Theory and Application of the Information Bottleneck Method

Theory and Application of the Information Bottleneck Method

Editors

Jan Lewandowsky
Gerhard Bauch

Basel • Beijing • Wuhan • Barcelona • Belgrade • Novi Sad • Cluj • Manchester

Editors
Jan Lewandowsky
Communication Systems
Fraunhofer FKIE
Wachtberg
Germany

Gerhard Bauch
Institute of Communications
Hamburg University of Technology
Hamburg
Germany

Editorial Office
MDPI
St. Alban-Anlage 66
4052 Basel, Switzerland

This is a reprint of articles from the Special Issue published online in the open access journal *Entropy* (ISSN 1099-4300) (available at: www.mdpi.com/journal/entropy/special_issues/ information_bottleneck_method).

For citation purposes, cite each article independently as indicated on the article page online and as indicated below:

Lastname, A.A.; Lastname, B.B. Article Title. *Journal Name* **Year**, *Volume Number*, Page Range.

ISBN 978-3-7258-0462-7 (Hbk)
ISBN 978-3-7258-0461-0 (PDF)
doi.org/10.3390/books978-3-7258-0461-0

Contents

Editorial

Theory and Application of the Information Bottleneck Method

Jan Lewandowsky [1,*] and Gerhard Bauch [2]

1. Fraunhofer Institute for Communication, Information Processing and Ergonomics, Fraunhoferstraße 20, 53343 Wachtberg, Germany
2. Institute of Communications, Hamburg University of Technology, Eißendorfer Straße 40, 21073 Hamburg, Germany; bauch@tuhh.de
* Correspondence: jan.lewandowsky@fkie.fraunhofer.de; Tel.: +49-228-9435-731

Citation: Lewandowsky, J.; Bauch, G. Theory and Application of the Information Bottleneck Method. *Entropy* **2024**, *26*, 187. https://doi.org/10.3390/e26030187

Received: 20 February 2024
Accepted: 21 February 2024
Published: 22 February 2024

In 1999, Naftali Tishby et al. introduced a powerful information theoretical framework called the information bottleneck method [1]. Conceptually, the aim of this method was to obtain a compressed representation T of an observed random variable Y by mapping Y onto T. In this setup, compression is caused by the rate–distortion concept of minimizing the mutual information $I(Y; T)$. The plain minimization of $I(Y; T)$, however, would destroy all information stored in Y. While classical rate–distortion theory defines limits on a distortion measure, typically with respect to the observation Y, the information bottleneck method introduces the concept of relevant information. Therefore, the fundamental idea of the information bottleneck method is to determine which information should be preserved under the invoked compression. This relevant information is defined through another random variable X. Hence, preserving the mutual information $I(X; T)$ while minimizing $I(Y; T)$ is the fundamental principle of the information bottleneck method.

The seemingly theoretical setup was proven to be very useful for various practical applications over the last few years. Its utilization evolved and covers too many different fields to list them here exhaustively. Examples include machine learning, deep learning, neuroscience, multimedia and image processing, data processing, source coding, channel coding and information processing. In addition, the theoretical backgrounds of the method, its generalizations and algorithms to find optimum mappings of Y onto T have become fruitful research topics. It should be noted that similar problems have been previously treated by Witsenhausen, Wyner and Ziv in the 1970s. However, this was on a rather abstract level and the information bottleneck with the involved mutual information measures as presented by Tishby and his colleagues was not as clearly carved out.

Today, more than 20 years after the publication of [1], researchers still conduct very promising research on the theory and the applications of the information bottleneck method in various disciplines of science and engineering. In light of this, we initiated a Special Issue to enrich the literature with topical theoretical and applied works oriented around the information bottleneck method. Therefore, we encouraged authors to submit their latest results on the theory and the applications of the information bottleneck method. We intentionally did not narrow the scope to a certain field of science or engineering. As a result, our Special Issue impressively illustrates the extensive theory and broad applicability of the information bottleneck framework.

In the following, we provide an overview of the published articles:

- Motivated by the information bottleneck and information-distortion systems, Parker and Dimitrov study the mathematical structure of information-based distortion-annealing problems in contribution 1. They investigate the bifurcations of solutions of certain degenerate constrained optimization problems related to the information bottleneck setup. Similarly, contribution 1 contributes to characterizing the local bifurcation structure of information bottleneck-type problems.
- Agmon leverages the information bottleneck's relations with the rate–distortion theory to provide deep insights into its general solution structure in contribution 2. Suboptimal solutions are seen to collide or exchange optimality at bifurcations of the

rate–information curve. By exploiting the dynamics of the optimal trade-off curve, a means to classify and handle bifurcations is presented. This understanding of bifurcations is used to propose a novel and surprisingly accurate numerical information bottleneck algorithm.

- Charvin, Volpi and Polani investigate the extent to which one can use existing compressed information bottleneck representations to produce new ones with a different granularity in contribution 3. First, they consider the notion of successive refinement, where no information needs to be discarded for this transition. For some specific information bottleneck problems, they derive successive refinability analytically and provide a tool to investigate it for discrete variables. Going further, they also quantify the loss of information optimality induced by several-stage processing in information bottleneck setups.

- Dikshtein, Ordentlich and Shamai introduce and study the double-sided information bottleneck problem in contribution 4, which is closely related to the biclustering domain. For jointly Gaussian and doubly symmetric binary sources in the double-sided information bottleneck setup, they provide insights on optimum solutions. They also explore a Blahut–Arimoto-like alternating maximization algorithm to find solutions for double-sided information bottleneck problems.

- Deng and Jia use the information bottleneck concept to deal with out-of-distribution generalization for classification tasks in contribution 5. In this context, they analyze failure situations of the information bottleneck invariant risk minimization principle and propose a new method, termed the counterfactual supervision-based information bottleneck, to overcome them. The effectiveness of their method is demonstrated empirically.

- Lyu, Aminian and Rodrigues use information bottleneck-inspired techniques to investigate the learning process of neural networks in contribution 6. They argue that the mutual information measures involved in the information bottleneck setup are difficult to estimate in this context. Therefore, they replace them with more tractable ones, i.e., the mean-squared error and the cross entropy. The resulting information bottleneck-inspired principle is used to study the learning dynamics of neural networks.

- Moldoveanu and Zaidi study distributed inference and learning over networks in contribution 7. They develop a framework to combine the information of observed features in a distributed information bottleneck setup with distributed observed nodes and a fusion node that conduct inference. Their experiments underline the advantages of their proposed scheme with respect to other distributed learning techniques.

- Steiner, Aminu and Kühn consider the optimization of distributed source coding in sensor networks in contribution 8. They investigate communication protocols in an extension of the so-called "extended chief executive officer problem setup". In this extension, the involved sensor nodes are allowed to communicate. The sensor nodes are optimized greedily and it is shown that their cooperation improves the performance significantly.

- Toledo, Venezian and Slonim revisit the sequential information information bottleneck (sIB) algorithm in contribution 9. Implementation aspects are discussed and the performance of their optimized information bottleneck algorithm is evaluated. The proposed implementation provides a trade-off between quality and speed that outperforms the considered reference algorithms. The novel sIB implementation is publicly available to ease further research on the information bottleneck method.

- Monsees, Griebel, Herrmann, Wübben, Dekorsy and Wehn study quantized decoders for low-density parity-check codes which are designed using the information bottleneck principle of maximizing the preserved relevant information under quantization in contribution 10. Such decoders allow for coarse quantization with minimum performance losses. A novel criterion for the required bit resolution in reconstruction–computation–quantization decoders is derived. Moreover, a comparison with a min-sum decoder implementation for throughput towards 1 Tb/s in fully depleted silicon-on-insulator technology is carried out.

- Contribution 11 describes the application of the information bottleneck method to quantized signal processing problems in communication receivers. For this purpose, contribution 11 summarizes recent ideas from various works to use the method for low-complexity quantized channel decoding, detection and channel estimation. In addition, novel results on a strongly quantized receiver chain, including channel estimation, detection and channel decoding, illustrate the ability to achieve optimum performance despite strong quantization with the proposed information bottleneck receiver design.

We would like to thank all of the aforementioned authors for their outstanding contributions to this Special Issue. We are also especially grateful to the reviewers. The articles in this Special Issue illustrate the current, highly versatile research directions related to the information bottleneck method, from relatively mathematical concepts and theories to their applications in practical engineering. We believe that research on the information bottleneck method will continue to play a very important role in the future. We look forward to further research work related to this fascinating method.

Conflicts of Interest: The authors declare no conflicts of interest.

List of Contributions

1. Parker, A.E.; Dimitrov, A.G. Symmetry-Breaking Bifurcations of the Information Bottleneck and Related Problems. *Entropy* **2022**, *24*, 1231.
2. Agmon, S. The Information Bottleneck's Ordinary Differential Equation: First-Order Root-Tracking for the IB. *Entropy* **2023**, *25*, 1370.
3. Charvin, H.; Volpi, N.C.; Polani, D. Exact and Soft Successive Refinement of the Information Bottleneck *Entropy* **2023**, *25*, 1355.
4. Dikshtein, M.; Ordentlich, O.; Shamai, S. The Double-Sided Information Bottleneck Function. *Entropy* **2022**, *24*, 1321.
5. Deng, B.; Jia, K. Counterfactual Supervision-Based Information Bottleneck for Out-of-Distribution Generalization. *Entropy* **2023**, *25*, 193.
6. Lyu, Z.; Aminian, G.; Rodrigues, M.R.D. On Neural Networks Fitting, Compression, and Generalization Behavior via Information-Bottleneck-like Approaches. *Entropy* **2023**, *25*, 1063.
7. Moldoveanu, M.; Zaidi, A. In-Network Learning: Distributed Training and Inference in Networks. *Entropy* **2023**, *25*, 920.
8. Steiner, S.; Aminu, A.D.; Kuehn, V. Distributed Quantization for Partially Cooperating Sensors Using the Information Bottleneck Method. *Entropy* **2022**, *24*, 438.
9. Toledo, A.; Venezian, E.; Slonim, N. Revisiting Sequential Information Bottleneck: New Implementation and Evaluation. *Entropy* **2022**, *24*, 1132.
10. Monsees, T.; Griebel, O.; Herrmann, M.; Wübben, D.; Dekorsy, A.; Wehn, N. Minimum-Integer Computation Finite Alphabet Message Passing Decoder: From Theory to Decoder Implementations towards 1 Tb/s. *Entropy* **2022**, *24*, 1452.
11. Lewandowsky, J.; Bauch, G.; Stark, M. Information Bottleneck Signal Processing and Learning to Maximize Relevant Information for Communication Receivers. *Entropy* **2022**, *24*, 972.

Reference

1. Tishby, N.; Pereira, F.C.; Bialek, W. The Information Bottleneck Method. In Proceedings of the 37th Allerton Conference on Communication and Computation, Monticello, NY, USA, 22–24 September 1999; pp. 368–377.

Article

Symmetry-Breaking Bifurcations of the Information Bottleneck and Related Problems

Albert E. Parker [1] and Alexander G. Dimitrov [2,*]

[1] Center for Biofilm Engineering, Department of Mathematical Sciences, Montana State University, Bozeman, MT 59717, USA
[2] Department of Mathematics and Statistics, Washington State University Vancouver, Vancouver, WA 98686, USA
* Correspondence: alex.dimitrov@wsu.edu

Abstract: In this paper, we investigate the bifurcations of solutions to a class of degenerate constrained optimization problems. This study was motivated by the Information Bottleneck and Information Distortion problems, which have been used to successfully cluster data in many different applications. In the problems we discuss in this paper, the distortion function is not a linear function of the quantizer. This leads to a challenging annealing optimization problem, which we recast as a fixed-point dynamics problem of a gradient flow of a related dynamical system. The gradient system possesses an S_N symmetry due to its invariance in relabeling representative classes. Its flow hence passes through a series of bifurcations with specific symmetry breaks. Here, we show that the dynamical system related to the Information Bottleneck problem has an additional spurious symmetry that requires more-challenging analysis of the symmetry-breaking bifurcation. For the Information Bottleneck, we determine that when bifurcations occur, they are only of pitchfork type, and we give conditions that determine the stability of the bifurcating branches. We relate the existence of subcritical bifurcations to the existence of first-order phase transitions in the corresponding distortion function as a function of the annealing parameter, and provide criteria with which to detect such transitions.

Keywords: information bottleneck; optimization; annealing; gradient flow; bifurcations; symmetry

Citation: Parker, A.E.; Dimitrov, A.G. Symmetry-Breaking Bifurcations of the Information Bottleneck and Related Problems. *Entropy* **2022**, *24*, 1231. https://doi.org/10.3390/e24091231

Academic Editors: Jan Lewandowsky and Gerhard Bauch

Received: 29 June 2022
Accepted: 29 August 2022
Published: 2 September 2022

Publisher's Note: MDPI stays neutral with regard to jurisdictional claims in published maps and institutional affiliations.

1. Introduction

This paper analyzes bifurcations of solutions to constrained optimization problems of the form

$$\max_{q \in \Delta} F(q, \beta) = \max_{q \in \Delta} \left(\sum_{i=1}^{N} f(q^i, \beta) \right) \tag{1}$$

as a function of a scalar parameter β and a quantizer or classifier $q = (q^1, \ldots, q^N)$ with $q^i \in \Re^K$. The real-valued function f is sufficiently smooth, and Δ is the constraint space of valid quantizers, a convex set of discrete probabilities (simplices).

This type of problem arises in Rate Distortion Theory [1,2], Deterministic Annealing [3] and biclustering [4]. The specific motivations for the abstract problem formulation given in (1) are the Information Bottleneck [5] and Information Distortion [6] functions

$$\max_{q \in \Delta} F(q, \beta) = \max_{q \in \Delta} (D(q) - \beta I(Y; T)). \tag{2}$$

These were proposed in [5,7] to analyze the Markov chain $X \to Y \to T$ in which $X \to Y$, characterized by a probability $p(X, Y)$, is the original system of interest, characterized by its mutual information $I(X; Y)$, and T is a simplification (quantized version of) Y. Here we work mainly with discrete versions of Y and T, with cardinalities $|Y| = K$ and $|T| = N$. Typically $N << K$. $I(Y; T)$ is the mutual information between the K objects in Y and the

N clusters in T. The goal is to cluster K objects in Y into N clusters in T given inputs X such that the function F is maximized in $[q^i]_j$; the probability that the jth element of Y is classified as being a member of the cluster with label $i \in T$. We call such a set of conditional probabilities a *stochastic quantizer*, or just a quantizer, to relate to the vector quantization literature [8]. The annealing parameter $\beta \in [0, \infty)$.

It has been shown that finding hard-clustering solutions to (2) is NP-complete (combinatorial search) when $D(q)$ is the mutual information $I(X; T)$ [9], as in the Information Bottleneck [5,10,11] and the Information Distortion [7,12,13] methods. Information Bottleneck (IB) approaches are gaining in penetration into multiple scientific and engineering domains [14–18]. As they typically involve the nonlinear optimization problem (2), there is need for optimization methods for such problems that can avoid the rise in complexity implied by the NP-complete hard-clustering solutions [9]. Originally, Tishby et al. [5] approached this problem with an algorithm inspired by the Blahut–Arimoto approach to solving Rate-Distortion types of problems ([2], Chapter 10). The "self-consistent" equations in [5] optimize both the quantizer and the "relevance" distribution $p(x|t)$. However, unlike the classic Blahut–Arimoto algorithm, which can guarantee convergence to a unique solution to its iterative scheme because of the convex geometry of the two state spaces, the "self-consistent" equations have no such guarantee due to the more-complicated geometry of *three* convex sets over which the optimization is performed, as also noted in [5]. Accordingly, in this work, we use the original optimization problem (2) over a single variable: the quantizer (conditional probability) $q(t|y)$. It may be possible that a related Blahut–Arimoto style optimization coupled to the bifurcation structure of its gradient flow discussed here can lead to additional insights into this problem, but we consider this beyond the scope of this particular manuscript.

We have investigated the structure of soft-clustering annealing-type methods that reach the hard-clustering solution in the limit of the annealing parameter [19,20] through a series of bifurcations. A bifurcation in this context is a point that is a solution (q^*, β^*) to (2) such that the number of solutions to (2) changes in a small neighborhood of (q^*, β^*). Because a bifurcation corresponds to a point at which some of the objects Y have just been classified, in the IB literature, a bifurcation is usually referred to as a phase transition. One of the goals of this and related work is to understand why annealing-type algorithms, such as the original optimization heuristics in [5,10], work as well as they do. This can help with designing further optimization heuristics and can assess how close those can get to the global solutions to IB problems. We believe that this amalgamation of optimization theory and dynamical systems theory, as stated in [19,20], can provide a solid foundation with which to address such optimization challenges.

Because of the form (1) of F, it possesses certain symmetries. That is, the value of $F(q, \beta)$ does not change (is invariant) under arbitrary permutations of the vectors q^i. In other words, F is S_N-invariant. The form (1) further implies that the Hessian $d_q^2 F(q)$ is block diagonal with blocks $\{d_{q^i}^2 f(q^i)\}_{i=1}^N$. These conditions are met by the Information Distortion function [6],

$$F_H(q, \beta) = H(T|Y) + \beta I(X, T), \tag{3}$$

where $H(T|Y)$ is the entropy, and by the cost function used in the original IB method [5],

$$F_{IB}(q, \beta) = -I(Y, T) + \beta I(X, T), \tag{4}$$

which is the focus of this manuscript. Both the Information Distortion and Informaton Bottleneck problems have the form given in (1) and (2). Importantly, $d^2 F_{IB}(q)$ has a "perpetual kernel" since each block $d^2 f(q^i)$ has the eigenpair $(0, q^i)$ for every q [20]. In other words, the Hessian $d^2 F$ is singular for every q and every value of β. This makes bifurcation detection challenging because bifurcations can usually be detected by identifying isolated singularities of $d^2 F$. This degeneracy is a consequence of the translational symmetry of

F_{IB}: if $\boldsymbol{k} \in \ker d_q^2 F_{IB}(q^*)$, then $F_{IB}(q^*) = F_{IB}(q^* + t\boldsymbol{k})$ for all $t \in \Re$ such that $q^* + t\boldsymbol{k} \in \Delta$. At bifurcations of solutions to (4), the translational symmetry never breaks.

To better understand bifurcations of solutions to problems of the form (1), which includes the problems (3) and (4), we consider the gradient flow

$$\begin{pmatrix} \dot{q} \\ \dot{\lambda} \end{pmatrix} = \nabla \mathcal{L}(q, \lambda, \beta)$$

Equilibria of this flow correspond to critical points of (1), where $\mathcal{L}$ is the Lagrangian with respect to the constraints imposed by Δ, and λ is the vector of Lagrange multipliers.

Previous work showed that when $d^2 F$ is generically non-singular, as occurs for the Information Distortion (3), then there are isolated singularities of $d^2 \mathcal{L}$ that indicate possible bifurcations of solutions to (1). In this case, an $M > 1$-dimensional $\ker d^2 F$ necessitates an $M - 1$-dimensional $\ker d^2 \mathcal{L}$, which admits a bifurcation of solutions to (1) where symmetry breaks from S_M to $S_m \times S_n$ for every $m, n > 0$ such that $m + n = M$ [21].

Here we allow $d^2 F$ and $d^2 \mathcal{L}$ to be singular for every $q \in \Delta$, as occurs for the Information Bottleneck (4). That is, the perpetual kernel for $d^2 F$ implies that $d^2 \mathcal{L}$ also has a perpetual kernel $\ker d^2 \mathcal{L} = \mathcal{K}_p(q)$, which means that the eigenvalue crossing condition that must occur at a bifurcation (i.e., $d^2 \mathcal{L}$ must have a zero eigenvalue at a bifurcation) [20] is never satisfied in $\mathcal{K}_p$. There are a few challenges due to the existence of the perpetual kernel (i.e., degeneracy) of the Information Bottleneck that we address in this paper. First, detecting bifurcations may be problematic because one cannot simply monitor the determinant of either $d^2 F$ or $d^2 \mathcal{L}$. Second, the standard theory that assures the existence of bifurcating branches, the Equivariance Branching Lemma, cannot be applied directly. Lastly, the spaces that contain the bifurcating solutions are always at least two-dimensional, which makes tracking the bifurcating solutions problematic.

Here we address two of these three challenges. We show that at a bifurcation, new eigenvalue(s) of $d^2 F_{IB}$ and $d^2 \mathcal{L}$ must cross zero, causing $\ker d^2 \mathcal{L}$ to expand so that $\ker d^2 \mathcal{L}(q^*) = \mathcal{K}_p \cup \mathcal{K}^*$, where $\mathcal{K}^*$ is the span of the eigenvectors with crossing eigenvalues. Instead of detecting bifurcations by the expensive process of monitoring the expansion of $\ker d^2 \mathcal{L}$ (from $\mathcal{K}_p$ to $\mathcal{K}_p \cup \mathcal{K}^*$), we give a simple way to check the eigenvalue crossing condition for annealing problems $F = G(q) + \beta D(q)$ as in (2) [20]. We prove the existence of the bifurcating branches by adapting the standard proof for the Equivariant Branching Lemma. This newly developed theory guarantees that bifurcating branches exist in $\mathcal{K}^*$, are generically pitchforks, and that symmetry breaks from S_M to $S_m \times S_n$. Additionally, we give conditions to check whether the pitchforks are subcritical or supercritical, and how stability of the bifurcating branches relates to optimality in the optimization problem (1).

2. Bifurcation Analysis

2.1. Equivariant Branching Lemma

The Equivariant Branching Lemma relates the subgroup structure of a symmetry group Γ with the existence of symmetry-breaking bifurcating branches of equilibria of $\dot{x} = f(x, \beta)$. Observe that we present a version that does not require absolute irreducibility. For a proof see [22] p. 83.

Theorem 1 (Equivariant Branching Lemma). *Let f be a smooth function $f : V \times \Re \to V$ that is Γ-equivariant for a compact Lie group Γ and a Banach space V. Let Σ be an isotropy subgroup of Γ with $\dim \mathrm{Fix}(\Sigma) = 1$. Suppose that $\mathrm{Fix}(\Gamma) = \{0\}$ and the crossing condition $d_{\beta x}^2 f(0, 0) x_0 \neq 0$ for $x_0 \in \mathrm{Fix}(\Sigma)$. Then there exists a unique smooth solution branch $(t x_0, \beta(t))$ to $f = 0$ with isotropy subgroup Σ.*

For an arbitrary Γ-equivariant system where bifurcation occurs at (x^*, β^*), the requirement in Theorem 1 that the bifurcation occurs at the origin is accomplished by a translation. Assuring that the Jacobian vanishes, $d_x f(0, 0) = 0$, can be effected by restricting and projecting the system onto the kernel of the Jacobian. This transform is called the Liapunov–Schmidt reduction (see [23]).

The Equivariant Branching Lemma does not directly apply to yield bifurcating branches for the problem (1) at q for which d^2F is singular for the following reasons:

- $\mathcal{K}_p$ and $\mathcal{K}^*$ have independent bases, which implies that each is invariant to the action of S_N, and so the decomposition $\ker d^2\mathcal{L}(q^*) = \mathcal{K}_p \times \mathcal{K}^*$ shows that S_N does not act absolutely irreducibly on $\ker d^2F(q^*)$, but it does act absolutely irreducibly on each of these disjoint subspaces separately. This is why we present a version of the Equivariant Branching Lemma that does not require absolute irreducibility.
- The Liapunov–Schmidt reduction onto $\ker d^2\mathcal{L}(q^*)$ is clear, but not onto $\mathcal{K}^*$.
- $\mathrm{Fix}(S_m \times S_n) \cap \ker d^2\mathcal{L}(q^*)$ is two-dimensional with basis

$$\{(n\boldsymbol{v},\ldots,n\boldsymbol{v},-m\boldsymbol{v},\ldots,-m\boldsymbol{v}),(n\boldsymbol{y},\ldots,n\boldsymbol{y},-m\boldsymbol{y},\ldots,-m\boldsymbol{y})\},$$

where $\boldsymbol{v},\boldsymbol{y} \in \Re^K$.

We address these issues in the manuscript and show that a small modification of the Equivariant Branching Lemma allows for similar analysis to be successfully applied to Information Bottleneck-style problems such as (2) with minimal modifications to the original algorithm from [20].

2.2. A Gradient Flow

We now lay the groundwork necessary to determine the bifurcations of local solutions to (1)

$$\max_{q \in \Delta} F(q, \beta),$$

where $F = \sum_{i=1}^{N} f(q^i, \beta)$, which includes as a special case the Information Distortion (3) and Information Bottleneck (4) problems. The convex set of discrete conditional probabilities is

$$\Delta := \left\{ q \in \Re^{NK} \mid \sum_{i=1}^{N} q_k^i = 1 \ \forall\, k : 1 \leq k \leq K \text{ and } q_k^i \geq 0 \ \forall\, i,k \right\}.$$

Due to the form of F, it has the following properties:

1. $F(q, \beta)$ is an S_N-invariant, real-valued function of q, where the action of S_N on q permutes the component vectors $q^i, i = 1, \ldots, N$, of $q \in \Delta$.
2. The $NK \times NK$ Hessian $d_q^2 F(q, \beta)$ is block diagonal, where the ith $K \times K$ block is $d^2 f(q^i)$.

The Lagrangian of (1) with respect to the equality constraints from Δ is

$$\mathcal{L}(q, \lambda, \beta) = F(q, \beta) + \sum_{k=1}^{K} \lambda_k \left(\sum_{i=1}^{N} q_k^i - 1 \right). \tag{5}$$

The scalar λ_k is the Lagrange multiplier for the constraint $\sum_{i=1}^{N} q_k^i - 1 = 0$, and $\lambda \in \Re^K$ is the vector of Lagrange multipliers $\lambda = (\lambda_1, \lambda_2, \ldots, \lambda_K)^T$. The gradient of the Lagrangian in (5) is

$$\nabla \mathcal{L} := \nabla_{q,\lambda} \mathcal{L}(q, \lambda, \beta) = \begin{pmatrix} \nabla_q \mathcal{L} \\ \nabla_\lambda \mathcal{L} \end{pmatrix},$$

where $\nabla_q \mathcal{L} = \nabla F(q, \beta) + \Lambda$ and $\Lambda = (\lambda^T, \lambda^T, \dots \lambda^T)^T \in \mathbf{R}^{NK}$. The gradient $\nabla_\lambda \mathcal{L}$ is a vector of K constraints

$$\nabla_\lambda \mathcal{L} = \begin{pmatrix} \Sigma_i q_1^i - 1 \\ \Sigma_i q_2^i - 1 \\ \vdots \\ \Sigma_i q_K^i - 1 \end{pmatrix}.$$

Let J be the Jacobian of $d_q \nabla_\lambda \mathcal{L}$

$$J := d_q \nabla_\lambda \mathcal{L} = \underbrace{(\ I_K \quad I_K \quad \cdots \quad I_K\ )}_{N \text{ blocks}}. \tag{6}$$

Observe that J has full row rank. The Hessian of (5) with respect to the vector $\begin{pmatrix} q \\ \lambda \end{pmatrix} \in \mathfrak{R}^{NK+K}$ is

$$d^2 \mathcal{L}(q) := d^2 \mathcal{L}(q, \lambda, \beta) = \begin{pmatrix} d^2 F(q, \beta) & J^T \\ J & \mathbf{0} \end{pmatrix}, \tag{7}$$

where $\mathbf{0}$ is $K \times K$. The $NK \times NK$ matrix $d^2 F(q) := d_q^2 F(q, \beta)$ is the block diagonal Hessian of F with $K \times K$ blocks $\{d^2 f(q^i, \beta)\}_{i=1}^N$.

The dynamical system whose equilibria are stationary points of (1) is the gradient flow of the Lagrangian

$$\begin{pmatrix} \dot{q} \\ \dot{\lambda} \end{pmatrix} = \nabla \mathcal{L}(q, \lambda, \beta) \tag{8}$$

for $\mathcal{L}$ as defined in (5) and $\beta \in [0, \infty)$. The equilibria of (8) are points $\begin{pmatrix} q^* \\ \lambda^* \end{pmatrix} \in \mathbf{R}^{NK+K}$ where

$$\nabla \mathcal{L}(q^*, \lambda^*, \beta) = 0.$$

The Jacobian of this system is the Hessian $d^2 \mathcal{L}(q, \lambda, \beta)$ from (7).

Remark 1. *By the theory of constrained optimization [24], the equilibria (q^*, λ^*, β) of (8) where $d^2 F(q^*, \beta)$ is negative definite on $\ker J$ are local solutions of (1). Conversely, if (q^*, β) is a local solution of (1), then there exists a vector of Lagrange multipliers λ^* so that (q^*, λ^*, β) is an equilibrium of (8) (this necessary requirement is called the Karush–Kuhn–Tucker conditions) such that $d^2 F(q^*, \beta)$ is non-positive definite on $\ker J$.*

2.3. Equilibria with Symmetry

Next, we categorize the equilibria of (8) according to their symmetries, which allows us to determine when to expect symmetry-breaking bifurcations.

Let $q \in \text{Fix}(S_M)$ for some $1 \leq M \leq N$. Then there exists a partition of $\{1, 2, \dots, N\}$ into the sets $\mathcal{U}$ and $\mathcal{R}$, where $|\mathcal{U}| = M$, so that $q^i = q^j$ if and only if $i, j \in \mathcal{U}$. Clearly, $d^2 F$ has M identical blocks, $\{d^f(q^i)\}_{i \in \mathcal{U}}$.

To ease the notation, and without loss of generality, we set

$$\mathcal{U} := \{1, \dots, M\} \quad \text{and} \quad \mathcal{R} := \{M+1, \dots, N\}.$$

To distinguish between the blocks of $d^2 F$, we write

$$B := d^2 f(q^i) \text{ for } 1 \leq i \leq M \text{ and } R_i := d^2 f(q^i) \text{ for } M+1 \leq i \leq N. \tag{9}$$

As mentioned in the introduction, we assume that for each $q \in \Delta$, each block $d^2 f(q^i)$ always has at least a one-dimensional kernel with basis vector(s) which depend on q. Thus, $\dim \ker d^2 F \geq N$. At an equilibrium of $(q^*, \lambda^*, \beta^*)$ of (8) where $q \in \text{Fix}(S_M)$, we consider the following three cases:

1. $\dim \ker d^2 F(q^*) > N + 1$;
2. $\dim \ker d^2 F(q^*) = N + 1$;
3. $\dim \ker d^2 F(q^*) = N$.

We will show that the first case necessitates a symmetry-breaking bifurcation (Theorem 3). In the second case, there is no bifurcation (Corollary 1). Finally, in the third case, we expect a saddle node [21], a symmetry-preserving bifurcation.

We are able to distinguish between the three cases above by considering which blocks of $d^2 F(q^*)$ have kernels that have more than one dimension. This motivates the following definition.

Definition 1. *An equilibrium $(q^*, \lambda^*, \beta^*)$ of (8) is M-singular (or, equivalently, q^* is M-singular) if:*

1. *$q \in \text{Fix}(S_M)$ so that $q^i = q^j$ for every $1 \leq i, j \leq M$.*
2. *For B, the M block(s) of the Hessian defined in (9), $\ker B$ has dimension 2 with basis vectors $\boldsymbol{v}, \boldsymbol{y} \in \Re^K$. $\boldsymbol{v}$ is associated with the crossing eigenvalues, and $\boldsymbol{y}$ is associated with the constant zero eigenvalue of B.*
3. *The $N - M$ block(s) of the Hessian $\{R_i\}_{i \in \Re}$, defined in (9), each have a one-dimensional kernel with basis vector $\boldsymbol{z}(i) \in \Re^K$.*
4. *The vectors $\boldsymbol{v}, \boldsymbol{y}$ and $\{\boldsymbol{z}(i)\}$ are linearly independent.*
5. *The matrix*

$$A := B \sum_{i=M+1}^{N} R_i^- + M I_K \tag{10}$$

is nonsingular. R_i^- is the Moore–Penrose inverse of R_i. When $M = N$, we define $A := N I_K$.

We wish to emphasize that we showed in [21] that requirements 2–5 in Definition 1 hold generically.

A straightforward calculation shows that every block of the Hessian $d^2 F$ of the Information Bottleneck cost function (2) is singular for every (q, β), and the basis for $\ker d^2 f(q^i)$ is $\boldsymbol{y} = q^i$ for $1 \leq i \leq M$ and $\boldsymbol{z}(i) = q^i$ for $M + 1 \leq i \leq N$ (Lemma 42 in [25]), which assures that these vectors are linearly independent, as in Definition 1.4. At a bifurcation, the kernels of the identical blocks B expand by $\boldsymbol{v}$ as in Definition 1.2. Using the notation above, $\boldsymbol{y} = q^i$ for each $i \in \mathcal{U}$, and $\boldsymbol{z}(i) = q^i$ for each $i \in \mathcal{R}$.

2.4. The Kernel at a Bifurcation

The equilibria of (8) change their stability with β, and hence change the solutions to (1). The changes of stability are determined by the kernel of $d^2 \mathcal{L}(q^*)$ at a bifurcation point q^*. In this section we show that for any $q \in \text{Fix}(S_M)$ with $M > 1$, $d^2 \mathcal{L}(q^*)$ has a perpetual kernel $\mathcal{K}_p$ that is at least $M - 1$ dimensional. The zero eigenvalues associated with the eigenvectors in $\mathcal{K}_p$ remain constant, so that at a bifurcation point $(q^*, \lambda^*, \beta^*)$ of (8) where q^* is M-singular, new eigenvalues of $d^2 \mathcal{L}$ must cross zero. Thus, the kernel expands, and the bifurcating directions exist in an "expanded" kernel of $d^2 \mathcal{L}(q^*)$, $\ker d^2 \mathcal{L}(q^*) = \mathcal{K}^* \times \mathcal{K}_p$.

We determine a basis for $\ker d^2 \mathcal{L}$ at an M-singular q^* when $M > 1$. If q is 1-singular with a trivial isotropy group (i.e., no symmetery), then $d^2 \mathcal{L}(q^*)$ is non-singular—$\mathcal{K}_p$ disappears. First, we ascertain a basis for $\ker d^2 F(q^*)$.

Recall that in the preliminaries, when $\boldsymbol{x} \in \Re^{NK}$, we defined $\boldsymbol{x}^j \in \mathbf{R}^K$ to be the jth vector component of $\boldsymbol{x}$. We now define the linearly independent vectors $\{\boldsymbol{v}_i\}_{i=1}^{M}$, $\{\boldsymbol{y}_i\}_{i=1}^{M}$, and $\{\boldsymbol{z}_k\}_{k=M+1}^{N}$ in $\Re^{NK}$ by

$$\boldsymbol{v}_i^j := \left\{ \begin{array}{l} \boldsymbol{v} \text{ if } 1 \leq i = j \leq M \\ \boldsymbol{0} \text{ otherwise} \end{array} \right. \quad \boldsymbol{y}_i^j := \left\{ \begin{array}{l} \boldsymbol{y} \text{ if } 1 \leq i = j \leq M \\ \boldsymbol{0} \text{ otherwise} \end{array} \right. , \quad \boldsymbol{z}_k^j := \left\{ \begin{array}{l} \boldsymbol{z}(i) \text{ if } M+1 \leq j = k \leq N \\ \boldsymbol{0} \text{ otherwise} \end{array} \right. \quad (11)$$

where $\boldsymbol{0} \in \Re^K$, and $\boldsymbol{v}$ and $\boldsymbol{y}$ are defined in Definition 1.2. For example, if $M = 2$ and $N = 3$, then $\boldsymbol{v}_1 := (\boldsymbol{v}^T, \boldsymbol{0}, \boldsymbol{0})^T$ and $\boldsymbol{v}_2 := (\boldsymbol{0}, \boldsymbol{v}^T, \boldsymbol{0})^T$.

Due to the block diagonal form of $d^2F(q^*)$, it is easy to see that the $N + M$ vectors defined in (11) form a basis for $\ker d^2F(q^*)$.

Now, let

$$V_i = \begin{pmatrix} \boldsymbol{v}_i \\ 0 \end{pmatrix} - \begin{pmatrix} \boldsymbol{v}_M \\ 0 \end{pmatrix}, \quad Y_i = \begin{pmatrix} \boldsymbol{y}_i \\ 0 \end{pmatrix} - \begin{pmatrix} \boldsymbol{y}_M \\ 0 \end{pmatrix}, \quad Z_k = \begin{pmatrix} \boldsymbol{z}_k \\ 0 \end{pmatrix} - \begin{pmatrix} \boldsymbol{z}_N \\ 0 \end{pmatrix} \quad (12)$$

for $i = 1, \ldots, M - 1$ and $M + 1 \leq k \leq N - 1$ where $\boldsymbol{0} \in \Re^K$. From (7), it is easy to see that these three sets of vectors are in $\ker d^2\mathcal{L}(q^*)$. The next theorem shows that $\{V_i\}_{i=1}^{M-1} \cup \{Y_i\}_{i=1}^{M-1}$ are a basis for $\ker d^2\mathcal{L}(q^*)$. This natural partition of the basis vectors shows that $\ker d^2\mathcal{L}(q^*)$ can be written as $\ker d^2\mathcal{L}(q^*) = \mathcal{K}_p \times \mathcal{K}^*$. According to Definition 1, the "perpetual kernel" corresponding to constant zero eigenvalues of $d^2\mathcal{L}(q^*)$ is generated by

$$\mathcal{K}_p = < \{Y_i\}_{i=1}^{M-1} > .$$

The part of the kernel that arises at a bifurcation corresponding to eigenvalues crossing zero is

$$\mathcal{K}^* = < \{V_i\}_{i=1}^{M-1} > .$$

The vectors $\{Z_k\}$ do not contribute to $\ker d^2\mathcal{L}(q^*)$.

Theorem 2. *If q^* is M-singular for $1 < M \leq N$, then $\{V_i\} \cup \{Y_i\}$ from (12) are a basis for $\ker d^2\mathcal{L}(q^*)$.*

Proof. To show that $\{V_i\}_{i=1}^{M-1} \cup \{Y_i\}_{i=1}^{M-1}$ span $\ker d^2\mathcal{L}(q^*)$, let $\boldsymbol{k} \in \ker d^2\mathcal{L}(q^*)$ and decompose it as

$$\boldsymbol{k} = \begin{pmatrix} \boldsymbol{k}_F \\ \boldsymbol{k}_J \end{pmatrix} \quad (13)$$

where $\boldsymbol{k}_F$ is $NK \times 1$, and $\boldsymbol{k}_J$ is $K \times 1$. Hence,

$$d^2\mathcal{L}(q^*, \lambda^*, \beta)\boldsymbol{k} = \begin{pmatrix} d^2F(q^*, \beta^*) & J^T \\ J & 0 \end{pmatrix} \begin{pmatrix} \boldsymbol{k}_F \\ \boldsymbol{k}_J \end{pmatrix} = 0$$
$$\implies d^2F(q^*, \beta)\boldsymbol{k}_F = -J^T\boldsymbol{k}_J$$
$$J\boldsymbol{k}_F = \boldsymbol{0}. \quad (14)$$

Now, from (6) and the fact that d^2F is block diagonal, we have

$$\begin{pmatrix} d^2f(q^1) & 0 & \cdots & 0 \\ 0 & d^2f(q^2) & \cdots & 0 \\ \vdots & \vdots & & \vdots \\ 0 & 0 & \cdots & d^2f(q^N) \end{pmatrix} \boldsymbol{k}_F = - \begin{pmatrix} \boldsymbol{k}_J \\ \boldsymbol{k}_J \\ \vdots \\ \boldsymbol{k}_J \end{pmatrix} . \quad (15)$$

We set

$$\boldsymbol{k}_F := (\boldsymbol{x}_1^T \ \boldsymbol{x}_2^T \ \cdots \ \boldsymbol{x}_N^T)^T, \quad (16)$$

and using the notation from (9), then (15) implies

$$
\begin{aligned}
B\boldsymbol{x}_i &= -\boldsymbol{k}_J \text{ for } 1 \le i \le M \\
R_i\boldsymbol{x}_i &= -\boldsymbol{k}_J \text{ for } M+1 \le i \le N.
\end{aligned}
\tag{17}
$$

It follows that $\boldsymbol{x}_i = R_i^- B\boldsymbol{x}_1$ for every $M+1 \le i \le N$. By (14), we have that $\sum_{i=1}^N \boldsymbol{x}_i = \boldsymbol{0}$, and so

$$
\begin{aligned}
&\textstyle\sum_{i=1}^M \boldsymbol{x}_i + \sum_{i=M+1}^N \boldsymbol{x}_i = \boldsymbol{0} \\
\implies &\textstyle\sum_{i=1}^M \boldsymbol{x}_i + \sum_{i=M+1}^N R_i^- B\boldsymbol{x}_1 + = \boldsymbol{0}.
\end{aligned}
$$

By (17), for every $1 \le i \le M$, $\boldsymbol{x}_i$ can be written as $\boldsymbol{x}_i = \boldsymbol{x}_p + d_i\boldsymbol{v} + e_i\boldsymbol{y}$, where $\boldsymbol{x}_p \in \mathrm{range}(B)$, $d_\eta, e_\eta \in \Re$, and $\boldsymbol{v}$ and $\boldsymbol{y}$ are the basis vectors of $\ker B$ from Definition 1.2. Thus,

$$
B\sum_{i=1}^M (\boldsymbol{x}_p + d_i\boldsymbol{v} + e_i\boldsymbol{y}) \quad + \quad B\sum_{i=M+1}^N R_i^- B(\boldsymbol{x}_p + d_1\boldsymbol{v} + e_1\boldsymbol{y}) = \boldsymbol{0}
$$

$$
\Leftrightarrow \quad (B\sum_{i=M+1}^N R_i^- + MI_K)B\boldsymbol{x}_p = \boldsymbol{0}
$$

$$
\Leftrightarrow \quad B\boldsymbol{x}_p = \boldsymbol{0}
$$

since $A = B\sum_{i=M+1}^N R_i^- + MI_K$ is nonsingular. This shows that $\boldsymbol{x}_p = \boldsymbol{0}$. Therefore, $\boldsymbol{x}_i = d_i\boldsymbol{v} + e_i\boldsymbol{y}$ for every $1 \le i \le M$. Now (17) shows that $\boldsymbol{k}_J = \boldsymbol{0}$, and so $\boldsymbol{x}_i \in \ker R_i$ for $M+1 \le i \le N$, which implies that

$$
\boldsymbol{x}_i = c_i\boldsymbol{z}(i) \text{ for } M+1 \le i \le N.
$$

Hence, $\boldsymbol{k} = \begin{pmatrix} \boldsymbol{k}_F \\ \boldsymbol{0} \end{pmatrix}$, where $\boldsymbol{k}_F^i = \begin{cases} d_i\boldsymbol{v} + e_i\boldsymbol{y} \text{ if } 1 \le i \le M \\ c_i\boldsymbol{z}(i) \text{ if } M+1 \le i \le N \end{cases}$, from which it follows that

$$
J\boldsymbol{k}_F = \sum_{i=1}^N \boldsymbol{x}_i = \sum_{i=1}^M d_i\boldsymbol{v} + \sum_{i=1}^M e_i\boldsymbol{y} + \sum_{i=M+1}^N c_i\boldsymbol{z}(i) = \boldsymbol{0}.
\tag{18}
$$

Linear independence (Definition 1.4) implies that $\sum d_i = \sum e_i = d_i = 0$. Thus, $\boldsymbol{k}_F = \sum_{i=1}^{M-1} d_i(\boldsymbol{v}_i - \boldsymbol{v}_M) + \sum_{i=1}^{M-1} e_i(\boldsymbol{y}_i - \boldsymbol{y}_M)$. Therefore, the linearly independent vectors $\{V_i\} = \{\begin{pmatrix} \boldsymbol{v}_i - \boldsymbol{v}_M \\ 0 \end{pmatrix}\}$ and $\{Y_i\} = \{\begin{pmatrix} \boldsymbol{y}_i - \boldsymbol{y}_M \\ 0 \end{pmatrix}\}$ span $\ker d^2\mathcal{L}(q^*)$. $\square$

Corollary 1. *If q^* is 1-singular and has isotropy group equal to the identity, then $d^2\mathcal{L}(q^*)$ is nonsingular.*

Proof. If q is 1-singular, then $d^2F(q^*)$ has a single block B with a two-dimensional kernel. The other $N-1$ blocks $\{R_i\}$ are distinct with one-dimensional kernels. By constructing the vectors as in (11), we see that $\dim \ker d^2F(q^*) = N+1$ with basis vectors $\boldsymbol{v}_1, \boldsymbol{y}_1, \{\boldsymbol{z}_i\}_{i=2}^N$. Now, following the proof of Theorem 2, we take an arbitrary $\boldsymbol{k} \in \ker d^2\mathcal{L}(q^*, \lambda, \beta)$, and then decompose $\boldsymbol{k}$ as in (13) and (16). The proof to Theorem 2 holds for the present case up until, and including (18). Linear independence now shows that $d_i = e_i = c_i = 0$, which implies that $\boldsymbol{k} = \boldsymbol{0}$. $\square$

Remark 2. *The independent bases given for $\mathcal{K}_p$ and $\mathcal{K}^*$ in Theorem 2 imply that each is invariant to the action of S_N, and so the decomposition $\ker d^2\mathcal{L}(q^*) = \mathcal{K}_p \times \mathcal{K}^*$ shows that S_N does not act absolutely irreducibly on $\ker d^2F(q^*)$. That is, by definition,*

$$
d_x r(\boldsymbol{0}, \beta) \neq c(\beta)I_{2M-2}.
$$

The explicit bases show that $\mathcal{K}_p, \mathcal{K} \cong \{x \in \mathbf{R}^M : \sum[x]_i = 0\}$, which implies that S_M acts absolutely irreducibly on $\mathcal{K}_p$ and $\mathcal{K}^*$ [26]. Thus, $\mathcal{K}_p$ and $\mathcal{K}^*$ are each S_M-irreducible.

2.5. Liapunov–Schmidt Reduction

To show the existence of bifurcating branches from a bifurcation point $(q^*, \lambda^*, \beta^*)$ of equilibria of (8), the Equivariant Branching Lemma requires that the bifurcation is translated to $(\mathbf{0}, 0, 0)$ and that the Jacobian vanishes at bifurcation. To accomplish the former, consider

$$\mathcal{F}(q, \lambda, \beta) := \nabla \mathcal{L}(q + q^*, \lambda + \lambda^*, \beta + \beta^*).$$

To assure that the Jacobian vanishes, we restrict and project $\mathcal{F}$ onto $\ker d^2 \mathcal{L}(q^*)$ in a neighborhood of $(\mathbf{0}, 0, 0)$. This is the Liapunov–Schmidt reduction of $\mathcal{F}$ [23],

$$
\begin{aligned}
r \quad &: \quad \mathbf{R}^{M-1} \times \mathbf{R} \to \mathbf{R}^{M-1} \\
r(x, \beta) \quad &= \quad W^T (I - E) \mathcal{F}(Wx + U(Wx, \beta), \beta)
\end{aligned}
\tag{19}
$$

where $Wx + U(Wx, \beta) = \begin{pmatrix} q \\ \lambda \end{pmatrix}$. The $(NK + K) \times (NK + K)$ matrix $I - E$ is the projection matrix onto $\ker \mathcal{F}(\mathbf{0}, 0) = \ker d^2 \mathcal{L}(q^*)$ with $\ker(I - E) = \text{range } d^2 \mathcal{L}(q^*)$. W is the $(NK + K) \times (2M - 2)$ matrix whose columns are the basis vectors $\{V_i\} \cup \{Y_i\}$ of $\ker d^2 \mathcal{L}(q^*)$ from (12) so that Wx is a vector in $\ker d^2 \mathcal{L}(q^*)$. The vector function $U(Wx, \beta)$ is the component of (q, λ) that is in range $d^2 \mathcal{L}(q^*)$ such that $E\mathcal{F}(Wx + U(x, \beta), \beta) = \mathbf{0}, U(\mathbf{0}, 0) = \mathbf{0}$, and

$$d_x U(\mathbf{0}, 0) = \mathbf{0}. \tag{20}$$

The system defined by the Liapunov–Schmidt reduction, $\dot{x} = r(x, \beta)$, has a bifurcation of equilibria at $(x = \mathbf{0}, \beta = 0)$, which are in $1 - 1$ correspondence with equilibria of (8). However, the stability of these associated equilibria is not necessarily the same.

It is straightforward to verify the following derivatives ([23] p. 32), which we will require in the sequel. The $(2M - 2) \times (2M - 2)$ Jacobian of (19) is

$$d_x r(x, \beta) = W^T (I - E) d^2_{q,\lambda} \mathcal{L}(q + q^*, \lambda + \lambda^*, \beta + \beta^*)(W + d_x U(Wx, \beta)), \tag{21}$$

which shows that

$$d_x r(\mathbf{0}, 0) = \mathbf{0} \tag{22}$$

since $\ker(I - E) = \text{range } d^2 \mathcal{L}(q^*)$.

Our crossing condition at a bifurcation depends on the matrix of derivatives

$$\frac{\partial^2 r_i}{\partial \beta \partial x_j}(\mathbf{0}, 0) = d_\beta d^2 \mathcal{L}[w_i, w_j] - d^3 \mathcal{L}[w_i, w_j, L^- d_\beta \nabla \mathcal{L}] \tag{23}$$

where the derivatives of $\mathcal{L}$ are evaluated at $(q^*, \lambda^*, \beta^*)$, and L^- is the Moore–Penrose-generalized inverse [27] of $d^2 \mathcal{L}(q^*)$. The vectors $\{w_i\}_{i=1}^{2M-2}$ are the basis vectors of $\ker d^2 \mathcal{L}(q^*)$ from Theorem 2.

The $(2M - 2) \times (2M - 2) \times (2M - 2)$ three-dimensional array of second derivatives is

$$\frac{\partial^2 r_i}{\partial x_j \partial x_k}(\mathbf{0}, 0) = d^3 \mathcal{L}(q^*, \lambda^*, \beta^*)[w_i, w_j, w_k].$$

In [21], we showed that $\frac{\partial^2 r_i}{\partial x_j \partial x_k}(\mathbf{0}, 0) = 0$ whenever $i = j = k \leq M - 1$. In the present case, there are more zero entries since now the basis vectors $\{w_i\}$ are of two types: $w_i = V_i$ for $1 \leq i \leq M - 1$ (basis vectors of $\mathcal{K}^*$); or $w_i = Y_{i-M+1}$ for $M \leq i \leq 2M - 2$ (basis vectors of

$\mathcal{K}_p$, see (12)). We now consider the case when $i, j \leq M - 1$ and $k > M - 1$. All other cases are dealt with using a similar argument. Substituting in for $\boldsymbol{w}_i$ we have

$$
\begin{aligned}
\frac{\partial^2 r_i}{\partial x_j \partial x_k}(\boldsymbol{0},0) &= \sum_{\nu,\delta,\eta=1}^{N} \sum_{l,m,n=1}^{K} \frac{\partial^3 F(q^*,\beta^*)}{\partial q_l^\nu \partial q_m^\delta \partial q_n^\eta} [\boldsymbol{v}_i - \boldsymbol{v}_M]_l^\nu [\boldsymbol{v}_j - \boldsymbol{v}_M]_m^\delta [\boldsymbol{y}_{k-M+1} - \boldsymbol{y}_M]_n^\eta \\
&= \sum_{l,m,n=1}^{K} \frac{\partial^3 f(q^\nu *,\beta^*)}{\partial q_l^\nu \partial q_m^\nu \partial q_n^\nu} \left(\delta_{ij(k-M+1)} [\boldsymbol{v}]_l [\boldsymbol{v}]_m [\boldsymbol{y}]_n - [\boldsymbol{v}]_l [\boldsymbol{v}]_m [\boldsymbol{y}]_n \right). \quad (24)
\end{aligned}
$$

The vectors $\boldsymbol{v}$ and $\boldsymbol{y}$ are defined in (2). An immediate consequence of this calculation is that $\frac{\partial^2 r_i}{\partial x_j \partial x_k}(\boldsymbol{0},0) = 0$ whenever $i = j = k - M + 1$. Thus, similar arguments show that $\frac{\partial^2 r_i}{\partial x_j \partial x_k}(\boldsymbol{0},0) = 0$ whenever:

- $i = j = k$;
- $i - M + 1 = j = k, \quad i = j - M + 1 = k, \quad i = j = k - M + 1$;
- $i - M + 1 = j - M + 1 = k, \quad i - M + 1 = j = k - M + 1, \quad i = j - M + 1 = k - M + 1$.

Further, we get four different "cubes" of identical entries in the 3-D array. They are:

- For $i, j, k \leq M - 1$, not all equal, the value of the cube is

$$
- \sum_{l,m,n=1}^{K} \frac{\partial^3 f(q^\nu *,\beta^*)}{\partial q_l^\nu \partial q_m^\nu \partial q_n^\nu} [\boldsymbol{v}]_l [\boldsymbol{v}]_m [\boldsymbol{v}]_n;
$$

- For $i, j \leq M - 1$, not both equal, and $j > M - 1$, the value of the cube is

$$
- \sum_{l,m,n=1}^{K} \frac{\partial^3 f(q^\nu *,\beta^*)}{\partial q_l^\nu \partial q_m^\nu \partial q_n^\nu} [\boldsymbol{v}]_l [\boldsymbol{v}]_m [\boldsymbol{y}]_n;
$$

- For $i \leq M - 1$ and $j, k > M - 1$, not both equal, the value of the cube is

$$
- \sum_{l,m,n=1}^{K} \frac{\partial^3 f(q^\nu *,\beta^*)}{\partial q_l^\nu \partial q_m^\nu \partial q_n^\nu} [\boldsymbol{v}]_l [\boldsymbol{y}]_m [\boldsymbol{y}]_n;
$$

- For $i, j, k > M - 1$, not all equal, the value of the cube is

$$
- \sum_{l,m,n=1}^{K} \frac{\partial^3 f(q^\nu *,\beta^*)}{\partial q_l^\nu \partial q_m^\nu \partial q_n^\nu} [\boldsymbol{y}]_l [\boldsymbol{y}]_m [\boldsymbol{y}]_n.
$$

The points above will prove useful when proving that $d^2 r(\boldsymbol{0},0) = \boldsymbol{0}$.

The four-dimensional array of third derivatives of r is

$$
\begin{aligned}
\frac{\partial^3 r_i}{\partial x_j \partial x_k \partial x_l}(\boldsymbol{0},0) = d^4 \mathcal{L}[\boldsymbol{w}_i, \boldsymbol{w}_j, \boldsymbol{w}_k, \boldsymbol{w}_l] \quad &- \quad d^3 \mathcal{L}[\boldsymbol{w}_i, \boldsymbol{w}_j, L^- d^3 \mathcal{L}[\boldsymbol{w}_k, \boldsymbol{w}_l]] \\
&- \quad d^3 \mathcal{L}[\boldsymbol{w}_i, \boldsymbol{w}_k, L^- d^3 \mathcal{L}[\boldsymbol{w}_j, \boldsymbol{w}_l]] \\
&- \quad d^3 \mathcal{L}[\boldsymbol{w}_i, \boldsymbol{w}_l, L^- d^3 \mathcal{L}[\boldsymbol{w}_j, \boldsymbol{w}_k]] \quad (25)
\end{aligned}
$$

where the derivatives of $\mathcal{L}$ are evaluated at $(q^*, \lambda^*, \beta^*)$, and L^- is the Moore–Penrose-generalized inverse [27] of $d^2 \mathcal{L}(q^*)$.

Since $\ker d^2 \mathcal{L}(q^*)$ is not absolutely irreducible, but $\mathcal{K}^*$ is, one might try to define a Liapunov–Schmidt reduction by restricting and projecting $\nabla \mathcal{L}$ onto $\mathcal{K}^*$. One issue with projecting the reduction onto $\mathcal{K}^*$ is how to define the projection matrix E so that

$$
E\mathcal{F} = 0 \text{ and } (I - E)\mathcal{F} = 0 \text{ if and only if } \mathcal{F} = 0
$$

holds and $E\, d_{\boldsymbol{x}} r(\mathbf{0}, 0)$ is non-singular in range (E) so that the Implicit Function Theorem assures the restriction $(q, \lambda) = W\boldsymbol{x} + U(W\boldsymbol{x}, \beta)$, where $U(W\boldsymbol{x}) \in$ range $(d^2\mathcal{L}(q^*))$, and $W\boldsymbol{x} \in \mathcal{K}^*$ instead of $W\boldsymbol{x} \in \ker d^2\mathcal{L}(q^*)$ as in (19) [23]. Simply ignoring the space $\mathcal{K}_p$ by considering $U \in$ range $(d^2\mathcal{L}(q^*))$ and $W\boldsymbol{x} \in \mathcal{K}^*$ amounts to setting $W\boldsymbol{x} = k^* + k_p$ and $k_p = 0$. Since $W\boldsymbol{x} + U$ is still embedded in the larger $\Re^{NK+K}$, which contains $\mathcal{K}_p$, then derivatives are affected by the implicit $k_p = 0$ constraint. This constraint $P_{\mathcal{K}_p}(q, \lambda) = k^* + U$ is nonlinear (and may not even be tractable) since $\mathcal{K}_p$ depends on q, where $P_{\mathcal{K}_p}$ is a projection matrix that depends on q (see Theorem 7).

2.6. Isotropy Subgroups $S_m \times S_n$ of S_N

The decomposition $\ker d^2\mathcal{L}(q^*) = \mathcal{K}_p \times \mathcal{K}^*$ shows that $\mathrm{Fix}(S_m \times S_n) \cap \ker d^2\mathcal{L}(q^*)$ is two-dimensional with basis vectors

$$\{(n\boldsymbol{y}^T, \ldots, n\boldsymbol{y}^T, -m\boldsymbol{y}^T, \ldots, -m\boldsymbol{y}^T)^T, (n\boldsymbol{v}^T, \ldots, n\boldsymbol{v}^T, -m\boldsymbol{v}^T, \ldots, -m\boldsymbol{v}^T)^T\}.$$

Restricted to $\mathcal{K}^*$, these isotropy subgroups $S_m \times S_n$ of S_M have one-dimensional fixed point spaces. This assures that we can use Theorem 1. We have the following Lemma.

Lemma 1. *Let $M = m + n$ such that $M > 1$ and $m, n > 0$. Let $\mathcal{U}_m$ be a set of m classes, and let $\mathcal{U}_n$ be a set of n classes such that $\mathcal{U}_m \cap \mathcal{U}_n = \varnothing$ and $\mathcal{U}_m \cup \mathcal{U}_n = \{1, \ldots, M\}$. Now define $\hat{\boldsymbol{u}}_{(m,n)} \in \Re^{NK}$ such that*

$$\hat{\boldsymbol{u}}^i_{(m,n)} = \left\{ \begin{array}{ll} n\boldsymbol{v} & \text{if } i \in \mathcal{U}_m \\ -m\boldsymbol{v} & \text{if } i \in \mathcal{U}_n \\ \mathbf{0} & \text{otherwise} \end{array} \right.$$

where $\boldsymbol{v}$ is defined as in Definition 1.2, and let

$$\boldsymbol{u}_{(m,n)} = \left(\begin{array}{c} \hat{\boldsymbol{u}}_{(m,n)} \\ \mathbf{0} \end{array} \right) \tag{26}$$

where $\mathbf{0} \in \mathbf{R}^K$. Then the isotropy subgroup of $\boldsymbol{u}_{(m,n)}$ is $\Sigma_{(m,n)} \subset \Gamma_{\mathcal{U}}$ such that $\Sigma_{(m,n)} \cong S_m \times S_n$, where S_m permutes $\boldsymbol{u}^i$ when $i \in \mathcal{U}_m$, and S_n permutes $\boldsymbol{u}^i$ when $i \in \mathcal{U}_n$. The fixed point space of $\Sigma_{(m,n)}$ restricted to $\mathcal{K}^* \subset d^2\mathcal{L}(q^*)$ is one dimensional.

2.7. Bifurcating Branches

Theorem 3. *Let $(q^*, \lambda^*, \beta^*)$ be an equilibrium of (8) such that q^* is M-singular for $1 < M \leq N$, and the crossing condition*

$$d_\beta d^2\mathcal{L}[\boldsymbol{u}, \boldsymbol{u}] - d^3\mathcal{L}[\boldsymbol{u}, \boldsymbol{u}, L^- d_\beta \nabla \mathcal{L}] \neq 0$$

is satisfied. Then there exists bifurcating solutions, $\left(\begin{array}{c} q^* \\ \lambda^* \\ \beta^* \end{array} \right) + \left(\begin{array}{c} t\boldsymbol{u}_{(m,n)} \\ \beta(t) \end{array} \right)$, *where $\boldsymbol{u}_{(m,n)} \in \mathcal{K}^*$ is defined in (26), for every pair (m, n) such that $M = m + n$, each with an isotropy group isomorphic to $S_m \times S_n$.*

Proof. We mimic the proof of the Equivariant Branching Lemma. Let $\boldsymbol{u} := \boldsymbol{u}_{(m,n)} \in \mathrm{Fix}(S_m \times S_n) \cap \mathcal{K}^*$ and let V be a matrix with columns composed of the $M - 1$ vectors $\{V_i\}$. Thus, there exists $\boldsymbol{x}_0 \in \Re^{M-1}$ so that $\boldsymbol{u} = V\boldsymbol{x}_0$. Since $r(\mathrm{Fix}(S_m \times S_n) \cap \mathcal{K}^*) \subseteq \mathrm{Fix}(S_m \times S_n) \cap \mathcal{K}^*$ (for every $\sigma \in S_m \times S_n$, $r(V\boldsymbol{x}) = r(\sigma V\boldsymbol{x})$ ($\boldsymbol{u} \in \mathrm{Fix}(S_M \times S_n)$ that equals $\sigma r(V\boldsymbol{x})$ (by equivariance)), then $r(t\boldsymbol{x}_0, \beta) = h(t, \beta)\boldsymbol{x}_0$, where r is the Liapunov–Schmidt reduction (19), and h is a polynomial in t.

Since $\mathcal{K}^*$ is S_M-irreducible, then $\mathrm{Fix}(S_M) \cap \mathcal{K}^* = \{\mathbf{0}\}$ (otherwise, $\sigma\boldsymbol{x} = \boldsymbol{x}$ for some $\boldsymbol{x} \in \mathcal{K}^*$ for every $\sigma \in S_M$, which implies that $\mathrm{span}(\boldsymbol{x})$ is an invariant subspace of $\mathcal{K}^*$).

Now [22] p. 75 shows that $r(\mathbf{0}, \beta) = \mathbf{0}$, and so $h(0, \beta) = 0$, from which it follows that $h(t, \beta) = tk(t, \beta)$. Thus,

$$r(t\mathbf{x}_0, \beta) = tk(t, \beta)\mathbf{x}_0. \tag{27}$$

Differentiating with respect to t yields

$$d_{\mathbf{x}} r(t\mathbf{x}_0, \beta)\mathbf{x}_0 = (k(t, \beta) + td_t k(t, \beta))\mathbf{x}_0, \tag{28}$$

from which it follows that

$$k(t, \beta)\mathbf{x}_0 = d_{\mathbf{x}} r(t\mathbf{x}_0, \beta)\mathbf{x}_0 - td_t k(t, \beta)\mathbf{x}_0,$$

and so $k(0,0) = 0$. Furthermore, we see that $d_\beta k(0,0)\mathbf{x}_0 = d^2_{\mathbf{x},\beta} r(0,0)\mathbf{x}_0 \neq \mathbf{0}$ by assumption (see (23)). This shows that $d_\beta k(0,0)$ is a non-zero eigenvalue of $d_{\mathbf{x}} r(t\mathbf{x}_0, \beta)$ with associated eigenvector $\mathbf{x}_0$. By the Implicit Function Theorem, $k(t, \beta) = 0$ has a non-zero unique solution for $\beta = \beta(t)$. $\square$

2.8. The Crossing Condition for Annealing Problems

We next determine how to check the crossing condition in Theorem 3 when F is an annealing problem, as in (2)

$$F(q, \beta) = H(q) + \beta D(q).$$

First, we show that the crossing condition can be checked in terms of the Hessian of the function D. Furthermore, when G is strictly concave on $\mathrm{span}(\{\mathbf{v}_i\})$, then the crossing condition is always satisfied, and every singularity is a bifurcation.

Theorem 4. *The crossing condition*

$$d_\beta d^2 \mathcal{L}[\mathbf{u}, \mathbf{u}] - d^3 \mathcal{L}[\mathbf{u}, \mathbf{u}, L^- d_\beta \nabla \mathcal{L}] \neq 0$$

given in Theorem 3 is satisfied for M-singular q for $M > 1$ if $d^2 D(q)$ is either positive or negative definite on $\mathrm{span}(\{\mathbf{v}_i\})$.

Proof. Let $\mathbf{x}_0 \in \Re^{2M-2}$ so that $\mathbf{u} = W\mathbf{x}_0 \in \mathrm{Fix}(S_m \times S_n) \cap \mathcal{K}^*$. Multiplying Equation (21) on the left by $\mathbf{x}_0^T$ and on the right by $\mathbf{x}_0$ yields

$$\mathbf{x}_0^T d_{\mathbf{x}} r(\mathbf{0}, \beta)\mathbf{x}_0 = \mathbf{u}^T d^2_{q,\lambda} \mathcal{L}(q^*, \lambda^*, \beta + \beta^*)(I_{NK+K} + d_{\mathbf{w}} U(\mathbf{0}, \beta))\mathbf{u}. \tag{29}$$

By Theorem 2, an arbitrary $\mathbf{u} \in \mathcal{K}^*$ can be written as $\mathbf{u} = \begin{pmatrix} \hat{\mathbf{u}} \\ \mathbf{0} \end{pmatrix}$, where $\hat{\mathbf{u}} \in \mathrm{span}(\{\mathbf{v}_i\}) \subset$ $\ker d^2 F(q^*, \beta^*)$. Substituting this into (29) and observing that $d^2 F(q^*, \beta + \beta^*) = d^2 G(q^*) + (\beta + \beta^*)d^2 D(q^*) = d^2 F(q^*, \beta^*) + \beta d^2 D(q^*)$ yields

$$\mathbf{x}_0^T d_{\mathbf{x}} r(\mathbf{0}, \beta)\mathbf{x}_0 = \beta \begin{pmatrix} \hat{\mathbf{u}}^T d^2 D(q^*) & \mathbf{0}^T \end{pmatrix}(I_{NK+K} + \partial_{\mathbf{w}} U(\mathbf{0}, \beta))\begin{pmatrix} \hat{\mathbf{u}} \\ \mathbf{0} \end{pmatrix}.$$

Differentiating with respect to β, evaluating at $\beta = 0$, and using (20) yields

$$\mathbf{x}_0^T d^2_{\mathbf{x},\beta} r(\mathbf{0},0)\mathbf{x}_0 = \hat{\mathbf{u}}^T d^2 D(q^*)\hat{\mathbf{u}}, \tag{30}$$

which must be non-zero since we assume that $d^2 D(q)$ is either positive or negative definite on $\mathrm{span}(\{\mathbf{v}_i\})$. $\square$

From (30), we can get an expression for ζ, the eigenvalue of $d^2_{\boldsymbol{x},\beta}r(0,0)$ with eigenvector $\boldsymbol{x}_0$. Substituting $d^2_{\boldsymbol{x},\beta}r(0,0)\boldsymbol{x}_0 = \zeta\boldsymbol{x}_0$ and observing that $\boldsymbol{x}_0^T\boldsymbol{x}_0 = \boldsymbol{x}_0^TW^TW\boldsymbol{x}_0 = \hat{\boldsymbol{u}}^T\hat{\boldsymbol{u}}$ yields

$$\zeta = \frac{\hat{\boldsymbol{u}}^T d^2 D(q^*)\hat{\boldsymbol{u}}}{||\hat{\boldsymbol{u}}||^2}. \tag{31}$$

The requirement that $d^2 D(q)$ is either positive or negative definite on $\mathrm{span}(\{\boldsymbol{v}_i\})$ holds when $d^2 G(q^*)$ is either negative or positive definite, respectively, on $\mathrm{span}(\{\boldsymbol{v}_i\})$.

Lemma 2. *Let $d^2 F(q^*,\beta^* \neq 0)$ be singular where q^* is M-singular such that $d^2 G(q^*)$ is negative (or positive) definite on $\mathrm{span}(\{\boldsymbol{v}_i\})$. Then $d^2 D(q^*)$ is positive (or negative) definite on $\mathrm{span}(\{\boldsymbol{v}_i\})$.*

Proof. If $\boldsymbol{u} \in \mathrm{span}(\{V_i\}) \subset \ker d^2 F(q^*)$, then $\boldsymbol{u}^T d^2 G(q^*)\boldsymbol{u} + \beta^*\boldsymbol{u}^T d^2 D(q^*)\boldsymbol{u} = 0$. Since $\boldsymbol{u}^T d^2 G(q^*)\boldsymbol{u} < 0$, then $\boldsymbol{u}^T d^2 D(q^*)\boldsymbol{u} > 0$. $\square$

These results are important for the Information Bottleneck problem (2), where $d^2 G(q) = -d^2 I(Y;Z)$ is only non-positive definite on $\ker d^2 F(q^*)$, but is negative definite on $\mathrm{span}(\{\boldsymbol{v}_i\})$. Thus, every singularity of the Information Bottleneck with $\ker d^2\mathcal{L}(q^*) = \mathcal{K}^* \times \mathcal{K}_p$ is a bifurcation point. The space $\mathcal{K}_p$ does not contain bifurcating branches since the crossing condition is never satisfied there: for $\boldsymbol{u} \in \mathcal{K}_p$, $\hat{\boldsymbol{u}}^T d^2 G(q)\hat{\boldsymbol{u}} + \beta\hat{\boldsymbol{u}}^T d^2 D(q)\hat{\boldsymbol{u}} = 0 + 0$ (by Lemma 42 in [25]), and so (Theorem 109, [25]) $\zeta = \frac{\hat{\boldsymbol{u}}^T d^2 D(q)\hat{\boldsymbol{u}}}{||\hat{\boldsymbol{u}}||} = 0$.

2.9. Bifurcation Type

Suppose that a bifurcation occurs at (q^*,λ^*,β^*), where q^* is M-singular. This section examines the type of bifurcation from which emanate the branches

$$\left(\begin{pmatrix} q^* \\ \lambda^* \end{pmatrix} + t\boldsymbol{u}, \beta^* + \beta(t) \right),$$

whose existence is guaranteed by Theorem 3.

As we showed in [21], the derivative $\beta'(0) \neq 0$ indicates a transcritical bifurcation. If $\beta'(0) = 0$, then the bifurcation is degenerate, and if $\beta''(0) \neq 0$, then we have a pitchfork-like bifurcation. Further, $t\beta'(t) < 0$ for small t indicates a subcritical bifurcating branch, and $t\beta'(t) > 0$ for small t indicates a supercritical bifurcating branch.

Expressions for $\beta'(0)$ and $\beta''(0)$ are derived as follows. Differentiating $k(t,\beta) = 0$ from (27) yields

$$d_t k(t,\beta(t)) + d_\beta k(t,\beta(t))\beta'(t) = 0, \tag{32}$$

so that $\beta'(t) = -\frac{d_t k(t,\beta(t))}{d_\beta k(t,\beta(t))}$. Differentiating (28) with respect to t and then evaluating at $t = 0$ shows that

$$\beta'(0) = \frac{-d^2_{\boldsymbol{x}}r(0,0)[\boldsymbol{x}_0,\boldsymbol{x}_0,\boldsymbol{x}_0]}{2||\boldsymbol{x}_0||^2\zeta} \tag{33}$$

where $d^2_{\boldsymbol{x}}r(0,0)[\boldsymbol{x}_0,\boldsymbol{x}_0,\boldsymbol{x}_0] = \sum_{i,j,k} \frac{\partial^2 r}{\partial[\boldsymbol{x}]_i\partial[\boldsymbol{x}]_j\partial[\boldsymbol{x}]_k}(0,0)[\boldsymbol{x}_0]_i[\boldsymbol{x}_0]_j[\boldsymbol{x}_0]_k$ (see (24)). As shown in the proof to Theorem 3, $\zeta = d_\beta k(0,0)$ is the non-zero eigenvalue of $d^2_{\boldsymbol{x},\beta}r(0,0)$ with eigenvector $\boldsymbol{x}_0$.

This expression is similar to the one given in [22] p. 90. The numerator can be calculated via (24). In [21], we showed that $\beta'(0) = 0$. We have the same result in the present case.

Theorem 5. *If q^* is M-singular for $1 < M \leq N$, then all of the bifurcating branches guaranteed by Theorem 3 are degenerate, i.e., $\beta'(0) = 0$.*

Proof. To show that the numerator of (33) $d_x^2 r(\mathbf{0},0) = \mathbf{0}$, expand r_i, the ith component of r, about $\mathbf{x} = \mathbf{0}$,

$$
\begin{aligned}
r_i(\mathbf{x}, \beta) &= r_i(\mathbf{0}, \beta) + d_x r_i(\mathbf{0}, \beta)^T \mathbf{x} + \mathbf{x}^T d_x^2 r_i(\mathbf{0}, \beta) \mathbf{x} + \mathcal{O}(\mathbf{x}^3) \\
&= d_x r_i(\mathbf{0}, \beta)^T \mathbf{x} + \mathbf{x}^T d_x^2 r_i(\mathbf{0}, \beta) \mathbf{x} + \mathcal{O}(\mathbf{x}^3),
\end{aligned}
$$

and so

$$
r_i(\mathbf{x}, 0) = \mathbf{x}^T d_x^2 r_i(\mathbf{0}, 0) \mathbf{x} + \mathcal{O}(\mathbf{x}^3).
$$

Applying the equivariance relation $Ar(\mathbf{x}, 0) = r(A\mathbf{x}, 0)$, where A is any element of the group isomorphic to S_M that acts on r in $\mathbf{R}^{M-1}$, and equating the quadratic terms yields

$$
A \begin{pmatrix} \mathbf{x}^T d_x^2 r_1 \mathbf{x} \\ \mathbf{x}^T d_x^2 r_2 \mathbf{x} \\ \vdots \\ \mathbf{x}^T d_x^2 r_{M-1} \mathbf{x} \end{pmatrix} = \begin{pmatrix} \mathbf{x}^T A^T d_x^2 r_1 A\mathbf{x} \\ \mathbf{x}^T A^T d_x^2 r_2 A\mathbf{x} \\ \vdots \\ \mathbf{x}^T A^T d_x^2 r_{M-1} A\mathbf{x} \end{pmatrix}.
$$

By (24), the diagonal $\frac{\partial^2 r_i}{\partial x_i \partial x_i}(\mathbf{0}, 0) = 0$ for each i as well as for all of the "multi-diagonals". This shows that $\frac{\partial^2 r_i}{\partial x_j \partial x_k}(\mathbf{0}, 0) = 0$ for every i, j, k (see Theorem 124 in [25]). $\square$

When $\beta'(0) = 0$, we need to compute $\beta''(0)$ to determine whether a branch is subcritical or supercritical. Differentiating (32) and setting $t = 0$ shows that $\beta''(0) = -\frac{d_t^2 k(0,0)}{d_\beta k(0,0)}$. Differentiating (28) twice and solving for $d_t^2 k(0,0)$ shows that

$$
\beta''(0) = \frac{-d_x^3 r(\mathbf{0}, 0)[\mathbf{x}_0, \mathbf{x}_0, \mathbf{x}_0, \mathbf{x}_0]}{3\|\mathbf{x}_0\|^2 \varsigma} \tag{34}
$$

where $W\mathbf{x}_0 = \mathbf{u} = \mathbf{u}_{(m,n)}$. Use Equation (25) to calculate the numerator, and $\varsigma = d_\beta k(\mathbf{0}, 0)$ is the non-zero eigenvalue of $d_{x,\beta}^2 r(0,0)$ with eigenvector $\mathbf{x}_0$, for which we give an explicit expression in (31) when F is an annealing problem.

If $\beta''(0) \neq 0$, which we expect to be true generically, then Theorem 5 shows that the bifurcation guaranteed by Theorem 3 is pitchfork-like.

2.10. Stability and Optimality

The next Theorem relates the stability of equilibria (q^*, λ^*, β) in the flow (8) with optimality of q^* in Problem (1). In particular, if a bifurcating branch corresponds to an eigenvalue of $d^2 \mathcal{L}(q^*)$ changing from negative to positive, then the branch consists of stationary points (q^*, β^*) that are not solutions of (1). Positive eigenvalues of $d^2 \mathcal{L}(q^*)$ do not necessarily show that q^* is not a solution of (1) (see Remark 1). For example, see page 668 of [21]. A proof of this theorem is given in [21].

Theorem 6. *For each bifurcating branch guaranteed by Theorem 3, $\mathbf{u}$ is an eigenvector of $d^2 \mathcal{L}(\begin{pmatrix} q^* \\ \lambda^* \end{pmatrix} + t\mathbf{u}, \beta^* + \beta(t))$ for sufficiently small t. Furthermore, if the corresponding eigenvalue is positive, then the branch consists of unstable stationary points that are not solutions to (1).*

2.11. Structure of the Symmetry Projection

The matrix $P_R(q^*)$ that projects $(q, \lambda) \in \Re^{NK+K}$ onto range $(d^2 \mathcal{L}(q^*)) \times \mathcal{K}^*$ by annihilating $\mathcal{K}_p$ is important for numerical computations for equilibria of IB, since we may want to take each equilibrium found by Newton's method and take out any part in $\mathcal{K}_p$. P_R is written as a function of q since its constitutive vectors $\mathbf{y}$ (from Definition 1) depend on q. The following theorems clarify the structure of this projection.

Theorem 7. $P_R(q) = I - P_{\mathcal{K}_p}(q)$, where $P_{\mathcal{K}_p} = \begin{pmatrix} A & 0 \\ 0 & 0 \end{pmatrix}$. P_R and $P_{\mathcal{K}_p}$ are $(NK + K) \times (NK + K)$. The matrix A is $NK \times NK$ with N^2 blocks, $\{A_{ij}\}_{i,j=1}^{N}$, of size $K \times K$, defined by

$$A_{i,j} = \begin{cases} (M-1)\boldsymbol{y}\boldsymbol{y}^T & \text{if } 1 \leq i = j \leq M \\ -\boldsymbol{y}\boldsymbol{y}^T & \text{if } 1 \leq i \neq j \leq M \\ 0 & \text{otherwise} \end{cases}$$

For example, if $M = N = 3$, then

$$P_R = I - \begin{pmatrix} 2\boldsymbol{y}\boldsymbol{y}^T & -\boldsymbol{y}\boldsymbol{y}^T & -\boldsymbol{y}\boldsymbol{y}^T & 0 \\ -\boldsymbol{y}\boldsymbol{y}^T & 2\boldsymbol{y}\boldsymbol{y}^T & -\boldsymbol{y}\boldsymbol{y}^T & 0 \\ -\boldsymbol{y}\boldsymbol{y}^T & -\boldsymbol{y}\boldsymbol{y}^T & 2\boldsymbol{y}\boldsymbol{y}^T & 0 \\ 0 & 0 & 0 & 0 \end{pmatrix} = I - \begin{pmatrix} (N-1) & -1 & -1 & 0 \\ -1 & (N-1) & -1 & 0 \\ -1 & -1 & (N-1) & 0 \\ 0 & 0 & 0 & 0 \end{pmatrix} \otimes \boldsymbol{y}\boldsymbol{y}^T.$$

Proof. Theorem 2 gives the basis of $\mathcal{K}_p$ as $\{Y_i\}_{i=1}^{M-1}$. Let Y be the $(NK + K) \times (M - 1)$ matrix whose columns are the vectors $\{Y_i\}$. For example, if $M = 3$ and $N = 4$, then

$$Y = \begin{pmatrix} \boldsymbol{y} & 0 \\ 0 & \boldsymbol{y} \\ -\boldsymbol{y} & -\boldsymbol{y} \\ 0 & 0 \\ 0 & 0 \end{pmatrix}.$$ Thus, the matrix that projects onto $\mathcal{K}_p$ is $P_{\mathcal{K}_p} = Y(Y^TY)^{-1}Y^T$,

and the projection matrix onto $\text{range}(d^2\mathcal{L}(q^*))$ is $P_R = I - P_{\mathcal{K}_p}$. Direct multiplication of $Y(Y^TY)^{-1}Y^T$, with an appeal to Lemma 34 in [25] to compute the inverse, shows that $P_{\mathcal{K}_p} = \frac{1}{N\boldsymbol{y}^T\boldsymbol{y}} \begin{pmatrix} A & 0 \\ 0 & 0 \end{pmatrix}$. Dropping the constant yields the result. $\square$

For the Information Bottleneck, the matrix P_R is easy to calculate, since $\boldsymbol{y} = q^i$ for any $i \in \mathcal{U}$. For example, when $q = q_{\frac{1}{N}}$, then $\boldsymbol{y}^T\boldsymbol{y} = \frac{K}{N^2}$ and $\boldsymbol{y}\boldsymbol{y}^T = \frac{1}{N^2}\boldsymbol{1}$, and so

$$P_{\mathcal{K}_p} = \frac{1}{NK} \begin{pmatrix} (N-1) & -1 & \cdots & -1 & 0 \\ -1 & (N-1) & \cdots & -1 & 0 \\ \vdots & \vdots & & \vdots & \vdots \\ -1 & -1 & \cdots & (N-1) & 0 \\ 0 & 0 & 0 & 0 & 0 \end{pmatrix} \otimes \boldsymbol{1}$$

where $\boldsymbol{1}$ is a $K \times K$ matrix of 1s. Thus,

$$P_R = I_{NK+K} - \begin{pmatrix} (N-1) & -1 & \cdots & -1 & 0 \\ -1 & (N-1) & \cdots & -1 & 0 \\ \vdots & \vdots & & \vdots & \vdots \\ -1 & -1 & \cdots & (N-1) & 0 \\ 0 & 0 & 0 & 0 & 0 \end{pmatrix} \otimes \boldsymbol{1}.$$

Theorem 8. *The symmetry group S_M commutes with the matrix P_R, which projects onto $\Re^{NK+K} \setminus \mathcal{K}_p$.*

Proof. Let $P := P_R$ be the matrix that projects onto $\text{range } d^2\mathcal{L}(q^*) \times \mathcal{K}^* = \Re^{NK+K} \setminus \mathcal{K}_p$. Since $\Re^{NK+K} = \text{range } d^2\mathcal{L}(q^*) \times \mathcal{K}^* \times \mathcal{K}_p$, then any $x \in \Re^{NK+K}$ can be decomposed in the respective subspaces as $x = r + k^* + k_p$. Let σ be an arbitrary permutation matrix in S_M. Then $\sigma P \begin{pmatrix} q \\ \lambda \end{pmatrix} = \sigma P(r + k^* + k_p) = \sigma(r + v)$. Since $\text{range } d^2\mathcal{L}(q^*)$, $\mathcal{K}^*$ and $\mathcal{K}_p$ are all S_M invariant; then $\sigma(r + v) \in \text{range } d^2\mathcal{L}(q^*) \times \mathcal{K}^*$ implies that $\sigma(r + v) = P\sigma(r + v)$, and $\sigma z \in \mathcal{K}_p$ implies that $P\sigma(r + v) = P\sigma(r + v + z)$. Thus, $\sigma Px = P\sigma x$. $\square$

2.12. Visualizations of Sample Results

We illustrate these structures numerically. In [7], we introduced the toy "Four-blob" probability distribution $p(x, y)$ shown in Figure 1.

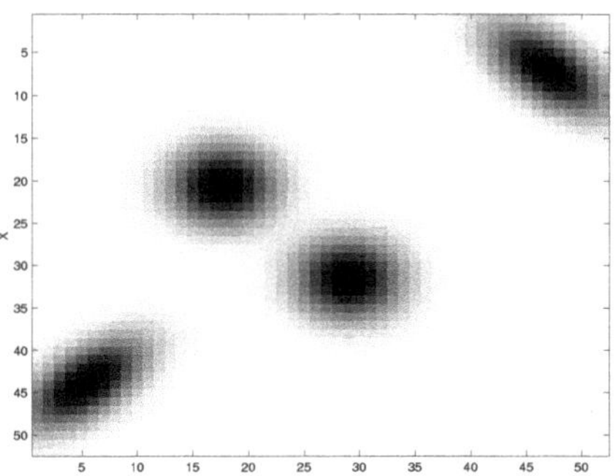

Figure 1. The probability distribution $p(x, y)$ for the "Four-blob" toy problem for a system of interest $X \to Y$. We use this probability to illustrate some results of the bifurcation analysis reported here.

For the Information Distortion problem (3) [7,12,13] and the synthetic dataset composed of a mixture of four Gaussians (Figure 1), we determined the bifurcation structure of solutions to (3) by annealing in β and finding the corresponding stationary points to (1). A typical run of the derived gradient dynamical system tends to follow the main bifurcation branch $S_K \to S_{K-1}$ from the fully symmetric uniform quantizer $q_{\frac{1}{N}}$ ($N = 4$ here) to the fully resolved deterministic quantizer (hard clustering) seen at the end in Figure 2. The permutation symmetry is also obvious there—the value of the cost function does not change if the classes along the vertical axis in T are permuted/relabeled. The uniform quantizer $q_{\frac{1}{N}}$ (Item 1 in the figure) plays a special role in the formulation (3), as it is the *unique* solution to the problem for $\beta = 0$ as the maximum entropy solution of $\max_q H(T|Y)$. Its loss of stability at the first bifurcation for increasing β can hence be determined analytically and the first bifurcation structure characterized completely. Because of the "perpetual kernel" of the cost function in (4), the uniform quantizer is just one of a continuous set of "uninformative" quantizers for the IB problem (4): all $\{q(t|y) : q(t|y) = f(t)\}$, having constant probability of assignment of each y to class t, but the assignment weight can be different for different classes. Such a structure does not change the value of the cost function in the IB problem (4) (but does change it for (3), which hence does not have this degeneracy). We address the degeneracy of the IB optimization by projecting onto the subspace that has the correct symmetry (i.e., just the uniform quantizer $q_{\frac{1}{N}}$ in this case), as outlined in Remark 2.

A more-thorough structure of the bifurcation diagram, using the analysis presented above, is shown in Figure 3.

Similar to the results we presented in [28], the close-up of the bifurcation at $\beta \approx 1.038706$ in Figure 3B shows a subcritical bifurcating branch (a first-order phase transition) that consists of stationary points of Problem (1). By projecting the Hessian $\Delta_q(G(q^*) + \beta D(q^*))$ onto each of the kernels referenced in Theorem 6, we determined that the points on this subcritical branch are **not** solutions of (1), and yet they **are** solutions of (2).

Furthermore, observe that Figure 3B indicates that a saddle-node bifurcation occurs at $\beta \approx 1.037479$. That this is indeed the case was proved in [21]. In fact, for any problem of the form (2), these are the only two types of bifurcations to be expected: pitchfork and saddle-node.

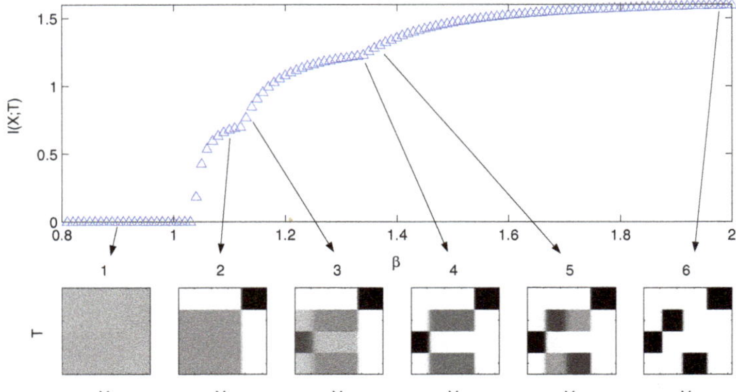

Figure 2. The bifurcations of the solutions (q^*, β) to the Information Distortion problem (3). For the mixture of 4 well-separated Gaussians shown in Figure 1, the behavior of $D(q) = I(X; T)$ as a function of β is shown in the top panel, and some of the solutions $q^*(T|Y)$ are shown in the bottom panels. Item 1 shows the uniform quantizer $q_{\frac{1}{N}}$, assigning equal probability of each $y \in Y$ to belong to one of the four clusters in T. Subsequent items 2–5 point to a set of partially resolved quantizations, in which subsets of Y are assigned with high probability to one (2) or more (3–5) classes (dark colors, close to 1), while other subsets are still unresolved (gray levels), albeit as a higher probability than $q_{\frac{1}{N}}$ (darker gray, as some of the classes are excluded after being resolved for another subset). Item 6 shows an almost fully resolved quantizer at sufficiently high β. They become fully resolved (deterministic; $q(t|y) = 1$ or 0) as $\beta \to \infty$ (not shown).

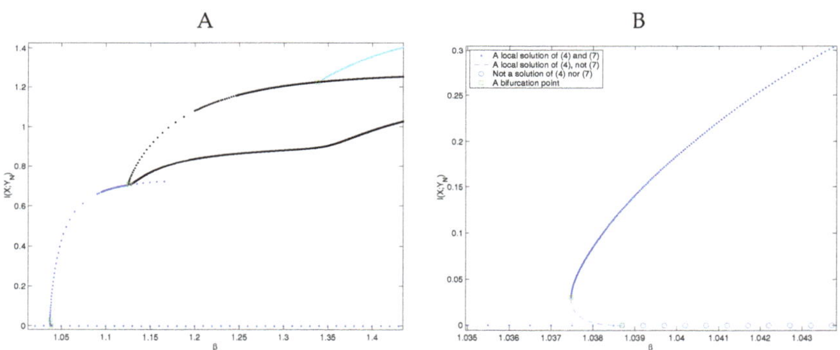

Figure 3. (**A**) The bifurcation structure of stationary points of the Information Distortion problem (3), a problem of form (2). We found these points by annealing in β and finding stationary points for Problem (1) using the algorithm presented in [28]. A square indicates where a bifurcation occurs. (**B**) A close-up of the subcritical bifurcation at $\beta \approx 1.038706$, indicated by a square. Observe the subcritical bifurcating branch, and the subsequent saddle-node bifurcation at $\beta \approx 1.037479$, indicated by another square. We applied Theorem 6 to show that the subcritical bifurcating branch is composed of quantizers that are solutions of (3) but not of (1).

3. Conclusions and Discussion

The main goal of this contribution was to show that information-based distortion-annealing problems such as (2) have an interesting mathematical structure. The most interesting aspects of that mathematical structure are driven by the symmetries present in the cost functions—their invariance to actions of the permutation group S_N, represented as relabeling of the reproduction classes. Such a structure would hold for any biclustering problem [4] that relies on the intrinsic interaction of a pair of variables for unsupervised clustering. The second mathematical structure that we used successfully was bifurcation theory, which allowed us to identify and study the discrete points at which the character of the cost function changed. The combination of those two tools in [20] allowed us to explicitly compute the value of the annealing parameter β at which the initial maximum at the uniform quantizer $q_{\frac{1}{N}}$ of (1) loses stability. We concluded that for a fixed system $C \to Y$ characterized by $p(X, Y)$, this value is the same for both problems, that it does not depend on the number of elements of the reproduction variable T, and that it is always greater than 1. We further introduced an eigenvalue problem that links the critical values of β and q for bifurcations, or phase transitions, branching off arbitrary intermediate solutions.

Even though the cost functions F_{IB} (4) and F_H (3) have similar properties, they also differ in some important aspects. We have shown that the function F_{IB} is degenerate since its constitutive functions $I(X; Y)$ and $I(X; T)$ are not strictly convex in q. That introduces additional invariances and singularities that are always preserved, which makes phase transitions more difficult to detect (e.g., the "uninformative quantizers" $q(t|y) = f(t)$ only) and post-transition directions more difficult to determine. In contrast, F_H is strictly convex except at points of phase transitions. The theory we developed here allows us to identify bifurcation directions and determine their stability. Despite the presence of a high-dimensional null space at bifurcations, the symmetries restrict the allowed transition dimensions to multiple co-dimension 1 transitions, all related by group transformations. We achieved that here with three main results. Theorem 8 extended the Equivariant Branching Lemma 1 to the Information Bottleneck case with additional translation invariance. Theorem 4 identified specific conditions at which a bifurcation of the gradient flow (8) occurs. This condition is computable analytically for the initial bifurcation off the uniform quantizer $q_{\frac{1}{N}}$ and with numeric continuation for subsequent bifurcation. Finally, in Section 2.9, we provided checks for the types of bifurcations that occur, giving conditions to detect saddle-node and pitchfork bifurcations and to determine whether pitchforks are supercritical (second-order phase transitions) or subcritical (leading to first-order phase transitions discontinuous in β). The combination of the three results, together with our previous results in [20], completely characterize the local bifurcation structure of Information Bottleneck-type problems with or without the added translation symmetry.

Despite the further development of the bifurcation formalism for IB presented her, there are still open questions that this manuscript did not resolve. In particular, we still cannot confirm or reject the conjecture that the set of S_K symmetric soft-clustering branches connected through symmetry-breaking bifurcations leads to the global hard-clustering optima at $\beta \to \infty$ (multiple equivalent solutions connected by the permutation symmetry of the problem). We believe this is partially due to a discrepancy between practical observations and theoretical results. In particular, we and other practitioners [29,30] note that the only observed symmetry-breaking bifurcations during optimization are of the kind $S_M \to S_{M-1}$, while the theory allows for arbitrary $S_M \to S_m \times S_n$ bifurcations. The latter are known to happen and be stable in other biological systems and circumstances [26,31]. This suggests a research approach of comparing and contrasting the different systems that possess the same S_N symmetry and symmetry-breaking bifurcations to lead to breakthroughs in this application to optimization in the Information Bottleneck problem.

An additional open problem involves the use of continuous variables, already noted in [5] and explored further in [32,33]. This approach, while important for many real-world problems, involves the application of additional mathematical tools, namely Calculus of Variations [34], which further increases the complexity of an otherwise already complex problem. These difficulties are illustrated in a pair of papers [35,36] that use the continuous formulation. They do present some significant results on conditions of learnability, but both papers manage to only get bounds on β under which learnability (optimal solutions beyond the "uninformative" quantizer) can be achieved. This is possibly due to the presence of continuous spectra in covariance operators of continuous quantizers, something that we avoid by focusing on finite spaces. As a consequence, here and in prior work [20], we show specific values for β for the initial bifurcation from the uniform quantizer, which supports nontrivial clustering. We consider formulation with continuous variables beyond the scope of this manuscript, but look forward to the development of additional techniques to incorporate this important case in the bifurcation framework presented here. Regardless of such developments, any practical problem with numeric optimization will involve discretization of the continuous variables, which effectively converts a continuous problem to the discrete state discussed here.

Author Contributions: Conceptualization, A.G.D.; Formal analysis, A.G.D. and A.E.P.; Investigation, A.G.D. and A.E.P.; Writing–original draft, A.G.D. and A.E.P. All authors have read and agreed to the published version of the manuscript.

Funding: This research received no external funding.

Institutional Review Board Statement: Not applicable.

Informed Consent Statement: Not applicable.

Data Availability Statement: Not applicable.

Conflicts of Interest: The authors declare no conflict of interest.

References

1. Gray, R.M. *Entropy and Information Theory*; Springer: Berlin/Heidelberg, Germany, 1990.
2. Cover, T.; Thomas, J. *Elements of Information Theory*; Wiley Series in Communication; Wiley: New York, NY, USA, 1991.
3. Rose, K. Deteministic Annealing for Clustering, Compression, Classification, Regression, and Related Optimization Problems. *Proc. IEEE* **1998**, *86*, 2210–2239. [CrossRef]
4. Madeira, S.C.; Oliveira, A.L. Biclustering algorithms for biological data analysis: A survey. *IEEE/ACM Trans. Comput. Biol. Bioinform.* **2004**, *1*, 24–45. [CrossRef] [PubMed]
5. Tishby, N.; Pereira, F.C.; Bialek, W. The information bottleneck method. In *37th Annual Allerton Conference on Communication, Control, and Computing*; University of Illinois: Champaign, IL, USA, 1999.
6. Dimitrov, A.G.; Miller, J.P.; Aldworth, Z.; Gedeon, T.; Parker, A.E. Analysis of neural coding through quantization with an information-based distortion measure. *Netw. Comput. Neural Syst.* **2003**, *14*, 151–176. [CrossRef]
7. Dimitrov, A.G.; Miller, J.P. Neural coding and decoding: Communication channels and quantization. *Netw. Comput. Neural Syst.* **2001**, *12*, 441–472. [CrossRef]
8. Gersho, A.; Gray, R.M. *Vector Quantization and Signal Compression*; Kluwer Academic Publishers: New York, NY, USA, 1992.
9. Mumey, B.; Gedeon, T. Optimal mutual information quantization is NP-complete. In Proceedings of the Neural Information Coding (NIC) Workshop, Snowbird, UT, USA, 1–4 March 2003.
10. Slonim, N.; Tishby, N. Agglomerative Information Bottleneck. In *Advances in Neural Information Processing Systems*; Solla, S.A., Leen, T.K., Müller, K.R., Eds.; MIT Press: Cambridge, MA, USA, 2000; Volume 12, pp. 617–623.
11. Slonim, N. The Information Bottleneck: Theory and Applications. Ph.D. Thesis, Hebrew University, Jerusalem, Israel, 2002.
12. Dimitrov, A.G.; Miller, J.P. Analyzing sensory systems with the information distortion function. In *Proceedings of the Pacific Symposium on Biocomputing 2001*; Altman, R.B., Ed.; World Scientific Publishing Co.: Singapore, 2000.
13. Gedeon, T.; Parker, A.E.; Dimitrov, A.G. Information Distortion and Neural Coding. *Can. Appl. Math. Q.* **2003**, *10*, 33–70.
14. Slonim, N.; Somerville, R.; Tishby, N.; Lahav, O. Objective classification of galaxy spectra using the information bottleneck method. *Mon. Not. R. Astron. Soc.* **2001**, *323*, 270–284. [CrossRef]
15. Bardera, A.; Rigau, J.; Boada, I.; Feixas, M.; Sbert, M. Image segmentation using information bottleneck method. *IEEE Trans. Image Process.* **2009**, *18*, 1601–1612. [CrossRef]
16. Aldworth, Z.N.; Dimitrov, A.G.; Cummins, G.I.; Gedeon, T.; Miller, J.P. Temporal encoding in a nervous system. *PLoS Comput. Biol.* **2011**, *7*, e1002041. [CrossRef]
17. Buddha, S.K.; So, K.; Carmena, J.M.; Gastpar, M.C. Function identification in neuron populations via information bottleneck. *Entropy* **2013**, *15*, 1587–1608. [CrossRef]
18. Lewandowsky, J.; Bauch, G. Information-optimum LDPC decoders based on the information bottleneck method. *IEEE Access* **2018**, *6*, 4054–4071. [CrossRef]
19. Parker, A.E.; Dimitrov, A.G.; Gedeon, T. Symmetry breaking in soft clustering decoding of neural codes. *IEEE Trans. Inf. Theory* **2010**, *56*, 901–927. [CrossRef]
20. Gedeon, T.; Parker, A.E.; Dimitrov, A.G. The mathematical structure of information bottleneck methods. *Entropy* **2012**, *14*, 456–479. [CrossRef]
21. Parker, A.E.; Gedeon, T. Bifurcations of a class of S_N-invariant constrained optimization problems. *J. Dyn. Differ. Equ.* **2004**, *16*, 629–678. [CrossRef]
22. Golubitsky, M.; Stewart, I.; Schaeffer, D.G. *Singularities and Groups in Bifurcation Theory II*; Springer: New York, NY, USA, 1988.

23. Golubitsky, M.; Schaeffer, D.G. *Singularities and Groups in Bifurcation Theory I*; Springer: New York, NY, USA, 1985.
24. Nocedal, J.; Wright, S.J. *Numerical Optimization*; Springer: New York, NY, USA, 2000.
25. Parker, A.E. Symmetry Breaking Bifurcations of the Information Distortion. Ph.D. Thesis, Montana State University, Bozeman, MT, USA, 2003.
26. Golubitsky, M.; Stewart, I. *The Symmetry Perspective: From Equilibrium to Chaos in Phase Space and Physical Space*; Birkhauser Verlag: Boston, MA, USA, 2002.
27. Schott, J.R. *Matrix Analysis for Statistics*; John Wiley and Sons: New York, NY, USA, 1997.
28. Parker, A.; Gedeon, T.; Dimitrov, A. Annealing and the rate distortion problem. In *Advances in Neural Information Processing Systems 15*; Becker, S.T., Obermayer, K., Eds.; MIT Press: Cambridge, MA, USA, 2003; Volume 15, pp. 969–976.
29. Dimitrov, A.G.; Cummins, G.I.; Baker, A.; Aldworth, Z.N. Characterizing the fine structure of a neural sensory code through information distortion. *J. Comput. Neurosci.* **2011**, *30*, 163–179. [CrossRef] [PubMed]
30. Schneidman, E.; Slonim, N.; Tishby, N.; de Ruyter van Steveninck, R.R.; Bialek, W. Analyzing neural codes using the information bottleneck method. In *Advances in Neural Information Processing Systems*; MIT Press: Cambridge, MA, USA, 2003; Volume 15.
31. Stewart, I. Self-Organization in evolution: A mathematical perspective. *Philos. Trans. R. Soc.* **2003**, *361*, 1101–1123. [CrossRef] [PubMed]
32. Chechik, G.; Globerson, A.; Tishby, N.; Weiss, Y. Information bottleneck for Gaussian variables. In Proceedings of the Advances in Neural Information Processing Systems 16 (NIPS 2003), Vancouver, BC, Canada, 8–13 December 2003.
33. Chechik, G.; Globerson, A.; Tishby, N.; Weiss, Y. Information Bottleneck for Gaussian Variables. *J. Mach. Learn. Res.* **2005**, *6*, 165–188.
34. Gelfand, I.M.; Fomin, S.V. *Calculus of Variations*; Dover Publications: Mineola, NY, USA 2000.
35. Wu, T.; Fischer, I.; Chuang, I.L.; Tegmark, M. Learnability for the information bottleneck. In Proceedings of the Uncertainty in Artificial Intelligence, PMLR, Virtual, 3–6 August 2020; pp. 1050–1060.
36. Ngampruetikorn, V.; Schwab, D.J. Perturbation theory for the information bottleneck. *Adv. Neural Inf. Process. Syst.* **2021**, *34*, 21008–21018.

Article

The Information Bottleneck's Ordinary Differential Equation: First-Order Root Tracking for the Information Bottleneck

Shlomi Agmon

School of Computer Science and Engineering, The Hebrew University of Jerusalem, Jerusalem 9190401, Israel; shlomi.agmon@mail.huji.ac.il

Abstract: The Information Bottleneck (IB) is a method of lossy compression of relevant information. Its rate-distortion (RD) curve describes the fundamental tradeoff between input compression and the preservation of relevant information embedded in the input. However, it conceals the underlying dynamics of optimal input encodings. We argue that these typically follow a piecewise smooth trajectory when input information is being compressed, as recently shown in RD. These smooth dynamics are interrupted when an optimal encoding changes qualitatively, at a *bifurcation*. By leveraging the IB's intimate relations with RD, we provide substantial insights into its solution structure, highlighting caveats in its finite-dimensional treatments. Sub-optimal solutions are seen to collide or exchange optimality at its bifurcations. Despite the acceptance of the IB and its applications, there are surprisingly few techniques to solve it numerically, even for finite problems whose distribution is known. We derive anew the IB's first-order Ordinary Differential Equation, which describes the dynamics underlying its optimal tradeoff curve. To exploit these dynamics, we not only detect IB bifurcations but also identify their type in order to handle them accordingly. Rather than approaching the IB's optimal tradeoff curve from sub-optimal directions, the latter allows us to follow a solution's trajectory along the optimal curve under mild assumptions. We thereby translate an understanding of IB bifurcations into a surprisingly accurate numerical algorithm.

Keywords: Information Bottleneck; bifurcations; ordinary differential equation; numerical approximation

Citation: Agmon, S. The Information Bottleneck's Ordinary Differential Equation: First-Order Root Tracking for the Information Bottleneck. *Entropy* **2023**, *25*, 1370. https://doi.org/10.3390/e25101370

Academic Editors: Jan Lewandowsky and Gerhard Bauch

Received: 16 May 2023

Revised: 8 August 2023

Accepted: 9 August 2023

Published: 22 September 2023

1. Introduction

The Information Bottleneck (IB) describes the fundamental tradeoff between the compression of information on an *input* X to the preservation of relevant information on a hidden *reference* variable Y. Formally, let X and Y be random variables defined, respectively, on finite *source* and *label alphabets* $\mathcal{X}$ and $\mathcal{Y}$, and let $p_{Y|X}(y|x)p_X(x)$ be their joint probability distribution, or $p(y|x)p(x)$ for short (without loss of generality, $p(x) > 0$ for every $x \in \mathcal{X}$ and so $p_{Y|X}$ is well-defined). One seeks [1] to maximize the information $I(Y; \hat{X})$ over all Markov chains $Y \longleftrightarrow X \longleftrightarrow \hat{X}$, subject to a constraint on the mutual information $I(X; \hat{X}) := \mathbb{E}_{p(\hat{x}|x)p(x)} \log \frac{p(\hat{x}|x)}{p(\hat{x})}$,

$$I_Y(I_X) := \max_{p(\hat{x}|x)} \left\{ I(Y; \hat{X}) : I(X; \hat{X}) \leq I_X \right\} . \tag{1}$$

The latter maximization is over conditional probability distributions or *encoders* $p(\hat{x}|x)$. The graph of $I_Y(I_X)$ is the *IB curve*. We write $T := |\hat{\mathcal{X}}|$, for a codebook or *representation alphabet* $\hat{\mathcal{X}}$. An encoder $p(\hat{x}|x)$ which achieves the maximum in (1) is *IB optimal* or simply *optimal*.

Written in a Lagrangian formulation $\mathcal{L} := I(X; \hat{X}) - \beta\, I(Y; \hat{X})$ with $\beta > 0$ (normalization constraints omitted for clarity), [1] showed that a necessary condition for extrema in (1) is that the *IB Equations* hold. Namely,

$$p(\hat{x}|x) = \frac{p(\hat{x})}{Z(x,\beta)} \exp\left\{-\beta\, D_{KL}\left[p(y|x)\|p(y|\hat{x})\right]\right\},\tag{2}$$

$$p(y|\hat{x}) = \sum_x p(y|x)p(x|\hat{x}),\quad \text{and}\tag{3}$$

$$p(\hat{x}) = \sum_x p(\hat{x}|x)p(x).\tag{4}$$

In these, $Z(x,\beta) := \sum_{\hat{x}} p(\hat{x}) \exp\left\{-\beta D_{KL}[p(y|x)\|p(y|\hat{x})]\right\}$ is the *partition function*, D_{KL} is the Kullback–Leibler divergence, $D_{KL}[p\|q] := \sum_i p(i)\log p(i)/q(i)$, and $p(x|\hat{x})$ in (3) is defined by the Bayes rule $p(\hat{x}|x)p(x)/p(\hat{x})$. The IB Equations (2)–(4) are a necessary condition for an extremum of $\mathcal{L}$ also when it is considered as a functional in three independent families of normalized distributions $\{p(\hat{x}|x)\}$, $\{p(y|\hat{x})\}$ and $\{p(\hat{x})\}$, ref. [1] (Section 3.3), rather than in $\{p(\hat{x}|x)\}$ alone. While satisfying them is necessary to achieve the curve (1), it is not sufficient. Indeed, Equations (2)–(4) have solutions that do not achieve curve (1), and so are *sub-optimal*. This results in sub-optimal IB curves, which intersect or *bifurcate* as the multiplier β varies (see Section 3.4 in [1]).

Iterating over the IB Equations (2)–(4) is essentially Blahut–Arimoto's algorithm variant for the IB (BA-IB) due to [1], brought below for reference. While the minimization problem (1) can be solved exactly in special cases [2] (Section IV), exact solutions of an arbitrary finite IB problem whose distribution is known are usually obtained nowadays using BA-IB; see [3] (Section 3) for a survey of other computation approaches. Write BA_β for a single iteration of BA-IB. Since BA_β encodes an iteration over the IB Equations (2)–(4), then an encoder $p(\hat{x}|x)$ is its fixed point, $BA_\beta[p(\hat{x}|x)] = p(\hat{x}|x)$, if and only if it satisfies the IB Equations. Or equivalently, if $p(\hat{x}|x)$ is a root of the *IB operator*,

$$F := Id - BA_\beta,\tag{5}$$

in a manner similar to [4]. We shall then call it an *IB root*. Agmon et al. [4] used a similar formulation of rate-distortion (RD) and its relations in [5] to the IB, to show that BA-IB suffers from *critical slowing down* near *critical points*, where the marginal $p(\hat{x})$ of a representor $\hat{x}$ in an optimal encoder vanishes gradually. That is, the number of BA-IB iterations required until convergence increases dramatically as one approaches such points.

Formulating fixed points of an iterative algorithm as operator roots can also be leveraged for computational purposes in a constrained optimization problem, as noted recently by [6] for RD. Indeed, let $F(\cdot,\beta)$ be a differentiable operator on $\mathbb{R}^n$ for some $n > 0$, $F : \mathbb{R}^n \times \mathbb{R} \to \mathbb{R}^n$, where β is a (real) constraint parameter. Suppose now that (x,β) is a root of F,

$$F(x,\beta) = 0,\tag{6}$$

such that $x = x(\beta)$ is a differentiable function of β. Write $D_x F := \left(\frac{\partial}{\partial x_j}F_i\right)_{i,j}$ for its Jacobian matrix, and $D_\beta F := \left(\frac{\partial}{\partial \beta}F_i\right)_i$ for its vector of partial derivatives with respect to β. The point (x,β) of evaluation is omitted whenever understood. As is often discussed along with the Implicit Function Theorem, e.g., [7], applying the multivariate chain rule to $F(x(\beta),\beta)$ in (6) yields an implicit ordinary differential equation (ODE)

$$D_x F\, \frac{dx}{d\beta} = -D_\beta F,\tag{7}$$

for the roots of F. Plugging in explicit expressions for the first-order derivative tensors $D_x F$ and $D_\beta F$, one can specialize (7) to a particular setting, which allows one to compute the *implicit derivatives* $\frac{dx}{d\beta}$ numerically. While [6] discovered the *RD ODE* this way, they showed that (7) can be generalized to arbitrary order under suitable differentiability assumptions. Namely, they showed that the derivatives $\frac{d^l x}{d\beta^l}$ implied by $F = 0$ (6) can be computed via a recursive formula, for an arbitrary-order $l > 0$. By specializing this with the higher derivatives of Blahut's algorithm [8], they obtained a family of numerical algorithms for following the path of an optimal RD root (Part I there).

In this work, we specialize the implicit ODE (7) to the IB. Namely, we plug into (7) the first-order derivatives of the IB operator $Id - BA_\beta$ (5) to obtain the *IB ODE*, and then use it to reconstruct the path of an optimal IB root, in a manner similar to [6]. This is not to be confused with the gradient flow (of arbitrary encoders) towards an optimal root at a fixed β value, described in [9] (Equation (6)) by an ODE, which is a different optimization approach. In contrast, the implicit Equation (7) describes how a root evolves *with* β. So, in principle, one may compute an optimal IB root once and then follow its evolution along the IB curve (1). While the discovery of the IB ODE is due to [10], we derive it here anew in a form that is better suited for computational (and other) purposes, especially when there are fewer possible labels $\mathcal{Y}$ than input symbols $\mathcal{X}$, as often is the case. To that end, we consider several natural choices of a coordinate system for the IB in Section 2 and compare their properties. This allows us to make an apt choice for the ODE's variable x in (7). In Section 3, we present the IB ODE in these coordinates (Theorem 1). This enables one to numerically compute the first-order implicit derivatives at an IB root, if it can be written as a differentiable function in β. So long as an optimal root remains differentiable, a simple way to reconstruct its trajectory is by taking small steps at a direction determined by the IB ODE. This is *Euler's method* for the IB. The error accumulated by Euler's method from the true solution path is roughly proportional to the step size, when small enough. For comparison, reverse deterministic annealing [11] with BA-IB is nowadays common for computing IB roots. The dependence of its error on the step size is roughly the same as in Euler's method. This is discussed in Section 4, where we combine Euler's method with BA-IB to obtain a modified numerical method whose error decreases at a faster rate than either of the above.

However, the differentiability of optimal IB roots breaks where the solution changes qualitatively. Such a point is often called a phase transition in the IB literature, or a *bifurcation*—namely, a point where there is a change in the problem's number of solutions; e.g., [12] (Section 2.3) for basic definitions. As noted already by Tishby et al. in [1], their existence in the IB stems from restricting the cardinality of the representation alphabet $\hat{\mathcal{X}}$. Since IB roots are the solutions of the fixed-point Equations (2)–(4), then the gap between achieving the IB curve (1) to merely satisfying these Equations lies in understanding the solution structure of the IB operator (5), or equivalently its bifurcations. While IB bifurcations were analyzed in several works, including [9,13,14] and others, little is known about the practical value of understanding them. In [15,16] it was shown that they correspond to the onset of learning new classes, and in [4] that they inflict a hefty computational cost to BA-IB. Following [6], this work demonstrates that understanding bifurcations can be translated to a new numerical algorithm to solve the IB. To that end, merely detecting a bifurcation along a root's path does not suffice. Rather, it is also necessary to identify its type, as this allows one to handle the bifurcation accordingly. One can then continue following the path dictated by the IB ODE.

Almost all of the literature on IB bifurcations is based on a perturbative approach, in a manner similar to [17] (Section IV.C). That is, suppose that the first variation

$$\frac{\partial}{\partial \epsilon} \mathcal{L}[p(\hat{x}|x) + \epsilon \Delta p(\hat{x}|x); \beta]\Big|_{\epsilon=0} \tag{8}$$

of the IB Lagrangian $\mathcal{L}$ vanishes, for every perturbation $\Delta p(\hat{x}|x)$. This condition is necessary for extremality and implies [1] the IB Equations (2)–(4). Then, $(p(\hat{x}|x), \beta)$ is said to be a *phase transition* only if there exists a particular direction $\Delta q(\hat{x}|x)$ at which $p(\hat{x}|x)$ can be perturbed without affecting the Lagrangian's value to second order,

$$\frac{\partial^2}{\partial \epsilon^2} \mathcal{L}[p(\hat{x}|x) + \epsilon \Delta q(\hat{x}|x); \beta]\Big|_{\epsilon=0} = 0 . \tag{9}$$

For finite IB problems, condition (8) boils down to requiring that the gradient of $\mathcal{L}$ vanishes, while condition (9) is equivalent to requiring that its Hessian matrix has a non-trivial kernel (as both are conditions on the directional derivatives, e.g., [18]). The works [9,14–16] take such an approach, while [13] focuses on one type of IB bifurcations.

While a perturbative approach is common in analyzing phase transitions, it has several shortcomings when applied to the IB, as noted by [10]. First, the IB's Lagrangian $\mathcal{L}$ is constant on a linear manifold of encoders $p(\hat{x}|x)$ [9] (Section 3.1), and so condition (9) leads to false detections. While this was considered there and in its sequel [19] by giving subtle conditions on the nullity of the second variation in (9), in practice it is difficult to tell whether a particular direction $\Delta q(\hat{x}|x)$ is in the kernel due to a bifurcation or due to other reasons, as they note. Second, note that a finite IB problem can be written as an *infinite* RD problem [20]. As discussed in Section 5, representing an IB root by a finite-dimensional vector leads to inherent subtleties in its computation. Among other things, these may well result in a bifurcation *not* being detectable under certain circumstances (Section 5.3). To our understanding, many of the difficulties that hindered the understanding of IB bifurcations throughout the years are, in fact, artifacts of finite dimensionality. Third, conditions (8) and (9) do not suffice to reveal the type of the bifurcation, information which is necessary for handling it when following a root's path. While [19] (Section 2.9) give conditions for identifying the type, these partially agree with our findings and do not suggest a straightforward way for handling a bifurcation.

Rather than imposing conditions on the scalar functional $\mathcal{L}$, our approach to IB bifurcations follows that of [6] for RD. That is, we rely on the fact that the IB's local extrema are fixed points of an iterative algorithm, and so they also satisfy a vector equation $F = \mathbf{0}$ (6). We shall now consider a toy problem to motivate our approach. *"Bifurcation Theory can be briefly described by the investigation of problem (6) in a neighborhood of a root where $D_x F$ is singular"* [21]. Indeed, recall that if $D_x F$ is non-singular at a root (x_0, β_0), then by the Implicit Function Theorem (IFT), there exists a function $x(\beta)$ through the root, $x(\beta_0) = x_0$, which satisfies $F\big(x(\beta), \beta\big) = \mathbf{0}$ (6) at the vicinity of β_0. The function $x(\beta)$ is then not only unique at some neighborhood of (x_0, β_0), but further, $x(\beta)$ inherits the differentiability properties of F [21] (I.1.7). In particular, if the operator F is real-analytic in its variables—as with the IB operator (5)—then so is its root $x(\beta)$. While a bifurcation can occur only if $D_x F$ is singular, singularity is not sufficient for a bifurcation to occur. For example, the roots of the operator

$$F(x, y; \beta) := (x - \beta, 0) \tag{10}$$

on $\mathbb{R}^2$ consist of the vertical line $x = \beta$, $\{(\beta, y) : y \in \mathbb{R}\}$, for every $\beta \in \mathbb{R}$. For a fixed y, each such root is real-analytic in β. However, one cannot deduce this directly from the IFT, as the Jacobian $\begin{pmatrix} 1 & 0 \\ 0 & 0 \end{pmatrix}$ of F (10) is always singular. Note, however, that in this particular example, the x coordinate alone suffices to describe the problem's dynamics, and so its y coordinate is redundant. One can ignore the y coordinate by considering the "reduction" $\tilde{F}(x; \beta) := x - \beta$ of F to $\mathbb{R}^1$. Further, discarding y also removes or *mods-out* the direction $\begin{pmatrix} 0 \\ 1 \end{pmatrix}$ from $\ker D_x F$, which does not pertain to a bifurcation in this case. This results in the non-singular Jacobian matrix (1) of $\tilde{F}$, and so it is now possible to invoke the IFT on the reduced problem. The root guaranteed by the IFT can always be considered in $\mathbb{R}^2$ by putting back a redundant y coordinate at some fixed value. In [6], a similarly defined *reduction* of finite RD problems was used to show that their dynamics are piecewise real-analytic under mild assumptions.

The intuition behind our approach is similar to [20] (Section III), who observed that *"in the IB one can also get rid of irrelevant variables <u>within</u> the model"*. Nevertheless, the details differ. Mathematically, we consider the *quotient* V/W of a vector space V by its subspace W. Elements of V are identified in the quotient if they differ by an element of W: $v_1 \sim v_2 \Leftrightarrow v_1 - v_2 \in W$, for $v_1, v_2 \in V$. This way, one "mods-out" W, collapsing it to a single point in the quotient vector space V/W. The resulting problem is smaller and so easier to handle, whether for theoretical or practical purposes (although not needed for our purposes, this can be made precise in terms of the *tangent space* of a differentiable manifold; cf., Section 3 in [22]). This is how the one-dimensional vector space ker $D_x F$ in our toy example (10) was reduced to the trivial ker $D_x \tilde{F} = \{0\}$. However, one needs to understand the solution structure, for example, to ensure that the directions in W are not due to a bifurcation. We note in passing that V/W has a simple geometric interpretation as the translations of W in V, in a manner reminiscent of its better-known counterparts of quotient groups and rings; e.g., [23] (Section 10.2). To keep things simple, however, we shall not use quotients explicitly. Instead, the reader may simply consider the sequel as a removal of redundant coordinates, for we shall only remove coordinates that the reader does not care about anyway, as in the above toy example.

To achieve this approach, one needs to consider the IB in a coordinate system that permits a simple reduction as in (10), and to understand its solution structure. We achieve these in Section 5 by exploiting two properties of the IB which are often overlooked. First, proceeding with the coordinates' exchange of Section 2, the intimate relations [5,20] of the IB with RD suggest a "minimally sufficient" coordinate system for the IB, just as the x axis is for problem (10). *Reducing* an IB root to these coordinates is a natural extension of reduction in RD [6]. Reduction of IB roots facilitates a clean treatment of IB bifurcations. These are roughly divided into *continuous* and *discontinuous* bifurcations, in Sections 5.2 and 5.3, respectively. While understanding continuous bifurcations is straightforward, the IB's relations with RD allow us to understand the discontinuous bifurcation examples of which we are aware as a *support switching bifurcation* in RD, by leveraging [6] (Section 6). A second property is the analyticity of the IB operator (5), which stems from the analyticity of the IB Equations (2)–(4). By building on the first property, analyticity leads us to argue that the Jacobian of the IB operator (5) is generally non-singular (Conjecture 1) when considered in reduced coordinates as above. As an immediate consequence, the dynamics underlying the IB curve (1) are piecewise real-analytic in β, in a manner similar to RD. Indeed, the fact that there exist dynamics underlying the IB curve (1) in the first place can arguably be attributed to analyticity (see the discussion following Conjecture 1). Combining both properties sheds light on several subtle yet important practical caveats in solving the IB (Section 5.3) due to using finite-dimensional representations of its roots. These subtleties are compatible with our numerical experience. The results here suggest that, unlike RD, the IB is inherently infinite-dimensional, even for finite problems.

Finally, Section 6 combines the modified Euler method of Section 4 with the understanding of IB bifurcations in Section 5, to obtain an algorithm for following the path of an optimal IB root, in Section 6.1. That is, First-order Root Tracking for the IB (IBRT1). For simplicity, we focus mainly on continuous IB bifurcations, as these are the ones most often encountered in practice (see Section 6.3 on the algorithm's handling of discontinuous bifurcations). The resulting approximations in the information plane are surprisingly close to the true IB curve (1), even on relatively sparse grids (i.e., with large step sizes), as seen in Figure 1. See Section 6.2 for the numerical results underlying the latter. The reasons for this are discussed in Section 6.3, along with the algorithm's basic properties. Unlike BA-IB, which suffers from an increased computational cost near bifurcations, our IBRT1 algorithm suffers from a reduced accuracy there, in a manner similar to root tracking for RD [6].

With that, we note that there are standard techniques in Bifurcation Theory for handling a non-trivial kernel of $D_x F$ at a root. For example, the *Lyapunov–Schmidt reduction* replaces the high-dimensional problem $F = 0$ (6) on $\mathbb{R}^n$ by a smaller but equivalent problem $\Phi = 0$, where $\Phi(\cdot, \beta)$ maps vectors in the (right) kernel of $D_x F$ to vectors in its left kernel. To achieve this, it separates the kernel and non-kernel directions of the problem, essentially handling each in turn; e.g., [21] (Theorem I.2.3) or [24] (Section 9.7). This technique is generic, as it does not rely on any particular property of the problem at hand. As such, it is considerably more involved than removing redundant coordinates, which requires an understanding of the solution structure. In contrast, reduction in the IB is straightforward. For the purpose of following a root's path, carrying on with redundant kernel directions is burdensome, computationally expensive, and sensitive to approximation errors. Applying

Lyapunov–Schmidt to our toy problem (10), for instance, reduces $F = \mathbf{0}$ (6) to choosing a continuously differentiable function Φ on the y-axis there (which is obtained by first solving for $x = \beta$; see the proof of Theorem I.2.3 in [21] for details). However, since y is redundant in this example, then solving for Φ can provide no useful information on the dynamics of its roots. In [19], a variant of the Lyapunov–Schmidt reduction was used to consider IB bifurcations due to symmetry breaking. While our findings are in partial agreement with theirs for continuous IB bifurcations, they differ for discontinuous bifurcations (see Sections 5.2 and 5.3).

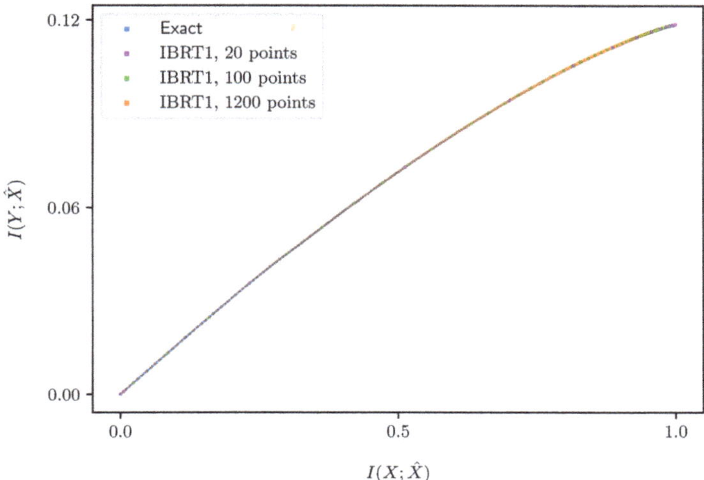

Figure 1. The approximate IB curves yielded by our algorithm, based on the IB ODE (16). Our First-order Root Tracking algorithm for the IB (IBRT1) of Section 6.1 was used to approximate the optimal IB roots of a binary symmetric channel with crossover probability 0.3 and a uniform source, BSC(0.3), for several grid densities. The points in the information plane yielded from these approximations are plotted on top of the problem's exact solution (see Appendix E). Despite the algorithm's approximation errors (Section 6.2), the approximate curves it yields are visually indistinguishable from the true IB curve (1), even on relatively few grid points. The reasons for this are discussed below (Section 6.3).

Notations

Vectors are written in boldface $\mathbf{x}$, and scalars in regular font x. A distribution p pertaining to a particular multiplier value β of the IB Lagrangian $\mathcal{L}$ is denoted with a subscript, p_β. Blahut–Arimoto's algorithm for the IB (BA-IB) is brought below as Algorithm 1, with a single iteration over the IB Equations (2)–(4) (in steps 1.4–1.8) denoted BA_β. The *probability simplex* on a set S is denoted $\Delta[S]$ (see Section 5.1). The *support* of a probability distribution p on S is supp $p := \{s \in S : \ p(s) \neq 0\}$. The *source, label*, and *representation* alphabets of an IB problem are denoted $\mathcal{X}, \mathcal{Y}$, and $\hat{\mathcal{X}}$, respectively; we write $T := |\hat{\mathcal{X}}|$. δ denotes Dirac's delta function, $\delta_{i,j} = 1$ if $i = j$, and zero otherwise.

Algorithm 1 Blahut–Arimoto for the Information Bottleneck (BA-IB), [1].

1: **function** BA-IB$(p_0(\hat{x}|x); p_{Y|X}\, p_X, \beta)$

Input:

    An initial encoder $p_0(\hat{x}|x)$, a problem definition $p(y|x)p(x)$, and $\beta > 0$.

Output:

    A fixed point $p(\hat{x}|x)$ of the IB Equations (2)–(4).

2:      Initialize $i \leftarrow 0$.

3:      **repeat**

4:         $p_i(\hat{x}) \leftarrow \sum_x p_i(\hat{x}|x)p(x)$

5:         $p_i(x|\hat{x}) \leftarrow p_i(\hat{x}|x)p(x)/p_i(\hat{x})$

6:         $p_i(y|\hat{x}) \leftarrow \sum_x p(y|x)p_i(x|\hat{x})$

7:         $Z_i(x, \beta) \leftarrow \sum_{\hat{x}} p_i(\hat{x}) \exp\left\{ -\beta\, D_{KL}\left[p(y|x)||p_i(y|\hat{x})\right] \right\}$

8:         $p_{i+1}(\hat{x}|x) \leftarrow \frac{p_i(\hat{x})}{Z_i(x,\beta)} \exp\left\{ -\beta\, D_{KL}\left[p(y|x)||p_i(y|\hat{x})\right] \right\}$

9:         $i \leftarrow i + 1$

10:     **until** convergence.

11: **end function**

---

2. Coordinates Exchange for the IB

Just as a point in the plane can be described by different coordinate systems, so can IB roots. As demonstrated recently by [6] for the related rate-distortion theory, picking the right coordinates matters when analyzing its bifurcations. The same holds also for the IB. Our primary motivations for exchanging coordinates are to reduce computational costs and to mod-out irrelevant kernel directions, as explained in Section 1. In this Section, we discuss three natural choices of a coordinate system for parameterizing IB roots and the reasoning behind our choice for the sequel before setting to derive the IB ODE in the following Section 3. This work is complemented by the later Section 5.1, which facilitates a transparent analysis of IB bifurcations.

IB roots have been classically parameterized in the literature by (direct) encoders $p(\hat{x}|x)$, following [1]. Considering the BA-IB Algorithm 1 reveals two other natural choices, illustrated by Equation (11) below. First, an encoder $p(\hat{x}|x)$ determines a *cluster marginal* $p(\hat{x})$ and an *inverse encoder* $p(x|\hat{x})$, via steps 4 and 5 of Algorithm 1 (denoted 1.4 and 1.5, for short), respectively. These can be interpreted geometrically as $p(\hat{x})$-weighted points $q_{\hat{x}}(x)$ in the simplex $\Delta[\mathcal{X}]$ of X, so long as they are well-defined, $\forall \hat{x}\ p(\hat{x}) \neq 0$. No more than $|\mathcal{X}| + 1$ points in the simplex are required to represent an IB root [2]. The latter is readily seen to analyze the IB in these coordinates, although it pre-dates [1] and has generally escaped broader attention. Second, an inverse encoder determines a *decoder* $p(y|\hat{x})$, via step 6. Along with the cluster marginal, $\big(p(y|\hat{x}), p(\hat{x})\big)$ can be similarly interpreted as $p(\hat{x})$-weighted points $r_{\hat{x}}(y)$ in the simplex $\Delta[\mathcal{Y}]$ of Y. This choice of coordinates is implied already by Theorem 5 in [1]. Cycling around Equation (11), a decoder $\big(p(y|\hat{x}), p(\hat{x})\big)$ determines via steps 7 and 8 a new encoder, which may differ from the one with which we have started. For notational simplicity, we shall usually write $\big(p(y|\hat{x}), p(\hat{x})\big)$ rather than $\big(r_{\hat{x}}(y), p(\hat{x})\big)$ for decoder coordinates (similarly for inverse encoder coordinates).

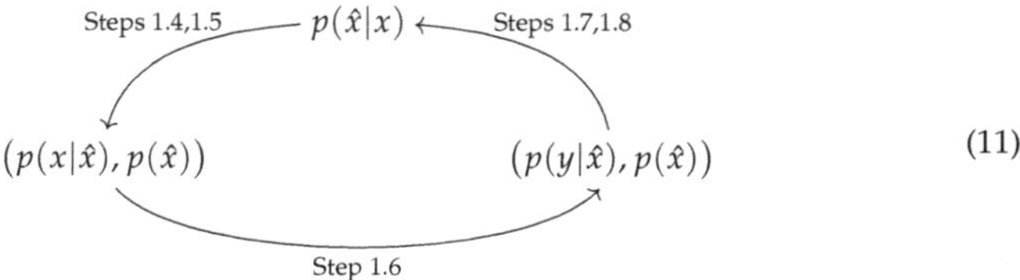

$$\tag{11}$$

The above allows us to define *three* BA operators as the composition of three consecutive maps in Equation (11), encoding an iteration of Algorithm 1. When starting at an encoder $p(\hat{x}|x)$, its output is a newly defined encoder. Similarly, when starting at one of the other two vertices, it sends an inverse encoder $\big(p(x|\hat{x}), p(\hat{x})\big)$ or a decoder pair $\big(p(y|\hat{x}), p(\hat{x})\big)$ to a newly defined one. By abuse of notation, we denote all three compositions by BA_β, with the choice of coordinate system mentioned accordingly. Indeed, these are representations of a single BA-IB iteration in three different coordinate systems, and so may be considered as distinct representations of the same operator. For completeness, BA_β in decoder coordinates is spelled out explicitly in Equation (A1) in Appendix A. A newly defined encoder (or inverse encoder or decoder) at a cycle's completion need not generally equal the one at which we have started. These are equal precisely at IB roots, when the IB Equations (2)–(4) hold. Therefore, the choice of a coordinate system does *not* matter then, and so moving around Equation (11) from one vertex to another yields different parameterizations of the same root, at least when $\forall \hat{x}\ p(\hat{x}) \neq 0$. In particular, this shows that the inverse encoders $q_{\hat{x}}$ in $\Delta[\mathcal{X}]$ of an IB root are in bijective correspondence with its decoders $r_{\hat{x}}$ in $\Delta[\mathcal{Y}]$, an observation which shall come in handy in Section 5.

Next, we consider how well each of these coordinate systems can serve for following the path of an IB root. The minimal number of symbols $\hat{x}$ needed to write down an IB root typically varies with the constraints, cf., [1] (Section 3.4) or [2] (Section II.A). Therefore, inverse encoder and decoder coordinates are better suited than encoder coordinates for considering the dynamics of a root with β, as they allow us to consider its evolution via a varying number of points in a fixed space, $\Delta[\mathcal{X}]$ or $\Delta[\mathcal{Y}]$, respectively. Indeed, a direct encoder $p(\hat{x}|x)$ can be interpreted geometrically as a point in the $|\mathcal{X}|$-fold product $\Delta[\hat{\mathcal{X}}]^{\mathcal{X}}$ of simplices $\Delta[\hat{\mathcal{X}}]$ [9] (Section 2). So, if a particular symbol $\hat{x}'$ is not in use anymore, $p(\hat{x}') = 0$, then one is forced to choose between replacing $\Delta[\hat{\mathcal{X}}]$ by a smaller space $\Delta[\hat{\mathcal{X}} \setminus \{\hat{x}'\}]$ or carrying on with a redundant symbol $\hat{x}'$. The latter leads to non-trivial kernels in the IB due to duplicate clusters (e.g., Section 3.1 there), making it difficult to tell whether a particular kernel direction pertains to a bifurcation (or to a "perpetual kernel" [9,19]). In contrast, when considered in decoder coordinates, for example, an IB root is nothing but $p(\hat{x})$-weighted paths $r_1, \ldots, r_T$ in $\Delta[\mathcal{Y}]$, with $\beta \mapsto r_{\hat{x}}(\beta)$ a path for each $\hat{x}$. And so, once a symbol $\hat{x}'$ is not needed anymore, then one can discard the path $r_{\hat{x}'}$ without replacing the underlying space $\Delta[\mathcal{Y}]$. This permits the clean treatment of IB bifurcations in Section 5.

The computational cost of solving a first-order ODE as in (7) numerically in $\frac{dx}{d\beta}$ depends on $\dim x$. Much of this cost is due to computing a linear pre-image under $D_x F$, which is of order $O(\dim x)^3$ [25] (Section 28.4); cf., Section 6. Representing an IB root on T clusters in encoder coordinates requires $|\mathcal{X}| \cdot T$ dimensions (ignoring normalization constraints), in inverse encoder coordinates $(|\mathcal{X}| + 1) \cdot T$ dimensions, and in decoder coordinates $(|\mathcal{Y}| + 1) \cdot T$ dimensions. Thus, the computational cost is lowest in decoder coordinates, at least when there are fewer possible labels $\mathcal{Y}$ than input symbols $\mathcal{X}$.

A priori, one might expect that derivatives with respect to β vanish when the solution barely changes, regardless of the choice of coordinate system. For example, at a very large "$\beta = \infty$" value, an obvious IB root is the diagonal encoder (setting $\hat{\mathcal{X}} := \mathcal{X}$ and $p(\hat{x}|x) := \delta_{x,\hat{x}}$), as can be seen by a direct examination of the IB Equations (2)–(4). It consists of one IB cluster of weight (or *mass*) $p(x)$ at $p_{Y|X=x} \in \Delta[\mathcal{Y}]$ for each $x \in \mathcal{X}$, and so one might expect that it would barely change so long as β is very large. However, the logarithmic derivative $\frac{d \log p_\beta(\hat{x}|x)}{d\beta}$ in encoder coordinates need *not* vanish even when the derivatives $\frac{d \log p_\beta(y|\hat{x})}{d\beta}$ and $\frac{d \log p_\beta(\hat{x})}{d\beta}$ in decoder coordinates do (see Section 3 on logarithmic coordinates), as seen to the right of Figure 2. Indeed, given the derivative in decoder coordinates, one can exchange it to encoder coordinates by

$$\frac{d \log p_\beta(\hat{x}|x)}{d\beta} = J_{\text{dec}}^{\text{enc}} \frac{d \log p_\beta(y'|\hat{x}')}{d\beta} + J_{\text{mrg}}^{\text{enc}} \frac{d \log p_\beta(\hat{x}')}{d\beta}$$
$$- D_{KL}[p(y|x)||p_\beta(y|\hat{x})] + \sum_{\hat{x}''} p_\beta(\hat{x}''|x) D_{KL}[p(y|x)||p_\beta(y|\hat{x}'')] , \quad (12)$$

where $J_{\text{dec}}^{\text{enc}}$ and $J_{\text{mrg}}^{\text{enc}}$ are the two coordinate exchange Jacobian matrices of orders $(T \cdot |\mathcal{X}|) \times (T \cdot |\mathcal{Y}|)$ and $(T \cdot |\mathcal{X}|) \times T$, respectively, given by Equations (A68) and (A70) in

Appendix B.4.2. And so, $\frac{d \log p_\beta(\hat{x}|x)}{d\beta}$ would often be non-zero even if both $\frac{d \log p_\beta(y|\hat{x})}{d\beta}$ and $\frac{d \log p_\beta(\hat{x})}{d\beta}$ vanish. This unintuitive behavior of the derivative in encoder coordinates is due to the explicit dependence of the IB's encoder Equation (2) on β. This dependence is the source of the last two terms in Equation (12) (see Equation (A73)). The comparison between encoder and inverse encoder coordinates can be seen to be similar. See Appendix B.4 for further details.

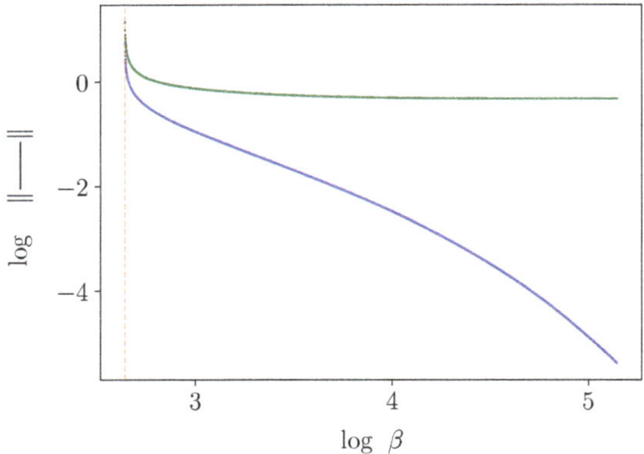

Figure 2. Derivatives' norm by coordinate system, for the exact solution of BSC(0.3) with a uniform source, as in Figure 1; see Appendix E. The derivative's L_2-norm is plotted in green for encoder coordinates and blue for decoder coordinates. The solution barely changes at high β values, and so the derivative in *decoder* coordinates is smaller (see main text). Nevertheless, the derivative in *encoder* coordinates does not vanish then, due to Equation (12). At low β values, however, the derivative in either coordinate system may generally be large. Both vanish to the left of the bifurcation in this problem (dashed red vertical), as the solution there is trivial (single-clustered). The derivatives diverge near the bifurcation (to its right) regardless of the coordinate system, as might be expected by the implicit ODE (7)—see also Section 6.1.

In light of the above, we proceed with decoder coordinates in the sequel.

3. Implicit Derivatives at an IB Root and the IB's ODE

We now specialize the implicit ODE (7) (of Section 1) to the IB, using the decoder coordinates of the previous Section 2. This allows us to compute first-order implicit derivatives at an IB root (Theorem 1) with remarkable accuracy, under one primary assumption—that the root is a differentiable function of β. While differentiability breaks at IB bifurcations (Section 5), this allows us to reconstruct a solution path from its local approximations in the following Section 4, so long as it holds.

To simplify calculations, we take the logarithm $\left(\log p(y|\hat{x}), \log p(\hat{x}) \right)$ of the decoder coordinates of Section 2 as our variables. Exchanging the BA_β operator to log-decoder coordinates is immediate, by writing $\log BA_\beta [\exp \left(\log p(y|\hat{x}) \right), \exp \left(\log p(\hat{x}) \right)]$. For short, we denote it $BA_\beta[\log p(y|\hat{x}), \log p(\hat{x})]$ when in these coordinates, by abuse of notation. Similarly, exchanging the IB ODE (below) back to non-logarithmic coordinates is immediate, via $\frac{d}{d\beta} \log p = \frac{1}{p} \frac{d}{d\beta} p$. In Section 6, we shall assume that $p(\hat{x})$ never vanishes. To ensure that taking logarithms is well-defined, we require that no decoder coordinate $p(y|\hat{x})$ vanishes (while it may have a well-defined derivative $\frac{d}{d\beta} p(y|\hat{x})$ even with a vanished coordinate, calculation details would differ). A sufficient condition for that is that $p(y|x) > 0$ for every x and y (Lemma A1 in Appendix A).

Next, define a variable $x \in \mathbb{R}^{T \cdot (|\mathcal{Y}|+1)}$ as the concatenation of the vector $\left(\log p_\beta(y|\hat{x}) \right)_{y \in \mathcal{Y}, \hat{x} \in \hat{\mathcal{X}}}$ with $\left(\log p_\beta(\hat{x}) \right)_{\hat{x} \in \hat{\mathcal{X}}}$. Differentiating $\partial/\partial \log p$ with respect to log-probabilities is given by $p \cdot \frac{\partial}{\partial p}$, by the chain rule (setting $u := \log p$, $\frac{df(p)}{du} = \frac{df}{dp}\frac{dp}{du}$, or equivalently $\frac{df}{d\log p} = p \cdot \frac{df}{dp}$; see Appendix B.1 for a gentler treatment). This gives meaning to the Jacobian matrix $D_x(\cdot)$ with respect to our logarithmic variable x. The Jacobian $D_{\log p(y|\hat{x}),\log p(\hat{x})}BA_\beta$ of a single Blahut–Arimoto iteration in these log-decoder coordinates is a square matrix of order $T \cdot (|\mathcal{Y}| + 1)$. Its $(T \cdot |\mathcal{Y}|) \times (T \cdot |\mathcal{Y}|)$ upper-left block (below) corresponds to perturbations of BA's output log-decoder $\log p(y|\hat{x})$ due to varying an input log-decoder $\log p(y'|\hat{x}')$. Since we prime input but not output coordinates, this is to say that the *columns* of this block are indexed by pairs $(y', \hat{x}')$ and its *rows* by $(y, \hat{x})$ (one could also enumerate $\mathcal{Y} := \{y_1, \ldots, y_{|\mathcal{Y}|}\}$ and $\hat{\mathcal{X}} := \{\hat{x}_1, \ldots, \hat{x}_T\}$ explicitly, replacing $(y, \hat{x})$ and $(y', \hat{x}')$ throughout by $(y_i, \hat{x}_j)$ and $(y_k, \hat{x}_l)$, respectively, with $i, k = 1, \ldots, |\mathcal{Y}|$ and $j, l = 1, \ldots, T$). Its $(T \cdot |\mathcal{Y}|) \times T$ upper-right block corresponds to perturbations in BA's output log-decoder $\log p(y|\hat{x})$ due to varying an input log-marginal $\log p(\hat{x}')$. That is, its columns are indexed by $\hat{x}'$ and rows by $(y, \hat{x})$. Similarly, for the bottom-left and bottom-right blocks, of respective sizes $T \times (T \cdot |\mathcal{Y}|)$ and $T \times T$. See (A25) ff., in Appendix B.2, and the end-result at Equation (A44) there. Explicitly, when evaluated at an IB root $\big(\log p(y|\hat{x}), \log p(\hat{x})\big)$, BA's Jacobian matrix is given by

$$D_{\log p(y|\hat{x}),\log p(\hat{x})}BA_\beta[\log p(y|\hat{x}), \log p(\hat{x})] =$$

$$\left(\begin{array}{c|c} \beta \cdot \sum_{\hat{x}'',y''} \left(\delta_{\hat{x}'',\hat{x}'} - \delta_{\hat{x},\hat{x}'} \right) \cdot \left[1 - \frac{\delta_{y'',y}}{p_\beta(y|\hat{x})} \right] C(\hat{x}, \hat{x}''; \beta)_{y',y''} & (1-\beta) \cdot \sum_{y''} \left[1 - \frac{\delta_{y'',y}}{p_\beta(y|\hat{x})} \right] B(\hat{x}, \hat{x}'; \beta)_{y''} \\ \hline \beta \cdot \left[\delta_{\hat{x},\hat{x}'}\, p_\beta(y'|\hat{x}) - B(\hat{x}, \hat{x}'; \beta)_{y'} \right] & (1-\beta) \cdot \left[\delta_{\hat{x},\hat{x}'} - A(\hat{x}, \hat{x}'; \beta) \right] \end{array} \right) \qquad (13)$$

where $\delta_{i,j} = 1$ if $i = j$ and is 0 otherwise. As mentioned above, primed coordinates y' and $\hat{x}'$ index the columns, and un-primed coordinates y and $\hat{x}$ the rows. Indices y'' and $\hat{x}''$ with more than a single prime are summation variables. A, B, and C are a scalar, a vector, and a matrix, each involving two IB clusters. They are defined by

$$A(\hat{x}, \hat{x}'; \beta) := \sum_{x''} p_\beta(\hat{x}'|x'')p_\beta(x''|\hat{x}) \,,$$

$$B(\hat{x}, \hat{x}'; \beta)_y := \sum_{x''} p(y|x'')p_\beta(\hat{x}'|x'')p_\beta(x''|\hat{x}) \,, \quad \text{and} \qquad (14)$$

$$C(\hat{x}, \hat{x}'; \beta)_{y,y'} := \sum_{x''} p(y|x'')p(y'|x'')p_\beta(\hat{x}'|x'')p_\beta(x''|\hat{x}) \,.$$

In these, y indexes B and the rows of C, y' the columns of C, and x'' is a summation variable. These $(\hat{x}, \hat{x}')$-labeled tensors have only $|\mathcal{Y}|$ entries along each axis, thanks to our choice of decoder coordinates. A and B can be expressed in terms of C via some obvious relations; see Equation (A32) and below in Appendix B.2. Appendix B.1 elaborates on the mathematical subtleties involved in calculating the Jacobian (13). See also Equation (A45) in Appendix B.2 for an implementation-friendly form of (13).

Together with $D_\beta BA_\beta$ (Equations (A58) and (A57) in Appendix B.3), we have both of the first-order derivative tensors of BA_β in log-decoder coordinates. This allows us to specialize the implicit ODE (7) (of Section 1) to the IB, in terms of our variable x. By abuse of notation, we write $\big(\log p_\beta(y|\hat{x}), \log p_\beta(\hat{x})\big)_{y,\hat{x}}$ for its $|\mathcal{Y}| \cdot T + T$ coordinates, and similarly for its derivatives vector v (15) below.

Theorem 1 (The IB's ODE). *Let $\big(p(y|\hat{x}), p(\hat{x})\big)$ be an IB root, and suppose that it can be written as a differentiable function $\beta \mapsto \big(p_\beta(y|\hat{x}), p_\beta(\hat{x})\big)$ in β. If none of its coordinates vanish, then the vector*

$$v := \left(\frac{d\log p_\beta(y|\hat{x})}{d\beta}, \frac{d\log p_\beta(\hat{x})}{d\beta} \right)_{y,\hat{x}} \qquad (15)$$

of its implicit logarithmic derivatives is well-defined and satisfies an ordinary differential equation in β,

$$\left(I - D_{\log p(y|\hat{x}),\log p(\hat{x})} BA_{\beta}\right) v = \begin{pmatrix} -\sum_{x,\hat{x}''} \left[1 - \frac{p(y|x)}{p_{\beta}(y|\hat{x})}\right] \cdot \left[\delta_{\hat{x},\hat{x}''} - p_{\beta}(\hat{x}''|x)\right] p_{\beta}(x|\hat{x}) \, D_{KL}\left[p(y|x)||p_{\beta}(y|\hat{x}'')\right] \\ \\ \sum_{x,\hat{x}''} \left[\delta_{\hat{x},\hat{x}''} - p_{\beta}(\hat{x}''|x)\right] p_{\beta}(x|\hat{x}) \, D_{KL}\left[p(y|x)||p_{\beta}(y|\hat{x}'')\right] \end{pmatrix} \tag{16}$$

where I is the identity matrix of order $T \cdot (|\mathcal{Y}| + 1)$, and the Jacobian matrix $D_{\log p(y|\hat{x}),\log p(\hat{x})} BA_{\beta}$ at the given IB root is given by Equation (13). The right-hand side of (16) is indexed as in (15), by $(y, \hat{x})$ at its top and $\hat{x}$ at its bottom coordinates.

While the IB ODE was discovered by [10], it is derived here anew in log–decoder coordinates due to the considerations in Section 2. It is analogous to the RD ODE, due to [6]; Corollary 1 and around (in Section 5.1) provides a relation between these two ODEs. We emphasize that the first assumption of Theorem 1, that the IB root is a differentiable function of β, is essential. It consists of two parts: (i) that the root can be written as a function of β, and (ii) that this function is differentiable. These are precisely the assumptions needed to compute the first-order implicit multivariate derivative v (15) at the given root [6] (Section 2.1). Continuous IB bifurcations violate (ii) (Section 5.2), while discontinuous ones violate (i) (Section 5.3). In contrast, the requirement that no coordinate vanishes is a technical one, due to our choice of logarithmic coordinates.

It is not necessary for the Jacobian of the IB operator (5) (to the left of (16)) to be non-singular in order to solve the IB ODE numerically. Nevertheless, non-singularity of the Jacobian will follow from the sequel (see Conjecture 1 in Section 5). With that, the derivatives $v = \frac{d}{d\beta}\left(\log p_{\beta}(y|\hat{x}), \log p_{\beta}(\hat{x})\right)$ (15) computed numerically from the IB ODE (16) at an exact root are remarkably accurate, as demonstrated in Figure 3. As in RD [6], calculating implicit derivatives numerically loses its accuracy when approaching a bifurcation because the Jacobian is increasingly ill-conditioned there. For comparison, the BA-IB Algorithm 1 also loses its accuracy near a bifurcation. This is a consequence of BA's critical slowing down [4], just as with its corresponding RD variant.

Each coordinate of $\left(p(y|\hat{x}), p(\hat{x})\right)$ is treated by the IB ODE (16) as an independent variable. However, the normalization of $p(y|\hat{x})$ imposes one constraint per cluster $\hat{x}$ (and one for the normalization of $p(\hat{x})$). Thus, one might expect the behavior of BA's Jacobian (13) to be determined by fewer than $T \cdot (|\mathcal{Y}| + 1)$ coordinates, at least qualitatively. This intuition is justified by the following Lemma 1, which allows us to consider the kernel of the IB operator (5) by a smaller and simpler matrix S; see Appendix C for its proof.

Lemma 1. *Given an IB root as above, define a square matrix of order $T \cdot |\mathcal{Y}|$ by*

$$S_{(y,\hat{x}),(y',\hat{x}')} := \sum_{x} p_{\beta}(x|\hat{x}) \left[\beta \cdot \frac{p(y|x)}{p_{\beta}(y|\hat{x})} + (1 - 2\beta)\right] p(y'|x) \left[\delta_{\hat{x},\hat{x}'} - p_{\beta}(\hat{x}'|x)\right] . \tag{17}$$

Then, the nullity of the Jacobian $I - D_{\log p(y|\hat{x}),\log p(\hat{x})} BA_{\beta}$ of the IB operator (5) equals that of $I - S$, where I is the identity matrix (of the respective order), and S is defined by (17),

$$\dim \ker(I - S) = \dim \ker\left(I - D_{\log p(y|\hat{x}),\log p(\hat{x})} BA_{\beta}\right) . \tag{18}$$

Specifically, write $v := \left(v_{y,\hat{x}}\right)_{y,\hat{x}}$ for a left eigenvector which corresponds to $1 \in \mathrm{eig}\, S$. Then, there is a bijective correspondence between the left kernels at both sides of (18), mapping

$$v \mapsto (v, u) , \tag{19}$$

where $u := (u_{\hat{x}})_{\hat{x}}$ is defined by $u_{\hat{x}} := \frac{1-\beta}{\beta} \cdot \sum_{y} v_{y,\hat{x}}.$

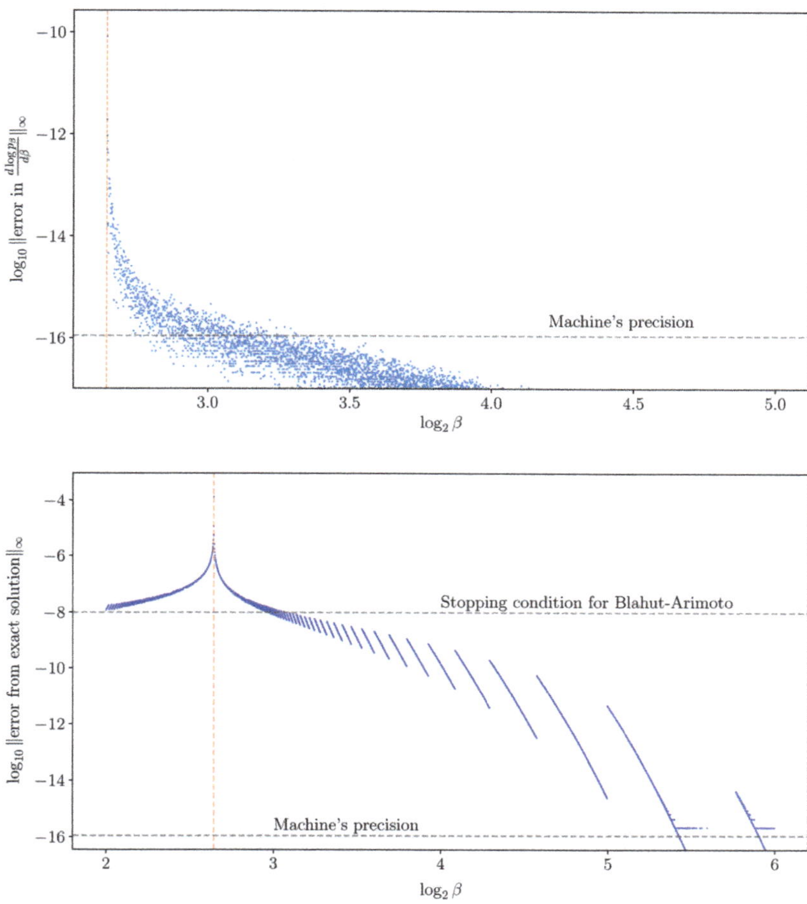

Figure 3. The implicit derivatives computed from the IB ODE (16) are very accurate, as is the BA-IB Algorithm 1. However, both lose their accuracy near a bifurcation. To verify their accuracy, we compared both to the exact solutions of BSC(0.3) with a uniform source (see Appendix E). (**Top**): Derivatives were computed at the problem's exact solution using the IB ODE (16) and compared to the problem's exact derivatives. These are accurate beyond the machine's precision, except when approaching the bifurcation (red vertical), since the Jacobian of the IB operator (5) is ill-conditioned there. (**Bottom**): The L_∞-errors of the solutions produced by the BA-IB Algorithm 1, with a 10^{-8} stopping condition, and uniform initial conditions. Error is measured from the true direct encoder to avoid biases due to clusters of low mass. Both plots are as in Figure 2.3 of [6].

In addition to offering a form more transparent than BA's Jacobian in (13), Lemma 1 also reduces the computational cost of testing $I - D_{\log p(y|\hat{x}),\log p(\hat{x})} BA_\beta$ (16) for singularity, by using the smaller $I - S$ (17) in its place. This makes it easier to detect upcoming bifurcations (see Conjecture 1 in Section 5). Further, one can verify directly that the IB ODE (16) indeed follows the right path. Indeed, if the ODE is non-singular, then, by the Implicit Function Theorem, there is (locally) a unique IB root, which is a differentiable function of β. And so, there is a unique solution path for a numerical approximation to follow. Finally, we note that a relation similar to (18) holds also for eigenvalues of $D_{\log p(y|\hat{x}),\log p(\hat{x})} BA_\beta$ (13) other than 1. This can be seen either empirically or by tracing the proof of Lemma 1.

In Section 5, we shall proceed with this line of thought of removing redundant coordinates. In the following Section 4, we turn to reconstruct a solution path from implicit derivatives at a point, with bifurcations ignored for now.

4. A Modified Euler Method for the IB

We follow the path of a given IB root away from bifurcation by using its implicit derivatives computed from the IB ODE (16), of Section 3. We follow the classic Euler method for simplicity, modifying it slightly to get the most out of the calculated derivatives. Improvements using more sophisticated numerical methods are left to future work. The detection and handling of IB bifurcations are deferred to the following Section 5, and thus are ignored in this section.

Let $\frac{dx}{d\beta} = f(x, \beta)$ and $x(\beta_0) = x_0$ define an initial value problem. In numerical approximations of ordinary differential equations (ODEs), the *Euler method* for this problem is defined by setting

$$x_{n+1} := x_n + \Delta\beta \cdot f(x_n, \beta_n) \,, \tag{20}$$

where $\beta_{n+1} := \beta_n + \Delta\beta$, and $|\Delta\beta|$ is the *step size*. The *global truncation error* $\max_n \|x_n - x(\beta_n)\|_\infty$ is the largest error of the approximations x_n from the true solutions $x(\beta_n)$. A numerical method for solving ODEs is said to be *of order d* if its global truncation error is of order $O(|\Delta\beta|^d)$, for step sizes $|\Delta\beta|$ small enough. Euler's method error analysis is a standard result, provided as Theorem 2 below. See [26] (Theorem 212A) or [27] (Theorem 2.4), for example. It shows that Euler's method (20) is of order $d = 1$, under mild assumptions, as demonstrated in Figure 4. The immediate generalization of (20) using derivatives until order d is *Taylor's method*, which is a method of order d.

Theorem 2 (Euler's method error analysis). *Let an initial value problem be defined on $[\beta_0, \beta_f]$ by $\frac{dx}{d\beta} = f(x, \beta)$ as above (with x_0 allowed to deviate from $x(\beta_0)$), and suppose that f satisfies the Lipschitz condition with some constant $L > 0$. Namely, $\|f(x, \beta) - f(x', \beta)\|_\infty \leq L \cdot \|x - x'\|_\infty$ for every x, x' and $\beta \in [\beta_0, \beta_f]$.*

Then, Euler's method (20) global truncation error satisfies

$$\max_{\beta_0 \leq \beta_n \leq \beta_f} \|x_n - x(\beta_n)\|_\infty \leq e^{(\beta_f - \beta_0)L} \|x_0 - x(\beta_0)\|_\infty + \frac{e^{(\beta_f - \beta_0)L} - 1}{L} \cdot \frac{1}{2}|\Delta\beta| \max_{\beta_0 \leq \beta \leq \beta_f} \left\| \frac{d^2 x(\beta)}{d\beta^2} \right\|_\infty . \tag{21}$$

Specializing Euler's method to our needs, replace x in (20) above by the log-decoder coordinates of an IB root, as in Section 3. So long as an IB root $p_\beta := \left(p_\beta(y|\hat{x}), p_\beta(\hat{x})\right)$ is a differentiable function of β in the vicinity of β_n, it can be approximated by

$$\log p_{\beta_{n+1}}(y|\hat{x}) \approx \log p_{\beta_n}(y|\hat{x}) + \Delta\beta \cdot \left. \frac{d\log p_\beta(y|\hat{x})}{d\beta} \right|_{p_{\beta_n}} \quad \text{and}$$

$$\log p_{\beta_{n+1}}(\hat{x}) \approx \log p_{\beta_n}(\hat{x}) + \Delta\beta \cdot \left. \frac{d\log p_\beta(\hat{x})}{d\beta} \right|_{p_{\beta_n}} , \tag{22}$$

where $\frac{d\log p_\beta(y|\hat{x})}{d\beta}$ and $\frac{d\log p_\beta(\hat{x})}{d\beta}$ are calculated from the IB ODE (16). Thus, applying (22) repeatedly, we obtain an Euler method for the IB. We shall take only negative steps $\Delta\beta < 0$ when approximating the IB, due to reasons explained in Section 5.3 (after Proposition 1). In contrast to the BA-IB Algorithm 1, Euler's method (22) can be used to interpolate intermediate points, yielding a piecewise linear approximation of the root.

The problem of tracking an operator's root belongs in general to a family of hard-to-solve numerical problems—known as *stiff*—if the problem has a bifurcation [6] (Section 7.2). See [26] or [27] for example on stiff differential equations. Stopping early in the vicinity of a bifurcation restricts the computational difficulty and permits convergence guarantees. Early stopping in the IB shall be handled later, in Section 5.2. [6] (Theorem 5) proves that

Euler's method convergence guarantees (Theorem 2) hold for the closely related Euler method for RD with early stopping. While Euler's method may inadvertently switch between solution branches of the IB ODE (16), the latter guarantees ensure that it indeed follows the true solution path between bifurcations, if the step size $|\Delta\beta|$ is small enough and initializing close enough to the true solution (see Sections 5 and 6.3 on the distinction between IB bifurcations and singularities of the IB ODE (16)). Although we do not dive into these details for brevity, we note that similar convergence guarantees can also be proven here. Alternatively, Euler's method can be ensured to follow the true solution path by noting that an optimal IB root is (strongly) stable when negative steps $\Delta\beta < 0$ are taken; these details are deferred to Section 6.3, as they depend on Section 5.

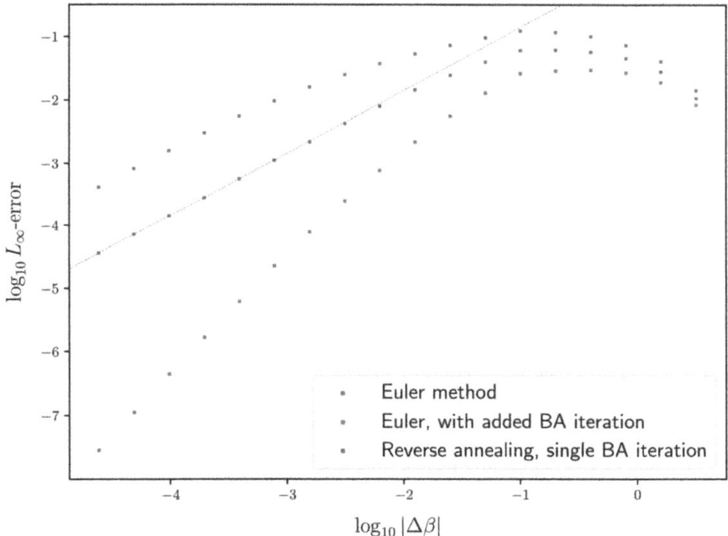

Figure 4. Error by step-size for a vanilla Euler method using the IB ODE (16), and with an added BA-IB iteration at each step, for $\mathrm{BSC}(0.3)$ with a uniform source (Appendix E). The linear regression (dashed black) of the third leftmost markers for the vanilla Euler method is of slope 0.99 ($R^2 \simeq 1$), matching the theory's prediction almost perfectly. A similar regression (not shown) for Euler's method with a single added BA iteration is of nearly double slope 1.93. For comparison, reverse deterministic annealing with a single BA iteration at each grid point yields a slope of 0.91 in this example. Taking a larger (pre-determined) number of iterations at each grid point pushes the error downwards, as expected. Yet, the resulting slopes approach 1 as the number of iterations is increased (not shown). See main text and Appendix D for details. The error was calculated as the supremum of the pointwise errors as in Figure 3, over the interval $[\beta_c + \frac{1}{10}, \beta_0]$ which contains no bifurcation. Each method was initialized with the exact solution at $\beta_0 = 2^5$, with $\Delta\beta = -\frac{103}{32}$ halved between consecutive markers.

Following the discussion in Section 2, there is a subtle disadvantage in choosing decoder coordinates as our variables compared to the other two coordinate systems there. Indeed, recall that the IB is defined as a maximization over Markov chains $Y \longleftrightarrow X \longleftrightarrow \hat{X}$. An (arbitrary) encoder $p(\hat{x}|x)$ defines a joint probability distribution $p(\hat{x}|x)p(y|x)p(x)$ which is Markov. An inverse encoder pair also similarly defines a Markov chain. In contrast, an arbitrary decoder pair $\left(p(y|\hat{x}), p(\hat{x})\right)$ need *not* necessarily define a Markov chain. Rather, by invoking the error analysis of Euler's method, one can see that Markovity is approximated at an increasingly improved quality as the step-size $|\Delta\beta|$ in (22) becomes smaller. To enforce Markovity, we shall perform a single BA iteration (in decoder coordinates) after each Euler method step. This ensures that the newly generated decoder pair satisfies the Markov condition, as it is now generated from an encoder.

As a side effect, adding a single BA-IB iteration after each Euler method step improves the approximation's quality significantly. By linearizing BA_β around a fixed point, one can show that deterministic annealing with a fixed number of BA iterations per grid point is a first-order method. Thus, deterministic annealing may arguably be considered a first-order method, as is with Euler's method. A similar argument shows that adding a single BA iteration after each Euler method step yields a second-order method. However, while a larger number of added BA iterations obviously improves the approximation's quality, it does not improve the method's order. See Appendix D for an approximate error analysis. The predicted orders are in good agreement with the ones found empirically, shown in Figure 4. We note that while [6] did not attempt an added BA iteration, they do discuss a variety of other improvements to root tracking (see Section 3.4 in [6]).

5. On IB Bifurcations

For the IB Equations (2)–(4) to exhibit a bifurcation, it is necessary that the Jacobian of the IB operator (5) be singular, as illustrated by Figure 5. However, a priori singularity is not sufficient to detect a bifurcation (cf., Section 3.1 in [9]), nor does this allow one to distinguish between bifurcations of different types. At an IB root, singularities of the IB ODE (16) (Section 3) coincide with those of $Id - BA_\beta$ (5) (in log-decoder coordinates). Thus, in order to be able to exploit the IB ODE (16), we shall now take a closer look into IB bifurcations. These can be broadly classified into two types: where an optimal root is continuous in β and where it is not. As noted after Theorem 1, each type violates an assumption necessary to compute implicit derivatives. Sections 5.2 and 5.3 provide the means to identify bifurcations, distinguish between their types, and handle them accordingly, mainly for continuous bifurcations. To facilitate the discussion, Section 5.1 considers the IB as a rate-distortion problem, following [20] and others. This allows us to leverage recent insights on RD bifurcations [6], while suggesting a "minimally sufficient" choice of coordinates for the IB. The latter permits a clean treatment of continuous IB bifurcations in Section 5.2. Viewing the IB as an infinite-dimensional RD problem facilitates the understanding of its discontinuous bifurcations, which in turn highlight subtleties in its finite-dimensional coordinate systems (of Section 2). These provide insight into the IB and are also of practical implications (Section 5.3), and so are necessary for our algorithms in Section 6.

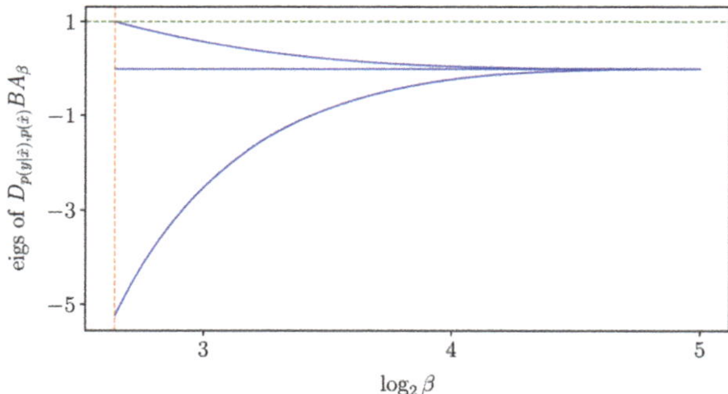

Figure 5. While the Jacobian $D_{\log p(y|\hat{x}),\log p(\hat{x})}(Id - BA_\beta)$ must be singular at a bifurcation, this does not suffice to identify its type. The Jacobian eigenvalues of BA_β (13) with respect to log-decoder coordinates are plotted for BSC(0.3) with a uniform source, as in Figure 1; see Appendix E for its exact solution. An eigenvalue reaches one (dashed green) precisely at the bifurcation (dashed red vertical), as expected by Conjecture 1 in Section 5.1. In particular, the Jacobian is increasingly ill-conditioned when approaching the bifurcation, as noted in Figure 3 (top). While this allows one to detect the bifurcation, identifying its type is necessary for handling it.

5.1. The IB as a Rate-Distortion Problem

We now explore the intimate relation between the IB and RD, following [5,20]. This leads to a "minimally sufficient" coordinate system for the IB, thereby completing the work of Section 2. In this coordinate system, results [6] on the dynamics of RD roots are readily considered in the IB context. This leads to Conjecture 1, that the IB operator (5) in these coordinates is typically non-singular. The discussion here facilitates the treatment of IB bifurcations in the following Sections 5.2 and 5.3.

First, recall a few definitions. A *rate distortion problem* on a *source alphabet* $\mathcal{X}$ and a *reproduction alphabet* $\hat{\mathcal{X}}$ is defined by a *distortion measure* $d : \mathcal{X} \times \hat{\mathcal{X}} \to \mathbb{R}_{\geq 0}$ (a non-negative function on $\mathcal{X} \times \hat{\mathcal{X}}$ with no further requirements—see Section 2.2 in [28]) and a source distribution $p_X(x)$. One seeks the minimal rate $I(X; \hat{X})$ subject to a constraint D on the expected distortion $\mathbb{E}[d(x, \hat{x})]$ [29,30],

$$R(D) := \min_{p(\hat{x}|x)} \left\{ I(X; \hat{X}) : \mathbb{E}_{p(\hat{x}|x)p_X(x)}[d(x, \hat{x})] \leq D \right\}, \tag{23}$$

known as the *rate-distortion curve*. The minimization is over *test channels* $p(\hat{x}|x)$. A test channel that attains the RD curve (23) is called an *achieving distribution*. We say that an RD problem is *finite* if both of the alphabets $\mathcal{X}$ and $\hat{\mathcal{X}}$ are finite. Using Lagrange multipliers for (23) with $I(X; \hat{X}) + \beta\, \mathbb{E}[d(x, \hat{x})]$ (normalization omitted for clarity), one obtains a pair of fixed-point equations

$$p(\hat{x}|x) = \frac{p(\hat{x})e^{-\beta\, d(x,\hat{x})}}{\sum_{\hat{x}} p(\hat{x})e^{-\beta\, d(x,\hat{x})}} \quad \text{and} \quad p(\hat{x}) = \sum_{x} p(\hat{x}|x)p(x) \tag{24}$$

in the marginal $p(\hat{x})$ and test channel $p(\hat{x}|x)$, similar to the IB Equations (2) and (4). Iterating over these is Blahut's algorithm for RD [8], denoted BA_{β}^{RD} here. As with the IB (1), β parameterizes the slope of the optimal curve (23) also for RD. See [28] or [31] for an exposition of rate-distortion theory.

We clarify a definition needed to rewrite the IB as an RD problem. We define the *simplex* $\Delta[S]$ on a (possibly infinite) set S as the collection of finite formal convex combinations $\sum_s a_s \cdot s$ of elements of S. That is, as the S-indexed vectors $(a_s)_{s \in S}$ (equivalently, as functions mapping each s in S to a real number a_s) that satisfy $\sum_s a_s = 1$ and $a_s \geq 0$, with a_s non-zero for only finitely many elements s (the *support* of $(a_s)_s$). Addition and multiplication are defined pointwise, as in $\sum_s a_s \cdot s + \sum_s b_s \cdot s = \sum_s (a_s + b_s) \cdot s$. $\Delta[S]$ is closed under finite convex combinations because the sum of finitely supported vectors is finitely supported. When taking $S = \{e_1, \ldots, e_n\}$ the standard basis vectors $(e_i)_j = \delta_{i,j}$ of $\mathbb{R}^n$, then one can identify the formal operations with those in $\mathbb{R}^n$, reducing the simplex $\Delta[S]$ to its usual definition. We write r for an element of $\Delta[\mathcal{Y}]$. In particular, an element of $\Delta\big[\Delta[\mathcal{Y}]\big]$ is merely a finite convex combination $\sum_{\hat{x}} p(\hat{x})r_{\hat{x}}$ of *distinct* probability distributions $r_{\hat{x}}(y) \in \Delta[\mathcal{Y}]$ on $\mathcal{Y}$ (note that $\Delta[S]$ is a set). When setting $\hat{\mathcal{X}} \subset \Delta[\mathcal{Y}]$ to be a finite subset of distributions, $|\hat{\mathcal{X}}| < \infty$, then $\Delta[\hat{\mathcal{X}}]$ is a special case of the decoder coordinates of Section 2 (unlike $\Delta[\hat{\mathcal{X}}]$ here, the decoder coordinates of Section 2 are *not* required to have their clusters r distinct).

Now, let a finite IB problem be defined by a joint probability distribution $p_{Y|X}\, p_X$, as in Section 1. To write it down as an RD problem [5,20], define the *IB distortion measure* by

$$d_{IB}(x, r) := D_{KL}\Big[p_{Y|X=x}||r\Big], \tag{25}$$

for $x \in \mathcal{X}$, $r \in \Delta[\mathcal{Y}]$, and $p_{Y|X=x} \in \Delta[\mathcal{Y}]$ the conditional probability distribution at $X = x$. The distortion measure d_{IB} (25) and p_X define an RD problem on the continuous reproduction alphabet $\hat{\mathcal{X}} := \Delta[\mathcal{Y}]$. Minimizing the IB Lagrangian $\mathcal{L}$ (in Section 1) is equivalent to minimizing the Lagrangian of this RD problem [20] (Theorem 5). That is, the IB is a rate-distortion problem when considered in these coordinates. IB clusters $r \in \Delta[\mathcal{Y}]$ assume the role of RD reproduction symbols, while an IB root (considered now as an RD root) is equivalently described either by the probabilities of each cluster—namely, by a point in $\Delta[\Delta[\mathcal{Y}]]$—or, by a test channel $p(r|x)$. The astute reader might notice that the IB Equations (2) and (4) are then equivalent to RD's fixed-point Equations (24), with the decoder Equation (3) implied by the IB's Markovity. The IB's Y-information $I(Y; \hat{X})$ equals the expected distortion $\mathbb{E}[d_{IB}(x, \hat{x})]$ in (23) up to a constant [20] (Section 5), and so is linear in the test channel $p(r|x)$. Unlike the finite-dimensional coordinate systems of Section 2, this definition of the IB entails no subtleties due to finite dimensionality, such as duplicate clusters (see more below). However, while it allows us to spell out the IB explicitly as an RD problem, handling an infinite reproduction alphabet is difficult for practical purposes. Since no more than $|\mathcal{X}| + 1$ reproduction symbols are needed to write down an IB root [2], this motivates one to consider the IB's *local* behavior, with clusters fixed.

So instead, one may require the reproduction symbols of d_{IB} (25) to be in a list $(r_{\hat{x}})_{\hat{x} \in \hat{\mathcal{X}}}$ indexed by some finite set $\hat{\mathcal{X}}$, with each $r_{\hat{x}}$ in $\Delta[\mathcal{Y}]$ (the elements $r_{\hat{x}_1}, \ldots, r_{\hat{x}_T}$ need *not* be distinct a priori). This defines a finite RD problem, for which d_{IB} (25) is merely an $|\mathcal{X}|$-by-T matrix. Yet, placing identical clusters in the list $(r_{\hat{x}})_{\hat{x}}$ inadvertently introduces degeneracy to the matrix d_{IB} (25), as discussed below. In [5] (Section 6), $(r_{\hat{x}})_{\hat{x}}$ is taken to be the decoders defined by a given encoder $p(\hat{x}|x)$, as in Equation (11) (Section 2). We shall then refer to d_{IB} (25) as *the distortion matrix defined by* $p(\hat{x}|x)$. When $p_{\beta_0}(\hat{x}|x)$ is an optimal IB root then the problem (d_{IB}, p_X) defined by it is called *the tangent RD problem*. Indeed, its RD curve (23) coincides with the IB curve (1) at this point (since an optimal choice of IB clusters is already encoded into d_{IB} (25), then solving the IB boils down to finding the clusters' optimal weights $p(\hat{x})$, which is an RD problem). However, the curves differ outside this point since IB clusters usually vary with β, while the distortion of the tangent problem was defined at $p_{\beta_0}(\hat{x}|x)$ and so is fixed. By definition (1), it follows that the IB curve is the lower envelope of the curves of its tangent RD problems [5] (Corollary 2). We note that a similar construction can also be carried out in inverse encoder coordinates, cf., [2].

Regardless of the formulation used to rewrite the IB as an RD problem, the associated RD problem has an expected distortion $\mathbb{E}[d_{IB}]$ of $I(X; Y) - I(\hat{X}; Y)$ at an IB root (Section 5 in [20] and Lemma 8 in [5]). That is, the IB is a method of lossy compression that strives to preserve the relevant information $I(\hat{X}; Y)$. Due to the Markov condition, information on Y is available only through X. Thus, one may intuitively consider the IB as a lossy compression method of the information on Y that is embedded in X. These intimate relations between the IB and RD suggest that studying bifurcations in either context could be leveraged to understand the other. Bifurcations in finite RD problems are discussed at length in [6] (Section 6). To facilitate the study of IB bifurcations in the sequel (Sections 5.2 and 5.3) using results from RD, we need a "minimally-sufficient" coordinate system for the IB.

Consider an IB root in decoder coordinates as finitely many $p(\hat{x})$-weighted points $r_{\hat{x}}(y)$ in $\Delta[\mathcal{Y}]$, as in Section 2. Exchanging *to* decoder coordinates (Equation (11) there) is well-defined as long as there are no zero-mass clusters, $\forall \hat{x}\ p(\hat{x}) \neq 0$. Yet, even then, the points $r_{\hat{x}}$ in $\Delta[\mathcal{Y}]$ yielded by BA's steps 4 through 6 (Algorithm 1) need *not* be distinct. Namely, they may yield identical clusters $r_{\hat{x}} = r_{\hat{x}'}$ at distinct indices $\hat{x} \neq \hat{x}'$. This leads to a discussion of structural symmetries of the IB (its degeneracies), which is not of use for our purposes; cf., [9]. To avoid such subtleties, we shall say that an IB root is *reduced* if it has no zero-mass clusters, $\forall \hat{x}\ p(\hat{x}) \neq 0$, and all its clusters are distinct, $\hat{x} = \hat{x}' \Leftrightarrow r_{\hat{x}} = r_{\hat{x}'}$. A root that is not reduced is called *degenerate* or *degenerately represented*. An IB root can be reduced by removing clusters of zero mass and merging identical clusters of distinct indices—see our reduction algorithm in Section 5.2 below. It is straightforward to see from the IB Equations (2)–(4) that reduction preserves the property of being an IB root. Similarly, reducing a root does not change its location in the information plane. So, a root achieves the IB curve (1) if and only if its reduction does. Therefore, reduction decreases the dimension in which the problem is considered while preserving all its essential properties. This allows us to represent an IB root on the smallest number of clusters possible—its *effective cardinality*—by factoring out the IB's structural symmetries. See also [13] (2.3 in Chapter 7), upon which this definition is based.

While the purpose of reduction is to mod-out redundant kernel coordinates (Section 1), it highlights the differences between the various IB definitions found in the literature, bringing to light a subtle caveat of finite dimensionality. To see this, note that reduction could have been defined above in terms of the other coordinate systems of the IB. Its definition in inverse encoder coordinates is nearly identical to that above, while defining it in encoder coordinates is a straightforward exercise. Since the coordinate systems of Section 2 are equivalent at an IB root (without zero-mass clusters), the precise definition does not matter then. Each of these parameterizations encodes the coordinates $r(y)$ of a root's clusters r using a finite-dimensional vector x (note Equation (11)). This enables one to represent duplicate clusters $\hat{x} \neq \hat{x}'$ with $r_{\hat{x}} = r_{\hat{x}'}$, and obliges one to choose the order in which clusters are being encoded into the coordinates of x. A finite-dimensional representation x of an IB root is invariant to interchanging clusters $\hat{x} \neq \hat{x}'$ precisely when they are identical, $r_{\hat{x}} = r_{\hat{x}'}$. The IB's functionals (e.g., its X- and Y-information) are invariant to any cluster permutation; cf., [9,19]. Both of these structural symmetries result from using a finite-dimensional parameterization, with the former eliminated by reduction. In contrast, the elements of $\Delta[\mathcal{Y}]$ are distinct by definition (since $\Delta[\mathcal{Y}]$ is a set), and so parameterizing the IB by points in $\Delta[\Delta[\mathcal{Y}]]$ does not permit identical clusters. An element $\sum_r p(r) r$ of $\Delta[\Delta[\mathcal{Y}]]$ assigns a probability mass $p(r)$ to *every* point r in $\Delta[\mathcal{Y}]$, with only finitely many points r supported. Thus, it implicitly encodes all the entries $r(y)$ of every probability distribution $r \in \Delta[\mathcal{Y}]$ in a "one size fits all" approach, giving no room for the choices above. This leads us to argue that the IB's structural symmetries are *not* an inherent property but rather an artifact of using its finite-dimensional representations. This is best understood in the context of discontinuous bifurcations, in Section 5.3 below. For comparison, both of the IB formulations [2,20] do not impose an a priori restriction on the number of clusters. The latter does not enable one to encode duplicate clusters, while the former does. The formulation [1] ignores these subtleties altogether, and [9,19] consider the IB on a pre-determined number of possibly duplicate clusters.

In rate-distortion, the *reduction* of a finite RD problem is defined similarly [6] (Section 3.1), by removing a symbol $\hat{x}$ from the reproduction alphabet $\hat{\mathcal{X}}$ and its column $d(\cdot, \hat{x})$ from the distortion matrix once it is not in use anymore (of zero mass). A distortion matrix d is *non-degenerate* if its columns are distinct, $d(\cdot, \hat{x}) \neq d(\cdot, \hat{x}')$ for all $\hat{x} \neq \hat{x}'$. Non-degeneracy arises naturally when considering the RD problem tangent to a given IB root $p(\hat{x}|x)$. Indeed, the distortion matrix d_{IB} (25) defined by $p(\hat{x}|x)$ has duplicate columns if the root has identical clusters, while the other direction holds under mild assumptions (if the $|\mathcal{X}|$ vectors $p_{Y|X=x}$ span $\mathbb{R}^{|\mathcal{Y}|}$, then $D_{KL}[p_{Y|X=x}||r_{\hat{x}}] = D_{KL}[p_{Y|X=x}||r_{\hat{x}'}]$ for all x implies that $r_{\hat{x}} = r_{\hat{x}'}$). Under these assumptions, the distortion matrix induced by an IB root $p(\hat{x}|x)$ is reduced and non-degenerate precisely when $p(\hat{x}|x)$ is a reduced IB root.

Reduction in RD provides the means to show that the dynamics underlying the RD curve (23) are piecewise analytic in β [6], under mild assumptions. Just as in definition (5) of the IB operator, [4] (Equation (5)) similarly define the *RD operator* $Id - BA_\beta^{RD}$ in terms of Blahut's algorithm for RD [8]. By using their Theorem 1, [6] (Section 3.1) observed that reducing a finite RD problem to the support of a given RD root mods-out redundant kernel coordinates if the distortion measure is finite and non-degenerate (the *support* of $p(\hat{x})$ is defined by $\operatorname{supp} p(\hat{x}) := \{\hat{x} : \ p(\hat{x}) > 0\}$). That is, the Jacobian $D(Id - BA_\beta^{RD})$ of the RD operator on the reduced problem is then non-singular (in the right coordinate system—see therein), just as with our toy problem (10) in Section 1. By the Implicit Function Theorem, there is therefore a unique RD root of the reduced problem through the given one; this root is real-analytic in β (details there). Considering this for the RD problem tangent to a reduced IB root immediately yields the following:

Corollary 1. *Let $p_{\beta_0}(\hat{x}|x)$ be a reduced IB root of a finite IB problem defined by $p_{Y|X}\, p_X$, such that the matrix $p_{Y|X}$ is of rank $|\mathcal{Y}|$. Then, near β_0, there is a unique function continuous in β, which is a root of the tangent RD problem through $p_{\beta_0}(\hat{x}|x)$; it is real-analytic in β.*

Corollary 1 shows that the local approximation of an IB problem (the roots of its tangent RD problem) is guaranteed to be as well-behaved as one could hope for, provided that the IB is viewed in the right coordinate system. Note, however, that the RD root through $p_{\beta_0}(\hat{x}|x)$ of the tangent problem does *not* in general coincide with the IB root outside of β_0 since the IB distortion d_{IB} (25) varies along with the clusters that define it. However, when the IB clusters are fixed, then one might expect that the Jacobian (13) of BA_β in log-decoder coordinates would be the same as the Jacobian of its RD variant. Indeed, the Jacobian matrix of BA_β^{RD} is the $T \times T$ bottom-right sub-block of the Jacobian (13) of BA_β, up to a multiplicative factor. For this, see Equations (5) and (6) in [4], Equations (14) and (13) in Section 3, and (A25) in Appendix B.2.

As in RD, we argue that reduction in the IB also provides the means to show that the dynamics underlying the optimal curve (1) are piecewise analytic in β. Corollary 1 concludes that, under mild assumptions, through every reduced IB root passes a unique real-analytic RD root. However, its crux is that the Jacobian of the RD operator $Id - BA_\beta^{RD}$ is non-singular at a reduced root. Due to the IB's close relations with RD, and since reduction in the IB is a natural extension of reduction in RD, we argue that the same is also to be expected of the IB operator $Id - BA_\beta$ (5) in decoder coordinates. To see this, note that IB roots are finitely supported [2] (Lemma 2.2(i)), and so one may take finitely supported probability distributions $\Delta\big[\Delta[\mathcal{Y}]\big]$ for the IB's optimization variable. Thus, the IB's BA_β operator in decoder coordinates (of Section 2) may be considered as an operator on $\Delta\big[\Delta[\mathcal{Y}]\big]$. Next, consider the RD problem defined by p_X and d_{IB} (25) on the continuous reproduction alphabet $\Delta[\mathcal{Y}]$, as in [20]. This defines on $\Delta\big[\Delta[\mathcal{Y}]\big]$ also the BA operator BA_β^{RD} for RD. Now that both BA operators are considered on an equal footing, we note the following. First, while BA_β^{RD} iterates over the IB Equations (2) and (4), its IB variant BA_β iterates also over the decoder Equation (3) (plug the IB distortion measure d_{IB} (25) into the Equations (24) defining BA_β^{RD} to see this). The latter Equation (3) is a necessary condition for $Y \to X \to \hat{X}$ to be Markov, and so can be understood as an enforcement of Markovity (in contrast, an arbitrary triplet $(Y, X, \hat{X})$ of random variables only satisfies $p(y|\hat{x}) = \sum_x p(y|x, \hat{x})p(x|\hat{x})$). That is, IB roots are RD roots with an extra constraint. Second, by Theorem 1 in [4], reducing $Id - BA_\beta^{RD}$ from the continuous reproduction alphabet $\Delta[\mathcal{Y}]$ to a root of finite support renders it non-singular, under mild assumptions. This suggests that reducing $Id - BA_\beta$ (5) from $\Delta\big[\Delta[\mathcal{Y}]\big]$ to a root's effective cardinality should also render it non-singular, due to the similarity between these operators, and since reduction in the IB is a natural extension of reduction in RD. In line with the discussion of Section 1 on reduction, we therefore state the following:

Conjecture 1. *The Jacobian matrix $I - D_{\log p(y|\hat{x}), \log p(\hat{x})} BA_\beta$ at (16) of the IB operator (5) in log-decoder coordinates is non-singular at reduced IB roots so long as it is well-defined, except perhaps at points of bifurcation.*

The intuition behind this conjecture stems from analyticity, as follows. The IB operator $Id - BA_\beta$ (5) is real-analytic, since each of the Equations 1.4–1.8 defining it (in the BA-IB Algorithm 1) is real-analytic in its variables. For a root x_0 of a real-analytic operator F, one might expect that, in general, (i) no roots other than x_0 exist in its vicinity and that (ii) $D_x F|_{x_0}$ has no kernel. That is, unless the operator is degenerate at x_0 in some manner or x_0 is a bifurcation. To see this, recall [32] (Section IX.3) that a real-valued function F_i in $x \in \mathbb{R}^n$ is *real-analytic* in some open neighborhood of x_0 if it is a power series in $x = (x_1, \dots, x_n)$, within some radius of convergence (although a strictly positive radius is needed, we omit these details for clarity). For every practical purpose, one may replace F_i by a polynomial in $(x_1, \dots, x_n)$ when x is close enough to the base-point x_0, by truncating the power series. Viewed this way, a root of an operator $F(x) = \big(F_1(x), \dots, F_n(x)\big)$ is nothing but a solution of n polynomial equations in n variables. However, a square polynomial system typically has only isolated roots, which is (i). This is best understood in terms of Bézout's Theorem; see [33] (6 in IV.4) for example. For (ii), a vector v is in $\ker D_x F$

precisely when it is orthogonal to each of the gradients ∇F_i. However, ∇F_i is the vector of the first-order monomial coefficients of $x_1, \ldots, x_n$ in F_i. In a general position, these n coefficient vectors $\nabla F_1, \ldots, \nabla F_n$ are linearly independent, and so v must vanish as claimed. If F is degenerate such that $F_i = F_j$ for particular $i \neq j$, for example, then both points fail, of course. See also Section I.2 of [34] for (i) and (ii). This intuition accords with the comments of [28] (Section 2.4) on RD: *"usually, each point on the rate distortion curve [...] is achieved by a unique conditional probability assignment. However, if the distortion matrix exhibits certain form of symmetry and degeneracy, there can be many choices of [a minimizer]"*. Indeed, the fact that the dynamics underlying the RD curve (23) are piecewise real-analytic [6] (under mild assumptions) can be similarly understood to stem from the analyticity of the RD operator $Id - BA_\beta^{RD}$.

Subject to Conjecture 1, a Jacobian eigenvalue of the IB operator (5) must vanish gradually as one approaches a bifurcation, causing the critical slowing down of BA-IB [4] (observe that BA's Jacobian (13) is continuous in the root at which it is evaluated). When an IB root traverses a bifurcation in which its effective cardinality decreases, then it is not reduced anymore. One can then handle the bifurcation by reducing the root anew. To ensure proper handling by the bifurcation's type, we consider the latter closely in Sections 5.2 and 5.3 below. In a nutshell, following the IB's ODE (16) along with a proper handling of its bifurcations is the idea behind our root-tracking algorithm (in Section 6), for approximating the IB numerically.

Conjecture 1 is compatible with our numerical experience. However, we leave its proof to future work. To that end, one could examine closely the smaller matrix S (17) (of Lemma 1 in Section 3), for example. However, even if Conjecture 1 were violated, then one could detect that easily by inspecting the Jacobian's eigenvalues. Conjecture 1 also implies that IB roots are locally unique outside of bifurcations when presented in their reduced form. Non-uniqueness of optimal roots is detectable by inspecting the Jacobian's eigenvalues—see Corollary 3 in Section 5.3 and the discussion following it. See also Section 6.3 in [6] for the respective discussion in RD. With that, most of the results in Sections 5.2 and 5.3 below do not depend on the validity of Conjecture 1.

5.2. Continuous IB Bifurcations: Cluster Vanishing and Cluster Merging

Following [10], we consider the evolution of IB roots which are a continuous function of β. By representing an IB root in its reduced form (Section 5.1), it is evident that there are two types of *continuous* IB bifurcations. We provide a practical heuristic (Algorithm 2) for identifying and handling such bifurcations. The discussion here is complemented by Section 5.3 below, which considers the case where continuity does not hold.

The evolution of an IB root in β obeys the ODE (16) as long as it can be written as a differentiable function in β, as in Theorem 1. Considering the root in decoder coordinates, this amounts to an evolution of a T-tuple of points $r_{\hat{x}}$ in $\Delta[\mathcal{Y}]$ and their weights $p(\hat{x})$. These typically traverse the simplex smoothly as the constraint β is varied, as demonstrated in Figure 6. We now consider two cases where this evolution does not obey the ODE (16), due to violating differentiability.

Consider an optimal IB root in its reduced form (see Section 5.1). Namely, consider the reduced form of a root that achieves the IB curve (1). Suppose that its decoders $r_{\hat{x}}$ and weights $p(\hat{x})$ are continuous in β. Then, a qualitative change in the root can occur only if either (i) two (or more) of its clusters collide or (ii) the marginal probability $p(\hat{x})$ of a cluster $\hat{x}$ vanishes. In either case, the minimal number of points in $\Delta[\mathcal{Y}]$ required to represent the root decreases. That is, its effective cardinality decreases (a qualitative change where the effective cardinality increases is obtained by merely reversing the dynamics in β). We call the first a *cluster-merging bifurcation* and the second a *cluster-vanishing bifurcation*, or *continuous bifurcations* collectively. Both types were observed already in [17] (Section IV.C) in the related setting of RD problems with a continuous source alphabet. Among the two, cluster-vanishing bifurcations are more frequent in practice than cluster merging. This can be understood by considering cluster trajectories in the simplex. In a general position, one might expect clusters to seldom be at the same "time" and place (that is, β and $r \in \Delta[\mathcal{Y}]$).

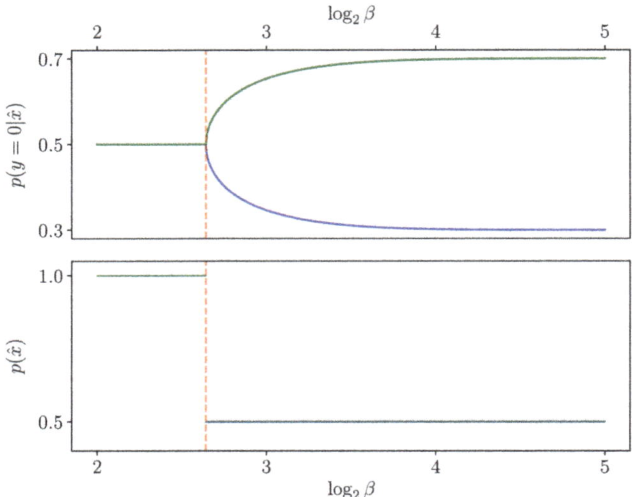

Figure 6. A cluster-merging bifurcation. The reduced form of the optimal IB root in decoder coordinates as a function of β, for the exact solution of BSC(0.3) with a uniform source, as in Figure 1 (see Appendix E). At high enough β, the root consists of two clusters (in green and blue), each of a marginal probability $\frac{1}{2}$. The clusters collide at $\beta_c = 6^{1/4}$ (dashed red vertical) and merge to one, yielding the trivial solution—a single cluster of probability 1 at p_Y. Carefully note that only a *single* IB root is plotted here, in its reduced form, with one cluster to the left of β_c and two to the right. The violation of clusters' differentiability at β_c can be observed visually (**top**), and the root is otherwise real-analytic in β, as can be deduced from Figure 5. Since the trivial solution is an IB root for every $\beta > 0$ (not shown), then β_c is indeed a bifurcation, where the trivial and non-trivial roots intersect. To see this, consider the degenerate form of the trivial solution on *two* copies of p_Y, each of probability $\frac{1}{2}$. The marginals $p(\hat{x})$ (**bottom**) appear to be discontinuous at β_c because the root was reduced before plotted (the latter degenerate form of the trivial root is *not* plotted to the left of β_c).

We argue that cluster merging and cluster vanishing are indeed bifurcations, where IB roots of distinct effective cardinalities collide and merge into one. We offer two ways to see this. First, using the inverse encoder formulation of the IB in [2] (Section II.A), one can consider an optimization problem in which the number of IB clusters is constrained explicitly (the inverse encoders of an IB root with no zero-mass clusters are in bijective correspondence with its decoders, as noted in Section 2, and so inverse encoder and decoder coordinates are interchangeable). By the arguments therein, the constrained problem has an optimal root (due to compactness), which achieves the optimal curve of the constrained problem. The latter curve must be sub-optimal if fewer clusters are allowed than needed to achieve the IB curve (1). Thus, whenever the effective cardinality of an optimal root (in the un-constrained problem) decreases, it must therefore collide with an optimal root of the constrained IB problem (by Corollary 3 in Section 5.3 below). This accords with [1] (Section 3.4), which describes IB bifurcations as a separation of optimal and sub-optimal IB curves according to their effective cardinalities. Second, consider the reduced form of an IB root at the point of a continuous bifurcation. Since its effective cardinality decreases there strictly, say from T_2 to T_1, then the root can be represented on T_1 clusters at the bifurcation itself. However, the Jacobian of the IB operator (5) in log-decoder coordinates is non-singular when represented on T_1 clusters, as discussed after Proposition 1 (in Section 5.3). Thus, by the Implicit Function's Theorem, there is a unique IB root on T_1 clusters through this point. It exists at both sides of the bifurcation (above and below the critical point). When represented on T_2 clusters, however, the latter intersects at the bifurcation with the root of effective cardinality T_2, and so the two roots collide and merge there to one. This argument is identical to [6] (Section 6.2), which proves that distinct RD roots collide and merge at cluster-vanishing bifurcations in RD.

At a continuous bifurcation, IB roots of distinct effective cardinalities collide and merge into one, as discussed above. Specifically, one root achieves the minimal value of the IB Lagrangian and so is stable, while the other root is sub-optimal. As we shall now elaborate, continuous IB bifurcations are thus *pitchfork bifurcations* (e.g., Section 3.4 in [35]), in accordance with [19]. Even though the optimal root is continuous in β (by assumption), its differentiability is violated at the point of bifurcation. This can be inferred from the comments following Theorem 1 and seen in Figure 6. Strictly speaking, several copies of the root of larger effective cardinality collide at a continuous bifurcation. When two clusters $r \neq r'$ collide in a cluster merging bifurcation, then the root itself is invariant to interchanging their coordinates after the collision but not before it, breaking the IB's first structural symmetry discussed in Section 5.1. Interchanging the coordinates of r and r' (and their marginals) before the collision yields two distinct copies of essentially the same root. For a cluster vanishing bifurcation, the IB's functionals (e.g., its X- and Y-information) do not depend on the coordinates $(r(y))_y$ of a vanished cluster r, rendering these redundant; cf., [9] (Section 3.1). Before the cluster r vanishes, there is one copy of the root for each index $\hat{x}$, with r placed at its $\hat{x}$ coordinates. Considered in reduced coordinates, these coincide to a single copy after the cluster vanishes. This breaks the IB's second structural symmetry.

With that, we note that cluster-vanishing bifurcations *cannot* be detected directly by standard local techniques (i.e., considering the derivative's kernel directions at the bifurcation point), whether considering the Hessian of the IB's loss function as in [9] or the Jacobian of the IB operator (5) as here. The technical reason for this is as follows, while the root cause underlying it is best understood in the context of discontinuous bifurcations (after Proposition 1 in Section 5.3). Observe that the $I(Y; \hat{X})$ and $I(X; \hat{X})$ functionals do not depend on the coordinates $(r(y))_y$ of clusters r of zero mass. Thus, the directions corresponding to these coordinates are always in the kernel regardless of whether evaluating at a bifurcation or not, and so cannot be used to detect a bifurcation (the direction corresponding to a cluster's marginal is useless when one does not know which coordinates $(r(y))_y$ to pick for r). Indeed, with its dynamics in β reversed, *"a new symbol grows continuously from zero mass"* in a cluster-vanishing bifurcation, as [17] (Section IV.C) comments in a related setting. It is then not clear a priori which point in $\Delta[\mathcal{Y}]$ should be chosen for the new symbol, rendering the perturbative condition at Equation (9) difficult to test. In accordance with this, Ref. [9] (Section 5) offers a perturbative condition for detecting arbitrary IB bifurcations, while ref. [13] (3.2 in Part III) offers a condition for detecting cluster-merging bifurcations by analyzing cluster stability. However, both conditions are equivalent (Appendix F), and so must detect the same type of bifurcations. In contrast, a cluster-splitting (or merging) bifurcation is straightforward to detect because the stability of a particular cluster $\hat{x}$ is a property of the root itself—see Appendix F and the references therein for details.

One may wonder whether bifurcations exist in the IB for the same reason as they do in RD. As in the IB, RD problems typically have many sub-optimal curves [6] (Section 6.1). While (continuous) bifurcations in the IB stem from restricting the effective cardinality [1] (Section 3.4), in RD they stem from the various restrictions that a reproduction alphabet has. For example, a reproduction alphabet $\hat{\mathcal{X}} := \{r_1, r_2, r_3\}$ of an RD problem may be restricted to the distinct subsets $\{r_1, r_2\}$ and $\{r_2, r_3\}$, usually yielding distinct sub-optimal RD curves (e.g., Figure 6.1 in [6]). In contrast to RD, the IB's distortion d_{IB} (25) defined by a root's clusters is determined a posteriori by the problem's solution rather than a priori by the problem's definition. As a result, both reasons for the existence of bifurcations coincide. To see this, consider the IB as an RD problem whose reproduction symbols $\hat{\mathcal{X}}$ are a finite subset of $\Delta[\mathcal{Y}]$ which is allowed to vary (i.e., as if defining the tangent RD problem anew at each β). Distinct restrictions of a reproduction alphabet $\hat{\mathcal{X}}$ can be forced to agree by altering the symbols themselves, so long as they are of the same size. For example, restricting the set $\{r_1, r_2, r_3\}$ of reproduction symbols to $\{r_1, r_2\}$ is the same as restricting it to $\{r_2, r_3\}$ instead, and then replacing r_3 with $r_1 \in \Delta[\mathcal{Y}]$ in the restricted problem (this is not to be confused with cluster permutations, which change the order in which clusters are listed but do *not* alter the symbols themselves).

The dynamical point of view above, considering an IB root as weighted points traversing $\Delta[\mathcal{Y}]$, offers a straightforward way to identify and handle continuous IB bifurcations. It is spelled out as our root-reduction Algorithm 2. For cluster-vanishing bifurcations, one can set a small threshold value $\delta_1 > 0$ and consider the cluster $\hat{x}$ as vanished if $p(\hat{x}) < \delta_1$ (Step 2.3), as in [6] (Section 3.1). Similarly, for cluster-merging bifurcations, one can set a small threshold $\delta_2 > 0$ and consider the clusters $\hat{x} \neq \hat{x}'$ to have merged if $\|r_{\hat{x}} - r_{\hat{x}'}\|_\infty < \delta_2$ (Step 2.9). A vanished cluster is then erased (and merged clusters replaced by one), resulting in an approximate IB root on fewer clusters. This not only identifies continuous IB bifurcations but also handles them, since the output of the root-reduction Algorithm 2 is a numerically reduced root, represented in its effective cardinality. To re-gain accuracy, we shall later invoke the BA-IB Algorithm 1 on the reduced root, as part of our root-tracking algorithm (in Section 6). We note that one should pick the thresholds δ_1 and δ_2 small enough to avoid false detections, and yet not too small so as to cause mis-detections. Mis-detections will be handled later, in Section 6.1, using a heuristic algorithm.

Algorithm 2 Root reduction for the IB

1: **function** REDUCE ROOT($p(y|\hat{x}), p(\hat{x}); \delta_1, \delta_2$)
Input:
    An approximate IB root $\big(p(y|\hat{x}), p(\hat{x})\big)$ in decoder coordinates,
    a cluster-mass threshold $0 < \delta_1 < 1$ and a cluster-merging threshold $0 < \delta_2 < 1$.
Output: An approximate IB root $\big(\tilde{p}(y|\hat{x}), \tilde{p}(\hat{x})\big)$ at its effective cardinality.
2:    **for** $\hat{x}$ **do**
3:        **if** $p(\hat{x}) < \delta_1$ **then**   ▷ Delete clusters of near-zero mass.
4:            delete the coordinates of $\hat{x}$, from $p(\hat{x})$ and $p(y|\hat{x})$.
5:        **end if**
6:    **end for**
7:    $p(\hat{x}) \leftarrow$ normalize $p(\hat{x})$ ▷ Preserve normalization, in case clusters were removed.

8:    **for** $\hat{x} \neq \hat{x}'$ **do**
9:        **if** $\|p(y|\hat{x}) - p(y|\hat{x}')\|_\infty < \delta_2$ **then**   ▷ Merge nearly identical points in $\Delta[\mathcal{Y}]$.
10:          $p(\hat{x}) \leftarrow p(\hat{x}) + p(\hat{x}')$
11:          delete the coordinates of $\hat{x}'$, from $p(\hat{x})$ and $p(y|\hat{x})$.
12:        **end if**
13:    **end for**

14:    **return** $\big(p(y|\hat{x}), p(\hat{x})\big)$
15: **end function**

Using the root-reduction Algorithm 2 allows one to stop early in the vicinity of a bifurcation when following the path of an IB root. As mentioned in Section 4, early stopping restricts the computational difficulty of root tracking [6]. Further, reducing the root *before* invoking BA-IB (Algorithm 1) allows us to avoid BA's critical slowing down [4], since reduction removes the nearly vanished Jacobian eigenvalues that pertain to the nearly vanished (or nearly merged) cluster(s), which are the cause of BA's critical slowing down. cf., Proposition 1 (Section 5.3) and the discussion around it. See also [6] (Figure 3.1(C) and Section 3.2) for the respective behavior in RD. Finally, we comment that the root-reduction Algorithm 2 can also be implemented in the other two coordinate systems of Section 2.

5.3. Discontinuous IB Bifurcations and Linear Curve Segments

In the previous Section 5.2, we considered continuous IB bifurcations—namely, when the clusters $r_{\hat{x}} \in \Delta[\mathcal{Y}]$ and weights $p(\hat{x})$ of an IB root are continuous functions of β. By exploiting the intimate relations between the IB and RD (Section 5.1), we now consider IB bifurcations where these *cannot* be written as a continuous function of β. In our experience, discontinuous bifurcations are infrequent in practice. However, the theory they evoke has several subtle consequences of practical implications important for computing IB roots (in Section 6). Though, perhaps more importantly, they oblige one to ask what *is* the IB? We start with several examples before diving into the theory; e.g., Figure 7.

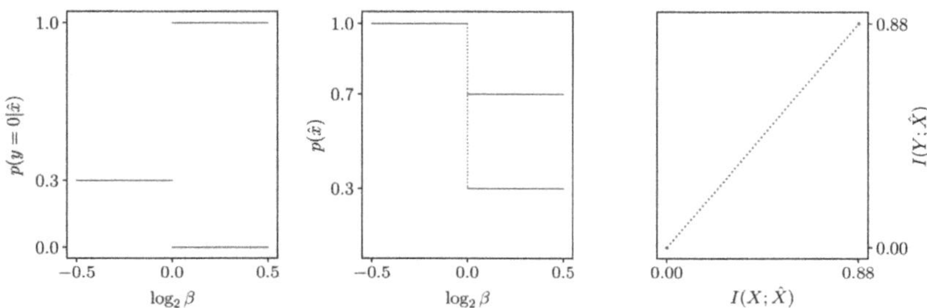

Figure 7. A discontinuous IB bifurcation at $\beta_c = 1$, of the problem defined by $p_{Y|X}\, p_X = \begin{pmatrix} 0.3 & \\ & 0.7 \end{pmatrix}$. (**Left**): to the left of β_c, the optimal solution is the trivial one, supported on the IB cluster p_Y. To the right it is supported on the boundary points $(1, 0)$ and $(0, 1)$ of $\Delta[\mathcal{Y}]$. (**Middle**): the marginals are constant, except at the point of bifurcation. Any convex combination of the trivial and non-trivial roots is optimal there (dotted). That is, this is a **support-switching bifurcation** as in RD [6] (Figure 6.2). (**Right**): the IB curve exhibits a linear segment of slope $1/\beta_c = 1$, connecting the image of the trivial solution in the information plane (bottom-left) to that of the non-trivial one (top-right). See comments in the main text.

The examples of discontinuous IB bifurcations of which we are aware can be understood in RD context as follows. Consider the IB as an RD problem on the continuous reproduction alphabet $\Delta[\mathcal{Y}]$, with IB roots parameterized by points in $\Delta[\Delta[\mathcal{Y}]]$ (see Section 5.1). In RD, the existence of linear curve segments is well-known [28]—e.g., Figure 2.7.6 in the latter and its reproduction in [6] (Figure 6.2). Section 6.5 in [6] offers an explanation of linear segments in terms of a *support-switching bifurcation*. Namely, a bifurcation where two RD roots of distinct supports exchange optimality at a particular multiplier value β_c. Both roots evolve smoothly in β while only exchanging optimality at the bifurcation. At β_c itself, every convex combination of these two roots is also an RD root. In particular, the optimal RD root *cannot* be written as a continuous function of β. The sudden emergence of an entire segment of roots at β_c can be understood by RD's convexity and analyticity properties, as follows. The RD curve (23) is parameterized by the slope $-\beta$ of its tangents [28] (Theorems 2.5.1 and 2.5.2). Above and below β_c, specifying the tangent's slope determines a curve-achieving distribution on the optimal root (the root whose curve is lower at this slope value). Equivalently, the lower convex envelope of these roots in the RD plane coincides with one root above β_c and with the other below it, as seen in Figure 8 (black). At β_c itself, specifying the slope determines a distribution on both roots. Thus, the convexity of the RD curve and of the set of achieving distributions implies a linear segment at β_c (Theorem 2.4.1 in [28] and Theorem 5 below). Finally, this behavior is possible due to analyticity, since the roots of a real-analytic operator $Id - BA_{\beta}^{RD}$ are either isolated (typical) or an algebraic curve (atypical) by Bézout's Theorem—see (i) in the discussion following Conjecture 1.

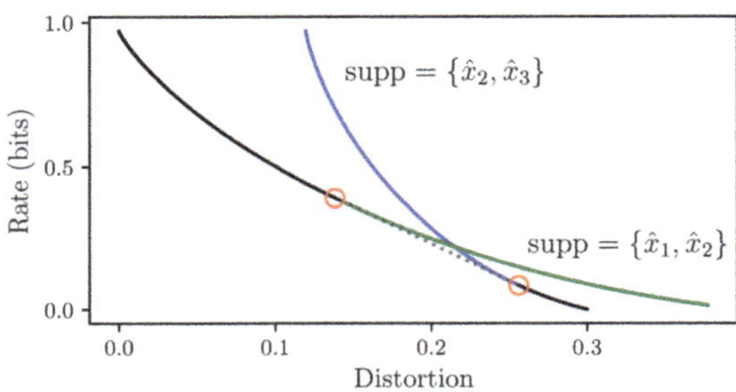

Figure 8. A support-switching bifurcation in RD, reproducing Figure 6.2(F) in [6] (details therein). The RD curve (23) (black) is the envelope of its tangents, parameterized by their slope $-\beta$, [28]. At high slopes, the envelope coincides with that of the problem restricted to the reproduction alphabet $\{\hat{x}_1, \hat{x}_2\}$ (green), and at low slopes with that restricted to $\{\hat{x}_2, \hat{x}_3\}$ (blue). At a critical slope $-\beta_c$, the tangent touches both curves (red circles). Convexity then implies a linear segment (dashed)—see main text.

For one example of linear curve segments in the IB, say that a matrix M *decomposes* if it can be written (non-trivially) as a block matrix by permuting its rows or columns. In light of the above, we have the following refinement of Theorem 2.6 in [2]:

Theorem 3. *The IB curve* (1) *has a linear segment at $\beta = 1$ if and only if the problem's definition $p_{Y|X}\, p_X$ decomposes.*

Recall that the slope of the IB curve is $1/\beta$ at a multiplier value β [1] (Equation (32)). Thus, Theorem 3 equates decomposable problems with linear curve segments of slope 1 (the slope cannot exceed one due to the data processing inequality). Figure 7 provides a simple decomposable example, exhibiting a support-switching bifurcation between its trivial and non-trivial roots. Non-decomposable examples also exist, exhibiting a support-switching bifurcation at lower slope values (higher critical β's). For example, a symmetric binary erasure channel exhibits a support-switching bifurcation [2] (Section IV.B), which is manifested by a linear segment of slope $1/\beta_c \leq 1$, for $\beta_c \geq 1$ (switching between the trivial root at p_Y and a bi-clustered root supported on $(\beta_c^{-1}, 1 - \beta_c^{-1}, 0)$ and $(0, 1 - \beta_c^{-1}, \beta_c^{-1}) \in \Delta[\mathcal{Y}]$; the linear segment of slope β_c^{-1} is Equation (4.8) there). See [2] (Section IV) for further examples. We argue that in the IB, support-switching bifurcations exhibit the same behavior as in RD. That is, two roots that evolve smoothly in β and exchange optimality at the bifurcation. While the sequel can justify this in general, there is a simple way to see this in practice. Namely, following the two roots of Figure 7 through the bifurcation by using BA-IB with deterministic annealing [11] (follow the trivial root of Figure 7 from left to right and the non-trivial one from right to left, through the bifurcation at $\beta_c = 1$ there). As deterministic annealing usually follows a solution branch continuously, this immediately reveals either root at the region where it is sub-optimal (not displayed).

A support-switching bifurcation evidently has similar characteristics to a *transcritical* bifurcation (e.g., Section 3.2 in [35]), though it should perhaps be classified as an *imperfect transcritical* since the roots do not intersect per se as in a classical transcritical. This extends the results of [19], who conclude that IB bifurcations *"are only of pitchfork type"* (Theorem 5 therein says that the bifurcations detected by their Theorem 3 are degenerate rather than transcritical, concluding that *"the bifurcation guaranteed by Theorem 3 is [generically] pitchfork-like"*). To see the reason for this discrepancy, note that they employ the mathematical machinery in [36] of bifurcations under symmetry. Since pitchfork bifurcations are *"common in physical problems that have a symmetry"* [35] (Section 3.4), then detecting only pitchforks by using the above machinery might not come as a surprise. Both [9] and its sequel [19] consider the IB's symmetry to interchanging the coordinates of identical clusters (Definition

1(1) in [19]). However, this is a structural symmetry of the IB which stems from representing IB roots by finite-dimensional vectors (Section 5.1), and is broken in continuous IB bifurcations (Section 5.2). On the other hand, discontinuous IB bifurcations need not break this symmetry, as can be seen by inspecting the roots of Figure 7 closely (the trivial solution to the left of β_c there may be given a degenerate bi-clustered representation, which is fully supported on p_Y but has a second cluster $r \neq p_Y$ of zero mass. Neither of its roots then possesses a symmetry to interchanging cluster coordinates, at either side of β_c).

A few convexity results from rate-distortion theory are needed to consider discontinuous bifurcations in general. These have subtle practical implications, which are of interest in their own right.

Theorem 4 (Theorem 2.4.2 in [28]). *The set of conditional probability distributions $p(\hat{x}|x)$ which achieve a point $(D, R(D))$ on the rate-distortion curve (23) is convex.*

Viewing the IB as an RD problem as in [20] immediately yields an identical result for the IB:

Corollary 2. *The set of IB encoders that achieve a point (I_X, I_Y) on the IB curve (1) is convex.*

The proof is provided below for completeness. We note that a version of Corollary 2 in inverse encoder coordinates can also be synthesized from the ideas leading to Theorem 2.3 in [2].

Proof of Corollary 2. Consider a finite IB problem $p_{Y|X} \, p_X$ as an RD problem (d_{IB}, p_X) on the continuous reproduction alphabet $\Delta[\mathcal{Y}]$, as defined by (25) in Section 5.1. As noted above, its encoders (or test channels) are conditional probability distributions $p(r|x)$, with $r \in \Delta[\mathcal{Y}]$, supported on finitely many coordinates (r, x).

Let $p_1(r|x)$ and $p_2(r|x)$ be encoders achieving a point (I_X, I_Y) on the IB curve (1). Define their support by supp $p(r|x) := $ supp $p(r)$, where $p(r)$ is defined from $p(r|x)$ via marginalization, as in (4). By Theorem 5 in [20], $p_1(r|x)$ and $p_2(r|x)$ may be considered as test channels achieving the curve (23) of the RD problem (d_{IB}, p_X). The reproduction symbols $r \in \Delta[\mathcal{Y}]$ supporting a convex combination $p_\lambda := \lambda \cdot p_1 + (1 - \lambda) \cdot p_2, 0 \leq \lambda \leq 1$, are contained in the the supports of p_1 and p_2: supp $p_\lambda \subseteq$ supp $p_1 \cup$ supp p_2. Therefore, p_λ is finitely supported. Although Berger's Theorem 4 assumes that the reproduction alphabet is finite, one can readily see that its proof works just as well when the distributions involved are finitely supported. Thus, by Theorem 4, p_λ achieves the above point on the RD curve (23). Since this point (I_X, I_Y) is on the IB curve (1), then p_λ is an optimal IB root. $\square$

The RD curve (23) is the envelope of lines of slope $-\beta$ and intercept $\min_{p(\hat{x}|x)} \left(I(X; \hat{X}) + \beta \, \mathbb{E}[d(x, \hat{x})] \right)$ along the R-axis, e.g., [28]. Thus, Theorem 4 can be generalized by considering the achieving distributions that pertain to a particular slope value rather than to a particular curve point $(D, R(D))$—see [6] (Section 6.3).

Theorem 5 (Theorem 20 in [6]). *For any $\beta > 0$ value, the set of distributions achieving the RD curve (23) that correspond to β is convex.*

As with Corollary 2, we immediately have an identical result for roots achieving the IB curve (1):

Corollary 3. *For any $\beta > 0$ value, the set of optimal IB encoders that correspond to β is convex.*

See also [2] (Section IV) for an argument in inverse encoder coordinates. In particular, note the duality technique leading to (b) and (c) in Theorem 4.1 there. This duality boils down to describing a compact convex set in the plane by its lines of support, as in the observation leading to Theorem 5. Commensurate with the IB being a special case of RD,

Corollary 3 can also be proven directly from the IB's definitions in direct encoder terms [37]. Note that the requirement that the IB root indeed achieves the curve is necessary. Otherwise, one could take convex combinations with the trivial IB root $p(r|x) = \delta_{r,p_Y}$ (which satisfies the IB Equations (2)–(4) for every $\beta > 0$, as one can verify directly). This yields absurd results, since the trivial root contains no information on either X or Y.

As in [6] (Section 6.3), the convexity of optimal IB roots (Corollary 3) has several important consequences. For one, unlike the (local) bifurcations we have considered so far, bifurcation theory also has *global bifurcations*. These are *"bifurcations that cannot be detected by looking at small neighborhoods of fixed points"* [12] (Section 2.3). From convexity, it immediately follows that

Corollary 4. *There are no global bifurcations in finite IB problems.*

Indeed, if at a given β value there exists more than one optimal root, then the Jacobian of the IB operator $Id - BA_\beta$ (5) must have a kernel vector pointing along the line connecting these optimal roots, by Corollary 3.

With that comes an important practical caveat. Corollaries 2 and 3 hold for the IB when parameterized by points in $\Delta[\Delta[\mathcal{Y}]]$. However, the above kernel vector (which exists due to convexity) may not be detectable if an IB root is improperly represented by a finite-dimensional vector. For example, consider the bifurcation in Figure 7, where a line segment at β_c connects the trivial (single-clustered) root to the 2-clustered root. Obviously, the bifurcation there cannot be detected by the Jacobian of the IB operator (5) when it is computed on $T = 1$ clusters (Jacobian of order $1 \cdot (|\mathcal{Y}| + 1)$). Indeed, the root of effective cardinality two cannot be represented on a single cluster, and so the line segment connecting it to the trivial root does not exist in a 1-clustered representation. This is demonstrated in Figure 9, which compares Jacobian eigenvalues at reduced representations to those at 2-clustered representations. The same reasoning gives the following necessary condition:

Proposition 1 (A necessary condition for detectability of IB bifurcations). *A bifurcation at β_c in a finite IB problem which involves roots of effective cardinalities T_1 and T_2 is detectable by a non-zero vector in $\ker(I - D_{\log p(y|\hat{x}),\log p(\hat{x})} BA_{\beta_c})$ only if the latter is evaluated at a representation on at least $\max\{T_1, T_2\}$ clusters.*

Indeed, suppose that $T_1 \lneq T_2$ (the conclusion is trivial if $T_1 = T_2$). By definition, a root of effective cardinality T_2 does not exist in representations with less than T_2 clusters. Thus, there is no bifurcation in a T-clustered representation if $T < T_2$, and so there is then nothing to detect. As a special case of this argument, note that Conjecture 1 (Section 5.1) implies that the Jacobian is non-singular in a T_1-clustered representation of the T_1-clustered root (namely, at its reduced representation). With that, we have observed numerically that the eigenvalues of $D_{\log p(y|\hat{x}),\log p(\hat{x})} BA_\beta$ do *not* depend on the representation's dimension if computed on strictly more clusters than the effective cardinality (which makes sense considering Theorem 2.6.1 in [28] or Lemma 2.2(i) in [2]). Rather, only the eigenvalues' multiplicities vary by dimension. We omit practical caveats on exchanging between the coordinate systems of Section 2 for brevity.

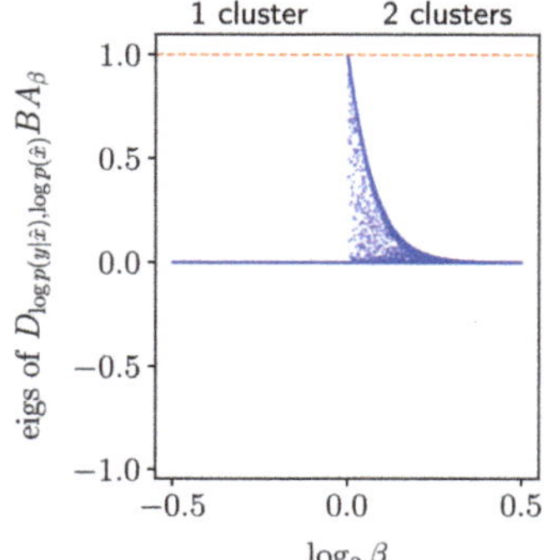
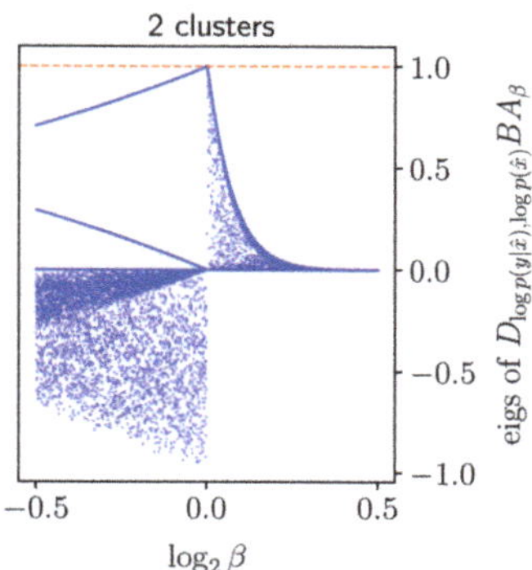

Figure 9. Bifurcations can be detected by BA_β's Jacobian only if computed on enough clusters. The approximate eigenvalues of $D_{\log p(y|\hat{x}),\log p(\hat{x})}BA_\beta$ are plotted by the representation's dimension for the problem in Figure 7. The eigenvalues are evaluated at solutions obtained by the BA-IB Algorithm 1 (stopping condition $= 10^{-9}$), initialized anew at random for each β. While the random initializations account for much of the eigenvalues' spread, they reveal the solution's behavior through its various approximations. Other factors which contribute to this spread are the degeneracy of the solutions (when $\beta < 1$, right panel), BA's loss of accuracy near the bifurcation (Figure 3 bottom), and the decoders' proximity to the simplex boundaries (see Equation (13)). (**Left**): when computed at reduced representations (on $T = 1$ clusters to the left, $T = 2$ to the right), then the eigenvalues at the trivial solution give no indication of the upcoming bifurcation (at $\beta < 1$), unlike the eigenvalues at the 2-clustered root ($\beta > 1$). (**Right**): the bifurcation's presence is clearly noticed also at the trivial solution ($\beta < 1$) when evaluated at its degenerate 2-clustered representations. Indeed, the trivial solution is then represented on the same number of clusters ($T = 2$) as the root to the right ($\beta > 1$)—see Proposition 1. However, due to the bifurcation, the eigenvalues' trajectories are not smooth at $\beta_c = 1$. **Both**: a similar dependency on the representation's dimension also exists in the other bifurcation examples in this paper (though without the eigenvalues' spread).

The discussion of discontinuous bifurcations naturally leads one to consider the IB as an RD problem on the continuous reproduction alphabet $\Delta[\mathcal{Y}]$, as in Corollaries 2 and 3, unlike its usual definitions in the literature. When considered this way, IB roots are merely paths $p(\beta)$ in $\Delta[\Delta[\mathcal{Y}]]$, following a piecewise smooth trajectory dictated by the IB ODE (16) (which may be considered as a non-autonomous ODE on $\Delta[\Delta[\mathcal{Y}]]$). Due to Conjecture 1 and the IFT, these paths are isolated outside bifurcations. Two (or more) roots may intersect in a continuous bifurcation. If one of the intersecting roots is optimal, then the other must be of a strictly smaller effective cardinality due to the arguments in Section 5.2. If two distinct roots are optimal simultaneously, then $\Delta[\Delta[\mathcal{Y}]]$ contains an entire segment of optimal IB roots, due to Corollary 3. Viewing the IB this way also highlights several subtleties in its calculation. First, parameterizing IB roots with $\Delta[\Delta[\mathcal{Y}]]$ avoids its structural symmetries (Section 5.1). Second, it shows that, a priori, it is possible to follow the path of an optimal root using local techniques (Corollary 4). Third, it highlights that one must compute on enough clusters to detect a bifurcation (Proposition 1). Though obvious in retrospect, this caveat was not given proper attention in the IB literature. Fourth, as we shall now see, cluster-vanishing bifurcations can only be detected by following an optimal root to its collision with a root of smaller effective cardinality. Fifth, this implies (below) that only negative step sizes $\Delta\beta < 0$ should be used to follow an optimal root.

The arguments above imply that cluster-vanishing bifurcations cannot be detected directly by considering kernel directions of the IB operator (5) at the bifurcation, as argued in Section 5.2. Indeed, consider a continuous bifurcation, where roots p_1 and p_2 of respective effective cardinalities $T_1 < T_2$ intersect. These are paths in $\Delta[\Delta[\mathcal{Y}]]$ that coincide at the bifurcation itself, $p_1(\beta_c) = p_2(\beta_c)$, and so in particular are of the same effective cardinality T_1 there. This is in contrast to the situation in Corollary 3, where two *distinct* roots are simultaneously optimal at β_c, leading to an entire segment of optimal roots. Asking whether a bifurcation is detectable amounts to considering the evaluation of $\ker D(Id - BA_\beta)$ at a finite-dimensional representation (or "projection") of p. The Jacobian $D(Id - BA_\beta)$ of the IB operator (5) is non-singular when evaluated on a T_1-clustered representation of $p_1(\beta_c)$ in log-decoder coordinates, as noted after Proposition 1. We argue that evaluating it on representations with more clusters $T \gneqq T_1$ does *not* allow one to detect the bifurcation (even if $T \geq T_2$). See Appendix H for a formal argument. Intuitively, this is because picking a degenerate representation amounts to duplicating clusters of the reduced representation or adding clusters of zero mass (see *reduction* in Section 5.1). Introducing degeneracies to a reduced root adds no information about the problem at hand.

Due to the above, cluster-vanishing bifurcations cannot be detected by following a root p_1 of effective cardinality T_1 through the bifurcation point, but only by following a root p_2 with $T_2 > T_1$ to its collision with p_1. As discussed after Conjecture 1 (Section 5.1), the Jacobian of $Id - BA_\beta$ in reduced log-decoder coordinates can then be used to indicate the upcoming collision of p_2 with p_1, in addition to the root-reduction Algorithm 2. The exact same arguments as above apply also to cluster-merging bifurcations. However, as noted in Section 5.2 (and Appendix F), the stability of a particular IB cluster $\hat{x}$ is a property of the root itself. Thus, these are detectable by standard local techniques at the point of bifurcation. Unlike continuous bifurcations, discontinuous bifurcations are inherently detectable due to the line segment in $\Delta[\Delta[\mathcal{Y}]]$ connecting the roots at the bifurcation (Corollary 3), as long as the IB root is represented on sufficiently many clusters (Proposition 1)—see Figure 9. These results make sense, considering that cluster-vanishing bifurcations are more frequent in practice than other types. Intuitively, branching from a suboptimal root p_1 to an optimal one p_2 is harder than the other way around, just as learning new relevant information is harder than discarding it. Cases where both directions are equally difficult are the exception, as one might expect. This is consistent with the later discussion in Section 6.3 on the stability of optimal IB roots (Appendix G).

When following the path of a reduced IB root (as in Section 4), one would like to ensure that its bifurcations are indeed detectable by BA's Jacobian. Due to the caveats involved in detecting bifurcations of either type, it is necessary to follow the path as the effective cardinality decreases rather than increases. As a result, we take only negative step sizes $\Delta\beta < 0$, since the effective cardinality of an optimal IB root cannot decrease with β. To see this, first note that the IB curve $I_Y(I_X)$ (1) is concave, and so its slope $1/\beta$ cannot increase with I_X. That is, β cannot decrease with I_X. Second, note that allowing more clusters cannot decrease the X-information $\sum_{\hat{x}} p(\hat{x}) H(p(x|\hat{x}))$ achieved by the IB's optimization variables. Indeed, a T-clustered variable $(p(x|\hat{x}), p(\hat{x}))$ (*not* necessarily a root) can always be considered as $(T+1)$-clustered, by adding a cluster of zero mass; cf., the construction of [2] (Section II.A). Thus, the effective cardinality of an optimal root cannot decrease as the constraint I_X on the X-information is relaxed. With both points combined, the effective cardinality cannot decrease with β, as argued. In contrast to the IB, we note that the behavior of RD problems is more complicated since the distortion of each reproduction symbol is fixed a priori; e.g., Example 2.7.3 and Problems 2.8–2.10 in [28].

Returning to discontinuous IB bifurcations, we proceed with the argument of Section 5.2 when continuity fails. That is, consider the reduced form of an optimal IB root, and suppose that either its decoders or its weights (or both) cannot be written as a continuous function of β at β_c. Write $r_{\hat{x}}^+$ and $r_{\hat{x}}^-$ for its distinct decoders as $\beta \to \beta_c^+$ and $\beta \to \beta_c^-$, respectively. Similarly, write $p^+(\hat{x})$ and $p^-(\hat{x})$ for its non-zero weights. Consider the tangent RD problem on the reproduction alphabet $\hat{\mathcal{X}} := \{r_{\hat{x}}^+\}_{\hat{x}} \cup \{r_{\hat{x}}^-\}_{\hat{x}} \subset \Delta[\mathcal{Y}]$, as in Section 5.1. See also [4] (Section V), upon which this argument is based. By construction, the IB coincides with its tangent RD problem at the two points $(r_{\hat{x}}^+, p^+(\hat{x}))$, and $(r_{\hat{x}}^-, p^-(\hat{x}))$. Since both points achieve the optimal curve at the same slope value $1/\beta_c$, then the linear segment of distributions connecting these points is also optimal, by Theorem 5. Alternatively, one could apply Corollary 3 directly to the IB problem. Either way, there exists a line segment of optimal IB roots, which pertain to the given slope value. In summary,

Theorem 6. *Let a finite IB problem have a discontinuous bifurcation at $\beta_c \geq 1$. Then, its IB curve (1) has a linear segment of slope $1/\beta_c$.*

Unless the decoder sets $\{r_{\hat{x}}^+\}_{\hat{x}}$ and $\{r_{\hat{x}}^-\}_{\hat{x}}$ are identical, then this is a support-switching bifurcation [6] (Section 6.5), as in Figure 7. *A priori*, the IB roots $(r_{\hat{x}}^+, p^+(\hat{x}))$ and $(r_{\hat{x}}^-, p^-(\hat{x}))$ *may* achieve the same point in the information plane, in which case the linear curve segment is of length zero. However, we are unaware of such examples. Yet, even if such bifurcations exist, they would be detectable by the Jacobian of BA-IB (when represented on enough clusters), subject to Conjecture 1.

6. First-Order Root Tracking for the Information Bottleneck

Gathering the results of Sections 2–5, we can now not only follow the evolution of an IB root along the first-order Equation (16), but can also identify and handle IB bifurcations. This is summarized by our First-order Root-Tracking algorithm for the IB (IBRT1) in Section 6.1, with some numerical results in Section 6.2. Section 6.3 discusses the basic properties of IBRT1, and mainly the surprising quality of approximations of the IB curve (1) that it produces, as seen in Figure 1. We focus on continuous bifurcations (Section 5.2), since these are far more frequent in our experience than discontinuous ones and are straightforward to handle (see Section 6.3 on the handling of discontinuous bifurcations).

6.1. The IBRT1 Algorithm 5

To assist the reader, we first present a simplified version of IBRT1 as Algorithm 3, with edge cases handled later by Algorithm 4—clarifications follow. When combined, these two form our IBRT1 Algorithm 5, specified below.

We now elaborate on the main steps of the Simplified First-order Root Tracking for the IB (Algorithm 3), following Root Tracking for RD [6] (Algorithm 3). Its purpose is to follow the path of a given IB root $p_{\beta_0}(\hat{x}|x)$ in a finite IB problem. The initial condition $p_{\beta_0}(\hat{x}|x)$ is required to be reduced and IB-optimal. Its optimality is needed below to ensure that the path traced by the algorithm is indeed optimal. The step-size $\Delta\beta$ is negative, for reasons explained in Section 5.3 (Proposition 1 ff.). The cluster mass and cluster merging thresholds are as in the root-reduction Algorithm 2 (Section 5.2).

Denote $\tilde{p}$ (step 3 of Algorithm 3) for the distributions generated from an encoder (see Equation (11) in Section 2). Algorithm 3 iterates over grid points $\tilde{p}$, with each **while** iteration generating the reduced form of the next grid point, as follows. On step 6, evaluate the IB ODE (16) at the current root $\tilde{p}$, solving the linear equations numerically. By Conjecture 1 (Section 5.1), the IB ODE has a unique numerical solution v if $\tilde{p}$ is a reduced root and not a bifurcation. Steps 7 and 8 approximate the root at the next grid point at $\beta + \Delta\beta$, by exponentiating Euler method's step (22) (Section 4). Normalization is enforced on step 9, since it is assumed throughout. Off-grid points can be generated by repeating steps 7 through 9 for intermediate $\Delta\beta$ values if desired. The approximate root at $\beta + \Delta\beta$ is reduced on step 11, by invoking the root-reduction Algorithm 2 (Section 5.2). Note that Algorithm 2 returns its input root unmodified unless reducing it numerically. If reduced, then the root is a vector of a lower dimension—either a cluster mass $p(\hat{x})$ has nearly vanished or distinct clusters have nearly merged. To re-gain accuracy, we invoke (on step 14) the Blahut–Arimoto Algorithm 1 for the IB until convergence, on the encoder defined at step 13 by the reduced root. Although BA-IB is invoked near a bifurcation, this does *not* incur a hefty computational cost due to its critical slowing down [4]—see comments at the bottom of Section 5.2. Invoking BA (on step 14) *before* reducing (on step 11) would have inflicted a hefty computational cost to BA-IB due to the nearby bifurcation. Finally, a single BA-IB iteration in decoder coordinates is invoked on the approximate root (step 17), whether reduced earlier or not. This enforces Markovity while improving the order of this method (see Section 4, and Figure 4 in particular). Algorithm 3 continues this way (step 4) until the approximate solution is trivial (single-clustered), or β is non-positive. In the IB, the trivial solution is always optimal for tradeoff values $\beta < 1$. However, here β plays the role of the ODE's independent variable instead. Thus, we allow Algorithm 3 to continue beyond $\beta = 1$, as long as $\beta > 0$, which is assumed throughout (the condition $\beta > |\Delta\beta|$ on step 4 ensures that the target β value of the *next* grid point is non-negative). This shall be useful for overshooting—see below.

Algorithm 3 Simplified First-order Root-Tracking for the IB

1: **function** SIBRT1$(p_{Y|X}\, p_X, \beta_0, p_{\beta_0}(\hat{x}|x); \Delta\beta, \delta_1, \delta_2)$
Input:
    An IB problem definition $p_{Y|X}\, p_X$ with $\forall x\, p_X(x) > 0$.
    A reduced IB-optimal root $p_{\beta_0}(\hat{x}|x)$ at β_0. A step size $\Delta\beta < 0$.
    Cluster-mass threshold δ_1 and cluster-merging threshold δ_2, with $0 < \delta_i < 1$.
Output: Approximations $\tilde{\boldsymbol{p}}_{\beta_n}$ of the optimal IB roots $\boldsymbol{p}_{\beta_n}$ at $\beta_n := \beta_0 + n\Delta\beta$.
2:     Initialize $\beta \leftarrow \beta_0$ and *results* $\leftarrow \{\}$.
3:     Initialize $\tilde{\boldsymbol{p}} := \big(\tilde{p}(\hat{x}|x), \tilde{p}(x|\hat{x}), \tilde{p}(y|\hat{x}), \tilde{p}(\hat{x})\big)$ from $p_{\beta_0}(\hat{x}|x)$, via Steps 1.4–1.6.
4:     **while** $\beta > |\Delta\beta|$ and $|\text{supp }\tilde{p}(\hat{x})| > 1$ **do**   ▷ See main text on stopping condition.
5:        Append $\tilde{\boldsymbol{p}}$ to *results*.
6:        $v := \left(\frac{d\log\tilde{p}(y|\hat{x})}{d\beta}, \frac{d\log\tilde{p}(\hat{x})}{d\beta}\right) \leftarrow$ solve the IB ODE (16) at $\tilde{\boldsymbol{p}}$.
7:        $\tilde{p}(y|\hat{x}) \leftarrow \tilde{p}(y|\hat{x})\exp\left(\Delta\beta \cdot \frac{d\log\tilde{p}(y|\hat{x})}{d\beta}\right)$
8:        $\tilde{p}(\hat{x}) \leftarrow \tilde{p}(\hat{x})\exp\left(\Delta\beta \cdot \frac{d\log\tilde{p}(\hat{x})}{d\beta}\right)$ ▷ Exponentiate the linear approximations (22).
9:        $\big(\tilde{p}(y|\hat{x}), \tilde{p}(\hat{x})\big) \leftarrow$ normalize $\big(\tilde{p}(y|\hat{x}), \tilde{p}(\hat{x})\big)$
10:       $old\_dim \leftarrow \dim\tilde{p}(\hat{x})$
11:       $\big(\tilde{p}(y|\hat{x}), \tilde{p}(\hat{x})\big) \leftarrow$ REDUCE ROOT$(\tilde{p}(y|\hat{x}), \tilde{p}(\hat{x}); \delta_1, \delta_2)$.   ▷ Algorithm 2.
12:       **if** $old\_dim \neq \dim\tilde{p}(\hat{x})$ **then**   ▷ Root was reduced due to bifurcation.
13:          $\tilde{p}(\hat{x}|x) \leftarrow$ the encoder defined by $\big(\tilde{p}(y|\hat{x}), \tilde{p}(\hat{x})\big)$, via Steps 1.7–1.8.
14:          $\tilde{\boldsymbol{p}} \leftarrow$ BA-IB$(\tilde{p}(\hat{x}|x); p_{Y|X}\, p_X, \beta + \Delta\beta)$.

              ▷ Ensure accuracy of the reduced root, using BA-IB Algorithm 1 till convergence.

15:       **end if**
16:       $\beta \leftarrow \beta + \Delta\beta$.
17:       $\tilde{\boldsymbol{p}} \leftarrow$ BA$_\beta(\tilde{p}(y|\hat{x}), \tilde{p}(\hat{x}))$ ▷ A single BA-IB iteration in decoder coordinates.
18:     **end while**
19:     Append $\tilde{\boldsymbol{p}}$ to *results*.
20:     **return** *results*.
21: **end function**

With that, there are caveats in Algorithm 3, which stem from passing too far or close to a bifurcation. For one, suppose that the error accumulated from the true solution is too large for a bifurcation to be detected. The approximations generated by the algorithm will then overshoot the bifurcation. Namely, it will proceed with more clusters than needed until the conditions for reduction are met later on (see Section 6.3 below), as demonstrated by the two sparse grids in Figure 10 (Section 6.2). For another, suppose that the current grid point $\tilde{\boldsymbol{p}}$ is too close to a bifurcation. This might happen due to a variety of numerical reasons, e.g., thresholds δ_1, δ_2 too small, or due to the particular grid layout. The coefficients matrix $I - D_{\log p(y|\hat{x}), \log p(\hat{x})} BA_\beta$ of the IB ODE (16) (which is the Jacobian of the IB operator (5)) would then be ill-conditioned, typically resulting in very large implicit numerical derivatives v on step 6; cf., Conjecture 1 ff. in Section 5.1. Any inaccuracy in v might then send the next grid point astray, derailing the algorithm from there on (e.g., inaccuracies due to the accumulated approximation error or due to the error caused by computing implicit derivatives in the vicinity of a bifurcation—see Figure 3 (top) in Section 3). Indeed, the derivatives $\frac{dx}{d\beta} = -(D_x F)^{-1} D_\beta F$ defined by the implicit ODE (7) are in general unbounded near a bifurcation of F (in our case, $D_x F$ is always non-singular outside bifurcations, due to Conjecture 1 and the use of reduced coordinates). This can be seen in Figure 2 (Section 2) for example, where the derivatives "explode" at the bifurcation's vicinity. See also [6] (Section 7.2) on the computational difficulty incurred by a bifurcation. While overshooting a bifurcation is not a significant concern for our purposes (see Section 6.3), passing too close to one is. The latter is important, especially when the step size $|\Delta\beta|$ is small. While decreasing $|\Delta\beta|$ generally improves the error of Euler's method, it also makes it easier for the approximations to come close to a bifurcation, thus potentially worsening the approximation dramatically if it derails. This motivates one to consider how singularities of the IB ODE (16) should be handled.

Algorithm 4 A heuristic for handling singularities of the IB ODE (16)

1: **function** HANDLE SINGULARITY($p_{Y|X}\, p_X$, $(\tilde{p}(y|\hat{x}), \tilde{p}(\hat{x}))$, v, β)
Input:
    An IB problem definition $p_{Y|X}\, p_X$, with $\forall x\ p_X(x) > 0$.
    An approximate root $(\tilde{p}(y|\hat{x}), \tilde{p}(\hat{x}))$ of the given problem, near a singularity of the IB ODE (16).
    Approximate numerical derivatives $v := \left(\frac{d\log \tilde{p}(y|\hat{x})}{d\beta}, \frac{d\log \tilde{p}(\hat{x})}{d\beta} \right)$ at the given root.
    The $\beta > 0$ value of the next (output) grid point.
Output: An approximate IB root $\tilde{p}$ at β on one fewer cluster.

2:     $\hat{x}', \hat{x}'' \leftarrow$ the two indices $\hat{x}$ of largest $\left\| \frac{d\log \tilde{p}(y|\hat{x})}{d\beta} \right\|_\infty$ value (norm of y-indexed vectors).

3:     $\tilde{p}(y|\hat{x}') \leftarrow \frac{1}{2} \cdot (\tilde{p}(y|\hat{x}') + \tilde{p}(y|\hat{x}''))$ ▷ Replace fastest-moving clusters by their mean.
4:     Erase $\hat{x}''$ from the decoder $\tilde{p}(y|\hat{x})$.
5:     $\tilde{p}(\hat{x}') \leftarrow \tilde{p}(\hat{x}') + \tilde{p}(\hat{x}'')$
6:     Erase $\hat{x}''$ from the marginal $\tilde{p}(\hat{x})$.
7:     $\tilde{p}(\hat{x}|x) \leftarrow$ the encoder generated from $(\tilde{p}(y|\hat{x}), \tilde{p}(\hat{x}))$, via Steps 1.7–1.8.

                    ▷ A new encoder on one cluster *less* than the input.

8:     $\tilde{p} \leftarrow$ BA-IB$(\tilde{p}(\hat{x}|x); p_{Y|X}\, p_X, \beta)$.   ▷ Re-gain accuracy, by the BA-IB Algorithm 1.
9:     **return** $\tilde{p}$
10: **end function**

Next, we elaborate on our heuristic for handling singularities of the IB ODE (16), Algorithm 4. The inputs of this heuristic are defined as in Algorithm 3. It starts with the assumption that the coefficients matrix $I - D_{\log p(y|\hat{x}),\log p(\hat{x})} BA_\beta$ of the IB ODE (16) is nearly singular at the current grid point $\tilde{p}$ due to a nearby bifurcation (although a priori the Jacobian $D_{\log p(y|\hat{x}),\log p(\hat{x})}(Id - BA_\beta)$ may be singular also due to other reasons, by Conjecture 1 it is non-singular at the approximations generated so far since they are assumed to be in their reduced form—see Section 5.1). As a result, the implicit derivatives v at $\tilde{p}$ are not to be used directly to extrapolate the next grid point, as explained above. Instead, we use them to identify the two fastest moving clusters, on step 2 of Algorithm 4 (while this can be refined to handle more than two fast-moving clusters at once, that is not expected to be necessary for typical bifurcations). These are replaced by a single cluster (steps 3 through 6), resulting in an approximate root on one fewer cluster. To re-gain accuracy, the BA-IB Algorithm 1 is then invoked (on step 8) on the encoder generated (on step 7) from the latter root, thereby generating the next grid point. If the fastest-moving clusters have merged (in the true solution) by the next grid point, then the output of Algorithm 4 will be an IB-optimal root if its input grid point is so. Namely, the branch followed by the algorithm remains an optimal one. Otherwise, if these clusters merge shortly after the next grid point, then Algorithm 4 yields a sub-optimal branch. However, optimality is re-gained shortly afterward since the sub-optimal branch collides and merges with the optimal one in continuous IB bifurcations (Section 5.3). Figure 10 below demonstrates Algorithm 4. cf., [6] (Section 3.2) on the similar heuristic in root tracking for RD, which may also lose optimality near a bifurcation and re-gain it shortly after.

Algorithm 5 First-order Root Tracking for the IB (IBRT1)

1: **function** IBRT1$(p_{Y|X}\, p_X, \beta_0, p_{\beta_0}(\hat{x}|x); \Delta\beta, \delta_1, \delta_2, \delta_3)$

Input:

    An IB problem definition $p_{Y|X}\, p_X$ with $\forall x\ p_X(x) > 0$.

    A reduced IB-optimal root $p_{\beta_0}(\hat{x}|x)$ at β_0. A step size $\Delta\beta < 0$.

    Thresholds $0 < \delta_1, \delta_2 < 1$ for the root-reduction Algorithm 2 (cluster mass and merging).

    A threshold $0 < \delta_3 < 1$ for eigenvalues' singularity.

Output: Approximations $\tilde{p}_{\beta_n}$ of the optimal IB roots p_{β_n} at $\beta_n := \beta_0 + n\Delta\beta$.

2:      Initialize $\beta \leftarrow \beta_0$ and $results \leftarrow \{\}$.

3:      Initialize $\tilde{p} := \big(\tilde{p}(\hat{x}|x), \tilde{p}(x|\hat{x}), \tilde{p}(y|\hat{x}), \tilde{p}(\hat{x})\big)$ from $p_{\beta_0}(\hat{x}|x)$, via Steps 1.4–1.6.

4:      **while** $\beta > |\Delta\beta|$ and $|\operatorname{supp} \tilde{p}(\hat{x})| > 1$ **do**

5:          Append $\tilde{p}$ to $results$.

6:          $v := \left(\frac{d\log \tilde{p}(y|\hat{x})}{d\beta}, \frac{d\log \tilde{p}(\hat{x})}{d\beta}\right) \leftarrow$ solve the IB ODE (16) at $\tilde{p}$.

7:          $eigs \leftarrow \operatorname{eig}(I - S)\big|_{\tilde{p}}$ ▷ Test ODE for singularity, using S (17) from Lemma 1.

8:          **if** $\big(\min_{v \in eigs} |v|\big) < \delta_3$ **then**   ▷ ODE is nearly singular.

9:              $\tilde{p} \leftarrow$ HANDLE SINGULARITY$(p_{Y|X}\, p_X, (\tilde{p}(y|\hat{x}), \tilde{p}(\hat{x})), v, \beta + \Delta\beta)$

                    ▷ Handle an otherwise undetected singularity using Algorithm 4.

10:         **else**

11:            $\tilde{p}(y|\hat{x}) \leftarrow \tilde{p}(y|\hat{x}) \exp\left(\Delta\beta \cdot \frac{d\log \tilde{p}(y|\hat{x})}{d\beta}\right)$

12:            $\tilde{p}(\hat{x}) \leftarrow \tilde{p}(\hat{x}) \exp\left(\Delta\beta \cdot \frac{d\log \tilde{p}(\hat{x})}{d\beta}\right)$

13:            $(\tilde{p}(y|\hat{x}), \tilde{p}(\hat{x})) \leftarrow$ normalize $(\tilde{p}(y|\hat{x}), \tilde{p}(\hat{x}))$

14:            $old\_dim \leftarrow \dim \tilde{p}(\hat{x})$

15:            $(\tilde{p}(y|\hat{x}), \tilde{p}(\hat{x})) \leftarrow$ REDUCE ROOT$(\tilde{p}(y|\hat{x}), \tilde{p}(\hat{x}); \delta_1, \delta_2)$.

16:            **if** $old\_dim \neq \dim \tilde{p}(\hat{x})$ **then**

17:               $\tilde{p}(\hat{x}|x) \leftarrow$ encoder defined from $(\tilde{p}(y|\hat{x}), \tilde{p}(\hat{x}))$, via Steps 1.7–1.8.

18:               $\tilde{p} \leftarrow$ BA-IB$(\tilde{p}(\hat{x}|x); p_{Y|X}\, p_X, \beta + \Delta\beta)$.

19:            **end if**

20:         **end if**

21:         $\beta \leftarrow \beta + \Delta\beta$.

22:         $\tilde{p} \leftarrow BA_\beta(\tilde{p}(y|\hat{x}), \tilde{p}(\hat{x}))$

23:      **end while**

24:      Append $\tilde{p}$ to $results$.

25:      **return** $results$.

26: **end function**

The heuristic Algorithm 4 is motivated by cluster-merging bifurcations. In these, the implicit derivatives are very large only at the coordinates $\frac{d\log p(y|\hat{x})}{d\beta}$ of the points colliding in $\Delta[\mathcal{Y}]$ (note that cluster masses barely change in the vicinity of a cluster merging, until the point of bifurcation itself). While intended for cluster-merging bifurcations, this heuristic works nicely in practice also for cluster-vanishing ones. To see why, note that one can always add a cluster of zero mass to an IB root without affecting the root's essential properties, regardless of its coordinates in $\Delta[\mathcal{Y}]$ (cf., Section 5.1 on reduction in the IB). Therefore, a numerical algorithm may, in principle, do anything with the coordinates $p(y|\hat{x}) \in \Delta[\mathcal{Y}]$ of a nearly vanished cluster $\hat{x}$, $p(\hat{x}) \simeq 0$, without affecting the approximation's quality too much. Thus, for numerical purposes, one may treat a cluster-vanishing bifurcation as a cluster-merging one. Conversely, in a cluster-merging bifurcation, a numerical algorithm may, in principle, zero the mass of one cluster while adding it to the remaining cluster, again without affecting the approximation's quality too much. To conclude, for numerical purposes, cluster vanishing is very similar to cluster merging. A variety of treatments between these extremities may be possible by a numerical algorithm. Empirically, we have observed that our ODE-based algorithm treats both as cluster-merging bifurcations. To our understanding, this is because our algorithm operates in decoder coordinates, unlike the BA-IB Algorithm 1, for example, which operates in encoder coordinates.

Finally, we combine the simplified root-tracking Algorithm 3 with the heuristic Algorithm 4 for handling singularities, yielding our IBRT1 Algorithm 5. It follows the lines of the simplified Algorithm 3, except that after solving for the implicit derivatives on step 6, we test the IB ODE (16) for singularity. To that end, we propose using the matrix S (17) (from Lemma 1 in Section 3), since its order $T \cdot |\mathcal{Y}|$ is smaller than the order $T \cdot (|\mathcal{Y}| + 1)$ of the ODE's coefficients matrix. This might make it computationally cheaper to test for singularity (on steps 7 and 8 of Algorithm 5). Our heuristic Algorithm 4 is invoked (on step 9) if the ODE (16) is found to be nearly singular, otherwise proceeding as in Algorithm 3.

6.2. Numerical Results for the IBRT1 Algorithm 5

To demonstrate the IBRT1 Algorithm 5, we present the numerical results used to approximate the IB curve in Figure 1 (Section 1)—see Section 6.3 below on the approximation quality and the algorithm's basic properties. This example was chosen both because it has an analytical solution (Appendix E) and because it allows one to get a good idea of the bifurcation handling added (in Section 6.1) on top of the modified Euler method (from Section 4).

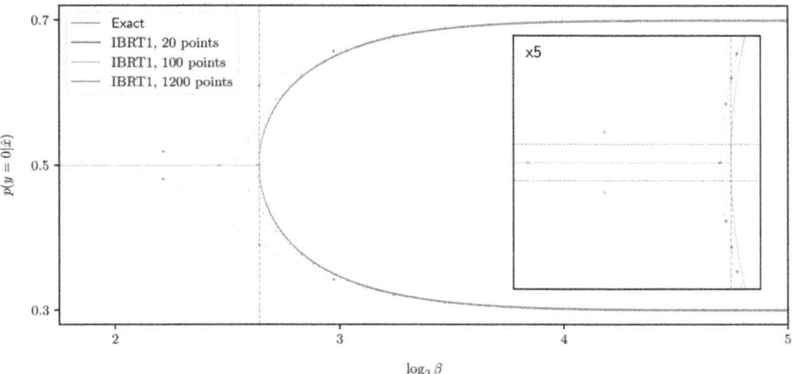

Figure 10. Clusters of the approximate IB roots generated by the IBRT1 Algorithm 5 for several step-sizes, on top of the exact solutions of $\mathrm{BSC}(0.3)$ with a uniform source (Appendix E). Carefully note that only a *single* IB root is plotted here; its two clusters merge at β_c (dashed red vertical), as seen in Figure 6 (Section 5.2). At 20 and 100 grid points, the approximations overshoot the bifurcation, terminating due to (approximate) cluster collision, while on 1200 grid points, the approximations pass too close to the bifurcation, terminating due to the nearby singularity. This can be seen in **the inset to the right**: The leftmost green marker has passed the cluster-merging threshold (dashed green lines), and so was numerically reduced to the trivial (single-clustered) solution by the root-reduction Algorithm 2. On the other hand, the orange markers to the right are still far from the cluster-merging threshold; the leftmost one was reduced by the singularity-handling heuristic Algorithm 4 since the IB ODE (16) is nearly singular there. Indeed, the numerical derivative is about five orders of magnitude larger there than at the algorithm's initial condition (see Figure 2) due to the bifurcation's proximity. The leftmost green and orange markers were drawn *after* the reductions took place. See main text and Section 6.1 for details, Figure 11 for errors, and Figure 1 (in Section 1) for the approximate IB curves. The marginals $p(\hat{x})$ are not shown, as these barely deviate from their true value in this problem. For each step-size $\Delta\beta$, the algorithm was initialized at the problem's exact solution at $\beta = 2^5$, with thresholds set to $\delta_i = 10^{-2}$, for $i = 1, 2, 3$. The lines connecting consecutive markers are for visualization only.

We discuss the numerical examples of this Section in light of the explanations provided in the previous Section 6.1. The error of the IBRT1 Algorithm 5 generally improves as the step-size $|\Delta\beta|$ becomes smaller, as expected. The single BA-IB iteration added to Euler's method (in Section 4) typically allows one to achieve the same error by using much fewer grid points, thus lowering computational costs. For example, the two denser grids in Figure 10 require about an order of magnitude fewer points to achieve the same error compared to Euler's method for the IB; this can be seen from Figure 4 (Section 4).

In sparse grids, the approximations often pass too far away from a bifurcation for the root-reduction Algorithm 2 to detect it. When overshooting it, the conditions for numerical reduction are generally met later on, as discussed in Section 6.3 below. Decreasing $|\Delta\beta|$ further often leads the approximations too close to a bifurcation, as can be seen in the densest grid of Figure 10. The implicit derivatives are typically very large at the proximity of a bifurcation, while the least accurate there (see Section 6.1). As these might send subsequent grid points off-track, the heuristic Algorithm 4 is invoked to handle the nearby singularity (see inset of Figure 10). As noted earlier, the computational difficulty in tracking IB roots (or root tracking in general) stems from the presence of a bifurcation, manifested here by large approximation errors in its vicinity. While the algorithm's error peaks at the bifurcation, it typically decreases afterward when overshooting, as seen in Figure 11. The reasons for this are discussed below in Section 6.3.

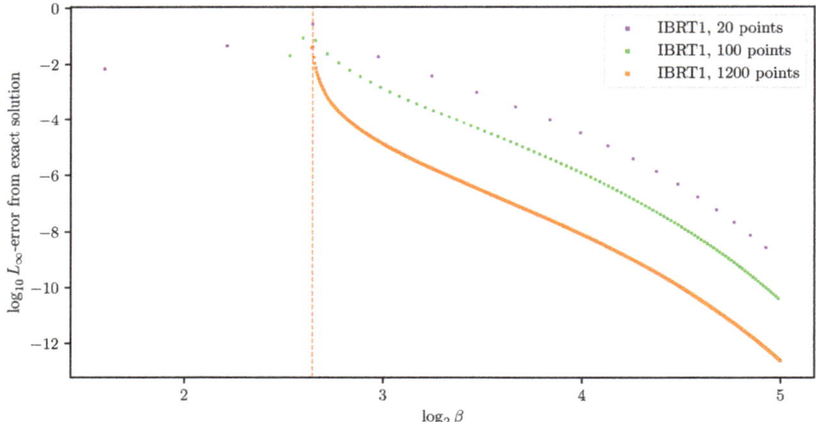

Figure 11. The error of the IBRT1 Algorithm 5 from the exact solution for several step-sizes. The figure shows the (log-) L_∞-error of the numerical approximations in Figure 10 from the exact solutions; the error is measured as in Figure 3 (bottom). Increasing the grid density decreases the error, as one might expect. While the error peaks at the bifurcation (dashed red vertical), it decreases afterward—see main text and Section 6.3 below. The rightmost marker for each grid density is missing since the initial error is zero.

6.3. Basic Properties of the IBRT1 Algorithm 5 and Why It Works

Apart from presenting the basic properties of the IBRT1 Algorithm 5, the primary purpose of this section is to understand why it approximates the problem's true IB curve (1) so well, despite its apparent errors in approximating the IB roots. While shown here only in Figures 1 and 10 (in Sections 1 and 6.2), this behavior is consistent in the few numerical examples that we have tested. We offer an explanation why this may be true in general.

To understand why the IBRT1 Algorithm 5 approximates the true IB curve (1) so well, we first explain why overshooting is not a significant concern, as noted earlier in Section 6.1. To that end, consider the implicit ODE (7)

$$\tfrac{dx}{d\beta} = -(D_x F)^{-1} D_\beta F \, ,$$

from Section 1. As long as $D_x F$ and $D_\beta F$ at its right-hand side are well-defined, it defines a vector field on the entire *phase space* of admissible x values, at least when $D_x F$ is non-singular. That is, even for x's which are *not* roots (6) of F. Ignoring several technicalities, the IB ODE (16) therefore defines a vector field also *outside* IB roots (although at a reduced root singularities of the IB ODE (16) coincide with IB bifurcations, the IB's vector field might a priori be singular elsewhere). Indeed, due to Conjecture 1, the Jacobian of the IB operator $Id - BA_\beta$ (5) is non-singular in the vicinity of a reduced root ($D_{\log p(y|\hat{x}),\log p(\hat{x})} BA_\beta$ (13) is continuous in the distributions defining it, and thus so are its eigenvalues, under mild assumptions—cf., Lemma A1 in Appendix A). Now, suppose that p_β is an optimal IB root, and consider a point $p' \neq p_\beta$ in its vicinity. An argument based on a strong notion of Lyapunov stability (in Appendix G) shows that p' flows along the IB's vector field towards p_β in regions that do not contain a bifurcation, though only if flowing in *decreasing* β as done by our IBRT Algorithm 5. An approximation p' would then be "pulled" towards the true root. Stability in decreasing (rather than increasing) β values is very reasonable, considering that p_β follows a path of decreasingly informative representations as β decreases. Indeed, all the paths to oblivion lead to one place—the trivial solution, whose representation in reduced coordinates is unique. As a result, a numerical approximation p' would gradually settle in the vicinity of the true root p_β as seen in Figures 10 and 11, so long as p_β does not change much and the step-size $|\Delta\beta|$ is small enough. While this explanation obviously breaks near a bifurcation, it does suggest that the approximation error should decrease when overshooting it (see Section 6.1), once the true reduced root has settled down. In a sense, overshooting is similar to being in the right place but at the wrong time.

The above suggests that the IBRT1 Algorithm 5 should generally approximate the true IB curve (1) well, despite its errors in approximating IB roots. To see this, note that while β^{-1} is the slope of the optimal curve (1) of the IB [1] (Equation (32)), for the IB ODE (16) it is merely an independent "time-like" variable. When solving for the optimal curve (1), one is not interested in an optimal root or in its β value, but rather in its image $\left(I(X; \hat{X}), I(Y; \hat{X})\right)$ in the information plane. As a result, achieving the optimal roots but with the wrong β values does yield the true IB curve (1), as required. This is the reason that the true curve (1) is achieved in Figure 1 (Section 1) even on sparse grids, despite the apparent approximation errors in Figures 10 and 11 (Section 6.2). With that, we expect the approximate IB curve produced by the IBRT1 Algorithm 5 to be of lesser quality when there are more than two possible labels y. To see why, note that the space $\Delta[\mathcal{Y}]$ traversed by approximate clusters is not one-dimensional when $|\mathcal{Y}| > 2$, and so it is possible to maneuver around the clusters of an optimal root.

Next, we briefly discuss the basic properties of the IBRT1 Algorithm 5. Its **computational complexity** is determined by the complexity of a single grid point. The latter is readily seen to be dominated by the complexity $O\left(T^2 \cdot |\mathcal{Y}|^2 \cdot (|\mathcal{X}| + T \cdot |\mathcal{Y}|)\right)$ of computing the coefficients matrix of the IB ODE (16) and of solving it numerically (on step 6). To that, one should add the complexity of the BA-IB Algorithm 1 each time a root is reduced. However, the critical slowing down of BA-IB [4] is avoided since we reduce the root before invoking BA-IB (see Section 5.2). The complexity is only linear in $|\mathcal{X}|$ thanks to the choice of decoder coordinates. Had we chosen one of the other coordinate systems in Section 2, then solving the ODE would have been cubic in $|\mathcal{X}|$ rather than linear (see there). The **computational difficulty** in following IB roots stems from the existence of bifurcations (Section 4), as it generally is with following an operator's root [6] (Section 7.2).

As noted in Section 4, **convergence guarantees** can be derived for Euler's method for the IB when away from bifurcation, in terms of the step-size $|\Delta\beta|$, in a manner similar to [6] (Theorem 5) for RD. These imply similar guarantees for the IBRT1 Algorithm 5, since adding a single BA-IB iteration in our modified Euler method improves its order (see there). These details are omitted for brevity, however.

For a numerical method of order $d > 0$ (see Section 4) with a fixed step-size $|\Delta\beta|$ and a fixed computational cost per grid point, the **cost-to-error tradeoff** is given by

$$error \propto cost^{-d} , \tag{26}$$

as in [6] (Equation (3.6)), when $|\Delta\beta|$ is small enough. See [26] for example. Figure 3.4 in [6] demonstrates for RD that methods of higher order achieve a better tradeoff, as expected, as in the fixed-order Taylor methods they employ. Since computing implicit derivatives of higher orders requires the calculation of many more derivative tensors of $Id - BA_\beta$ (5) than done here [6] (Section 2.2), we have used only first-order derivatives for simplicity. However, while the vanilla Euler method for the IB is of order $d = 1$, the discussion in Section 4 (and Figure 4 in particular) suggests that the order d of the modified Euler method used by the IBRT1 Algorithm 5 is nearly twice than that; cf., Section 6.2 and Appendix D.

With that, we comment on the behavior of the IBRT1 Algorithm 5 at **discontinuous bifurcations**. Consider the problem in Figure 7 (Section 5.3), for example. When Algorithm 5 follows the optimal 2-clustered root there, the Jacobian's singularity (in Figure 9) is detectable by it because the step size $\Delta\beta$ is negative (see the discussion in Section 5.3 there). Indeed, due to Conjecture 1 ff., the algorithm can detect discontinuous bifurcations in general. Whether a particular discontinuous bifurcation is detected by Algorithm 5 in practice depends on the details, of course, as with continuous bifurcations (e.g., on the threshold value δ_3 for detecting singularity and on the precise grid point layout). Indeed, the details may or may not cause a particular example to be detected by the conditions on steps 7 and 8 (in Algorithm 5). If missed, Algorithm 5 will continue to follow the 2-clustered root in Figure 7 to the left of the bifurcation, where it is sub-optimal, just as BA-IB with reverse deterministic annealing would. Once detected, though, one may wonder whether the heuristic Algorithm 4 works well also for discontinuous bifurcations. The example of Figure 7 has just one single-clustered root to the left of the bifurcation. Thus, the BA-IB Algorithm 1 invoked on step 8 (of Algorithm 4) must converge to it. However, there may generally be more than a single root of smaller effective cardinality to the left of the bifurcation, to which BA-IB may converge. The handling of discontinuous bifurcations is left to future work. Such handling is expected to be easier in the IB than in RD, since, in contrast to RD, the effective cardinality of an optimal IB root cannot decrease with β (bottom of Section 5.3). See Problems 2.8–2.10 in [28] for counter-examples in RD. This makes detecting discontinuous bifurcations easier in the IB and is also expected to assist with their handling.

We list the **assumptions** used along the way for reference. These are needed to guarantee the optimality of the IBRT1 Algorithm 5 at the limit of small step-sizes $|\Delta\beta|$, except at a bifurcation's vicinity. In Section 1, it was assumed without loss of generality that the input distribution p_X is of full support, $p(x) > 0$ for every x (otherwise, one may remove symbols x with $p_X(x) = 0$ from the source alphabet). The requirement $p(y|x) > 0$ was added in Section 3 as a sufficient technical condition for exchanging to logarithmic coordinates (Lemma A1 in Appendix A), and could perhaps be alleviated in alternative derivations. Together, these are equivalent to having a never-vanishing IB problem definition, $p(y|x)p(x) > 0$ for every x and y. The algorithm's initial condition is assumed to be a reduced and optimal IB root, since reduction is needed by Conjecture 1 in Section 5.1. Finally, the given IB problem is assumed to have only continuous bifurcations, except perhaps for its first (leftmost) one. While these assumptions are sufficient to guarantee optimality, we note that milder conditions might do in a particular problem.

7. Concluding Remarks

The IB is intimately related to several problems in adjacent fields [3], including coding problems, inference, and representation learning. Despite its importance, there are surprisingly few techniques to solve it numerically. This work attempts to fill this gap by exploiting the dynamics of IB roots.

The end result of this work is a new numerical algorithm for the IB, which follows the path of a root along the IB's optimal tradeoff curve (1). A combination of several novelties was required to achieve this goal. First, the dynamics underlying the IB curve (1) obeys an ODE [10]. Following the discussion around Conjecture 1 (in Section 5.1), the existence of such a dynamics stems from the analyticity of the IB's fixed-point Equations (2)–(4), thus typically resulting in piece-wise smooth dynamics of IB roots. Several natural choices of a coordinate system for the IB were considered, both for computational purposes and to facilitate a clean treatment of IB bifurcations below. The IB's ODE (16) was derived anew in appropriate coordinates, allowing an efficient computation of implicit derivatives at an IB root. Combining BA-IB with Euler's method yields a modified numerical method whose order is higher than either.

Second, one needs to understand where the IB ODE (16) is *not* obeyed, thereby violating the differentiability of an optimal root with respect to β. To that end, one not only needs to detect IB bifurcations but also needs to identify their type in order to handle them properly. Unlike standard techniques, our approach is to remove redundant coordinates, following root tracking for RD [6] (see Section 1). To achieve a reduction, we follow the arguably better definition of the IB in [20]. Namely, a finite IB problem is an RD problem on the continuous reproduction alphabet $\Delta[\mathcal{Y}]$. Therefore, the IB may be intuitively considered as a method of lossy compression of the information on Y embedded in X. Viewing a finite IB problem as an infinite RD problem suggests a particular choice of a coordinate system for the IB, which enables reduction in the IB; this extends reduction in RD [6]. Furthermore, this point of view highlights subtleties of finite dimensionality in computing representations of IB roots. To our understanding, these subtleties hindered the understanding of IB bifurcations throughout the years.

Combining the above allows us to translate an understanding of IB bifurcations to a new numerical algorithm for the IB (the IBRT1 Algorithm 5). There are several directions that one could consider to improve our algorithm. Near bifurcations, one could improve its handling of discontinuous bifurcations. While we used implicit derivatives only of the first order for simplicity, higher-order derivatives generally offer a better cost-to-error tradeoff when away from bifurcations. See also [6] (Section 3.4) on possible improvements for following an operator's root.

Funding: This work was partially funded by the Israel Science Foundation grant 1641/21.

Institutional Review Board Statement: Not applicable.

Informed Consent Statement: Not applicable.

Data Availability Statement: Not applicable.

Acknowledgments: The author is grateful to Or Ordentlich for helpful conversations and for his support, and to Noam and Dafna Agmon for their unwavering support throughout this journey. The author thanks the late Naftali Tishby for insightful conversations and Etam Benger for his involvement during the early stages of this work. The author thanks the reviewers for their helpful comments.

Conflicts of Interest: The author declares no conflict of interest.

Abbreviations

The following abbreviations are used in this manuscript:

IB	Information Bottleneck
RD	Rate Distortion
IFT	Implicit Function Theorem
ODE	Ordinary Differential Equation
BA	Blahut–Arimoto

Appendix A. The BA-IB Operator in Decoder Coordinates

For reference, we give an explicit expression for the BA-IB operator in decoder coordinates, defined in Section 2.

Denote by $\boldsymbol{p}_{Y|\hat{X}}$ and $\boldsymbol{p}_{\hat{X}}$ the vectors whose coordinates are $\left(p(y|\hat{x}), p(\hat{x})\right)$. We denote the evaluation of BA_β at this point by $BA_\beta[\boldsymbol{p}_{Y|\hat{X}}, \boldsymbol{p}_{\hat{X}}]$. Its output is again a decoder–marginal pair, whose coordinates are denoted, respectively, $BA_\beta[\boldsymbol{p}_{Y|\hat{X}}, \boldsymbol{p}_{\hat{X}}](y|\hat{x})$ and $BA_\beta[\boldsymbol{p}_{Y|\hat{X}}, \boldsymbol{p}_{\hat{X}}](\hat{x})$. Explicitly, BA_β in decoder coordinates is given by

$$
\begin{aligned}
BA_\beta[\boldsymbol{p}_{Y|\hat{X}}, \boldsymbol{p}_{\hat{X}}](y|\hat{x}) &:= \sum_x \frac{p(y|x)p(x)}{Z(x,\beta)} \exp\left\{ -\beta\, D_{KL}[p(y'|x)||p(y'|\hat{x})] \right\} \quad \text{and} \\
BA_\beta[\boldsymbol{p}_{Y|\hat{X}}, \boldsymbol{p}_{\hat{X}}](\hat{x}) &:= \sum_x \frac{p(\hat{x})p(x)}{Z(x,\beta)} \exp\left\{ -\beta\, D_{KL}[p(y'|x)||p(y'|\hat{x})] \right\},
\end{aligned}
\tag{A1}
$$

where $Z(x,\beta)$ is defined in terms of $p(y|\hat{x})$ and $p(\hat{x})$ as in the IB's encoder Equation (2) (Section 1).

The following lemma is handy when exchanging to logarithmic coordinates in Section 3.

Lemma A1. *Let $p(y|x)p(x)$ define a finite IB problem, such that $p(y|x) > 0$ for every x and y. Let $p(y|\hat{x})$ be the decoder of an IB root, and $\hat{x}'$ such that $p(\hat{x}') > 0$. Then $p(y|\hat{x}') > 0$ for every y.*

Proof of Lemma A1. This follows immediately from the IB's decoder Equation (3), since $p(x|\hat{x}')$ is a well-defined normalized conditional probability distribution if $p(\hat{x}') > 0$. $\square$

Appendix B. The First-Order Derivative Tensors of Blahut–Arimoto for the IB

We calculate the first-order derivative tensors of the Blahut–Arimoto operator BA_β in log-decoder coordinates (see Sections 2 and 3). Namely, its Jacobian matrix $D_{\log p(\hat{x}|x),\log p(\hat{x})} BA_\beta$, and the vector $D_\beta BA_\beta$ of its partial derivatives with respect to β. See also Appendix A for explicit formulae of BA_β in decoder coordinates.

While these are "just" differentiations, many subtleties are involved in getting the math right. For example, one needs to correctly identify the inputs and outputs of BA_β, when considered as an operator on log-decoder coordinates. For another, one must take special care as to which variable depends on which, and especially on which they do *not* depend, since multiple variables are involved. Above all, these calculations require a deep understanding of the chain rule. With that, a common caveat in such calculations is that the BA_β operator (and the equations defining it) should be differentiated *before* they are evaluated. While this is obvious for real functions, where $f'(3)$ stands for the derivative function of $f(x)$ evaluated at $x = 3$, for the BA_β operator, this might be obfuscated by the myriad of variables and variable dependencies that comprise it. Although calculating the derivative of BA_β (at an arbitrary point) first and only then evaluating at a fixed point might appear as a mere technical necessity, it is required by this work, for example, when considering the vector field defined by the IB operator (5) in Section 6.3. cf., [6] (Section 5), for the derivative tensors of Blahut's algorithm [8] for RD, of arbitrary order.

The subtitles involved in these differentiations are discussed in Appendix B.1, with the bulk of the calculations carried out in Appendix B.1.1. The latter are gathered and simplified in Appendix B.2 to obtain the Jacobian matrix $D_{\log p(\hat{x}|x),\log p(\hat{x})} BA_\beta$, and in Appendix B.3 to obtain the partial-derivatives vector $D_\beta BA_\beta$. The results provided here naturally depend on the choice of coordinate system. To compare results between log-decoder and log-encoder coordinates in Section 2 (e.g., in Figure 2), we derive in Appendix B.4 the coordinate-exchange Jacobians between these coordinate systems.

Appendix B.1. Calculation Setups and Partial Derivatives of Unnamed Functions

We explain the mathematical subtitles relevant to the sequel.

As we are interested in the derivatives of the Blahut–Arimoto Algorithm 1 for the IB (in Section 1), we shall follow its notation. Namely, distributions are subscripted i or $i + 1$ by the algorithm's iteration number. A subscript i is usually considered an input distribution, and a subscript $i + 1$ is usually considered an output distribution, e.g., $p_i(\hat{x})$ or $p_{i+1}(y|\hat{x})$. These need *not* be IB roots but rather are arbitrary distributions. On the other hand, a subscript β denotes a distribution of an IB root at a tradeoff value β, as in $p_\beta(y|\hat{x})$

for a root's decoders. To avoid subtleties due to zero-mass clusters, we usually assume in the sequel that $p_i(\hat{x}) \neq 0$ for every $\hat{x}$. cf., Sections 2 and 5.1 on *reduction* in the IB.

It is important to distinguish which variables are dependent and which are independent in a particular calculation, e.g., in Appendix B.3. Since this task is easier for a single real variable (as opposed to distributions, for example), we consider simplifications to the real case. Note that each of the Steps 1.4 through 1.8 defining the BA-IB Algorithm 1 yields a new distribution in terms of already-specified ones. These define *unnamed functions*, whose variables and values are probability distributions. For example, one could have formally defined $p_i(x|\hat{x})$ in 1.5 by the function

$$\mathcal{F}[p_i(\hat{x}|x), p_i(\hat{x})](x, \hat{x}) := p_i(\hat{x}|x)p(x)/p_i(\hat{x}) , \tag{A2}$$

where $p_i(\hat{x}|x)$ and $p_i(\hat{x})$ are the variables of $\mathcal{F}$, and its output is a conditional probability distribution, with x conditioned upon $\hat{x}$. As the input and representation alphabets $\mathcal{X}$ and $\hat{\mathcal{X}}$ are finite, $N := |\mathcal{X}|$ and $T := |\hat{\mathcal{X}}|$, the arguments $p_i(\hat{x}|x), p_i(\hat{x})$ and values $p_i(x|\hat{x})$ of $\mathcal{F}$ (A2) are merely real vectors. Thus, enumerating the variables $x_1, \ldots, x_N$ and $\hat{x}_1, \ldots, \hat{x}_T$ allows us to spell out (A2) by its coordinates,

$$\mathcal{F}[p_i(\hat{x}_1|x_1), p_i(\hat{x}_1|x_2), \ldots, p_i(\hat{x}_1|x_N), \ldots, p_i(\hat{x}_T|x_N), p_i(\hat{x}_1), \ldots, p_i(\hat{x}_T)](x, \hat{x}) := p_i(\hat{x}|x)p(x)/p_i(\hat{x}) . \tag{A3}$$

While (A3) is too cumbersome to work with, it does highlight that $\mathcal{F}$ is merely a vector of $N \cdot T$ real vector-valued functions, in $T + N \cdot T$ real variables. This allows us to use partial derivatives rather than their infinite-dimensional counterparts (namely, variational derivatives), as in

$$\frac{\partial \mathcal{F}[p_i(\hat{x}|x), p_i(\hat{x})]}{\partial p_i(\hat{x}_j|x_k)} := \lim_{h \to 0} \frac{\mathcal{F}[p_i(\hat{x}_1|x_1), \ldots, p_i(\hat{x}_j|x_k) + h, \ldots, p_i(\hat{x}_T)] - \mathcal{F}[\ldots, p_i(\hat{x}_j|x_k), \ldots]}{h} . \tag{A4}$$

This is the derivative of $\mathcal{F}$ (A3) with respect to a particular (j, k)-entry of its argument, by definition. However, to maintain a concise notation, we shall carry on with un-named function definitions, writing $\partial p_i(x|\hat{x})/\partial p_i(\hat{x}_j|x_k)$ for the partial derivative of (A2) rather than its explicit form (A4). If disoriented, the reader is encouraged to return to the definitions (A4).

We often exchange variables implicitly to logarithmic coordinates, as in Section 3. For example, $\frac{\partial \mathcal{F}[p_i(\hat{x}|x), p_i(\hat{x})]}{\partial \log p_i(\hat{x}_i|x_j)}$ is to be understood as exchanging variables to $u_i(\hat{x}, x) := \log p_i(\hat{x}|x)$, with $\mathcal{G}[u_i(\hat{x}, x), u_i(\hat{x})] := \mathcal{F}[\exp u_i(\hat{x}, x), \exp u_i(\hat{x})]$ now differentiated with respect to its variables $u_i(\hat{x}, x)$ and $u_i(\hat{x})$,

$$\frac{\partial \mathcal{F}[p_i(\hat{x}|x), p_i(\hat{x})]}{\partial \log p_i(\hat{x}_i|x_j)} = \frac{\partial \mathcal{F}[\exp u_i(\hat{x}, x), \exp u_i(\hat{x})]}{\partial u_i(\hat{x}, x)} =: \frac{\partial \mathcal{G}[u_i(\hat{x}, x), u_i(\hat{x})]}{\partial u_i(\hat{x}, x)} \tag{A5}$$

The output of $\mathcal{F}$ may similarly be exchanged to logarithmic coordinates, as in $\log \mathcal{F}[\exp u_i(\hat{x}, x), \exp u_i(\hat{x})]$.

To proceed, carefully note the dependencies between the various variables in a BA-IB iteration, in Steps 1.4 through 1.8. These are summarized compactly by the following diagram:

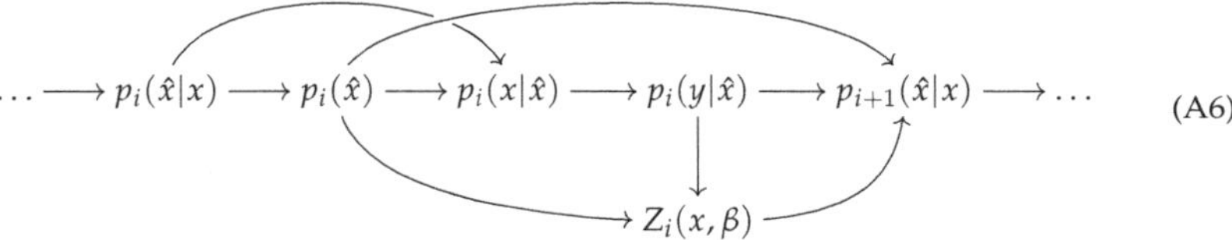

$$\ldots \longrightarrow p_i(\hat{x}|x) \longrightarrow p_i(\hat{x}) \longrightarrow p_i(x|\hat{x}) \longrightarrow p_i(y|\hat{x}) \longrightarrow p_{i+1}(\hat{x}|x) \longrightarrow \ldots \tag{A6}$$
$$\longrightarrow Z_i(x, \beta) \longrightarrow$$

by their order of appearance in the BA-IB Algorithm 1. This diagram proceeds to both sides by the iteration number i. Each node in (A6) serves both as a function of the nodes preceding it and as a variable for those succeeding it, and so it is a "function-variable".

To differentiate along the dependencies graph (A6), we shall need the multivariate chain rule

$$\frac{df}{dy} = \frac{\partial f}{\partial y} + \frac{\partial f}{\partial z}\frac{dz}{dy},$$

(A7)

for a function $f(y, z(y))$. As the dependencies graph (A6) involves multiple function variables, such as $z(y)$, we pause on the definition's subtleties. The partial derivative of a function g in several variables $x_1, \dots, x_N$ with respect to its i-th entry is defined by

$$\frac{\partial g}{\partial x_i} := \lim_{h \to 0} \frac{g(x_1, \dots, x_i + h, \dots, x_N) - g(x_1, \dots, x_i, \dots, x_N)}{h}.$$

(A8)

We emphasize that variables $x_1, \dots, x_{i-1}, x_{i+1}, \dots, x_N$ *other* than x_i are fixed when calculating $\frac{\partial g}{\partial x_i}$. And so, it makes no difference in (A8) whether or not they depend on x_i, as in $x_j = x_j(x_i)$ for $j \neq i$.

Next, suppose we would like to calculate how changing an input distribution affects some output distribution. This is relevant in Appendix B.2 for example, when considering how a change in a coordinate of an input decoder $p_i(y|\hat{x})$ or marginal $p_i(\hat{x})$ affects a particular coordinate of the output decoder or marginal. For exposition's simplicity, though, suppose that we would like to calculate how a change in the (k_1, k_2) coordinate $p_i(\hat{x}_{k_1}|x_{k_2})$ of an input encoder affects the (j_1, j_2) coordinate $p_{i+1}(\hat{x}_{j_1}|x_{j_2})$ of the output encoder. That is, deriving the rightmost node in (A6) with respect to a coordinate of the leftmost one,

$$\frac{d \log p_{i+1}(\hat{x}_{j_1}|x_{j_2})}{d \log p_i(\hat{x}_{k_1}|x_{k_2})},$$

(A9)

where we have exchanged to logarithmic coordinates to simplify calculations. To calculate (A9), one needs to apply the multivariate chain rule (A7) along all the possible dependencies of the output $\log p_{i+1}(\hat{x}_{j_1}|x_{j_2})$ on the input coordinate $\log p_i(\hat{x}_{k_1}|x_{k_2})$. This amounts to following all the paths in (A6) connecting these two nodes, summing the contributions of *every* possible path. For example, traversing from the input $p_i(\hat{x}_{k_1}|x_{k_2})$ rightwards at (A6) to $p_i(\hat{x})$, then downwards to $Z_i(x, \beta)$ and then to the output $p_{i+1}(\hat{x}_{j_1}|x_{j_2})$ yields the term

$$\frac{\partial \log p_i(\hat{x}'')}{\partial \log p_i(\hat{x}_{k_1}|x_{k_2})} \frac{\partial \log Z_i(x, \beta)}{\partial \log p_i(\hat{x}'')} \frac{\partial \log p_{i+1}(\hat{x}_{j_1}|x_{j_2})}{\partial \log Z_i(x, \beta)}$$

corresponding to this path, at particular x and $\hat{x}''$ coordinates. To collect the contribution from every intermediate function variable coordinate, we need to sum the latter over x and $\hat{x}''$. Writing down all such paths, one has for (A9),

$$\frac{d \log p_{i+1}(\hat{x}_{j_1}|x_{j_2})}{d \log p_i(\hat{x}_{k_1}|x_{k_2})}$$

$$= \frac{\partial \log p_i(\hat{x}'')}{\partial \log p_i(\hat{x}_{k_1}|x_{k_2})} \cdot \left\{ \frac{\partial \log p_{i+1}(\hat{x}_{j_1}|x_{j_2})}{\partial \log Z_i(x, \beta)} \cdot \frac{\partial \log Z_i(x, \beta)}{\partial \log p_i(\hat{x}'')} + \frac{\partial \log p_{i+1}(\hat{x}_{j_1}|x_{j_2})}{\partial \log p_i(\hat{x}'')} \right.$$

$$+ \left[\frac{\partial \log p_{i+1}(\hat{x}_{j_1}|x_{j_2})}{\partial \log Z_i(x, \beta)} \cdot \frac{\partial \log Z_i(x, \beta)}{\partial \log p_i(y|\hat{x})} + \frac{\partial \log p_{i+1}(\hat{x}_{j_1}|x_{j_2})}{\partial \log p_i(y|\hat{x})} \right] \cdot \frac{\partial \log p_i(y|\hat{x})}{\partial \log p_i(x'|\hat{x}')} \cdot \left. \frac{\partial \log p_i(x'|\hat{x}')}{\partial \log p_i(\hat{x}'')} \right\}$$

$$+ \left[\frac{\partial \log p_{i+1}(\hat{x}_{j_1}|x_{j_2})}{\partial \log Z_i(x, \beta)} \cdot \frac{\partial \log Z_i(x, \beta)}{\partial \log p_i(y|\hat{x})} + \frac{\partial \log p_{i+1}(\hat{x}_{j_1}|x_{j_2})}{\partial \log p_i(y|\hat{x})} \right] \cdot \frac{\partial \log p_i(y|\hat{x})}{\partial \log p_i(x'|\hat{x}')} \cdot \frac{\partial \log p_i(x'|\hat{x}')}{\partial \log p_i(\hat{x}_{k_1}|x_{k_2})}$$

(A10)

Repeated unbounded variables are understood to be summed over, as in Einstein's summation convention.

Appendix B.1.1. Differentiating along the Dependencies Graph

Next, we differentiate each edge in (the logarithm of) the dependency graph (A6). These are necessary to evaluate derivatives along dependency paths, which underlie the subsequent sections' calculations.

Step 1.4 in the BA-IB Algorithm 1 defines the cluster marginal in terms of the direct encoder,

$$\frac{\partial \log p_i(\hat{x})}{\partial \log p_i(\hat{x}'|x')} \overset{\text{Step 1.4}}{=} \frac{1}{p_i(\hat{x})} \sum_x p(x) \frac{\partial}{\partial \log p_i(\hat{x}'|x')} p_i(\hat{x}|x)$$

$$= \frac{1}{p_i(\hat{x})} \sum_x p(x) p_i(\hat{x}|x) \frac{\partial \log p_i(\hat{x}|x)}{\partial \log p_i(\hat{x}'|x')} \overset{\text{Step 1.5}}{=} p_i(x'|\hat{x}) \cdot \delta_{\hat{x},\hat{x}'} \quad \text{(A11)}$$

In the first and second equalities we have used the identity $\frac{\partial}{\partial x} y = y \frac{\partial}{\partial x} \log y$ for the differentiation of a function's logarithm, when y is a function of x.

Following the comments around the definition (A8) of a partial derivative, note that Step 1.5 defines the inverse encoder $\log p_i(x|\hat{x})$ as a function of the variables $\log p_i(\hat{x}|x)$ and $\log p_i(\hat{x})$ (and $p(x)$, which we ignore under differentiation). Thus, differentiating this equation with respect to an entry of the variable $\log p_i(\hat{x}|x)$ implies that the entries of the other variable $\log p_i(\hat{x})$ are held fixed, and vice versa. So, for the Bayes rule Step 1.5 we have

$$\frac{\partial \log p_i(x_{j_1}|\hat{x}_{j_2})}{\partial \log p_i(\hat{x})} = \frac{\partial}{\partial \log p_i(\hat{x})} \left[\log p_i(\hat{x}_{j_2}|x_{j_1}) - \log p_i(\hat{x}_{j_2}) \right] = -\delta_{\hat{x},\hat{x}_{j_2}} \quad \text{(A12)}$$

where $\log p_i(\hat{x}_{j_2}|x_{j_1})$ at the right-hand side is different from the variable $\log p_i(\hat{x})$ of differentiation, and so its partial derivative vanishes. Next, differentiating Step 1.5 with respect to a coordinate of its other variable $\log p_i(\hat{x}|x)$,

$$\frac{\partial \log p_i(x_{j_1}|\hat{x}_{j_2})}{\partial \log p_i(\hat{x}'|x')} = \frac{\partial \log p_i(\hat{x}_{j_2}|x_{j_1})}{\partial \log p_i(\hat{x}'|x')} - \frac{\partial \log p_i(\hat{x}_{j_2})}{\partial \log p_i(\hat{x}'|x')} = \delta_{x_{j_1},x'} \cdot \delta_{\hat{x}_{j_2},\hat{x}'} \quad \text{(A13)}$$

Using again the logarithmic derivative identity $\frac{\partial}{\partial x} y = y \frac{\partial}{\partial x} \log y$, by the decoder Step 1.6 we have

$$\frac{\partial \log p_i(y|\hat{x}'')}{\partial \log p_i(x_{k_1}|\hat{x}_{k_2})} = \frac{1}{p_i(y|\hat{x}'')} \sum_{x'''} p(y|x''') \frac{\partial}{\partial \log p_i(x_{k_1}|\hat{x}_{k_2})} p_i(x'''|\hat{x}'')$$

$$= \frac{1}{p_i(y|\hat{x}'')} \sum_{x'''} p(y|x''') p_i(x'''|\hat{x}'') \, \delta_{\hat{x}_{k_2},\hat{x}''} \cdot \delta_{x_{k_1},x'''} = \delta_{\hat{x}_{k_2},\hat{x}''} \cdot \frac{p(y|x_{k_1}) p_i(x_{k_1}|\hat{x}'')}{p_i(y|\hat{x}'')} \quad \text{(A14)}$$

Next, consider the KL-divergence term in the Definition 1.7 of the partition function Z_i,

$$\frac{\partial}{\partial \log p_i(y|\hat{x}'')} D_{KL}\left[p(y|x'') || p_i(y|\hat{x}) \right] = - \sum_{y'} p(y'|x'') \underbrace{\frac{\partial}{\partial \log p_i(y|\hat{x}'')} \log p_i(y'|\hat{x})}_{\delta_{\hat{x},\hat{x}''} \cdot \delta_{y,y'}} = -\delta_{\hat{x},\hat{x}''} \cdot p(y|x'') \quad \text{(A15)}$$

Since the partition Function 1.7 depends on the decoder $p_i(y|\hat{x})$ only via the KL-divergence,

$$\frac{\partial Z_i(x'',\beta)}{\partial \log p_i(y|\hat{x}'')} = \frac{\partial}{\partial \log p_i(y|\hat{x}'')} \sum_{\hat{x}} p_i(\hat{x}) \exp\big\{ -\beta\, D_{KL}\big[p(y|x'')\|p_i(y|\hat{x})\big]\big\}$$

$$= -\beta \sum_{\hat{x}} p_i(\hat{x}) \exp\big\{ -\beta\, D_{KL}\big[p(y|x'')\|p_i(y|\hat{x})\big]\big\} \frac{\partial}{\partial \log p_i(y|\hat{x}'')} D_{KL}\big[p(y|x'')\|p_i(y|\hat{x})\big]$$

$$\overset{(A15)}{=} \beta\, p_i(\hat{x}'') \exp\big\{ -\beta\, D_{KL}\big[p(y|x'')\|p_i(y|\hat{x}'')\big]\big\} p(y|x'')$$

$$\overset{\text{Step } 1.8}{=} \beta\, p_{i+1}(\hat{x}''|x'') Z_i(x'',\beta) p(y|x'') \tag{A16}$$

Hence,

$$\frac{\partial \log Z_i(x'',\beta)}{\partial \log p_i(y|\hat{x}'')} = \beta\, p_{i+1}(\hat{x}''|x'') p(y|x'') \tag{A17}$$

For the derivative of the partition function with respect to the marginal $p_i(\hat{x})$,

$$\frac{\partial Z_i(x,\beta)}{\partial \log p_i(\hat{x}')} \overset{\text{Step } 1.7}{=} \frac{\partial}{\partial \log p_i(\hat{x}')} \sum_{\hat{x}} p_i(\hat{x}) \exp\big\{ -\beta\, D_{KL}\big[p(y|x)\|p_i(y|\hat{x})\big]\big\}$$

$$= \sum_{\hat{x}} p_i(\hat{x}) \exp\big\{ -\beta\, D_{KL}\big[p(y|x)\|p_i(y|\hat{x})\big]\big\} \frac{\partial \log p_i(\hat{x})}{\partial \log p_i(\hat{x}')}$$

$$= \sum_{\hat{x}} p_i(\hat{x}) \exp\big\{ -\beta\, D_{KL}\big[p(y|x)\|p_i(y|\hat{x})\big]\big\} \cdot \delta_{\hat{x},\hat{x}'}$$

$$\overset{\text{Step } 1.8}{=} Z_i(x,\beta) \cdot p_{i+1}(\hat{x}'|x) \tag{A18}$$

where the second equality follows from the logarithmic derivative identity. Hence,

$$\frac{\partial \log Z_i(x,\beta)}{\partial \log p_i(\hat{x}')} = p_{i+1}(\hat{x}'|x) \tag{A19}$$

Finally, for the encoder Step 1.8,

$$\log p_{i+1}(\hat{x}'|x') := \log p_i(\hat{x}') - \log Z_i(x',\beta) - \beta\, D_{KL}\big[p(y|x')\|p_i(y|\hat{x}')\big] \tag{A20}$$

The first two terms to the right, $p_i(\hat{x})$ and $Z_i(x,\beta)$, take the role of a variable in Step 1.8. In contrast, we consider the last divergence term as a shorthand for summing over $p_i(y|\hat{x})$. Thus, the latter *is* a variable of (A20). With (A15), we thus have

$$\frac{\partial \log p_{i+1}(\hat{x}'|x')}{\partial \log p_i(y|\hat{x}'')} = \beta\, \delta_{\hat{x}',\hat{x}''} \cdot p(y|x') \,. \tag{A21}$$

For the other derivatives of the encoder Step 1.8,

$$\frac{\partial \log p_{i+1}(\hat{x}'|x')}{\partial \log Z_i(x'',\beta)} = -\frac{\partial \log Z_i(x',\beta)}{\partial \log Z_i(x'',\beta)} = -\delta_{x',x''} \tag{A22}$$

and

$$\frac{\partial \log p_{i+1}(\hat{x}|x)}{\partial \log p_i(\hat{x}')} = \frac{\partial \log p_i(\hat{x})}{\partial \log p_i(\hat{x}')} - \frac{\partial \log Z_i(x,\beta)}{\partial \log p_i(\hat{x}')} - \beta\frac{\partial D_{KL}\big[p(y|x)\|p_i(y|\hat{x})\big]}{\partial \log p_i(\hat{x}')} = \delta_{\hat{x},\hat{x}'} \tag{A23}$$

where the variable $p_i(\hat{x})$ of Step 1.8 differs from the variables Z_i and $p_i(y|\hat{x})$, on which the crossed-out terms depend.

We summarize the calculations of this subsection in the following diagram:

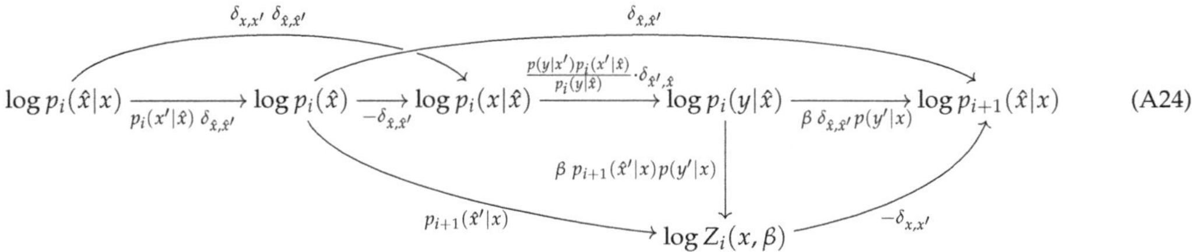

A differentiation variable is denoted with commas, at an arrow's source in this diagram. A coordinate of the function which we differentiate is written without commas, at an arrow's end, e.g.,

$$\log p_i(\hat{x}|x) \xrightarrow[\frac{\partial \log p_i(x|\hat{x})}{\partial \log p_i(\hat{x}'|x')} = \ldots]{} \log p_i(x|\hat{x})$$

Appendix B.2. The Jacobian Matrix of BA-IB in Log-Decoder Coordinates

By gathering the results of Appendix B.1.1 and following the lines of Appendix B.1, we calculate the Jacobian matrix (13) (in Section 3) of the Blahut–Arimoto operator BA_β in log-decoder coordinates, defined in Section 2.

The derivative of BA_β in decoder coordinates boils down to the four quantities: the effect $\frac{d \log p_{i+1}(y|\hat{x})}{d \log p_i(y'|\hat{x}')}$ that varying a coordinate $\log p_i(y'|\hat{x}')$ of an input cluster has on a coordinate $\log p_{i+1}(y|\hat{x})$ of an output cluster, the effect $\frac{d \log p_{i+1}(y|\hat{x})}{d \log p_i(\hat{x}')}$ that varying an input marginal coordinate $\log p_i(\hat{x}')$ has on a coordinate $\log p_{i+1}(y|\hat{x})$ of an output cluster, and so forth. And so, the Jacobian $D_{\log p(y|\hat{x}), \log p(\hat{x})} BA_\beta$ it is a block matrix,

$$\begin{pmatrix} \dfrac{d \log p_{i+1}(y|\hat{x})}{d \log p_i(y'|\hat{x}')} & \dfrac{d \log p_{i+1}(y|\hat{x})}{d \log p_i(\hat{x}')} \\[2ex] \hline \\[-1ex] \dfrac{d \log p_{i+1}(\hat{x})}{d \log p_i(y'|\hat{x}')} & \dfrac{d \log p_{i+1}(\hat{x})}{d \log p_i(\hat{x}')} \end{pmatrix} \tag{A25}$$

Its rows correspond to the output coordinates of BA_β. We index its upper rows by $y \in \mathcal{Y}$ and $\hat{x} \in \{1, \ldots, T\}$, while its lower rows are indexed by $\hat{x}$ alone. Similarly, its columns correspond to the input coordinates of BA_β. We index its leftmost columns by y' and $\hat{x}'$, and its rightmost columns by $\hat{x}'$ alone. Each block in (A25) consists of contributions along all the distinct paths connecting two vertices in the dependencies graph (A6). For example, the lower-left block in (A25) consists of the contributions along all the paths in (A6) connecting $p_i(y'|\hat{x}')$ to $p_{i+1}(\hat{x})$.

We now spell out the paths contributing to each block in (A25), with repeated dummy indices understood to be summed over. Afterward, we shall calculate the contributing paths explicitly, carrying out the summations. The upper-left block of (A25) consists of

$$\frac{d \log p_{i+1}(y|\hat{x})}{d \log p_i(y'|\hat{x}')} = \frac{\partial \log p_{i+1}(y|\hat{x})}{\partial \log p_{i+1}(x_1|\hat{x}_2)} \cdot \left[\frac{\partial \log p_{i+1}(x_1|\hat{x}_2)}{\partial \log p_{i+1}(\hat{x}_3)} \frac{\partial \log p_{i+1}(\hat{x}_3)}{\partial \log p_{i+1}(\hat{x}_4|x_5)} + \frac{\partial \log p_{i+1}(x_1|\hat{x}_2)}{\partial \log p_{i+1}(\hat{x}_4|x_5)} \right]$$

$$\cdot \left[\frac{\partial \log p_{i+1}(\hat{x}_4|x_5)}{\partial \log p_i(y'|\hat{x}')} + \frac{\partial \log p_{i+1}(\hat{x}_4|x_5)}{\partial \log Z_i(x_6, \beta)} \frac{\partial \log Z_i(x_6, \beta)}{\partial \log p_i(y'|\hat{x}')} \right] \tag{A26}$$

This Equation (A26) encodes the four paths connecting the vertex $p_i(y'|\hat{x}')$ to $p_{i+1}(y|\hat{x})$ in (A6). When accumulating the contributions in (A26), one must carefully sum only over repeated dummy indices that appear in the given term. For example, the two paths in (A26) which traverse the edge $\frac{\partial \log p_{i+1}(x_1|\hat{x}_2)}{\partial \log p_{i+1}(\hat{x}_4|x_5)}$ (pointing from $p_{i+1}(\hat{x}|x)$ to $p_{i+1}(x|\hat{x})$) do *not* involve a summation over $\hat{x}_3$. In contrast, the two paths involving $\frac{\partial \log p_{i+1}(x_1|\hat{x}_2)}{\partial \log p_{i+1}(\hat{x}_3)} \frac{\partial \log p_{i+1}(\hat{x}_3)}{\partial \log p_{i+1}(\hat{x}_4|x_5)}$ there do entail a summation over $\hat{x}_3$. This is relevant for the calculations below, as in (A31) for example.

Similarly, for the upper-right block of (A25),

$$
\begin{aligned}
\frac{d \log p_{i+1}(y|\hat{x})}{d \log p_i(\hat{x}')} &= \frac{\partial \log p_{i+1}(y|\hat{x})}{\partial \log p_{i+1}(x_1|\hat{x}_2)} \cdot \left[\frac{\partial \log p_{i+1}(x_1|\hat{x}_2)}{\partial \log p_{i+1}(\hat{x}_3)} \frac{\partial \log p_{i+1}(\hat{x}_3)}{\partial \log p_{i+1}(\hat{x}_4|x_5)} + \frac{\partial \log p_{i+1}(x_1|\hat{x}_2)}{\partial \log p_{i+1}(\hat{x}_4|x_5)} \right] \\
&\quad \cdot \left\{ \frac{\partial \log p_{i+1}(\hat{x}_4|x_5)}{\partial \log p_i(\hat{x}')} + \frac{\partial \log p_{i+1}(\hat{x}_4|x_5)}{\partial \log p_i(y_7|\hat{x}_8)} \frac{\partial \log p_i(y_7|\hat{x}_8)}{\partial \log p_i(x_9|\hat{x}_{10})} \frac{\partial \log p_i(x_9|\hat{x}_{10})}{\partial \log p_i(\hat{x}')} \right. \\
&\quad \left. + \frac{\partial \log p_{i+1}(\hat{x}_4|x_5)}{\partial \log Z_i(x_6,\beta)} \left[\frac{\partial \log Z_i(x_6,\beta)}{\partial \log p_i(\hat{x}')} + \frac{\partial \log Z_i(x_6,\beta)}{\partial \log p_i(y_7|\hat{x}_8)} \frac{\partial \log p_i(y_7|\hat{x}_8)}{\partial \log p_i(x_9|\hat{x}_{10})} \frac{\partial \log p_i(x_9|\hat{x}_{10})}{\partial \log p_i(\hat{x}')} \right] \right\}
\end{aligned}
\tag{A27}
$$

For the lower-left block of (A25),

$$
\frac{d \log p_{i+1}(\hat{x})}{d \log p_i(y'|\hat{x}')} = \frac{\partial \log p_{i+1}(\hat{x})}{\partial \log p_{i+1}(\hat{x}_1|x_2)} \left[\frac{\partial \log p_{i+1}(\hat{x}_1|x_2)}{\partial \log p_i(y'|\hat{x}')} + \frac{\partial \log p_{i+1}(\hat{x}_1|x_2)}{\partial \log Z_i(x_3,\beta)} \frac{\partial \log Z_i(x_3,\beta)}{\partial \log p_i(y'|\hat{x}')} \right]
\tag{A28}
$$

Last, for the lower-right block of (A25),

$$
\begin{aligned}
\frac{d \log p_{i+1}(\hat{x})}{d \log p_i(\hat{x}')} &= \frac{\partial \log p_{i+1}(\hat{x})}{\partial \log p_{i+1}(\hat{x}_1|x_2)} \cdot \left\{ \frac{\partial \log p_{i+1}(\hat{x}_1|x_2)}{\partial \log p_i(\hat{x}')} + \frac{\partial \log p_{i+1}(\hat{x}_1|x_2)}{\partial \log p_i(y_3|\hat{x}_4)} \frac{\partial \log p_i(y_3|\hat{x}_4)}{\partial \log p_i(x_5|\hat{x}_6)} \frac{\partial \log p_i(x_5|\hat{x}_6)}{\partial \log p_i(\hat{x}')} \right. \\
&\quad \left. + \frac{\partial \log p_{i+1}(\hat{x}_1|x_2)}{\partial \log Z_i(x_7,\beta)} \left[\frac{\partial \log Z_i(x_7,\beta)}{\partial \log p_i(\hat{x}')} + \frac{\partial \log Z_i(x_7,\beta)}{\partial \log p_i(y_3|\hat{x}_4)} \frac{\partial \log p_i(y_3|\hat{x}_4)}{\partial \log p_i(x_5|\hat{x}_6)} \frac{\partial \log p_i(x_5|\hat{x}_6)}{\partial \log p_i(\hat{x}')} \right] \right\}
\end{aligned}
\tag{A29}
$$

Next, by using the intermediate results summarized in (A24) (Appendix B.1.1), we calculate each of the four blocks of (A25) explicitly. For the upper-left block (A26), we have

$$
\begin{aligned}
\frac{d \log p_{i+1}(y|\hat{x})}{d \log p_i(y'|\hat{x}')} &= \frac{p(y|x_1) p_{i+1}(x_1|\hat{x})}{p_{i+1}(y|\hat{x})} \delta_{\hat{x},\hat{x}_2} \cdot \left[(-\delta_{\hat{x}_2,\hat{x}_3}) p_{i+1}(x_5|\hat{x}_3) \delta_{\hat{x}_3,\hat{x}_4} + \delta_{x_1,x_5} \delta_{\hat{x}_2,\hat{x}_4} \right] \\
&\quad \cdot \left[\beta \delta_{\hat{x}_4,\hat{x}'} p(y'|x_5) + (-\delta_{x_5,x_6}) \beta p_{i+1}(\hat{x}'|x_6) p(y'|x_6) \right]
\end{aligned}
\tag{A30}
$$

For clarity, we elaborate on each step needed to complete the calculation of the upper-left block (A26) while providing only the main steps for the other blocks. To carry out the summations over the dummy variables $x_1, \hat{x}_2, \hat{x}_3, \hat{x}_4, x_5$, and x_6 in (A30), we carefully sum only over repeated dummy indices, as explained after (A26). We carry out one summation at a time, starting with $\hat{x}_2$. This yields

$$
\begin{aligned}
\beta &\frac{p(y|x_1) p_{i+1}(x_1|\hat{x})}{p_{i+1}(y|\hat{x})} \cdot \left[-\delta_{\hat{x},\hat{x}_3} p_{i+1}(x_5|\hat{x}_3) \delta_{\hat{x}_3,\hat{x}_4} + \delta_{x_1,x_5} \delta_{\hat{x},\hat{x}_4} \right] \cdot \left[\delta_{\hat{x}_4,\hat{x}'} p(y'|x_5) - \delta_{x_5,x_6} \, p_{i+1}(\hat{x}'|x_6) p(y'|x_6) \right] \\
&= \beta \cdot \left[-\delta_{\hat{x},\hat{x}_3} p_{i+1}(x_5|\hat{x}_3) \delta_{\hat{x}_3,\hat{x}_4} + \delta_{\hat{x},\hat{x}_4} \frac{p(y|x_5) p_{i+1}(x_5|\hat{x})}{p_{i+1}(y|\hat{x})} \right] \cdot \left[\delta_{\hat{x}_4,\hat{x}'} p(y'|x_5) - \delta_{x_5,x_6} \, p_{i+1}(\hat{x}'|x_6) p(y'|x_6) \right] \\
&= \beta \cdot p_{i+1}(x_5|\hat{x}) \left[-\delta_{\hat{x},\hat{x}_4} + \delta_{\hat{x},\hat{x}_4} \frac{p(y|x_5)}{p_{i+1}(y|\hat{x})} \right] \cdot \left[\delta_{\hat{x}_4,\hat{x}'} p(y'|x_5) - \delta_{x_5,x_6} \, p_{i+1}(\hat{x}'|x_6) p(y'|x_6) \right] \\
&= -\beta \cdot p_{i+1}(x_5|\hat{x}) \left[1 - \frac{p(y|x_5)}{p_{i+1}(y|\hat{x})} \right] \cdot \left[\delta_{\hat{x},\hat{x}'} p(y'|x_5) - \delta_{x_5,x_6} \, p_{i+1}(\hat{x}'|x_6) p(y'|x_6) \right] \\
&= -\beta \cdot p(y'|x_5) p_{i+1}(x_5|\hat{x}) \left[1 - \frac{p(y|x_5)}{p_{i+1}(y|\hat{x})} \right] \cdot \left[\delta_{\hat{x},\hat{x}'} - p_{i+1}(\hat{x}'|x_5) \right] \\
&= -\beta \sum_x p(y'|x) p_{i+1}(x|\hat{x}) \cdot \left[1 - \frac{p(y|x)}{p_{i+1}(y|\hat{x})} \right] \cdot \left[\delta_{\hat{x},\hat{x}'} - p_{i+1}(\hat{x}'|x) \right]
\end{aligned}
\tag{A31}
$$

In the first equality above we carried out the summation over x_1, in the second over $\hat{x}_3$, in the third over $\hat{x}_4$, in the fourth over x_6, and in the fifth over x_5.

To simplify the notation, we replace summations over x with definitions as in Equation (14) (Section 3),

$$C(\hat{x}, \hat{x}'; i)_{y,y'} := \sum_x p(y|x)p(y'|x)p_i(\hat{x}'|x)p_i(x|\hat{x})$$

$$B(\hat{x}, \hat{x}'; i)_y := \sum_x p(y|x)p_i(\hat{x}'|x)p_i(x|\hat{x}) = \sum_{y'} C(\hat{x}, \hat{x}'; i)_{y,y'}$$

$$A(\hat{x}, \hat{x}'; i) := \sum_x p_i(\hat{x}'|x)p_i(x|\hat{x}) = \sum_y B(\hat{x}, \hat{x}'; i)_y \qquad \text{(A32)}$$

$$D(\hat{x}; i)_{y,y'} := \frac{1}{p_i(y|\hat{x})} \sum_x p(y|x)p(y'|x)p_i(x|\hat{x}) = \frac{1}{p_i(y|\hat{x})} \sum_{\hat{x}'} C(\hat{x}, \hat{x}'; i)_{y,y'}$$

and note that

$$\sum_{y',\hat{x}'} C(\hat{x}, \hat{x}'; i)_{y,y'} = p_i(y|\hat{x}) . \qquad \text{(A33)}$$

The quantities $A, B,$ and C involve two IB clusters. They are a scalar, a vector, and a matrix, respectively. The definition of D involves only one IB cluster and coincides with C_Y in [13] (3.2 in Part III). The relations to the right of (A32) show that each can be expressed in terms of $C(\hat{x}, \hat{x}'; i)_{y,y'}$. Equation (A33) shows that the latter can be rewritten as a right-stochastic matrix, up to trivial manipulations. As seen below, the Jacobian matrix (A25) of a BA-IB step in log-decoder coordinates can be computed in terms of the quantities in (A32). With the latter definitions (A32), (A31) can be rewritten as

$$\frac{d\log p_{i+1}(y|\hat{x})}{d\log p_i(y'|\hat{x}')} \overset{\text{(A31)}}{=} -\beta \sum_x \Big[\delta_{\hat{x},\hat{x}'}\, p(y'|x)p_{i+1}(x|\hat{x}) - p(y'|x)p_{i+1}(\hat{x}'|x)p_{i+1}(x|\hat{x})$$

$$- \delta_{\hat{x},\hat{x}'}\, \tfrac{1}{p_{i+1}(y|\hat{x})} p(y|x)p(y'|x)p_{i+1}(x|\hat{x}) + \tfrac{1}{p_{i+1}(y|\hat{x})} p(y'|x)p(y|x)p_{i+1}(\hat{x}'|x)p_{i+1}(x|\hat{x}) \Big]$$

$$\overset{\text{(A32)}}{=} -\beta \Big[\delta_{\hat{x},\hat{x}'}\, p_{i+1}(y'|\hat{x}) - B(\hat{x}, \hat{x}'; i+1)_{y'} - \delta_{\hat{x},\hat{x}'}\, D(\hat{x}; i+1)_{y,y'} + \tfrac{1}{p_{i+1}(y|\hat{x})} C(\hat{x}, \hat{x}'; i+1)_{y,y'} \Big]$$

$$= \beta \sum_{\hat{x}'',y''} \big(\delta_{\hat{x}'',\hat{x}'} - \delta_{\hat{x},\hat{x}'} \big) \Big(1 - \tfrac{\delta_{y'',y}}{p_{i+1}(y|\hat{x})} \Big) C(\hat{x}, \hat{x}''; i+1)_{y',y''} \qquad \text{(A34)}$$

The third equality above follows from (A33), the identities to the right of (A32), and simple algebra.

For the upper-right block (A27),

$$\frac{d\log p_{i+1}(y|\hat{x})}{d\log p_i(\hat{x}')} = \frac{p(y|x_1)p_{i+1}(x_1|\hat{x}_2)}{p_{i+1}(y|\hat{x}_2)} \delta_{\hat{x}_2,\hat{x}} \cdot \Big[\big(-\delta_{\hat{x}_2,\hat{x}_3} \big) p_{i+1}(x_5|\hat{x}_3)\delta_{\hat{x}_3,\hat{x}_4} + \delta_{x_1,x_5}\delta_{\hat{x}_2,\hat{x}_4} \Big]$$

$$\cdot \Big\{ \delta_{\hat{x}_4,\hat{x}'} + \beta\delta_{\hat{x}_4,\hat{x}_8} p(y_7|x_5)\frac{p(y_7|x_9)p_i(x_9|\hat{x}_8)}{p_i(y_7|\hat{x}_8)}\delta_{\hat{x}_8,\hat{x}_{10}}\big(-\delta_{\hat{x}_{10},\hat{x}'} \big)$$

$$+ \big(-\delta_{x_5,x_6} \big) \Big[p_{i+1}(\hat{x}'|x_6) + \beta p_{i+1}(\hat{x}_8|x_6)p(y_7|x_6)\frac{p(y_7|x_9)p_i(x_9|\hat{x}_8)}{p_i(y_7|\hat{x}_8)}\delta_{\hat{x}_8,\hat{x}_{10}}\big(-\delta_{\hat{x}_{10},\hat{x}'} \big) \Big] \Big\} \qquad \text{(A35)}$$

In a manner similar to (A31), summing over all ten dummy variables other than x_1 and x_5 yields

$$(1-\beta) \cdot \frac{p(y|x_1)p_{i+1}(x_1|\hat{x})}{p_{i+1}(y|\hat{x})} \cdot \big(\delta_{x_1,x_5} - p_{i+1}(x_5|\hat{x}) \big) \cdot \big(\delta_{\hat{x},\hat{x}'} - p_{i+1}(\hat{x}'|x_5) \big)$$

$$= (1-\beta) \cdot \Big(\tfrac{-1}{p_{i+1}(y|\hat{x})} \sum_x p(y|x)p_{i+1}(\hat{x}'|x)p_{i+1}(x|\hat{x}) + \sum_x p_{i+1}(\hat{x}'|x)p_{i+1}(x|\hat{x}) \Big)$$

$$= (1-\beta) \cdot \sum_x \Big(1 - \tfrac{p(y|x)}{p_{i+1}(y|\hat{x})} \Big) p_{i+1}(\hat{x}'|x)p_{i+1}(x|\hat{x}) \qquad \text{(A36)}$$

The two terms involving $\delta_{\hat{x},\hat{x}'}$ cancel out when summing over x_1 and x_5 at the first equality. Rewriting with the definitions (A32) of A and B further simplifies (A36) to

$$(1 - \beta) \cdot \left[A(\hat{x}, \hat{x}'; i+1) - \frac{1}{p_{i+1}(y|\hat{x})} B(\hat{x}, \hat{x}'; i+1)_y \right]$$
$$= (1 - \beta) \cdot \sum_{y''} \left[1 - \frac{\delta_{y'',y}}{p_{i+1}(y|\hat{x})} \right] B(\hat{x}, \hat{x}'; i+1)_{y''} \quad \text{(A37)}$$

For the lower-left block (A28),

$$\frac{d \log p_{i+1}(\hat{x})}{d \log p_i(y'|\hat{x}')} = p_{i+1}(x_2|\hat{x})\delta_{\hat{x},\hat{x}_1} \left[\beta\delta_{\hat{x}_1,\hat{x}'} p(y'|x_2) + (-\delta_{x_2,x_3})\beta p_{i+1}(\hat{x}'|x_3) p(y'|x_3) \right] \quad \text{(A38)}$$

Summing over dummy variables and simplifying yields

$$\beta \cdot \left[\delta_{\hat{x},\hat{x}'}\, p_{i+1}(y'|\hat{x}) - \sum_x p(y'|x) p_{i+1}(\hat{x}'|x) p_{i+1}(x|\hat{x}) \right] \quad \text{(A39)}$$

In terms of definitions (A32), this simplifies to

$$\beta \cdot \left[\delta_{\hat{x},\hat{x}'}\, p_{i+1}(y'|\hat{x}) - B(\hat{x}, \hat{x}'; i+1)_{y'} \right] \quad \text{(A40)}$$

Finally, for the lower-right block (A29),

$$\frac{d \log p_{i+1}(\hat{x})}{d \log p_i(\hat{x}')} = p_{i+1}(x_2|\hat{x})\delta_{\hat{x},\hat{x}_1} \cdot \left\{ \delta_{\hat{x}_1,\hat{x}'} + \beta\delta_{\hat{x}_1,\hat{x}_4} p(y_3|x_2) \frac{p(y_3|x_5)p_i(x_5|\hat{x}_4)}{p_i(y_3|\hat{x}_4)} \delta_{\hat{x}_4,\hat{x}_6}(-\delta_{\hat{x}_6,\hat{x}'}) \right.$$
$$\left. + (-\delta_{x_2,x_7}) \left[p_{i+1}(\hat{x}'|x_7) + \beta p_{i+1}(\hat{x}_4|x_7) p(y_3|x_7) \frac{p(y_3|x_5)p_i(x_5|\hat{x}_4)}{p_i(y_3|\hat{x}_4)} \delta_{\hat{x}_4,\hat{x}_6}(-\delta_{\hat{x}_6,\hat{x}'}) \right] \right\} \quad \text{(A41)}$$

This simplifies to

$$(1 - \beta) \left(\delta_{\hat{x},\hat{x}'} - \sum_x p_{i+1}(\hat{x}'|x) p_{i+1}(x|\hat{x}) \right) \quad \text{(A42)}$$

With definitions (A32), this can be written as

$$(1 - \beta) \left(\delta_{\hat{x},\hat{x}'} - A(\hat{x}, \hat{x}'; i+1) \right) \quad \text{(A43)}$$

Collecting the results from (A34), (A37), (A40), and (A43) back into (A25), BA's Jacobian in these coordinates is

$$\left(\begin{array}{c|c} \begin{array}{c} \beta \sum_{\hat{x}'',y''} \left(\delta_{\hat{x}'',\hat{x}'} - \delta_{\hat{x},\hat{x}'} \right) \\[4pt] \cdot \left(1 - \frac{\delta_{y'',y}}{p_{i+1}(y|\hat{x})} \right) C(\hat{x}, \hat{x}''; i+1)_{y',y''} \end{array} & (1 - \beta) \cdot \sum_{y''} \left[1 - \frac{\delta_{y'',y}}{p_{i+1}(y|\hat{x})} \right] B(\hat{x}, \hat{x}'; i+1)_{y''} \\[14pt] \hline \\ \beta \cdot \left[\delta_{\hat{x},\hat{x}'}\, p_{i+1}(y'|\hat{x}) - B(\hat{x}, \hat{x}'; i+1)_{y'} \right] & (1 - \beta) \left(\delta_{\hat{x},\hat{x}'} - A(\hat{x}, \hat{x}'; i+1) \right) \end{array} \right) \quad \text{(A44)}$$

When evaluated at an IB root, this is Equation (13) of Section 3. Equivalently, it can be written in the following form, which is more convenient for implementation:

$$\left(
\begin{array}{c|c}
\begin{aligned}
&\beta\left[B(\hat{x},\hat{x}';i+1)_{y'} - \delta_{\hat{x},\hat{x}'}\, p_{i+1}(y'|\hat{x})\right.\\
&\left. + \delta_{\hat{x},\hat{x}'}\, D(\hat{x};i+1)_{y,y'} - \tfrac{1}{p_{i+1}(y|\hat{x})} C(\hat{x},\hat{x}';i+1)_{y,y'}\right]
\end{aligned}
& (1-\beta)\cdot\left[A(\hat{x},\hat{x}';i+1) - \tfrac{1}{p_{i+1}(y|\hat{x})} B(\hat{x},\hat{x}';i+1)_y\right] \\[2ex]
\hline
\beta\cdot\left[\delta_{\hat{x},\hat{x}'}\, p_{i+1}(y'|\hat{x}) - B(\hat{x},\hat{x}';i+1)_{y'}\right]
& (1-\beta)\left(\delta_{\hat{x},\hat{x}'} - A(\hat{x},\hat{x}';i+1)\right)
\end{array}
\right) \tag{A45}$$

Appendix B.3. The Partial β-Derivatives of BA-IB in Log-Decoder Coordinates

We calculate the vector $D_\beta BA_\beta$ of partial derivatives of the BA_β operator in log-decoder coordinates (of Section 2), which appears at the right-hand side of the IB-ODE (16) (in Section 3).

To that end, we differentiate backward along the dependencies graph (A6) (in Appendix B.1) with respect to β, starting at the output coordinates $p_{i+1}(y|\hat{x})$ and $p_{i+1}(\hat{x})$ of BA_β. After differentiating, we mind our independent variables. Here, these are β, and the input coordinates $p_i(y|\hat{x})$ and $p_i(\hat{x})$ of BA_β. The differentiation of these with respect to β vanishes (except for $\frac{d\beta}{d\beta} = 1$), as they are independent. Finally, we compose the differentiations to obtain the effect $D_\beta BA_\beta$ of changing β on BA's output. We note that, in principle, one can differentiate the explicit formulae (A1) of BA_β in decoder coordinates (Appendix A) with respect to β. However, we find that to be cumbersome and far more error-prone than our approach, and so proceed in the spirit of the previous Appendix B.2.

We start by differentiating each of the equations defining the Blahut–Arimoto Algorithm 1 with respect to β, as if all its variables are dependent. For the cluster marginal Step 1.4,

$$\frac{d}{d\beta} p_i(\hat{x}) = \sum_x p(x) \frac{d}{d\beta} p_i(\hat{x}|x) \tag{A46}$$

For the inverse encoder Step 1.5,

$$\frac{d}{d\beta} p_i(x|\hat{x}) = \frac{p(x)}{p_i(\hat{x})} \frac{dp_i(\hat{x}|x)}{d\beta} - \frac{p_i(\hat{x}|x)p(x)}{p_i(\hat{x})^2} \frac{dp_i(\hat{x})}{d\beta} \tag{A47}$$

For the decoder Step 1.6,

$$\frac{d}{d\beta} p_i(y|\hat{x}) = \sum_x p(y|x) \frac{d}{d\beta} p_i(x|\hat{x}) \tag{A48}$$

For the KL-divergence,

$$\frac{d}{d\beta} D_{KL}\left[p(y|x)\|p_i(y|\hat{x})\right] = \frac{d}{d\beta} \sum_{y''} p(y''|x) \log \frac{p(y''|x)}{p_i(y''|\hat{x})} = -\sum_{y''} \frac{p(y''|x)}{p_i(y''|\hat{x})} \frac{d}{d\beta} p_i(y''|\hat{x}) \tag{A49}$$

And for its exponent,

$$\frac{d}{d\beta} \exp\left\{-\beta\, D_{x,\hat{x}}\right\} = -\left(D_{x,\hat{x}} + \beta \frac{dD_{x,\hat{x}}}{d\beta}\right)\cdot \exp\left\{-\beta\, D_{x,\hat{x}}\right\}$$

$$\stackrel{(A49)}{=} -\left(D_{x,\hat{x}} - \beta \sum_{y''} \frac{p(y''|x)}{p_i(y''|\hat{x})} \frac{d}{d\beta} p_i(y''|\hat{x})\right)\cdot \exp\left\{-\beta\, D_{x,\hat{x}}\right\} \tag{A50}$$

where we have written $D_{x,\hat{x}} := D_{KL}\left[p(y|x)\|p_i(y|\hat{x})\right]$ for short. Thus, for the partition function's Step 1.7, we have

$$\frac{d}{d\beta} Z_i(x,\beta) = \frac{d}{d\beta} \sum_{\hat{x}''} p_i(\hat{x}'') \exp\left\{ -\beta\, D_{x,\hat{x}''} \right\}$$

$$\stackrel{\text{(A50)}}{=} \sum_{\hat{x}''} \left(\frac{dp_i(\hat{x}'')}{d\beta} - p_i(\hat{x}'') D_{x,\hat{x}''} + \beta\, p_i(\hat{x}'') \sum_{y''} \frac{p(y''|x)}{p_i(y''|\hat{x}'')} \frac{d}{d\beta} p_i(y''|\hat{x}'') \right) \cdot \exp\left\{ -\beta\, D_{x,\hat{x}''} \right\} \quad \text{(A51)}$$

Finally, for the encoder Step 1.8 we have

$$\frac{d}{d\beta} p_{i+1}(\hat{x}|x) = \frac{d}{d\beta} \left(\frac{p_i(\hat{x}) e^{-\beta D_{x,\hat{x}}}}{Z_i(x,\beta)} \right)$$

$$\stackrel{\text{(A50)}}{=} \frac{p_i(\hat{x}) e^{-\beta D_{x,\hat{x}}}}{Z_i(x,\beta)} \left[\frac{1}{p_i(\hat{x})} \frac{dp_i(\hat{x})}{d\beta} - \left(D_{x,\hat{x}} - \beta \sum_{y''} \frac{p(y''|x)}{p_i(y''|\hat{x})} \frac{d}{d\beta} p_i(y''|\hat{x}) \right) - \frac{1}{Z_i(x,\beta)} \frac{dZ_i(x,\beta)}{d\beta} \right]$$

$$\stackrel{\text{Step 1.8}}{=} p_{i+1}(\hat{x}|x) \cdot \left[\frac{1}{p_i(\hat{x})} \frac{dp_i(\hat{x})}{d\beta} - \left(D_{x,\hat{x}} - \beta \sum_{y''} \frac{p(y''|x)}{p_i(y''|\hat{x})} \frac{d}{d\beta} p_i(y''|\hat{x}) \right) - \frac{1}{Z_i(x,\beta)} \frac{dZ_i(x,\beta)}{d\beta} \right] \quad \text{(A52)}$$

Next, picking β and the inputs $\log p_i(y|\hat{x})$ and $\log p_i(\hat{x})$ of BA_β as our independent variables, we compose the differentiations above to obtain $D_\beta BA_\beta$ at an output coordinate. That is, we seek $\frac{d}{d\beta} \log p_{i+1}(y|\hat{x})$ and $\frac{d}{d\beta} \log p_{i+1}(\hat{x})$. By the chain rule, we trace the dependencies graph (A6) (Appendix B.1) backwards, from the output nodes $p_{i+1}(y|\hat{x})$ and $p_{i+1}(\hat{x})$ back to the input nodes. The derivatives of the latter with respect to β vanish, as these are our independent variables.

Starting with a decoder output coordinate,

$$\frac{d}{d\beta} \log p_{i+1}(y|\hat{x}) = \frac{1}{p_{i+1}(y|\hat{x})} \frac{d}{d\beta} p_{i+1}(y|\hat{x}) \stackrel{\text{(A48)}}{=} \frac{1}{p_{i+1}(y|\hat{x})} \sum_x p(y|x) \frac{d}{d\beta} p_{i+1}(x|\hat{x})$$

$$\stackrel{\text{(A47)}}{=} \frac{1}{p_{i+1}(y|\hat{x})} \sum_x p(y|x) \left[\frac{p(x)}{p_{i+1}(\hat{x})} \frac{dp_{i+1}(\hat{x}|x)}{d\beta} - \frac{p_{i+1}(\hat{x}|x) p(x)}{p_{i+1}(\hat{x})^2} \frac{dp_{i+1}(\hat{x})}{d\beta} \right]$$

$$\stackrel{\text{(A46)}}{=} \sum_x \frac{p(y|x) p(x)}{p_{i+1}(y|\hat{x}) p_{i+1}(\hat{x})} \left[\frac{dp_{i+1}(\hat{x}|x)}{d\beta} - \frac{p_{i+1}(\hat{x}|x)}{p_{i+1}(\hat{x})} \sum_{x'} p(x') \frac{dp_{i+1}(\hat{x}|x')}{d\beta} \right]$$

$$\stackrel{\text{(A52)}}{=} \sum_x \frac{p(y|x) p(x)}{p_{i+1}(y|\hat{x}) p_{i+1}(\hat{x})} \left\{ p_{i+1}(\hat{x}|x) \cdot \left[\frac{1}{p_i(\hat{x})} \frac{dp_i(\hat{x})}{d\beta} - \left(D_{x,\hat{x}} - \beta \sum_{y''} \frac{p(y''|x)}{p_i(y''|\hat{x})} \frac{dp_i(y''|\hat{x})}{d\beta} \right) - \frac{1}{Z_i(x,\beta)} \frac{dZ_i(x,\beta)}{d\beta} \right] \right.$$
$$\left. - \frac{p_{i+1}(\hat{x}|x)}{p_{i+1}(\hat{x})} \sum_{x'} p(x') \left(p_{i+1}(\hat{x}|x') \cdot \left[\frac{1}{p_i(\hat{x})} \frac{dp_i(\hat{x})}{d\beta} - \left(D_{x',\hat{x}} - \beta \sum_{y''} \frac{p(y''|x')}{p_i(y''|\hat{x})} \frac{dp_i(y''|\hat{x})}{d\beta} \right) - \frac{1}{Z_i(x',\beta)} \frac{dZ_i(x',\beta)}{d\beta} \right] \right) \right\} \quad \text{(A53)}$$

Since $p_i(y|\hat{x})$ and $p_i(\hat{x})$ are independent input variables, their derivatives with respect to the independent variable β vanish, yielding

$$-\sum_x \frac{p(y|x) p(x)}{p_{i+1}(y|\hat{x}) p_{i+1}(\hat{x})} \left\{ p_{i+1}(\hat{x}|x) \cdot \left[D_{x,\hat{x}} + \frac{1}{Z_i(x,\beta)} \frac{dZ_i(x,\beta)}{d\beta} \right] \right.$$
$$\left. - \frac{p_{i+1}(\hat{x}|x)}{p_{i+1}(\hat{x})} \sum_{x'} p(x') p_{i+1}(\hat{x}|x') \cdot \left[D_{x',\hat{x}} + \frac{1}{Z_i(x',\beta)} \frac{dZ_i(x',\beta)}{d\beta} \right] \right\} \quad \text{(A54)}$$

To complete the calculation at (A53), note that the same argument can be used for two of the three summands in (A51), reducing it to

$$\frac{dZ_i(x,\beta)}{d\beta} = -\sum_{\hat{x}''} p_i(\hat{x}'') D_{x,\hat{x}''} e^{-\beta D_{x,\hat{x}''}} \quad \text{(A55)}$$

since $p_i(y|\hat{x})$ and $p_i(\hat{x})$ are considered as independent variables. Therefore,

$$\frac{d}{d\beta}\log p_{i+1}(y|\hat{x}) \overset{\substack{(A54)\\(A55)}}{=} -\sum_x \frac{p(y|x)p(x)}{p_{i+1}(y|\hat{x})p_{i+1}(\hat{x})}\left\{ p_{i+1}(\hat{x}|x)\cdot\left[D_{x,\hat{x}} - \sum_{\hat{x}''}\left(\frac{p_i(\hat{x}'')}{Z_i(x,\beta)}e^{-\beta\,D_{x,\hat{x}''}}\right)D_{x,\hat{x}''}\right]\right.$$
$$\left. -\frac{p_{i+1}(\hat{x}|x)}{p_{i+1}(\hat{x})}\sum_{x'}p(x')p_{i+1}(\hat{x}|x')\cdot\left[D_{x',\hat{x}} - \sum_{\hat{x}''}\left(\frac{p_i(\hat{x}'')}{Z_i(x',\beta)}e^{-\beta\,D_{x',\hat{x}''}}\right)D_{x',\hat{x}''}\right]\right\}$$

$$\overset{\text{Step 1.8}}{=} -\sum_x \frac{p(y|x)p(x)}{p_{i+1}(y|\hat{x})p_{i+1}(\hat{x})}\left\{ p_{i+1}(\hat{x}|x)\cdot\left[D_{x,\hat{x}} - \sum_{\hat{x}''}p_{i+1}(\hat{x}''|x)D_{x,\hat{x}''}\right]\right.$$
$$\left. -\frac{p_{i+1}(\hat{x}|x)}{p_{i+1}(\hat{x})}\sum_{x'}p(x')p_{i+1}(\hat{x}|x')\cdot\left[D_{x',\hat{x}} - \sum_{\hat{x}''}p_{i+1}(\hat{x}''|x')D_{x',\hat{x}''}\right]\right\}$$

$$\overset{\substack{\text{Step 1.5}\\\text{Step 1.6}}}{=} \sum_x p_{i+1}(x|\hat{x})D_{x,\hat{x}} - \sum_x \frac{p(y|x)}{p_{i+1}(y|\hat{x})}p_{i+1}(x|\hat{x})D_{x,\hat{x}}$$
$$+ \sum_{x,\hat{x}''}\frac{p(y|x)}{p_{i+1}(y|\hat{x})}p_{i+1}(\hat{x}''|x)p_{i+1}(x|\hat{x})D_{x,\hat{x}''} - \sum_{x,\hat{x}''}p_{i+1}(\hat{x}''|x)p_{i+1}(x|\hat{x})D_{x,\hat{x}''}$$
$$= \sum_x\left[1-\frac{p(y|x)}{p_{i+1}(y|\hat{x})}\right]p_{i+1}(x|\hat{x})D_{x,\hat{x}} - \sum_{x,\hat{x}''}\left[1-\frac{p(y|x)}{p_{i+1}(y|\hat{x})}\right]p_{i+1}(\hat{x}''|x)p_{i+1}(x|\hat{x})D_{x,\hat{x}''} \tag{A56}$$

At the second equality to the bottom we started with the third summand, then with the first, and only then with the third and fourth summands. And so,

$$\frac{d}{d\beta}\log p_{i+1}(y|\hat{x}) = \sum_{x,\hat{x}''}\left[1-\frac{p(y|x)}{p_{i+1}(y|\hat{x})}\right]\cdot\left[\delta_{\hat{x},\hat{x}''} - p_{i+1}(\hat{x}''|x)\right]\cdot p_{i+1}(x|\hat{x})D_{x,\hat{x}''} \tag{A57}$$

Next, consider a cluster marginal output coordinate,

$$\frac{d}{d\beta}\log p_{i+1}(\hat{x}) = \frac{1}{p_{i+1}(\hat{x})}\frac{d}{d\beta}p_{i+1}(\hat{x}) \overset{(A46)}{=} \frac{1}{p_{i+1}(\hat{x})}\sum_x p(x)\frac{d}{d\beta}p_{i+1}(\hat{x}|x)$$
$$\overset{(A52)}{=} \frac{1}{p_{i+1}(\hat{x})}\sum_x p(x)p_{i+1}(\hat{x}|x)\cdot\left[\frac{1}{p_i(\hat{x})}\frac{dp_i(\hat{x})}{d\beta} - \left(D_{x,\hat{x}} - \beta\sum_{y''}\frac{p(y''|x)}{p_i(y''|\hat{x})}\frac{d}{d\beta}p_i(y''|\hat{x})\right) - \frac{1}{Z_i(x,\beta)}\frac{dZ_i(x,\beta)}{d\beta}\right] \tag{A58}$$

Since $p_i(y|\hat{x})$ and $p_i(\hat{x})$ are independent variables, their derivatives with respect to β vanish, yielding

$$-\frac{1}{p_{i+1}(\hat{x})}\sum_x p_{i+1}(\hat{x}|x)p(x)\left[D_{x,\hat{x}} + \frac{1}{Z_i(x,\beta)}\frac{dZ_i(x,\beta)}{d\beta}\right]$$
$$\overset{(A55)}{=} -\frac{1}{p_{i+1}(\hat{x})}\sum_x p_{i+1}(\hat{x}|x)p(x)\left[D_{x,\hat{x}} - \sum_{\hat{x}''}\left(\frac{p_i(\hat{x}'')}{Z_i(x,\beta)}e^{-\beta\,D_{x,\hat{x}''}}\right)D_{x,\hat{x}''}\right]$$
$$\overset{\substack{\text{Step 1.8}\\\text{Step 1.5}}}{=} -\sum_{x,\hat{x}''}\left[\delta_{\hat{x},\hat{x}''} - p_{i+1}(\hat{x}''|x)\right]\cdot p_{i+1}(x|\hat{x})D_{x,\hat{x}''} \tag{A59}$$

Thus, for the marginals' coordinates, we have obtained

$$\frac{d}{d\beta}\log p_{i+1}(\hat{x}) = -\sum_{x,\hat{x}''}\left[\delta_{\hat{x},\hat{x}''} - p_{i+1}(\hat{x}''|x)\right]\cdot p_{i+1}(x|\hat{x})D_{x,\hat{x}''} \tag{A60}$$

When evaluated at an IB root, Equations (A57) and (A60) form, respectively, the decoder and marginal coordinates of $D_\beta BA_\beta$, which appears at the right-hand side of the IB ODE (16) (note the extra minus sign in the implicit ODE (7)).

*Appendix B.4. The Coordinate Exchange Jacobians between Log-Decoder
and Log-Encoder Coordinates*

Following the discussion in Section 2 on the pros and cons of each coordinate system,
we leverage the observations of Appendix B.1 in order to derive the coordinate exchange
Jacobians between the log-decoder and log-encoder coordinate systems. Exchanging
between the other coordinate system pairs adds little to the below and thus is omitted.

Given the encoder's logarithmic derivative $\frac{d}{d\beta} \log p_\beta(\hat{x}'|x')$, we would like to compute
from it the logarithmic derivative $\left(\frac{d}{d\beta} \log p_\beta(y|\hat{x}), \frac{d}{d\beta} \log p_\beta(\hat{x})\right)$ in decoder coordinates,
and vice versa. To that end, recall that an (arbitrary) encoder $p(\hat{x}'|\hat{x})$ determines a decoder–
marginal pair $\left(p(y|\hat{x}), p(\hat{x})\right)$ and vice versa (e.g., Equation (11) in Section 2). So, one can
follow the dependencies graph (A6) (in Appendix B.1) backward between these coordinate
systems to exchange the coordinates of an implicit derivative. For example, consider $p_i(y|\hat{x})$
and $p_i(\hat{x})$ as functions of the encoder $p_i(\hat{x}'|x')$ preceding it in the graph (A6). When at an
IB root, multiplying by the coordinates' exchange Jacobian yields

$$\frac{d\log p_\beta(y|\hat{x})}{d\beta} = \frac{d \log p_\beta(y|\hat{x})}{d \log p_\beta(\hat{x}'|x')} \frac{d\log p_\beta(\hat{x}'|x')}{d\beta} \quad \text{and} \tag{A61}$$

$$\frac{d\log p_\beta(\hat{x})}{d\beta} = \frac{d \log p_\beta(\hat{x})}{d \log p_\beta(\hat{x}'|x')} \frac{d\log p_\beta(\hat{x}'|x')}{d\beta}. \tag{A62}$$

Similarly, considering an encoder $p_\beta(\hat{x}|x)$ as a function of $p_\beta(y'|\hat{x}')$ and $p_\beta(\hat{x}')$,

$$\frac{d\log p_\beta(\hat{x}|x)}{d\beta} = \frac{d \log p_\beta(\hat{x}|x)}{d \log p_\beta(y'|\hat{x}')} \frac{d\log p_\beta(y'|\hat{x}')}{d\beta} + \frac{d \log p_\beta(\hat{x}|x)}{d \log p_\beta(\hat{x}')} \frac{d\log p_\beta(\hat{x}')}{d\beta} + \frac{\partial\log p_\beta(\hat{x}|x)}{\partial\beta}. \tag{A63}$$

The last term $\frac{\partial \log p_\beta(\hat{x}|x)}{\partial\beta}$ in (A63) stems from the fact that the encoder Step 1.8 depends
explicitly on β, unlike the marginal and decoder Steps 1.4 and 1.6. cf., the comments around
(A8) in Appendix B.1.

The matrices $\frac{d \log p_\beta(y|\hat{x})}{d \log p_\beta(\hat{x}'|x')}$ and $\frac{d \log p_\beta(\hat{x})}{d \log p_\beta(\hat{x}'|x')}$ for exchanging from encoder to decoder
coordinates follow from the chain rule, and are calculated in Appendix B.4.1 below,
at Equations (A64) and (A66). Similarly, the matrices $\frac{d \log p_\beta(\hat{x}|x)}{d \log p_\beta(y'|\hat{x}')}$ and $\frac{d \log p_\beta(\hat{x}|x)}{d \log p_\beta(\hat{x}')}$ and
the partial derivative $\frac{\partial \log p_\beta(\hat{x}|x)}{\partial\beta}$ for exchanging from decoder to encoder coordinates are
Equations (A68), (A70), and (A73), in Appendix B.4.2.

Appendix B.4.1. Exchanging from Encoder to Decoder Coordinates

An input encoder $p_i(\hat{x}'|x')$ determines a decoder $p_i(y|\hat{x})$ and a marginal $p_i(\hat{x})$. As
in previous subsections, we follow the dependencies graph (A6) along all the paths
between these.

Using diagram (A24) from Appendix B.1.1, for the marginal one has

$$\frac{d \log p_i(\hat{x})}{d \log p_i(\hat{x}'|x')} = p_i(x'|\hat{x}')\, \delta_{\hat{x},\hat{x}'}, \tag{A64}$$

while for the decoder,

$$\begin{aligned}
\frac{d \log p_i(y|\hat{x})}{d \log p_i(\hat{x}'|x')} &= \frac{\partial \log p_i(y|\hat{x})}{\partial \log p_i(x_1|\hat{x}_2)} \left[\frac{\partial \log p_i(x_1|\hat{x}_2)}{\partial \log p_i(\hat{x}'|x')} + \frac{\partial \log p_i(x_1|\hat{x}_2)}{\partial \log p_i(\hat{x}_3)} \frac{\partial \log p_i(\hat{x}_3)}{\partial \log p_i(\hat{x}'|x')} \right] \\
&= \frac{p(y|x_1) p_i(x_1|\hat{x}_2)}{p_i(y|\hat{x}_2)} \delta_{\hat{x}_2,\hat{x}} \left[\delta_{x_1,x'}\, \delta_{\hat{x}_2,\hat{x}'} - \delta_{\hat{x}_2,\hat{x}_3}\, p_i(x'|\hat{x}_3)\, \delta_{\hat{x}_3,\hat{x}'} \right]
\end{aligned} \tag{A65}$$

Summing over the three dummy variables as before, the latter simplifies to

$$\frac{d \log p_i(y|\hat{x})}{d \log p_i(\hat{x}'|x')} = \left[\frac{p(y|x')}{p_i(y|\hat{x})} - 1\right] p_i(x'|\hat{x}) \, \delta_{\hat{x},\hat{x}'} \, . \tag{A66}$$

Appendix B.4.2. Exchanging from Decoder to Encoder Coordinates

In the other way around, a decoder $p_i(y'|\hat{x}')$ and a marginal $p_i(\hat{x}')$ determine the subsequent encoder $p_{i+1}(\hat{x}|x)$. Using diagram (A24), one has

$$\begin{aligned}
\frac{d \log p_{i+1}(\hat{x}|x)}{d \log p_i(y'|\hat{x}')} &= \frac{\partial \log p_{i+1}(\hat{x}|x)}{\partial \log p_i(y'|\hat{x}')} + \frac{\partial \log p_{i+1}(\hat{x}|x)}{\partial \log Z_i(x_1)} \frac{\partial \log Z_i(x_1)}{\partial \log p_i(y'|\hat{x}')} \\
&= \beta \, \delta_{\hat{x},\hat{x}'} \, p(y'|x) - \delta_{x,x_1} \, \beta \, p_{i+1}(\hat{x}'|x_1) p(y'|x_1)
\end{aligned} \tag{A67}$$

Summing over the dummy variable x_1, this is the coordinates' exchange Jacobian $J_{\text{dec}}^{\text{enc}}$ mentioned in Section 2:

$$\frac{d \log p_{i+1}(\hat{x}|x)}{d \log p_i(y'|\hat{x}')} = \beta \, p(y'|x) \left[\delta_{\hat{x},\hat{x}'} - p_{i+1}(\hat{x}'|x)\right] \tag{A68}$$

Next, for the derivative with respect to the marginal,

$$\begin{aligned}
\frac{d \log p_{i+1}(\hat{x}|x)}{d \log p_i(\hat{x}')} =& \frac{\partial \log p_{i+1}(\hat{x}|x)}{\partial \log p_i(\hat{x}')} + \frac{\partial \log p_{i+1}(\hat{x}|x)}{\partial \log Z_i(x_1)} \frac{\partial \log Z_i(x_1)}{\partial \log p_i(\hat{x}')} \\
&+ \left[\frac{\partial \log p_{i+1}(\hat{x}|x)}{\partial \log Z_i(x_1)} \frac{\partial \log Z_i(x_1)}{\partial \log p_i(y_2|\hat{x}_3)} + \frac{\partial \log p_{i+1}(\hat{x}|x)}{\partial \log p_i(y_2|\hat{x}_3)}\right] \frac{\partial \log p_i(y_2|\hat{x}_3)}{\partial \log p_i(x_4|\hat{x}_5)} \frac{\partial \log p_i(x_4|\hat{x}_5)}{\partial \log p_i(\hat{x}')} \\
=& \delta_{\hat{x},\hat{x}'} - \delta_{x,x_1} \, p_{i+1}(\hat{x}'|x_1) \\
&+ \left[- \delta_{x,x_1} \, \beta \, p_{i+1}(\hat{x}_3|x_1) p(y_2|x_1) + \beta \, \delta_{\hat{x}_3,\hat{x}} \, p(y_2|x)\right] \frac{p(y_2|x_4) p_i(x_4|\hat{x}_3)}{p_i(y_2|\hat{x}_3)} \, \delta_{\hat{x}_3,\hat{x}_5} \cdot (-\delta_{\hat{x}_5,\hat{x}'})
\end{aligned} \tag{A69}$$

Summing over the five dummy variables, this is the coordinates' exchange Jacobian $J_{\text{mrg}}^{\text{enc}}$ from Section 2:

$$\frac{d \log p_{i+1}(\hat{x}|x)}{d \log p_i(\hat{x}')} = (1 - \beta) \left[\delta_{\hat{x},\hat{x}'} - p_{i+1}(\hat{x}'|x)\right] \tag{A70}$$

Finally, note that the encoder Step 1.8 depends on β explicitly, rather than indirectly only via its other variables. So, to calculate the partial derivative term $\frac{\partial \log p_{i+1}(\hat{x}|x)}{\partial \beta}$ in (A63), write as follows for $\log Z$:

$$\begin{aligned}
\frac{\partial}{\partial \beta} Z_i(x, \beta) &\overset{\text{Step 1.7}}{=} \sum_{\hat{x}} p_i(\hat{x}) \frac{\partial}{\partial \beta} \exp\left\{-\beta \, D_{KL}\left[p(y|x)||p_i(y|\hat{x})\right]\right\} \\
&= -\sum_{\hat{x}} p_i(\hat{x}) D_{KL}\left[p(y|x)||p_i(y|\hat{x})\right] \exp\left\{-\beta \, D_{KL}\left[p(y|x)||p_i(y|\hat{x})\right]\right\}
\end{aligned} \tag{A71}$$

Thus,

$$\begin{aligned}
\frac{\partial}{\partial \beta} \log Z_i(x, \beta) &= \frac{1}{Z_i(x, \beta)} \frac{\partial}{\partial \beta} Z_i(x, \beta) \\
&\overset{(A71)}{=} -\sum_{\hat{x}} \frac{p_i(\hat{x}) \exp\left\{-\beta \, D_{KL}\left[p(y|x)||p_i(y|\hat{x})\right]\right\}}{Z_i(x, \beta)} D_{KL}\left[p(y|x)||p_i(y|\hat{x})\right] \\
&\overset{\text{Step 1.8}}{=} -\sum_{\hat{x}} p_{i+1}(\hat{x}|x) D_{KL}\left[p(y|x)||p_i(y|\hat{x})\right] \, .
\end{aligned} \tag{A72}$$

And so, from the encoder Step 1.8, we have

$$\frac{\partial \log p_{i+1}(\hat{x}|x)}{\partial \beta} = \frac{\partial \log p_i(\hat{x})}{\partial \beta} - \frac{\partial \log Z_i(x,\beta)}{\partial \beta} - \frac{\partial}{\partial \beta}\left(\beta D_{KL}\left[p(y|x)||p_i(y|\hat{x})\right]\right)$$

$$\overset{(A72)}{=} \sum_{\hat{x}''} p_{i+1}(\hat{x}''|x) D_{KL}\left[p(y|x)||p_i(y|\hat{x}'')\right] - D_{KL}\left[p(y|x)||p_i(y|\hat{x})\right] \quad \text{(A73)}$$

where the term $\frac{\partial \log p_i(\hat{x})}{\partial \beta}$ vanishes since it is considered as an independent variable here..

Appendix C. Proof of Lemma 1, on the Kernel of the Jacobian of the IB Operator in Log-Decoder Coordinates

We prove Lemma 1 from Section 3, using the results of Appendix B.

In the first direction, suppose that $\left((v_{y,\hat{x}})_{y,\hat{x}}, (u_{\hat{x}})_{\hat{x}}\right)$ is a vector in the *left* kernel of the Jacobian of the IB operator (5) in log-decoder coordinates, $I - D_{\log p(y|\hat{x}),\log p(\hat{x})} B A_\beta$, as in (16) in Section 3. Using the Jacobian's implicit form (A25) (Appendix B.2), this is to say that

$$v_{y',\hat{x}'} = \sum_{y,\hat{x}} v_{y,\hat{x}} \frac{d\log p_{i+1}(y|\hat{x})}{d\log p_i(y'|\hat{x}')} + \sum_{\hat{x}} u_{\hat{x}} \frac{d\log p_{i+1}(\hat{x})}{d\log p_i(y'|\hat{x}')} \quad \text{and} \quad \text{(A74)}$$

$$u_{\hat{x}'} = \sum_{y,\hat{x}} v_{y,\hat{x}} \frac{d\log p_{i+1}(y|\hat{x})}{d\log p_i(\hat{x}')} + \sum_{\hat{x}} u_{\hat{x}} \frac{d\log p_{i+1}(\hat{x})}{d\log p_i(\hat{x}')} \quad \text{(A75)}$$

hold, for every y' and $\hat{x}'$. We spell out and manipulate these equations to obtain the desired result.

By the Jacobian's explicit form (A44) from Appendix B.2, Equation (A74) spells out as

$$v_{y',\hat{x}'} = \beta \cdot \sum_{y,\hat{x}} v_{y,\hat{x}} \sum_{\hat{x}'',y''} \left(\delta_{\hat{x}'',\hat{x}'} - \delta_{\hat{x},\hat{x}'}\right) \cdot \left(1 - \frac{\delta_{y'',y}}{p_{i+1}(y|\hat{x})}\right) C(\hat{x},\hat{x}'';i+1)_{y',y''}$$

$$+ \beta \cdot \sum_{\hat{x}} u_{\hat{x}}\left[\delta_{\hat{x},\hat{x}'}\, p_{i+1}(y'|\hat{x}) - B(\hat{x},\hat{x}';i+1)_{y'}\right], \quad \text{(A76)}$$

while the second Equation (A75) spells out as

$$u_{\hat{x}'} = (1-\beta) \cdot \sum_{y,\hat{x}} v_{y,\hat{x}} \sum_{y''} \left[1 - \frac{\delta_{y'',y}}{p_{i+1}(y|\hat{x})}\right] B(\hat{x},\hat{x}';i+1)_{y''}$$

$$+ (1-\beta) \cdot \sum_{\hat{x}} u_{\hat{x}}\left(\delta_{\hat{x},\hat{x}'} - A(\hat{x},\hat{x}';i+1)\right). \quad \text{(A77)}$$

Next, we expand and simplify each of the terms in (A76) and (A77), using the definition (A32) of A, B, and C from Appendix B.2.

For the first summand to the right of (A76),

$$\beta \cdot \sum_{y,\hat{x}} v_{y,\hat{x}} \sum_{\hat{x}'',y''} \left(\delta_{\hat{x}'',\hat{x}'} - \delta_{\hat{x},\hat{x}'}\right)\left(1 - \frac{\delta_{y'',y}}{p_{i+1}(y|\hat{x})}\right) C(\hat{x},\hat{x}'';i+1)_{y',y''}$$

$$\overset{(A32)}{=} \beta \cdot \sum_{y,\hat{x}} v_{y,\hat{x}} \sum_{\hat{x}'',y''} \left(\delta_{\hat{x}'',\hat{x}'} - \delta_{\hat{x},\hat{x}'}\right)\left(1 - \frac{\delta_{y'',y}}{p_{i+1}(y|\hat{x})}\right) \sum_{x} p(y'|x) p(y''|x) p_{i+1}(\hat{x}''|x) p_{i+1}(x|\hat{x}) \quad \text{(A78)}$$

We simplify each of the four addends to the right of (A78) while temporarily ignoring the β coefficient. For the $\delta_{\hat{x}'',\hat{x}'} \cdot 1$ term,

$$\sum_{y,\hat{x}} v_{y,\hat{x}} \sum_{\hat{x}'',y''} \delta_{\hat{x}'',\hat{x}'} \sum_{x} p(y'|x)p(y''|x)p_{i+1}(\hat{x}''|x)p_{i+1}(x|\hat{x})$$

$$= \sum_{y,\hat{x}} v_{y,\hat{x}} \sum_{x} p(y'|x)p_{i+1}(\hat{x}'|x)p_{i+1}(x|\hat{x}) \quad \text{(A79)}$$

For the $-\delta_{\hat{x}'',\hat{x}'} \cdot \frac{\delta_{y'',y}}{p_{i+1}(y|\hat{x})}$ term,

$$-\sum_{y,\hat{x}} v_{y,\hat{x}} \sum_{\hat{x}'',y''} \delta_{\hat{x}'',\hat{x}'}\delta_{y'',y} \sum_{x} \frac{1}{p_{i+1}(y|\hat{x})}\, p(y'|x)p(y''|x)p_{i+1}(\hat{x}''|x)p_{i+1}(x|\hat{x})$$

$$= -\sum_{y,\hat{x}} v_{y,\hat{x}} \sum_{x} \frac{1}{p_{i+1}(y|\hat{x})}\, p(y'|x)p(y|x)p_{i+1}(\hat{x}'|x)p_{i+1}(x|\hat{x}) \quad \text{(A80)}$$

For the $-\delta_{\hat{x},\hat{x}'} \cdot 1$ term,

$$-\sum_{y,\hat{x}} v_{y,\hat{x}} \sum_{\hat{x}'',y''} \delta_{\hat{x},\hat{x}'} \sum_{x} p(y'|x)p(y''|x)p_{i+1}(\hat{x}''|x)p_{i+1}(x|\hat{x})$$

$$= -\sum_{y,\hat{x}'} v_{y,\hat{x}'} \sum_{x} p(y'|x)p_{i+1}(x|\hat{x}') = -\sum_{y} v_{y,\hat{x}'}\, p_{i+1}(y'|\hat{x}') \quad \text{(A81)}$$

And for the last $-\delta_{\hat{x},\hat{x}'} \cdot \frac{-\delta_{y'',y}}{p_{i+1}(y|\hat{x})}$ term,

$$\sum_{y,\hat{x}} v_{y,\hat{x}} \sum_{\hat{x}'',y''} \delta_{\hat{x},\hat{x}'} \cdot \frac{\delta_{y'',y}}{p_{i+1}(y|\hat{x})} \sum_{x} p(y'|x)p(y''|x)p_{i+1}(\hat{x}''|x)p_{i+1}(x|\hat{x})$$

$$= \sum_{y} \frac{v_{y,\hat{x}'}}{p_{i+1}(y|\hat{x}')} \sum_{x} p(y'|x)p(y|x)p_{i+1}(x|\hat{x}') \quad \text{(A82)}$$

Collecting (A79)–(A82) back into (A78), we obtain

$$\beta \cdot \sum_{y,\hat{x}} v_{y,\hat{x}} \sum_{x} p(y'|x)p_{i+1}(\hat{x}'|x)p_{i+1}(x|\hat{x}) \left[1 - \frac{p(y|x)}{p_{i+1}(y|\hat{x})}\right]$$

$$+ \beta \cdot \sum_{y} v_{y,\hat{x}'} \frac{1}{p_{i+1}(y|\hat{x}')} \sum_{x} p(y|x)p(y'|x)p_{i+1}(x|\hat{x}') - \beta \cdot p_{i+1}(y'|\hat{x}') \sum_{y} v_{y,\hat{x}'} \quad \text{(A83)}$$

for the first summand to the right of (A76).

The second summand to the right of (A76) equals

$$\beta \cdot \sum_{\hat{x}} u_{\hat{x}} \left[\delta_{\hat{x},\hat{x}'}\, p_{i+1}(y'|\hat{x}) - B(\hat{x},\hat{x}';i+1)_{y'}\right]$$

$$\overset{\text{(A32)}}{=} \beta \cdot u_{\hat{x}'}\, p_{i+1}(y'|\hat{x}') - \beta \cdot \sum_{x} p(y'|x)p_{i+1}(\hat{x}'|x) \sum_{\hat{x}} u_{\hat{x}}\, p_{i+1}(x|\hat{x}) \quad \text{(A84)}$$

Combining (A83) and (A84), Equation (A76) is equivalent to

$$\frac{1}{\beta} \cdot v_{y',\hat{x}'} + p_{i+1}(y'|\hat{x}') \sum_{y} v_{y,\hat{x}'} - u_{\hat{x}'}\, p_{i+1}(y'|\hat{x}')$$

$$= \sum_{y,\hat{x}} v_{y,\hat{x}} \sum_{x} p(y'|x)p_{i+1}(\hat{x}'|x)p_{i+1}(x|\hat{x}) \left[1 - \frac{p(y|x)}{p_{i+1}(y|\hat{x})}\right]$$

$$+ \sum_{y} v_{y,\hat{x}'} \sum_{x} p(y'|x)p_{i+1}(x|\hat{x}') \frac{p(y|x)}{p_{i+1}(y|\hat{x}')} - \sum_{x} p(y'|x)p_{i+1}(\hat{x}'|x) \sum_{\hat{x}} u_{\hat{x}}\, p_{i+1}(x|\hat{x}) \quad \text{(A85)}$$

for any y' and $\hat{x}'$. Summing (A85) over y' and simplifying, we obtain

$$\frac{1}{\beta} \cdot \sum_y v_{y,\hat{x}'} - u_{\hat{x}'}$$

$$= \sum_{y,\hat{x}} v_{y,\hat{x}} \sum_x p_{i+1}(\hat{x}'|x) p_{i+1}(x|\hat{x}) \left[1 - \frac{p(y|x)}{p_{i+1}(y|\hat{x})} \right] - \sum_x p_{i+1}(\hat{x}'|x) \sum_{\hat{x}} u_{\hat{x}}\, p_{i+1}(x|\hat{x}) \quad (A86)$$

for any $\hat{x}'$.

Next, we expand and simplify Equation (A77). Using the definition (A32) of B, the first summand to its right can be written as

$$(1-\beta) \cdot \sum_{y,\hat{x}} v_{y,\hat{x}} \sum_x p_{i+1}(\hat{x}'|x) p_{i+1}(x|\hat{x}) \left[1 - \frac{p(y|x)}{p_{i+1}(y|\hat{x})} \right] . \quad (A87)$$

Similarly, the second summand to the right of (A77) can be written as

$$(1-\beta) \cdot \left[u_{\hat{x}'} - \sum_x p_{i+1}(\hat{x}'|x) \sum_{\hat{x}} u_{\hat{x}}\, p_{i+1}(x|\hat{x}) \right] . \quad (A88)$$

Combining (A87) and (A88), Equation (A77) can now be written explicitly,

$$\frac{\beta}{1-\beta} \cdot u_{\hat{x}'} = \sum_{y,\hat{x}} v_{y,\hat{x}} \sum_x p_{i+1}(\hat{x}'|x) p_{i+1}(x|\hat{x}) \left[1 - \frac{p(y|x)}{p_{i+1}(y|\hat{x})} \right] - \sum_x p_{i+1}(\hat{x}'|x) \sum_{\hat{x}} u_{\hat{x}}\, p_{i+1}(x|\hat{x}) \quad (A89)$$

for every $\hat{x}'$. Next, subtracting (A89) from (A86), we obtain

$$u_{\hat{x}} = \frac{1-\beta}{\beta} \cdot \sum_y v_{y,\hat{x}} \quad (A90)$$

for any $\hat{x}$.

Substituting (A90) into (A85) and using the decoder Step 1.6 to expand $p_{i+1}(y'|\hat{x}')$ there,

$$\frac{1}{\beta} \cdot v_{y',\hat{x}'} = \sum_{y,\hat{x}} v_{y,\hat{x}} \sum_x p_{i+1}(\hat{x}'|x) p(y'|x) p_{i+1}(x|\hat{x}) \left[\frac{2\beta-1}{\beta} - \frac{p(y|x)}{p_{i+1}(y|\hat{x})} \right]$$

$$- \sum_y v_{y,\hat{x}'} \sum_x p(y'|x) p_{i+1}(x|\hat{x}') \cdot \frac{2\beta-1}{\beta} + \sum_y v_{y,\hat{x}'} \sum_x p(y'|x) p_{i+1}(x|\hat{x}') \frac{p(y|x)}{p_{i+1}(y|\hat{x}')} \quad (A91)$$

Next, inserting $\sum_{\hat{x}} \delta_{\hat{x},\hat{x}'}$ into the sums on the last line,

$$\frac{1}{\beta} \cdot v_{y',\hat{x}'} = \sum_{y,\hat{x}} v_{y,\hat{x}} \sum_x p_{i+1}(\hat{x}'|x) p(y'|x) p_{i+1}(x|\hat{x}) \left[\frac{2\beta-1}{\beta} - \frac{p(y|x)}{p_{i+1}(y|\hat{x})} \right]$$

$$- \sum_{y,\hat{x}} v_{y,\hat{x}} \sum_x \delta_{\hat{x},\hat{x}'}\, p(y'|x) p_{i+1}(x|\hat{x}) \left[\frac{2\beta-1}{\beta} - \frac{p(y|x)}{p_{i+1}(y|\hat{x})} \right] \quad (A92)$$

Finally, this simplifies to

$$v_{y',\hat{x}'} = \sum_{y,\hat{x}} v_{y,\hat{x}} \sum_x p(y'|x) \left[\delta_{\hat{x},\hat{x}'} - p_{i+1}(\hat{x}'|x) \right] p_{i+1}(x|\hat{x}) \left[\beta \cdot \frac{p(y|x)}{p_{i+1}(y|\hat{x})} + (1-2\beta) \right] \quad (A93)$$

The latter is to say that $\left(v_{y,\hat{x}} \right)_{y,\hat{x}}$ is a left-eigenvector of the eigenvalue 1 of the matrix to the right. At an IB root, this is precisely the matrix S (17) from the Lemma's statement, as desired.

As a side note, we comment that Equations (A74) and (A75) also imply

$$\forall y \;\; \sum_{\hat{x}} v_{y,\hat{x}} = 0 \quad \text{and} \quad \sum_{\hat{x}} u_{\hat{x}} = 0\,, \quad (A94)$$

which can be seen by summing (A85) and (A89), respectively, over $\hat{x}'$, and simplifying.

In the other direction, let $v := (v_{y,\hat{x}})_{y,\hat{x}}$ be a left-eigenvector of the eigenvalue 1 of S (17). That is, assume that Equation (A93) holds. Define a vector $u := (u_{\hat{x}})_{\hat{x}}$ by Equation (A90). Reversing the algebra, (A93) is equivalent to (A91). Substituting (A90) into the latter yields back (A85), which is equivalent to the explicit form (A76) of Equation (A74). Next, summing (A85) over y' and simplifying yields (A86). Adding the latter to (A90) yields back (A89), which is equivalent to Equation (A77), the explicit form of (A75). To conclude, both of the Equations (A74) and (A75) hold, as claimed.

Appendix D. Approximate Error Analysis for Deterministic Annealing and for Euler's Method with BA

Complementing the results of Section 4, we provide an approximate error analysis for two computation methods for the IB: deterministic annealing and Euler's method combined with a fixed number of BA iterations.

First, we recap the linearization argument around Equation (10) in [4]. Denote repeated BA iterations initialized at p_0 by

$$p_{k+1} := BA_\beta[p_k] . \tag{A95}$$

Linearizing around a fixed-point p_β of BA,

$$BA[p_k] \simeq p_\beta + D\, BA_\beta|_{p_\beta} \cdot (p_k - p_\beta) , \tag{A96}$$

where $D\, BA_\beta|_{p_\beta}$ denotes the Jacobian matrix of BA_β evaluated at p_β. Rewriting in terms of the error $\delta p_k := p_k - p_\beta$ of the k-th iterate,

$$\delta p_{k+1} \simeq D\, BA_\beta|_{p_\beta} \cdot \delta p_k . \tag{A97}$$

Thus, to first order, repeated applications of BA_β reduce the initial error according to

$$\|\delta p_{k+1}\| \simeq \left\| \left(D\, BA_\beta|_{p_\beta} \right)^k \cdot \delta p_0 \right\| . \tag{A98}$$

Next, consider $k > 0$ applications of $BA_{\beta+\Delta\beta}$ to a root p_β at β. This is similar to deterministic annealing, but with a capped number of BA iterations. Plugging the initial error $\delta p_0 := p_\beta - p_{\beta+\Delta\beta} \simeq -\Delta\beta\, \frac{dp}{d\beta}|_\beta$ into Equation (A98) shows that this method is of the first order:

$$\|\delta p_{k+1}\| \simeq |\Delta\beta| \cdot \left\| \left(D\, BA_{\beta+\Delta\beta}|_{p_{\beta+\Delta\beta}} \right)^k \frac{dp}{d\beta}|_\beta \right\| . \tag{A99}$$

Finally, we combine BA with Euler's method for the IB, Equation (22). Consider $k > 0$ applications of $BA_{\beta+\Delta\beta}$ to the approximation $p_\beta + \Delta\beta\, \frac{dp}{d\beta}|_\beta$ produced by an Euler method step. Its initial error is

$$\delta p_0 := p_\beta + \Delta\beta\, \frac{dp}{d\beta}|_\beta - p_{\beta+\Delta\beta} = -\tfrac{1}{2}(\Delta\beta)^2\, \frac{d^2 p}{d\beta^2}|_{\beta'} , \tag{A100}$$

where the last equality follows from the second-order expansion $p_{\beta+\Delta\beta} = p_\beta + \Delta\beta\, \frac{dp}{d\beta}|_\beta + \tfrac{1}{2}(\Delta\beta)^2\, \frac{d^2 p}{d\beta^2}|_{\beta'}$, with $\beta' \in [\beta, \beta + \Delta\beta]$. Similar to before, plugging this into Equation (A98) shows that this method is of the second order:

$$\|\delta p_{k+1}\| \simeq \tfrac{1}{2}|\Delta\beta|^2 \cdot \left\| \left(D\, BA_{\beta+\Delta\beta}|_{p_{\beta+\Delta\beta}} \right)^k \frac{d^2 p}{d\beta^2}|_{\beta'} \right\| . \tag{A101}$$

Appendix E. An Exact Solution for a Binary Symmetric Channel

Define an IB problem by $Y \sim \mathrm{Bernoulli}(\frac{1}{2})$ and $X := Y \oplus Z$ for $Z \sim \mathrm{Bernoulli}(\alpha)$ independent of Y, $0 < \alpha < \frac{1}{2}$, where $\oplus$ denotes addition modulo 2. Explicitly, it is given by $p_{Y|X} = \begin{pmatrix} 1-\alpha & \alpha \\ \alpha & 1-\alpha \end{pmatrix}$ and $p_X = (\frac{1}{2}, \frac{1}{2})$. We synthesize exact solutions for this problem using Mrs. Gerber's Lemma [38] and by following [2].

Let $h(p) := -p \log p - (1-p) \log(1-p)$ be the binary entropy, with $h(0) := h(1) := 0$. It is injective on $[0, \frac{1}{2}]$, with a maximal value of $\log 2$ at $p = \frac{1}{2}$. So, its inverse function h^{-1} is well-defined on $[0, \log 2]$. Given a constraint $I_X \in [0, \log 2]$ on $I(\hat{X}; X)$, $I(\hat{X}; X) \leq I_X$, define a random variable $V \sim \mathrm{Bernoulli}(\delta)$ and set $\hat{X} := X \oplus V$, where δ is defined by $h(\delta) = \log 2 - I_X$ or equivalently in terms of h^{-1} by $\delta := h^{-1}(\log 2 - I_X)$. Explicitly, $p(\hat{x}|x) = \begin{pmatrix} 1-\delta & \delta \\ \delta & 1-\delta \end{pmatrix}$, with its rows indexed by $\hat{x}$ and columns by x. $\hat{X}$ is also a Bernoulli$(\frac{1}{2})$ variable since X is, and so

$$I(\hat{X}; X) = H(\hat{X}) - H(\hat{X}|X) = \log 2 - h(\delta) = I_X, \tag{A102}$$

showing that the constraint on $I(\hat{X}; X)$ holds. The chain $\hat{X} \to X \to Y$ of random variables is readily seen to be Markov. By Corollary 4 in [38], it follows that $I(\hat{X}; Y) \leq \log 2 - h(\alpha * \delta)$, where $a * b := a(1-b) + b(1-a)$. Finally, equality follows by Theorem 1 there. Thus, the above $p(\hat{x}|x)$ is IB-optimal.

The above defines an IB solution $p(\hat{x}|x)$ as a function of I_X. However, our numerical computations are phrased in terms of the IB's Lagrange multiplier β. To that end, note that [2] (Section IV.A) show that

$$\beta \cdot (1 - 2\alpha) \log \frac{1 - \alpha * \delta}{\alpha * \delta} = \log \frac{1 - \delta}{\delta}, \tag{A103}$$

and that the bifurcation of this problem occurs at

$$\beta_c = \frac{1}{(1 - 2\alpha)^2}. \tag{A104}$$

To conclude, we have $\beta = \beta(\delta)$ as a function of δ, $\delta = \delta(I_X)$ as a function of I_X, and the encoder $p(\hat{x}|x)$ as a function of δ. These functional dependencies are summarized as follows:

$$p(\hat{x}|x) \longrightarrow \delta \tag{A105}$$

where the variable at the tail of each arrow is a function of that at its head.

Writing $p = \big(p(\hat{x}|x)\big)_{\hat{x},x}$, its derivative with respect to β can be calculated by the chain rule:

$$\frac{dp}{d\beta} = \frac{d}{d\beta}\Big(p\big(\beta^{-1}(\delta)\big)\Big) = \frac{dp}{d\delta}\left(\frac{d\beta}{d\delta}\right)^{-1}, \tag{A106}$$

where we have applied the derivative of an inverse function $(f^{-1})' = 1/f'$ to $\beta(\delta)$ in (A103), to differentiate $\delta(\beta)$. From the argument around (A102), $\frac{dp}{d\delta} = \begin{pmatrix} -1 & 1 \\ 1 & -1 \end{pmatrix}$. While this yields an analytical expression for the derivative $\frac{dp}{d\beta}$, both of the terms to the right of (A106) are evaluated at $\delta(\beta)$, for a given β value. Although it is straightforward to compute $\delta(\beta)$ numerically from (A103), this entails numerical error, especially as δ approaches $1/2$ near the bifurcation. For the solution with respect to decoder coordinates, an immediate application of the Bayes rule shows that

$$p(\hat{x}) = \frac{1}{2} \quad \text{and} \quad p(y|\hat{x}) = \begin{pmatrix} \alpha * (1 - \delta) & \alpha * \delta \\ \alpha * \delta & \alpha * (1 - \delta) \end{pmatrix}, \tag{A107}$$

where the rows of $p(y|\hat{x})$ are indexed by y, and columns by $\hat{x}$. Along with $\frac{dp(y|\hat{x})}{d\delta} = (2\alpha - 1) \cdot \begin{pmatrix} 1 & -1 \\ -1 & 1 \end{pmatrix}$, its derivatives with respect to β follow as in (A106).

Appendix F. Equivalent Conditions for Cluster-Merging Bifurcations

We briefly discuss the equivalent conditions for cluster-merging bifurcations in the IB (Section 5.2) found in the literature.

Rose et al. [39] (Section 4) derive a condition for cluster-splitting phase transitions (Equation (17) there) in the context of fuzzy clustering. Following this, [13] (3.2 in Part III) derives an analogous condition for cluster splitting in the IB (Equation (12) there):

$$(I - \beta\, C_X(\hat{x}; \beta))u = 0 \,, \tag{A108}$$

where I is the identity. Namely, for a cluster $\hat{x}$ to split, it is necessary that $1/\beta$ be an eigenvalue of an $|\mathcal{X}|$-by-$|\mathcal{X}|$ matrix $C_X(\hat{x}; \beta)$, whose entries at an IB root are given by

$$C_X(\hat{x}; \beta)_{x,x'} := \sum_y \frac{p(y|x)p(y|x')p_\beta(x'|\hat{x})}{p_\beta(y|\hat{x})} \,. \tag{A109}$$

While the coefficients matrix (A109) for the IB differs from the one for fuzzy clustering, inter-cluster interactions are explicitly neglected in both derivations (see therein). Indeed, the definition (A109) of C_X involves the coordinates of cluster $\hat{x}$ alone, as one might expect when considering a root in either decoder or in inverse encoder coordinates (Section 2). Reversing the dynamics in β, condition (A108) characterizes cluster-merging bifurcations in the IB (Section 5.2).

In [13] it is noted that (A108) is closely related to the bifurcation analysis of [9]. The latter provides a condition to identify the critical β values of IB bifurcations, given in their Theorem 5.3. Indeed, their condition is equivalent to (A108), and therefore it also characterizes cluster-merging bifurcations. To see this, the necessary condition they give for a phase transition at β is that $1/\beta$ must be an eigenvalue of a matrix V (Equation (21) there). When written in our notation, this matrix is given by

$$V(\hat{x}; \beta)_{x,x'} := \sum_y \frac{p(x', y)p(x, y)p_\beta(\hat{x}|x)}{p_\beta(y, \hat{x})p(x')} \,. \tag{A110}$$

However, V (A110) is readily seen to be the transpose of C_X (A109), and so they have the same eigenvalues.

Appendix G. Lyapunov Stability of an Optimal IB Root

We provide the essential parts of a proof that an optimal IB root is Lyapunov uniformly asymptotically stable on closed intervals which do not contain a bifurcation when following the flow dictated by the IB's ODE (16) in *decreasing* β. Definitions for the below are as in [40] (see especially Section 4.2 there). See Section 6.3 for a discussion of the results below.

Let $p^*(\beta)$ be an optimal IB root. We start by rewriting it as an equilibrium of a non-autonomous ODE, as in [40] (Equation (4.1)). Consider the implicit ODE (7) $\frac{dp}{d\beta} = -(D_pF)^{-1}D_\beta F$, specialized to the IB by setting $F := Id - BA_\beta$ (5). Denote $\delta p := p - p^*$, for an arbitrary p. Subtracting the ODE at p from that at p^* yields a non-autonomous ODE in the error δp from the optimal root:

$$\frac{d\delta p}{d\beta} = (D_pF)^{-1}D_\beta F|_{p^*} - (D_pF)^{-1}D_\beta F|_{p^*+\delta p} \tag{A111}$$

This rewrites the given root p^* as an equilibrium $\delta p = 0$ of this ODE (A111), simplifying the below.

Next, we define a Lyapunov function for the flow of the equilibrium $\delta p = 0$ along the ODE (A111), when its dynamics in β is reversed. Consider the IB's Lagrangian $\mathcal{L}_\beta := I(X;\hat{X}) - \beta \cdot I(Y;\hat{X})$ as a functional in p, and let $\mathcal{L}_\beta^* := \mathcal{L}_\beta[p^*]$ be its optimal value at β. Then,

$$\left(\mathcal{L}_\beta - \mathcal{L}_\beta^*\right)(\delta p) \tag{A112}$$

is the desired Lyapunov function. Specifically, (i) $\mathcal{L}_\beta - \mathcal{L}_\beta^*$ is positive definite and (ii) $\frac{d}{d\beta}\left(\mathcal{L}_\beta - \mathcal{L}_\beta^*\right)$ is negative definite, when the dynamics in β are reversed. Theorem 4.1 in [40] then implies that $\delta p = 0$ is uniformly asymptotically stable [40] (Definition 4.6).

For (i), $\mathcal{L}_\beta - \mathcal{L}_\beta^*$ (A112) is immediately seen to be positive semi-definite from the definition of $\mathcal{L}_\beta^*$, up to technicalities ignored here (cf., Definition 4.7 in [40]). The results of Section 5.3 (after Proposition 1) imply that representing p in reduced log-decoder coordinates renders (A112) strictly positive definite. Indeed, $D(Id - BA_\beta)$ is non-singular in a reduced representation in these coordinates, as mentioned there, and so an optimal root p^* is locally unique. As for condition (ii), from the definition of $\mathcal{L}_\beta$ we have

$$\tfrac{d}{d\beta}\mathcal{L}_\beta = \tfrac{d}{d\beta}I(X;\hat{X}) - \beta\tfrac{d}{d\beta}I(Y;\hat{X}) - I(Y;\hat{X}) = -I(Y;\hat{X}) , \tag{A113}$$

where $\frac{d}{d\beta}I(X;\hat{X}) = \beta\frac{d}{d\beta}I(Y;\hat{X})$ in the last equality follows by direct calculations similar to those in the Appendix of [13] (Part III). Thus, for the β-derivative of (A112), we have

$$\tfrac{d}{d\beta}\left(\mathcal{L}_\beta - \mathcal{L}_\beta^*\right)(\delta p) = I(Y;\hat{X})|_{p^*} - I(Y;\hat{X})|_p . \tag{A114}$$

The latter is always positive semi-definite around p^*, since by definition (1) p^* yields the maximal Y-information subject to a constraint on the X-information. The same argument as above shows that it is strictly positive definite. Finally, reversing the dynamics in β leaves the ODE (A111) unaffected but flips the sign of (A114), rendering it negative definite as required.

Appendix H. Introducing Degeneracies to the IB Operator in Decoder Coordinates Is Uninformative at an Isolated Optimal Root

Suppose that the Jacobian of the IB operator $Id - BA_\beta$ (5) in log-decoder coordinates is non-singular at a reduced representation of an IB root, as in the argument following Proposition 1 (Section 5.3). We show that evaluating it on a non-reduced (degenerate) representation only permits one to detect kernel directions which are due to the selected degeneracy. Thus, it is inadequate for detecting bifurcations, as explained in Section 5.3.

Let $p \in \Delta[\Delta[\mathcal{Y}]]$ be an IB root of effective cardinality T_1. A T-clustered *representation* of a root (e.g., in decoder coordinates) is a function $\pi : \Delta[\Delta[\mathcal{Y}]] \to \mathbb{R}^{(|\mathcal{Y}|+1)\cdot T}$, defined on some neighborhood of p. In the other way around, one can consider the inclusion $i : \mathbb{R}^{(|\mathcal{Y}|+1)\cdot T} \to \Delta[\Delta[\mathcal{Y}]]$, defined in the obvious way on normalized decoder coordinates at some neighborhood of $\pi(p)$. Let π be a representation of p in its effective cardinality T_1, and $\tilde{\pi}$ a degenerate one on $T_2 > T_1$ clusters. These satisfy

$$\pi = reduc \circ \tilde{\pi} , \tag{A115}$$

where *reduc* is the reduction map (defined similar to the root-reduction Algorithm 2, by setting its thresholds to zero, $\delta_1 = \delta_2 = 0$, and replacing its strict inequalities with non-strict ones. Note that Algorithm 2 has a well-defined output for every input). In the other way around, one can pick a particular degenerating map *degen* (e.g., "split the third cluster to two copies of probability ratio 1:2"). Applying a particular degeneracy and then reducing is the identity,

$$reduc \circ degen = Id , \tag{A116}$$

though not the other way around. Let i and $\tilde{i}$ be the inclusions corresponding to π and $\tilde{\pi}$, respectively. Similar to (A115), degenerating a root has no effect before it is included into $\Delta[\Delta[\mathcal{Y}]]$,

$$i = \tilde{i} \circ degen \tag{A117}$$

Recall from Section 5.1 (before Conjecture 1) that BA_β in decoder coordinates may be considered as an operator on $\Delta[\Delta[\mathcal{Y}]]$. To summarize, we have the following diagram:

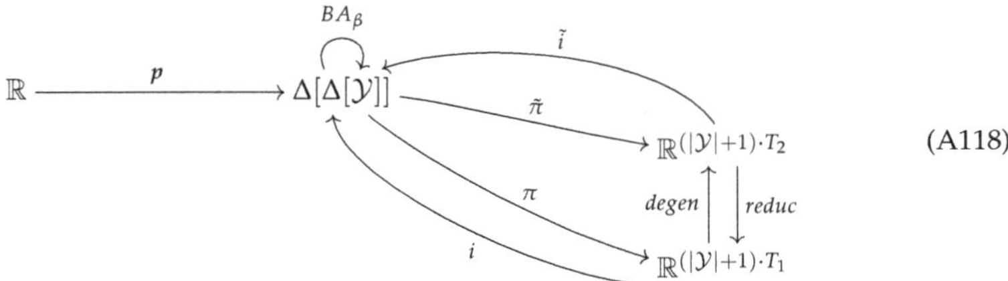

$$\tag{A118}$$

Next, consider the representations of the IB operator $Id - BA_\beta$ (5) on T_1 and T_2 clusters. These amount to pre-composing with the inclusions and post-composing with the representation maps. Denote by Id_i the identity operator on $\mathbb{R}^{(|\mathcal{Y}|+1)\cdot T_i}$. By identities (A115), (A116), and (A117), the T_1-clustered representation of $Id - BA_\beta$ (5) satisfies

$$Id_1 - \pi \circ BA_\beta \circ i = reduc \circ degen - reduc \circ \tilde{\pi} \circ BA_\beta \circ \tilde{i} \circ degen$$
$$= reduc \circ \left[Id_2 - \tilde{\pi} \circ BA_\beta \circ \tilde{i} \right] \circ degen \tag{A119}$$

Differentiating, by the chain rule we have

$$D\left(Id_1 - \pi \circ BA_\beta \circ i \right) = D(reduc) D\left(Id_2 - \tilde{\pi} \circ BA_\beta \circ \tilde{i} \right) D(degen) . \tag{A120}$$

Multiplying a matrix B from the left can only enlarge the kernel, $\dim \ker(AB) \geq \dim \ker B$, and so

$$\dim \ker D\left(Id_1 - \pi \circ BA_\beta \circ i \right) \geq \dim \ker \left(D\left(Id_2 - \tilde{\pi} \circ BA_\beta \circ \tilde{i} \right) D(degen) \right) . \tag{A121}$$

Since the left-hand side is evaluated at a reduced representation, it vanishes by assumption. Thus, $D\left(Id_2 - \tilde{\pi} \circ BA_\beta \circ \tilde{i} \right) D(degen)$ is of full rank, yielding

$$\ker D\left(Id_2 - \tilde{\pi} \circ BA_\beta \circ \tilde{i} \right) \subset (\operatorname{Im} D(degen))^{\perp} , \tag{A122}$$

where $(\operatorname{Im} D(degen))^{\perp}$ denotes the vectors tangent to the column space $\operatorname{Im} D(degen)$ of $D(degen)$. Stated differently, the Jacobian $D\left(Id_2 - \tilde{\pi} \circ BA_\beta \circ \tilde{i} \right)$ of a T_2-clustered representation of $Id - BA_\beta$ (5) can only detect directions in $(\operatorname{Im} D(degen))^{\perp}$, which are determined by the choice of the degenerating map $degen$, as argued.

For completeness, splitting a cluster $r \in \Delta[\mathcal{Y}]$ into two at some fixed ratio $0 < \lambda < 1$ is of the form $(r, p(r)) \mapsto (r, \lambda \cdot p(r), r, (1 - \lambda) \cdot p(r))$. Adding a pre-defined cluster $r' \in \Delta[\mathcal{Y}]$ of zero mass is constant $(\dots) \mapsto (\dots, r', 0)$ on the newly added coordinates. A general degeneracy map $degen$ is a composition of these, and is otherwise the identity map on unaffected coordinates.

References

1. Tishby, N.; Pereira, F.C.; Bialek, W. The Information Bottleneck Method. In Proceedings of the 37th Annual Allerton Conference on Communication, Control, and Computing, Monticello, IL, USA, 22–24 September 1999; pp. 368–377.
2. Witsenhausen, H.; Wyner, A. A Conditional Entropy Bound for a Pair of Discrete Random Variables. *IEEE Trans. Inf. Theory* **1975**, *21*, 493–501. [CrossRef]

3. Zaidi, A.; Estella-Aguerri, I.; Shamai, S. On the Information Bottleneck Problems: Models, connections, Applications and Information Theoretic Views. *Entropy* **2020**, *22*, 151. [CrossRef] [PubMed]
4. Agmon, S.; Benger, E.; Ordentlich, O.; Tishby, N. Critical Slowing Down Near Topological Transitions in Rate-Distortion Problems. In Proceedings of the 2021 IEEE International Symposium on Information Theory (ISIT), Melbourne, Australia, 12–20 July 2021; pp. 2625–2630. [CrossRef]
5. Gilad-Bachrach, R.; Navot, A.; Tishby, N. An Information Theoretic Tradeoff between Complexity and Accuracy. In *Learning Theory and Kernel Machines*; Springer: Berlin/Heidelberg, Germany, 2003. [CrossRef]
6. Agmon, S. Root Tracking for Rate-Distortion: Approximating a Solution Curve with Higher Implicit Multivariate Derivatives. *IEEE Trans. Inf. Theory* **2023**, *in press*.
7. De Oliveira, O. The Implicit and the Inverse Function Theorems: Easy Proofs. *Real Anal. Exch.* **2014**, *39*, 207–218. [CrossRef]
8. Blahut, R. Computation of Channel Capacity and Rate-Distortion Functions. *IEEE Trans. Inf. Theory* **1972**, *18*, 460–473. [CrossRef]
9. Gedeon, T.; Parker, A.E.; Dimitrov, A.G. The Mathematical Structure of Information Bottleneck Methods. *Entropy* **2012**, *14*, 456–479. [CrossRef]
10. Agmon, S. On Bifurcations in Rate-Distortion Theory and the Information Bottleneck Method. Ph.D. Thesis, The Hebrew University of Jerusalem, Jerusalem, Israel, 2022.
11. Rose, K.; Gurewitz, E.; Fox, G. A deterministic annealing approach to clustering. *Pattern Recognit. Lett.* **1990**, *11*, 589–594. [CrossRef]
12. Kuznetsov, Y.A. *Elements of Applied Bifurcation Theory*, 3rd ed.; Springer Science & Business Media: New York, NY USA, 2004; Volume 112. [CrossRef]
13. Zaslavsky, N. Information-Theoretic Principles in the Evolution of Semantic Systems. Ph.D. Thesis, The Hebrew University of Jerusalem, Jerusalem, Israel, 2019.
14. Ngampruetikorn, V.; Schwab, D.J. Perturbation Theory for the Information Bottleneck. *Adv. Neural Inf. Process. Syst.* **2021**, *34*, 21008–21018. [PubMed]
15. Wu, T.; Fischer, I.; Chuang, I.L.; Tegmark, M. Learnability for the Information Bottleneck. *PMLR* **2020**, *115*, 1050–1060. [CrossRef]
16. Wu, T.; Fischer, I. Phase Transitions for the Information Bottleneck in Representation Learning. In Proceedings of the Eighth International Conference on Learning Representations (ICLR 2020), Virtual Conference, 26 April–1 May 2020.
17. Rose, K. A Mapping Approach to Rate-Distortion Computation and Analysis. *IEEE Trans. Inf. Theory* **1994**, *40*, 1939–1952. [CrossRef]
18. Giaquinta, M.; Hildebrandt, S. *Calculus of Variations I*; Springer: Berlin/Heidelberg, Germany, 2004; Volume 310. [CrossRef]
19. Parker, A.E.; Dimitrov, A.G. Symmetry-Breaking Bifurcations of the Information Bottleneck and Related Problems. *Entropy* **2022**, *24*, 1231. [CrossRef] [PubMed]
20. Harremoës, P.; Tishby, N. The Information Bottleneck Revisited or How to Choose a Good Distortion Measure. In Proceedings of the 2007 IEEE International Symposium on Information Theory, Nice, France, 24–29 June 2007; pp. 566–570. [CrossRef]
21. Kielhöfer, H. *Bifurcation Theory: An Introduction with Applications to Partial Differential Equations*, 2nd ed.; Springer: New York, NY, USA, 2012. [CrossRef]
22. Lee, J.M. *Introduction to Smooth Manifolds*, 2nd ed.; Spinger: New York, NY, USA, 2012. [CrossRef]
23. Dummit, D.S.; Foote, R.M. *Abstract Algebra*, 3rd ed.; John Wiley & Sons, Inc.: Hoboken, NJ, USA, 2004.
24. Teschl, G. *Topics in Linear and Nonlinear Functional Analysis*; University of Vienna: Vienna, Austria, 2022. Available online: https://www.mat.univie.ac.at/~gerald/ftp/book-fa/fa.pdf (accessed on 20 December 2022).
25. Cormen, T.H.; Leiserson, C.E.; Rivest, R.L.; Stein, C. *Introduction to Algorithms*, 2nd ed.; MIT Press: Cambridge, MA, USA, 2001.
26. Butcher, J.C. *Numerical Methods for Ordinary Differential Equations*, 3rd ed.; John Wiley & Sons: Hoboken, NJ, USA, 2016. [CrossRef]
27. Atkinson, K.E.; Han, W.; Stewart, D. *Numerical Solution of Ordinary Differential Equations*; John Wiley & Sons: Hoboken, NJ, USA, 2009; Volume 108. [CrossRef]
28. Berger, T. *Rate Distortion Theory: A Mathematical Basis for Data Compression*; Prentice-Hall: Englewood Cliffs, NJ, USA, 1971.
29. Shannon, C.E. A Mathematical Theory of Communication. *Bell Syst. Tech. J.* **1948**, *27*, 379–423. [CrossRef]
30. Shannon, C.E. Coding Theorems for a Discrete Source with a Fidelity Criterion. *IRE Nat. Conv. Rec.* **1959**, *4*, 325–350.
31. Cover, T.M.; Thomas, J.A. *Elements of Information Theory*, 2nd ed.; John Wiley & Sons: Hoboken, NJ, USA, 2006; p. 748. [CrossRef]
32. Dieudonné, J. *Foundations of Modern Analysis*; Academic Press: Cambridge, MA, USA, 1969.
33. Gowers, T.; Barrow-Green, J.; Leader, I. *The Princeton Companion to Mathematics*; Princeton University Press: Princeton, NJ, USA, 2008.
34. Coolidge, J.L. *A Treatise on Algebraic Plane Curves*; Dover: Dover, NY, USA, 1959.
35. Strogatz, S.H. *Nonlinear Dynamics and Chaos: With Applications to Physics, Biology, Chemistry, and Engineering*, 2nd ed.; CRC Press: Boca Raton, FL, USA, 2018.
36. Golubitsky, M.; Stewart, I.; Schaeffer, D.G. *Singularities and Groups in Bifurcation Theory II*; Springer: New York, NY, USA, 1988. [CrossRef]
37. Benger, E. (The Hebrew University of Jerusalem, Jerusalem, Israel). Private communications, 2019.
38. Wyner, A.; Ziv, J. A Theorem on the Entropy of Certain Binary Sequences and Applications: Part I. *IEEE Trans. Inf. Theory* **1973**, *19*, 769–772. [CrossRef]

39. Rose, K.; Gurewitz, E.; Fox, G.C. Statistical Mechanics and Phase Transitions in Clustering. *Phys. Rev. Lett.* **1990**, *65*, 945. [CrossRef] [PubMed]
40. Slotine, J.J.E.; Li, W. *Applied Nonlinear Control*; Prentice Hall: Englewood Cliffs, NJ, USA, 1991; Volume 199.

Article

Exact and Soft Successive Refinement of the Information Bottleneck

Hippolyte Charvin *, Nicola Catenacci Volpi and Daniel Polani

School of Physics, Engineering and Computer Science, University of Hertfordshire, Hatfield AL10 9AB, UK;
n.catenacci-volpi@herts.ac.uk (N.C.V.); d.polani@herts.ac.uk (D.P.)
* Correspondence: h.charvin@herts.ac.uk

Abstract: The information bottleneck (IB) framework formalises the essential requirement for efficient information processing systems to achieve an optimal balance between the complexity of their representation and the amount of information extracted about relevant features. However, since the representation complexity affordable by real-world systems may vary in time, the processing cost of updating the representations should also be taken into account. A crucial question is thus the extent to which adaptive systems can *leverage the information content of already existing IB-optimal representations for producing new ones*, which target the same relevant features but at a different granularity. We investigate the information-theoretic optimal limits of this process by studying and extending, within the IB framework, the notion of *successive refinement*, which describes the ideal situation where no information needs to be discarded for adapting an IB-optimal representation's granularity. Thanks in particular to a new geometric characterisation, we analytically derive the successive refinability of some specific IB problems (for binary variables, for jointly Gaussian variables, and for the relevancy variable being a deterministic function of the source variable), and provide a linear-programming-based tool to numerically investigate, in the discrete case, the successive refinement of the IB. We then soften this notion into a *quantification* of the loss of information optimality induced by several-stage processing through an existing measure of unique information. Simple numerical experiments suggest that this quantity is typically low, though not entirely negligible. These results could have important implications for (*i*) the structure and efficiency of incremental learning in biological and artificial agents, (*ii*) the comparison of IB-optimal observation channels in statistical decision problems, and (*iii*) the IB theory of deep neural networks.

Keywords: information bottleneck; successive refinement; unique information; incremental learning; coarse-graining; Blackwell order; deep learning

Citation: Charvin, H.; Volpi, N.C.; Polani, D. Successive Refinement of the Information Bottleneck. *Entropy* **2023**, *25*, 1355. https://doi.org/10.3390/e25091355

Academic Editors: Jan Lewandowsky and Gerhard Bauch

Received: 25 July 2023
Revised: 8 September 2023
Accepted: 13 September 2023
Published: 19 September 2023

1. Introduction

1.1. Conceptualisation and Organisation Outline

Consider the problem, for an information-processing system, of extracting relevant information about a target variable Y within a correlated source variable X, under constraints on the cost of the information processing needed to do so—yielding a compressed representation T. This situation can be formalised in an information-theoretic language, where the information-processing cost is measured with the mutual information $I(X;T)$ between the source X and the representation T of it, while the relevancy about Y of the information extracted by T is measured by $I(Y;T)$. The problem thus becomes that of maximising the relevant information $I(Y;T)$ under bounded information-processing cost $I(X;T)$, i.e., we are interested in the *information bottleneck* (IB) problem [1,2], which, in primal form, can be formulated as

$$\underset{\substack{q(T|X): \\ T-X-Y,\, I(X;T)\leq\lambda}}{\arg\max}\quad I(Y;T). \tag{1}$$

Here, the trade-off parameter λ controls the bound on the permitted information-processing cost and thus, intuitively, the resulting representation's granularity. The Markov chain condition $T - X - Y$ ensures that any information that the bottleneck T extracts about the relevancy variable Y can only come from the source X. The solutions to (1) for varying λ trace the so-called *information curve*, i.e., the λ-parameterised curve

$$\left(I_\lambda(X;T), I_\lambda(Y;T)\right)_{\lambda \geq 0} \subseteq \mathbb{R}^2, \tag{2}$$

where $I_\lambda(X;T)$ and $I_\lambda(X;T)$ are defined by a bottleneck T of parameter λ (see the black curve in the first figure in Section 2 below). This curve indicates the informationally optimal bounds on the feasible trade-offs between relevancy $I(Y;T)$ and complexity $I(X;T)$ of the representation T. In this sense, the IB method provides a fundamental understanding of the *informationally optimal limits* of information-processing systems.

These limits are crucial for both understanding and building adaptive behaviour. For instance, choosing X to be an agent's past and Y to be its future leads it to extract the most relevant features of its environment [3–6]. More generally, the IB point of view on modelling embodied agents' representations has been leveraged for unifying efficient and predictive coding principles in theoretical neuroscience—at the level of single neurons [3,7–9] and neuronal populations [9–13]—but also for studying sensor evolution [14–16], the emergence of common concepts [17] and of spatial categories [18], the evolution of human language [19–21], or for implementing informationally efficient control in artificial agents [22–24]. This line of research brings increasing support to the hypothesis that, particularly for evolutionary reasons, biological agents are often poised close to optimality in the IB sense. It also provides a framework for both measuring and improving artificial agents' performance.

However, one aspect of the IB framework conflicts with a crucial feature of real-world systems: the informationally optimal limits that it describes only consider a given representation T *taken in isolation* from any other one in the system. This point of view *a priori* disregards the *relationship between representations*, which is crucial in real-world information-processing systems. Thus, it is crucial to consider the following question: does the relationship between a set of internal representations $T_1, \ldots, T_n$ impact their individual information optimality? In this paper, we are mostly interested in a specific kind of relationship: when $T_1, \ldots, T_n$ are successively produced in this order, and each new T_i builds on both the previous representation T_{i-1} and new information from the fixed source X to extract information about the fixed relevancy Y. This scenario formalises the *incorporation of information into already learned representations*—as is the case in developmental learning, or, more generally, any kind of learning process that goes through identifiable successive steps.

More precisely, consider an informationally bounded agent that extracts information about a relevant variable Y within an environment X. If the agent is informationally optimal, given an affordable complexity cost λ_1, it must maximise the relevant information that it extracts from the environment—resulting in a bottleneck representation T_1, i.e., a solution to (1) with parameter λ_1. Then, assume that at, a later stage, the complexity cost that the agent can afford increases to $\lambda_2 > \lambda_1$, while the goal is still to extract information about the same relevant feature Y within the same environment X. To keep being informationally optimal, the agent should thus update its representation so it becomes a bottleneck of parameter λ_2. Given this setting, the question we ask is: to which extent can the content learned into T_1 be leveraged for the production of T_2? It is indeed not intuitively clear that T_2 should keep all the information from T_1. An informal example is the fact that most pedagogical curricula teach knowledge via successive approximations, where, at a more advanced level, the content learned at the beginner level must sometimes be *unlearned* to successively proceed further, even though it was perfectly reasonable—in our language, informationally optimal—to deliver the first beginner sketch to students that would never progress to learn the expert level.

This question has been formalised, in the rate-distortion literature, with the notion of *successive refinement* (SR) [25–29], which, in short, refers to the situation where several-stage processing does not incur any loss of information optimality. More precisely, in the context outlined above, there is successive refinement if the processing cost of first producing a coarse bottleneck T_1 of parameter λ_1 and then refining it to a finer bottleneck T_2 of parameter $\lambda_2 > \lambda_1$ is no larger than the processing cost of directly producing a bottleneck T_2 of parameter λ_2 without any intermediary bottleneck T_1 (see Section 2.1 and Appendix B.2 for formal definitions). The aim of this work is to push the understanding of successive refinement in the IB framework [30–32] further, as well as to expand the analysis to a *quantification* of the lack of SR, in cases where the latter does not hold exactly. We start by leveraging general results in existing IB literature [33,34] to prove that successive refinement always holds for jointly Gaussian (X, Y), and when Y is a deterministic function of X. However, it is seems crucial, for further progress on more general scenarios, to design specifically tailored mathematical and numerical tools. In this regard, we provide two main contributions.

First, we present a simple geometric characterisation of SR, in terms of convex hulls of the decoder symbol-wise conditional probabilities $q(X|t)$, for t varying in the bottleneck alphabet $\mathcal{T}$. This characterisation is proven in the discrete case under an additional but mild assumption of injectivity of the decoder $q(X|T)$. This new point of view fits well with an ongoing convexity approach to the IB problem [35–39] and might thus help develop a new geometric perspective on the successive refinement of the IB. As an example, we use this geometric characterisation to prove that SR always holds for binary source X and binary relevancy Y. Moreover, this characterisation makes it straightforward to numerically assess, with a linear program checking convex hull inclusions, whether or not two discrete bottlenecks T_1 and T_2 achieve successive refinement. As we demonstrate with minimal numerical examples, this can help in investigating the SR structure of any given IB problem, i.e., how successive refinement depends on the particular combination of trade-off parameters λ_1 and λ_2.

Second, we soften [18] the traditional notion of successive refinement and study the *extent to which* several-stage processing incurs a loss of information optimality. More precisely, we propose to measure soft successive refinement with the *unique information* [40] (UI) that the coarser bottleneck T_1 holds about the source X, as compared to the finer one T_2. Explicitly, this UI is defined as the minimal value of $I_q(X; T_1|T_2)$ over all distributions $q := q(X, T_1, T_2)$ whose marginals $q(X, T_1)$ and $q(X, T_2)$ coincide with the corresponding bottleneck distributions (see Section 3.1 for details). As a first exploration of soft SR's qualitative features, we investigate the landscapes of unique information over trade-off parameters, for again some simple example distributions $p(X, Y)$. These landscapes seem to unveil a rich structure, which was largely hidden by the traditional notion of SR, that only distinguished between SR being present or absent. Among the general features suggested by these experiments, the most significant are that (i) soft SR seems strongly influenced by the trajectories of the decoders $q_\lambda(X|T)$ over λ; (ii) the UI often goes through sharp variations at the bifurcations [41–44] undergone by the bottlenecks (in a fashion compatible with the presence of discontinuities of either the UI itself, or its differential, with regard to trade-off parameters); and (iii) the loss of information optimality seems always small—more precisely, the global bound on the UI was observed to be typically one or two orders of magnitude lower than the system's globally processed information (see Section 3.2 for formal statements). These three conclusions are phenomenological and limited to our minimal examples, but they shed light on the kind of structure that can be investigated by further research. They also suggest the relevance that developing this theoretical framework might have for the scientific question that motivates it. In particular, the link with IB bifurcations and the overall small loss of information optimality would, if generalisable, have interesting consequences for the structure and efficiency of incremental learning.

As a side contribution, we draw along the paper formal equivalences between our framework and other notions proposed in the literature, thus making the formal framework also relevant to decision problems [40,45] and to the information-theoretic approach to deep learning [46]. This flexibility of interpretation stems from the fact that even though our formal framework crucially depends on the order of the bottleneck representations' trade-off parameters, it does not depend on the order in which these representations are produced. Thus, a sequence of bottlenecks can be equally well interpreted as produced from coarsest to finest—as is the case for the information incorporation interpretation outlined above—or from finest to coarsest—as is the case in feed-forward processing. This conceptual unity sheds light on the common formal structure shared by these diverse phenomena.

In the next Section 1.2, we review related work. After having established notations and recalled some general notions in Section 1.3, we formally introduce the notion of the successive refinement of the IB in Section 2.1, where we also prove successive refinability in the case of Gaussian vectors and deterministic channel $p(Y|X)$. We then present the convex hull characterisation in Section 2.2, before using it to prove successive refinement for the case of binary source and relevancy variables. The following Section 2.3 leverages the convex hull characterisation to gather some first insights from minimal experiments. These experiments suggest an intuition for defining soft successive refinement, which we formalise in Section 3.1 through a measure of unique information [40], where we provide theoretical motivations for our choice. This new measure is explored in Section 3.2 with additional numerical experiments that highlight the general features described above. The alternative interpretations of both exact and soft SR, in terms of decision problems and feed-forward deep neural networks, are developed in Sections 4.1 and 4.2, respectively. We then describe the limitations and potential future work in Section 5, and conclude in Section 6.

1.2. Related Work

The notion of successive refinement has been long studied in the rate-distortion literature [25–29]. However, classic rate-distortion theory [47] usually considers distortion functions defined on the random variables' *alphabets*, whereas the IB framework can be regarded as a rate-distortion problem only if one allows the distortion to be defined on the space of probability *distributions* [48]. Successive refinement thus needed to be adapted to the IB framework, which was achieved starting from various perspectives.

In [30,31], successive refinement is formulated within the IB framework. Then, Ref. [32] goes further by considering the informationally optimal limits of several-stage processing in general, without comparing it to single-stage processing. In both these works, the problem is initially defined in asymptotic coding terms, and only then given a single-letter characterisation. On the contrary, we will directly define successive refinement from a single-letter perspective. It turns out that our single-letter definition and the operational multi-letter definition from [30,31] are equivalent. The two latter works—as well as [32]—thus provide our single-letter definition with an operational interpretation that also formalises the intuition of an informationally optimal incorporation of information (see Proposition 1 and Appendix B.2).

Another notion named "successive refinement" as well can be found in [46]. This work, instead of modelling information incorporation, rather considers the successive processing of data along a feed-forward pipeline—which encompasses the example of deep neural networks. Fortunately, the "successive refinement" defined in [46] happens to encompass the notion we develop here; more precisely, in [46], the relevancy variable is allowed to vary across processing stages, but if we choose it to be always the same, then "successive refinement" as defined in [46] and "successive refinement" as defined here are formally equivalent (see Section 4.2). In other words, the situation considered in this paper is a particular case of [46], so our results, methods, and phenomenological insights are directly relevant to [46]. For instance, our proof of SR for binary X and Y (see Proposition 5) is a generalisation of Lemma 1 in [46], which proves SR when X is a Bernoulli variable of parameter $\frac{1}{2}$ and $p(Y|X)$ is a binary symmetric channel.

More generally speaking, the link between successive refinement and the IB theory of deep learning [49–56] has been noted since the inception of the latter research agenda [49], and, besides in [46], it was also further developed in [57]. Section 4.2 makes clear in which sense our results are relevant to this line of research. In particular, our minimal experiments suggest (if they are scalable to the much richer deep learning setting) that trained deep neural networks should lie close to IB bifurcations: i.e., if X is the network's input, Y the feature to be learnt and $L_1, \ldots, L_n$ the network's successive layers, the points $(I(X; L_i), I(Y, L_i))$ should lie close to points of the information curve corresponding to IB bifurcations. This feature was already suggested in [49,50], but for reasons not explicitly related to successive refinement. Note that while the phenomenon of IB bifurcations has been studied from a variety of perspectives (see, e.g., [41–44]), here, we adopt that of [43], which frames IB bifurcations as parameter values where the minimal number of symbols required to represent a bottleneck increases.

In [58], successive refinability is proved for discrete source X and relevancy $Y = X$. Our Proposition 3 generalises this result to either discrete or continuous source X, with relevancy Y being an arbitrary function of X, with a similar argument as that in [58].

In [33], links between the IB framework and renormalisation group theory are exhibited. Even though the questions addressed in the latter work are thus distinct from those addressed here, the Gaussian IB's *semigroup structure* defined and proven in [33] implies the successive refinability of Gaussian vectors (see Proposition 2, and see Appendix 2 for more details on the semigroup structure). This generalises Lemma 3 in [46], which proves SR when X and Y are jointly Gaussian, but each one-dimensional (see Section 4.2 for the relevance of [46] to our framework).

The geometric approach in which we propose to study the successive refinement of the IB is closely related to the convexity approach to the IB [35–39], which frames the IB problem as that of finding the lower convex hull of a well-chosen function. This formulation happens to fit neatly with our convex hull characterisation of successive refinement; we use it to apply the characterisation to proving successive refinability in the case of a binary source and relevancy. Moreover, it is worth noting that our convex hull characterisation makes successive refinement tightly related to the notion of *input-degradedness* [59], through which additional operational interpretations can be given to successive refinement, particularly in terms of randomised games.

The loss of information optimality induced by several-stage processing has already been studied in [60] (see next paragraph), but a quantification of it based on *soft Markovianity* was, to the best of our knowledge, only considered in [18]. Here, we take inspiration in the latter work to quantify soft successive refinement, but we explicitly address the problem that joint distributions over distinct bottlenecks are not uniquely defined. This leads us to use the *unique information* defined in [40] within the context of partial information decomposition [61–64] as our measure of soft SR. This unique information has tight links with the Blackwell order [45,65], which allows us in Section 4.1 to provide a second alternative interpretation of (exact and soft) successive refinement in terms of decision problems.

Ref. [60] proves the near-successive refinability of rate-distortion problems when the distortion measure is the squared error. However, the latter work's approach is different from ours in two respects. First, the distortion measures are different: in particular, as mentioned above, the IB distortion is defined over the space of probability distributions on symbols, unlike the squared error, which is defined on the space of symbols itself. Second, Ref. [60] quantifies the lack of SR as the respective differences between sequences of optimal rates (for given distortion sequences) of a several-stage processing system and the corresponding optimal rates (for the same distortions) of a single-stage processing system. Here, we quantify the lack of SR with a single quantity: the unique information defined by bottlenecks with different granularities. We are, at this stage, not aware of a link between this value of unique information and differences in one-stage and several-stage optimal rates.

1.3. Technical Preliminaries

In this section, we fix the notations and conventions that we will use along the paper and recall some general notions that we will need.

1.3.1. Notations and Conventions

The random variables are denoted by capital letters, e.g., X, their alphabets by calligraphic ones, e.g., $\mathcal{X}$, and their symbols by lower-case letters, e.g., x. Sometimes, we will mix upper- and lower-case notations to denote a family where some symbols vary, while others are fixed, e.g., $q(X|t) := \big(q(x|t)\big)_{x \in \mathcal{X}}$, or $q(x|T) := \big(q(x|t)\big)_{t \in \mathcal{T}}$. Throughout the whole paper, X is the fixed source and Y the fixed relevancy of the IB problem. The variable T defined by the solution $q(T|X)$ to the primal IB problem (1) is called a *primal* bottleneck. We use the same symbol T for *Lagrangian* bottlenecks, i.e., variables defined by solutions $q(T|X)$ to the Lagrangian bottleneck problem (see Equation (3) below). By "bottleneck" without further specification, we refer to either a primal or Lagrangian bottleneck. The fixed source-relevancy distribution is denoted $p(X, Y)$, and any distribution involving at least one bottleneck is denoted with the letter q, e.g., $q(X, Y, T)$. When it is necessary to make the trade-off parameter explicit, we index the corresponding objects by λ, e.g., $q_\lambda(T|X)$ or $I_\lambda(Y; T)$. Unless explicitly stated otherwise, the source X, relevancy Y, and any considered bottleneck T are defined as either all discrete or all continuous. Probability simplices, and sometimes some of their subsets are written using the generic symbol Δ; for instance, the source simplex is denoted by $\Delta_\mathcal{X}$.

Without loss of generality, we always restrict X, Y, and the bottleneck T to their respective supports so that, in particular, all the conditional distributions are unambiguously well-defined, both in the discrete and the continuous case.

We will denote by I_Y the function from $\mathbb{R}_+$ to $\mathbb{R}_+$ defined by $I_Y(\lambda) := I(Y; T)$, where T is a solution to the primal IB problem (1) for the parameter λ. The *information curve*, defined above in Equation (2), is thus also the graph of the function I_Y.

1.3.2. General Facts and Notions

The following properties of the IB framework will be useful [35,37]:

- A bottleneck must saturate the information constraint, i.e., solutions T to (1) must satisfy $I_\lambda(X; T) = \lambda$. In other words, the primal trade-off parameter is the complexity cost of the corresponding bottleneck.
- The function $I_Y : \lambda \mapsto I_\lambda(T; Y)$ is constant for $\lambda \geq H(X)$. We will thus always assume, without loss of generality, that $\lambda \in [0, H(X)]$.
- In the discrete case, choosing a bottleneck cardinality $|\mathcal{T}| = |\mathcal{X}| + 1$ is enough to obtain optimal solutions. Thus, we always assume, without loss of generality, that $|\mathcal{T}| \leq |\mathcal{X}| + 1$, where $|\mathcal{T}| < |\mathcal{X}| + 1$ might occur if needed to make T full support .

To compute bottleneck solutions, instead of directly solving the primal problem (1), following common practice, we will solve its Lagrangian relaxation [66]:

$$\underset{q(T|X)\,:\,T-X-Y,}{\arg\min} \quad I(X; T) - \beta\, I(Y; T), \tag{3}$$

where the complexity-relevancy trade-off is now parameterised by $\beta \geq 0$, which corresponds to the inverse of the information curve's slope [41]. As the information curve is known to be concave, the Lagrangian parameter β is an increasing function of the primal parameter $\lambda = I(X; T)$. Moreover, we can, without loss of generality, assume that $\beta \geq 1$ [43]. (Note that when the information curve is not strictly concave, the Lagrangian formulation does not allow one to obtain all the solutions to the primal problem [39,67]. However, in our simple numerical experiments, we always obtained strictly concave information curves.)

We will also need the following concepts [43]:

Definition 1. *Let T be a (primal or Lagrangian) discrete bottleneck. The* effective cardinality $k = k(T)$ *is the number of distinct pointwise conditional probabilities $q(X|t)$ for varying t.*

Definition 2. *A discrete (primal or Lagrangian) bottleneck T is a* canonical bottleneck, *or is in* canonical form, *if all the pointwise conditional probabilities $q(X|t)$ are distinct, i.e., equivalently, if $|\mathcal{T}| = k(T)$, where $k(T)$ is the effective cardinality of T.*

Our definition of effective cardinality, even though slightly different from the original one in [43], is equivalent to the latter for Lagrangian bottlenecks. And, importantly, every (primal or Lagrangian) bottleneck can be reduced to its canonical form by merging the symbols with identical $q(X|t)$ (see Appendix A.1 for more details). We will be particularly interested in the *change* of effective cardinality, which has been identified in [43] as characterising the bottleneck phase-transitions, or *bifurcations*.

In Figure 1, we present examples of bottleneck conditional distributions $q(X|T)$, visualised as the family of points $\{q(X|t), t \in \mathcal{T}\}$ on the source simplex $\Delta_{\mathcal{X}}$, where, here, $|\mathcal{X}| = 3$, and the bottleneck is computed with $|\mathcal{T}| = 3$ in both examples. However, in Figure 1 (left), there are only two distinct $q(X|t)$, so there must be two equal pointwise probabilities $q(X|t_1)$ and $q(X|t_2)$; thus, $k = 2$ and the canonical form of T is obtained by merging t_1 and t_2. On the contrary, in Figure 1 (right), there are three distinct $q(X|t)$, so, here, $k = 3$ and the bottleneck is already in canonical form.

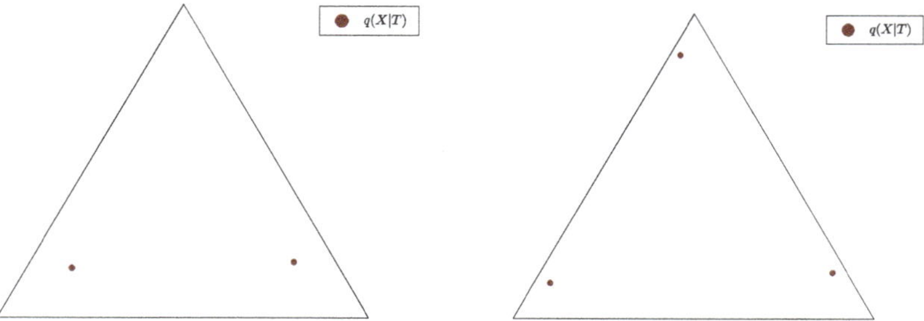

Figure 1. Examples of distributions $q(X|T)$, visualised as families of points $\{q(X|t),\ t \in \mathcal{T}\}$ on the source simplex $\Delta_{\mathcal{X}}$, where, here, $|\mathcal{X}| = 3$. Each of the triangle's vertices represents the Dirac probability of some $x \in \mathcal{X}$. The bottleneck's effective cardinality is $k = 2$ on the left and $k = 3$ on the right.

Eventually, the notions of *consistency* and *extension* will be crucial to us.

Definition 3. *Let $\mathcal{A} := \mathcal{A}_1 \times \cdots \times \mathcal{A}_m$ be a Cartesian product of (continuous or discrete) alphabets. For $C = \{c_1, \ldots, c_r\} \subseteq \{1, \ldots, m\}$ a subset of coordinates, we write*

$$\underset{c \in C}{\times} \mathcal{A}_c := \mathcal{A}_{c_1} \times \cdots \times \mathcal{A}_{c_r}.$$

For each $1 \le i \le n$, we consider a subset of coordinates C_i and a probability distribution q_i over $\times_{c \in C_i} \mathcal{A}_c$. The distributions $q_1, \ldots, q_n$ are said to be consistent *if, for every $1 \le i, j \le n$, the respective marginals of q_i and q_j on their common coordinates $\times_{c \in C_i \cap C_j} \mathcal{A}_c$ are equal.*

For instance, if T_1 and T_2 are two bottlenecks, they define consistent distributions $q_1(X, Y, T_1)$ and $q_2(X, Y, T_2)$ because, by definition, their respective marginals on their common coordinates $\mathcal{X} \times \mathcal{Y}$ are $q_1(X, Y) = q_2(X, Y) = p(X, Y)$.

Definition 4. *Let $\mathcal{A} := \mathcal{A}_1 \times \cdots \times \mathcal{A}_m$ be a Cartesian product of (continuous or discrete) alphabets, and $q_1, \ldots, q_n$ be consistent probability distributions over distinct but potentially overlapping coordinates of $\mathcal{A}$. A distribution q over the whole $\mathcal{A}$ is called an* extension *of the family of distributions $\{q_1, \ldots, q_n\}$ if it is consistent with each q_i.*

Consider bottlenecks $T_1, \ldots, T_n$ of same source X and relevancy Y for resp. parameters $\lambda_1, \ldots, \lambda_n$. They define a consistent family of distributions $\{q_{\lambda_i}(X, T_i), 1 \leq i \leq n\}$. One of the central mathematical objects of this work is the set of their extensions into *joint* distributions $q(X, T_1, \ldots, T_n)$:

Notation 1. *For given bottlenecks $T_1, \ldots, T_n$ of respective parameters $\lambda_1, \ldots, \lambda_n$, we denote by $\Delta_{\lambda_1, \ldots, \lambda_n}$ the set of extensions $q(X, T_1, \ldots, T_n)$ of the family of distributions $\{q_{\lambda_i}(X, T_i), 1 \leq i \leq n\}$.*

In general, for a fixed family of bottlenecks, there is a multitude of possible ways to extend them into a joint distribution; indeed, $\Delta_{\lambda_1, \ldots, \lambda_n}$ traces a polytope on the simplex $\Delta_{\mathcal{X} \times T_1 \times \cdots \times T_n}$ of joint distributions (see Appendix A in [40]). This feature is the formal version of our previous statement that the IB framework does not entirely specify the *relationship* between representations $T_1, \ldots, T_n$: it only constrains it through the set $\Delta_{\lambda_1, \ldots, \lambda_n}$. Questions about possible relationships between IB representations are thus questions about properties of the set $\Delta_{\lambda_1, \ldots, \lambda_n}$.

2. Exact Successive Refinement of the IB

2.1. Formal Framework and First Results

Here, we formally describe, within the IB framework, the rate-distortion-theoretic notion of *successive refinement* (SR) [25–27,29]. We propose a purely single-letter definition (i.e., we only consider single source, relevancy, and bottleneck variables), which makes the presentation simpler but still conveys the intuition of information incorporation. After having presented the notion of SR in the IB framework, we describe its Markov chain characterisation (see Proposition 1), which mirrors the characterisation of SR for classic rate-distortion problems [26], and makes our formulation equivalent to previous multi-letter operational definitions, which also formalise the intuition of information incorporation [30–32]. We then leverage this characterisation to prove SR in the case of Gaussian vectors and deterministic channel $p(Y|X)$.

Intuitively, there is successive refinement when a finer bottleneck T_2 does not discard any of the information extracted by a coarser bottleneck T_1. This can be imposed by requiring that $T_2 = (T_1, S_2)$ for some variable S_2, which encodes the "supplement" of information that "refines" T_1 into T_2. In the general case:

Definition 5. *Let $0 < \lambda_1 < \cdots < \lambda_n$, and a discrete or continuous $p(X, Y)$ be given. There is successive refinement (SR) for parameters $(\lambda_1, \ldots, \lambda_n)$ if there exist variables $(T_1, S_2, S_3, \ldots, S_n)$ such that*

- *T_1 is a bottleneck with parameter λ_1;*
- *For every $2 \leq i \leq n$, the variable $T_i := (T_{i-1}, S_i)$ is a bottleneck with parameter λ_i.*

Note that even though it does not appear explicitly in this definition, the relevancy variable Y is indeed crucial to it, as it defines what a bottleneck is (see Equation (1)). If the conditions of Definition 5 hold, we will also say that the IB problem defined by $p(X, Y)$ is $(\lambda_1, \ldots, \lambda_n)$-refinable. If bottlenecks $T_1, \ldots, T_n$ satisfy the definition's conditions, we will say that they *achieve* successive refinement, or, simply, that there is successive refinement between these bottlenecks. If there is successive refinement for all combinations $0 < \lambda_1 < \cdots < \lambda_n$ of trade-off parameters, we will say that the corresponding IB problem is successively refinable. Eventually, when it will be needed in later sections to contrast this notion with that of soft successive refinement, we will refer to it as *exact* successive refinement.

For instance, let $0 < \lambda_1 < \lambda_2$ and $\mathcal{X} = \mathcal{Y} = \{0,1\}$. We consider $Y := X \oplus Z$, where $\oplus$ denotes the modulo-2 addition, and X and Z are Bernouilli variables with parameters $\frac{1}{2}$ and a, respectively, for an arbitrary $0 \le a \le \frac{1}{2}$. In this case, it is proven in Lemma 1 of [46] that, for well-chosen binary variables S_1 and S_2, we have that X, S_1, and S_2 are mutually independent, and the variables $X \oplus S_1$ and $X \oplus S_1 \oplus S_2$ are bottlenecks of resp. parameters λ_1 and λ_2. Moreover, using the independence of S_2 with $(X, X \oplus S_1)$ and the assumed Markov chain $Y - X - X \oplus S_1 \oplus S_2$, a straightforward computation shows that to get a bottleneck of parameter λ_2, the variable $X \oplus S_1 \oplus S_2$ can be replaced by $(X \oplus S_1, S_2)$. Thus, here, the IB problem is (λ_1, λ_2)-refinable, where successive refinement is achieved by $T_1 := X \oplus S_1$ and $T_2 = (T_1, S_2)$.

It is helpful to visualise SR on the information plane, i.e., that on which lies the information curve. Indeed, successive refinement can be understood in terms of specific translations on the information plane: those resulting from concatenating an already existing variable T_{i-1} with a new variable S_i—let us call them "accumulative translations" because they result from a processing that does not discard any of the information already collected. Let us focus on the case $n = 2$ and first note that, whether or not (T_1, S_2) is a bottleneck, we have

$$I(X; T_1, S_2) = I(X; T_1) + I(X; S_2 | T_1),$$

and, similarly,

$$I(Y; T_1, S_2) = I(Y; T_1) + I(Y; S_2 | T_1).$$

In other words, the measure of both the complexity cost and relevance for (T_1, S_2) can be decomposed into the same measures first for T_1 and then for the "supplement" of information S_2, conditionally on the "already collected" information T_1. In Figure 2 (left and right), we first fix a coarse bottleneck T_1, understood here as a point $\big(I(X; T_1), I(Y; T_1)\big)$ on the information curve. Once T_1 is known, we supplement it with a new variable S_2, which incurs both an additional complexity cost $I(X; S_2 | T_1)$ and an additional relevant information gain $I(Y; S_2 | T_1)$. The question of successive refinement is that of whether the additional complexity cost can be leveraged enough for the resulting relevant information gain to take (T_1, S_2) "up to the information curve", i.e., to be such that $\big(I(X; T_1, S_2), I(Y; T_1, S_2)\big)$ is on the information curve. This is the case in Figure 2, right, and not the case in Figure 2, left. In short, there is successive refinement between two points on the information curve if and only if there exists an "accumulative translation" from the coarser one to the finer one.

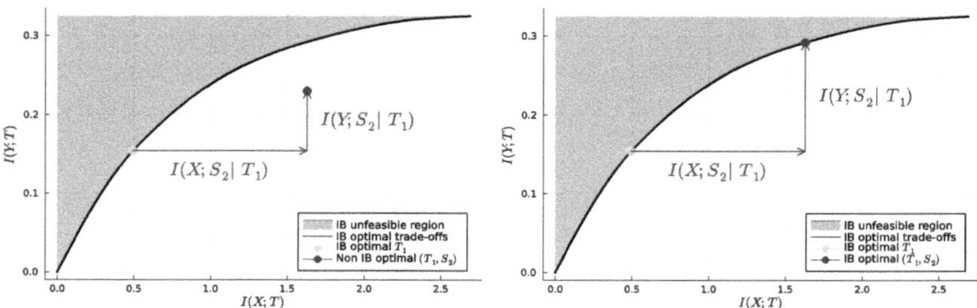

Figure 2. Successive refinement visualised on the information plane. On the left, adding the information from the variable S_2 (the supplement variable) is not efficient enough to achieve successive refinement. On the right, it is. See main text for details (the values of $I(X; S_2 | T_1)$ and $I(Y; S_2 | T_1)$ have been chosen arbitrarily to illustrate each case).

Let us now describe a more formal characterisation, where point (ii) will mirror the characterisation of SR for classic rate-distortion problems [26].

Proposition 1. *Let $0 < \lambda_1 < \cdots < \lambda_n$. The following are equivalent:*

(i) *There is successive refinement for parameters $(\lambda_1, \ldots, \lambda_n)$;*

(ii) *There exist bottlenecks $T_1, \ldots, T_n$, of common source X and relevancy Y, with respective parameters $\lambda_1, \ldots, \lambda_n$, and an extension $q(X, T_1, \ldots, T_n)$ of the $q_i := q_i(X, T_i)$, such that, under q, we have the Markov chain*

$$X - T_n - \cdots - T_1. \tag{4}$$

(iii) *There exist bottlenecks $T_1, \ldots, T_n$, of common source X and relevancy Y, with respective parameters $\lambda_1, \ldots, \lambda_n$, and an extension $q(Y, X, T_1, \ldots, T_n)$ of the $q_i := q_i(Y, X, T_i)$, such that, under q, we have the Markov chain*

$$Y - X - T_n - \cdots - T_1. \tag{5}$$

Proof. See Appendix B.1. It is relatively straightforward because we started directly from a single-letter definition. $\square$

Proposition 1 was already known to be a characterisation of SR of the IB [30–32]. However, as the latter references start from an operational problem in terms of asymptotic rates and distortions for multi-letter systems, here, Proposition 1 shows that our single-letter Definition 5 is equivalent to the operational definitions in [30–32]. See Appendix B.2 for more details.

Remark 1. *Crucially, the order of the indexing in (4) and (5) depends only on the order of the trade-off parameters $\lambda_1 < \cdots < \lambda_n$, and not on the order in which the bottlenecks T_i are produced, which is just the interpretation we started from. In particular, Proposition 1 makes equally legitimate the interpretation of bottlenecks produced from the finest one to the coarsest one, each new bottleneck thus implementing a further coarsening of the source X. This alternative interpretation renders successive refinement relevant to feed-forward processing, including in particular the Blackwell order (see Section 4.1) and deep neural networks (see Section 4.2). For ease of presentation, though, we will stick to the information incorporation interpretation along most of the paper.*

Moreover, from Proposition 1, we can leverage existing IB literature to prove the successive refinability of two specific settings. (For an explicit definition of what we mean, in Proposition 2, by successive refinement in the case of the *Lagrangian* IB problem, see Appendix B.3.)

Proposition 2. *If X, Y are jointly Gaussian vectors, then the Lagrangian IB problem defined by $p(X, Y)$ is $(\lambda_1, \ldots, \lambda_n)$-refinable for all $\lambda_1 < \cdots < \lambda_n$.*

This result is a direct consequence of a property named a *semigroup structure*, and is proven for the Gaussian IB framework in [33], which relates the latter framework with renormalisation group theory. The semigroup structure denotes, in short, the situation where iterating the operation of coarse graining a variable by computing a bottleneck—where, at each iteration, the previous bottleneck becomes the source of the next IB problem—still outputs a bottleneck for the original problem. This semigroup structure is a stronger property than successive refinement and, as it is satisfied in the Gaussian case, this implies the successive refinability of Gaussian vectors (see Appendix B.3 for more details). Beyond Proposition 2, this relationship between successive refinement and the semigroup structure hints at potentially interesting links between the composition of coarse-graining operators and successive refinement. In this respect, note that our numerical results below (see Sections 2.3 and 3.2) suggest that, for non-Gaussian vectors, successive refinement does not always hold and thus, *a fortiori*, that the semigroup structure might not always be satisfied in the IB framework—or at least not perfectly.

Eventually, in the case of deterministic channel $p(Y|X)$, an explicit solution to the IB problem (1) is known [34]: $T = Y$ with probability α, and $T = e$ with probability $1 - \alpha$, for e a dummy symbol, and some well-chosen $0 < \alpha < 1$. This specific solution allows one to address successive refinement for the deterministic case:

Proposition 3. *Let X be a discrete or continuous variable, and Y be a deterministic function of X. Then, the IB problem defined by $p(X, Y)$ is successively refinable for all trade-off parameters $\lambda_1 < \cdots < \lambda_n$.*

Proof. See Appendix B.4. A proof was already proposed, from an asymptotic coding perspective, for discrete X and $Y = X$, in [58]. We use a similar argument here. □

Note, though, that the solution used here to prove successive refinement is, as noted in [34], not very interesting: it is nothing more than an increasingly noisy version of Y. It is not clear whether or not there exists more interesting bottleneck solutions in the deterministic case, and if so, whether these other solutions are successively refinable. Proposition 3 will in any case be useful for our own purposes: we will use it to set aside the deterministic case in the proof of SR for binary X and Y (Proposition 5 below).

Until now, we used existing results from the IB literature that, even though not originally aimed at it, happen to yield interesting consequences for the problem of the successive refinement of the IB. However, it seems crucial, for further progress on the latter topic, to design specifically tailored mathematical and numerical tools. This is the purpose of the following sections of this paper; in particular, in the next section, we present a simple geometric characterisation of the IB's successive refinability.

2.2. The Convex Hull Characterisation and the Case $|\mathcal{X}| = |\mathcal{Y}| = 2$

In this section, we present our convex hull characterisation of successive refinement. We then show its relevance both to numerical computations—thanks to a linear program for checking the condition—and to proving new mathematical results—which we exemplify by proving, thanks to this new characterisation, the successive refinability of binary variables. Here, as in our subsequent numerical experiments in Section 2.3, we will focus on discrete variables and $n = 2$ processing stages, even though our results are thought of as a first step towards a generalisation to continuous variables and an arbitrary number of processing stages.

The convexity approach that we propose hinges upon changing the perspective on the IB problem (1) from an optimisation over the encoder channels $q(T|X)$ to an optimisation over the *decoder* channels $q(X|T)$; indeed, (1) can be equivalently presented as the "reversed" optimisation problem

$$\underset{\substack{(q(T),q(X|T))\,:\\ \sum_t q(t)q(X|t)=p(X)\\ T-X-Y,\ I(X;T)\leq\lambda}}{\arg\max}\quad I(Y;T). \tag{6}$$

Formulations (1) and (6) yield the same solutions because, through the Markov chain $T - X - Y$, the joint distribution $q(X, Y, T)$ is equivalently determined by specifying some $q(T|X)$ or specifying some pair $\big(q(T), q(X|T)\big)$ that satisfies the consistency condition $\sum_t q(t)q(X|t) = p(X)$. This condition says that the source distribution $p(X)$ must be retrievable as a convex combination of the $q(X|t)$, where the weights are given by the $q(t)$.

Moreover, this formulation leads to a crucial intuition concerning the relationship between successive refinement and the set $\mathcal{H}_T := Hull\{q(X|t), t \in \mathcal{T}\}$, where, for a set $E \subseteq \mathbb{R}^n$, we denote by $Hull(E)$ the convex hull of E, i.e., the set of points obtained as convex combinations of points in E. First, note that, for a bottleneck T, the set $\mathcal{H}_T$ is reduced to a single point if and only if T is independent from the source X. Conversely, $\mathcal{H}_T$ coincides with the whole source simplex $\Delta_{\mathcal{X}}$ if and only if T captures all the information from the source, i.e., if $I(X;T) = H(X)$. Generalising these extreme cases suggests the intuition that $\mathcal{H}_T$ describes the *information content* held by the bottleneck T about the source X. Now, let us recall that successive refinement from a coarse bottleneck T_1 to a finer bottleneck T_2 means intuitively that T_2 can be obtained without discarding any of the information extracted by T_1 about the source X; in other words, that the information content of T_1 about the source X *is included* in that of T_2. Combining this latter intuition with the one about $\mathcal{H}_T$ being the information content of a bottleneck T suggests the following characterisation of successive refinement:

$$Hull\{q(X|t_1), t_1 \in \mathcal{T}_1\} \subseteq Hull\{q(X|t_2), t_2 \in \mathcal{T}_2\}, \tag{7}$$

where T_1 and T_2 are bottlenecks of parameters $\lambda_1 < \lambda_2$, respectively. This condition is visualised in Figure 3. The characterisation indeed holds, at least for the discrete case and under a mild assumption of injectivity of the finer bottleneck's decoder:

Proposition 4. *Let $0 < \lambda_1 < \lambda_2$, and assume that $p(X, Y)$ is discrete.*

If there is successive refinement for parameters (λ_1, λ_2), then there exist bottlenecks T_1, T_2 of parameters λ_1, λ_2, respectively, such that the convex hull condition (7) is satisfied.

Conversely, if there exist bottlenecks T_1, T_2 of parameters λ_1, λ_2, respectively, such that the convex hull condition (7) holds and such that the decoder $q(X|T_2)$, seen as a probability transition matrix, is injective, then there is successive refinement for parameters (λ_1, λ_2). Moreover in this latter case, if T_1, T_2 are bottlenecks that achieve successive refinement, the extension $\tilde{q}(X, T_1, T_2)$ of $q(X, T_1)$ and $q(X, T_2)$ such that $X - T_2 - T_1$ holds is uniquely defined.

Proof. See Appendix B.5. The idea consists in translating the Markov chain characterisation $X - T_2 - T_1$ into the convex hull condition (7). The direct sense is straightforward. For the converse direction, observe that, even though as soon as (7) is satisfied it provides a joint distribution $\tilde{q}(X, T_1, T_2)$ that satisfies the Markov chain $X - T_2 - T_1$, it is not clear whether this distribution is consistent with $q(X, T_1)$. The potential problem stems from the fact that $\tilde{q}$ must be such that the channel $\tilde{q}(T_2|T_1)$ maps the marginal $q(T_1)$ to the marginal $q(T_2)$. The injectivity assumption, however, provides a sufficient condition for it to be the case. This assumption happens to also imply the uniqueness of the extension, among all those that satisfy the Markov chain $X - T_2 - T_1$. $\square$

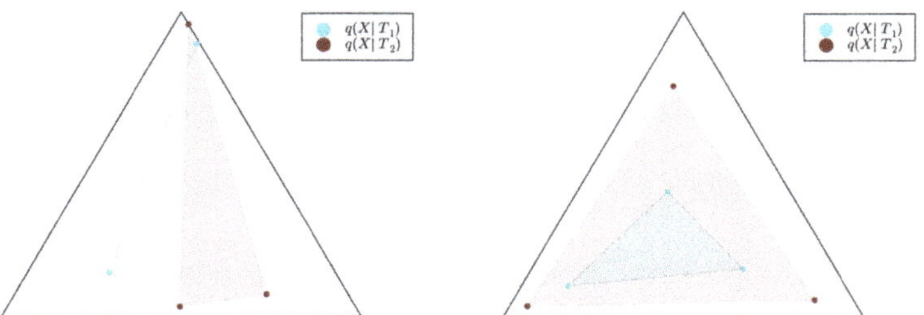

Figure 3. Illustration of the convex hull condition. The black triangle represents the source simplex $\Delta_{\mathcal{X}}$ with, here, $|\mathcal{X}| = 3$, and the pointwise bottleneck decoder probabilities $\{q(X|t), t \in \mathcal{T}\}$ are represented on it (in cyan for the coarser bottleneck T_1 and in red for the finer one T_2). The convex hull of the respective families of points are shaded with the corresponding color. On the left, the condition is not satisfied; on the right, it is.

Even though the injectivity assumption might seem restrictive, in practice, in our numerical experiments below (see Sections 2.3 and 3.2), we always found that the decoder channel $q(X|T_2)$ could be chosen as injective by reducing it to its effective cardinality (see Section 1.3)—a process that leaves the convex hull condition (7) unchanged because it leaves the points $q(X|t_2)$ unchanged. See also Appendix D for a conjecture that, if true, would simplify our convex hull characterisation in the case of a strictly concave information curve.

Remark 2. *The convex hull condition happens to be equivalent to the* input-degradedness *preorder on channels (see Proposition 1 in [59]). Even though we will not develop this point further when considering alternative interpretations of SR (Section 4), it is worth noting that, through input-degradedness, SR can be given additional operational interpretations, particularly in terms of randomised games (see Section IV-C in [59]).*

Our new characterisation provides a simple way of checking whether or not two bottlenecks T_1 and T_2 achieve SR. Recall that the Markov chain characterisation (Proposition 1, point (ii)) shows that SR is a feature of the space Δ_{q_1,q_2} of all extensions $q(X, T_1, T_2)$ of individual bottleneck distributions $q_1(X, T_1)$ and $q_2(X, T_2)$. While this set might, *a priori*, be difficult to study directly, our characterisation (7) reduces the problem to a simple geometric property relating only two explicitly given conditional distributions: $q(X|T_1)$ and $q(X|T_2)$. Moreover, note that (7) is equivalent to

$$\forall t_1 \in \mathcal{T}_1, \quad q(X|t_1) \in Hull\{q(X|t_2), t_2 \in \mathcal{T}_2\},$$

and that checking whether a point is in the convex hull of a finite set of other points can be cast as a linear programming problem [68]. As a consequence, one can bound the time complexity of checking condition (7) as $O(|\mathcal{X}| K)$, where K is the time complexity bound of a linear program with $2|\mathcal{X}| + 2$ variables and $3|\mathcal{X}| + 2$ constraints. As a consequence, using the bound on K proved in [69], the time complexity of checking (7) is no worse that $\tilde{O}(|\mathcal{X}|^{\omega+1} \log(\frac{|\mathcal{X}|}{\delta}))$, where $\omega \approx 2.38$ corresponds to the complexity of matrix multiplication, δ is the relative accuracy, and the $\tilde{O}(\cdot)$ notation hides polylogarithmic factors (see Appendix B.6 for details).

We deem this convex hull characterisation to be important for theory as well. Indeed, it reduces the question of successive refinement to a question about the structure of the trajectories, on the source probability simplex $\Delta_{\mathcal{X}}$, of the points $q_\lambda(X|t)$ for varying λ. Thus, any theoretical progress on the description of these bottleneck trajectories might lead to theoretical progress on the side of successive refinement. As a first step in this direction, we show that this geometric point of view helps to solve the question of SR in the case of a binary source and relevancy (This result generalises the already known fact that there is always successive refinement when X is a Bernoulli variable of parameter $\frac{1}{2}$ and $p(Y|X)$ is a binary symmetric channel (see Lemma 1 in [46] and see Section 4.2 for explanations on why the latter work's framework encompasses ours). Moreover, a potential generalisation of our result to an arbitrary number of processing stages is left to future work).

Proposition 5. *If $|\mathcal{X}| = |\mathcal{Y}| = 2$, then, for any discrete distribution $p(X, Y)$ and any trade-off parameters $\lambda_1 < \lambda_2$, the IB problem defined by $p(X, Y)$ is (λ_1, λ_2)-successively refinable.*

Proof. Let us here outline the proof presented in Appendix B.7. The case of deterministic $p(Y|X)$ was already dealt with in Proposition 3, so we can assume that $p(Y|X)$ is not deterministic. In this case, the IB problem with $|\mathcal{X}| = |\mathcal{Y}| = 2$ and $n = 2$ has been extensively studied in [35]. In short, the latter approach leverages the fact that a pair $(q(T), q(X|T))$ is a solution to the IB problem (6) if the convex combination of the points $F_\beta(q(X|t))$, with weights given by $q(T)$, achieves the lower convex envelope of the function F_β, where F_β is a well-chosen function on the source simplex $\Delta_{\mathcal{X}}$ and β is the information curve's inverse slope. This work, along with considerations from [37], which uses the same convexity approach, yields in particular that (i) the points $q_\beta(X|t)$ are the extreme points of a non-empty open segment uniquely defined by β, and (ii) this latter segments grows as a function of the inverse slope β and thus, by concavity, as a function of λ. This implies that the convex hull condition is always satisfied for $\lambda_1 < \lambda_2$. As point (i) also implies that, here, $q_{\lambda_2}(X|T_2)$ must be injective, Theorem 4 allows us to conclude the successive refinability for $n = 2$ processing stages. $\quad\square$

The proof of Proposition 5 exemplifies how our convex hull characterisation interlocks well with the convexity approach to the IB [35–39]. In this sense, our characterisation brings a new theoretical tool to the study of the successive refinement of the IB.

2.3. Numerical Results on Minimal Examples

In this section, we leverage our new convex hull characterisation to investigate successive refinement on minimal numerical examples, i.e., with discrete and low-cardinality distributions $p(X, Y)$. Our experiments suggest that, in general, successive refinement does not always hold exactly. However, they also highlight two other features: first, it seems that successive refinement is often shaped by IB bifurcations [41–44]. Second, even though successive refinement is often not satisfied exactly, visualisations suggest that it is often "close" to being satisfied. The formalisation of this latter intuition will be the topic of the next section.

We consider the Lagrangian form (3) of the IB problem (see Section 1.3). We compute solutions to it with the Blahut–Arimoto (BA) algorithm [1], combined with reverse deterministic annealing [19,70], starting from $\beta \approx \infty$ (i.e., in practice, $\beta \gg 1$) at the IB solution $T = X$ (we noticed that regular deterministic annealing sometimes yielded sub-optimal solutions because they followed sub-optimal branches at IB bifurcations [1,71], which was not the case for reverse annealing). We always obtained that $I(X; T)$ was a strictly increasing function of the Lagrangian parameter β, so it makes sense to index the solutions by $\lambda = I(X; T)$ rather than β; for instance, in this section and Section 3.2, we will write $q_\lambda(T|X)$ for our algorithm's output for a β such that $I(X; T) = \lambda$.

In all our numerical experiments, after reducing a bottleneck T to its canonical form (see Section 1.3), the decoder channel $q_\lambda(X|T)$ was injective. Therefore, thanks to Theorem 4, the convex hull condition (7) being satisfied here does imply successive refinement. In the remainder of the paper, we will thus use the convex hull condition as a proxy for numerically assessing successive refinement (see Appendix D for more details on what we mean by "proxy"). This condition can be investigated in two ways. First, for two distinct trade-off parameters $\lambda_1 < \lambda_2$, we can compute whether the convex hull condition (7) holds or not with the linear program described in Appendix B.6. Second, for $|\mathcal{X}| \leq 3$, we can visualise the whole trajectories, for varying λ, of the points $q_\lambda(X|t)$ on the source simplex $\Delta_\mathcal{X}$. As we will see, this yields interesting qualitative insights.

As a sanity check for our algorithm, we compute bottleneck solutions for binary X and Y, which we proved in Proposition 5 to be successively refinable for all trade-off parameters. We used the linear program to check the convex hull condition numerically for all pairs $\lambda_1 < \lambda_2$ and for distributions $p(X, Y)$ uniformly sampled on the joint probability simplex $\Delta_{\mathcal{X} \times \mathcal{Y}}$. We find that the convex hull condition is indeed always numerically satisfied.

Then, we study the case $|\mathcal{X}| = |\mathcal{Y}| = 3$, once again uniformly sampling example distributions $p(X, Y)$ on $\Delta_{\mathcal{X} \times \mathcal{Y}}$. Figures 4–6 show, for representative examples, visualisations of the trajectories over λ of the $q_\lambda(X|t)$ (left)—which we will refer to as the *bottleneck trajectories*—along with the corresponding computations of the convex hull condition as a function of λ_1 and $\lambda_2 \geq \lambda_1$ (right)—which we will refer to as the *SR patterns* (The correspodning $p(Y|X)$ are plotted in Appendix E, and $p(X)$ is shown in Figures 4–6 (left). The explicit $p(X, Y)$ corresponding to each of these paper's figures can be found at: https://gitlab.com/uh-adapsys/successive-refinement-ib/(accessed on 12 September 2023).

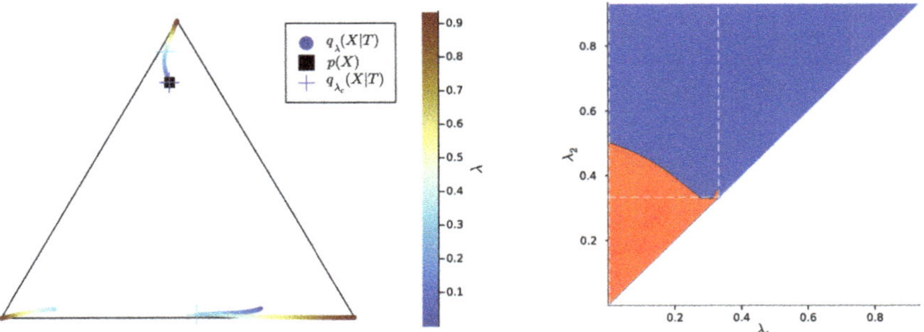

Figure 4. Left: bottleneck trajectories for an example distribution $p(X, Y)$ such that $|\mathcal{X}| = |\mathcal{Y}| = 3$, i.e., trajectory of $q_\lambda(X|T)$, represented as the family of points $\{q_\lambda(X|t),\ t \in \mathcal{T}\}$ on the source simplex $\Delta_\mathcal{X}$, as a function of $\lambda = I(X; T)$ (crosses: value of $q_{\lambda_c}(X|T)$ just before a symbol split at a critical parameter λ_c, where the crosses' color corresponds to the value of λ_c). The conditional distribution $q_\lambda(X|T)$ is defined by the single point $p(X)$ when $\lambda = 0$ (dark blue cross on the black square), or by two distinct points between the first and second symbol splits (dark blue to cyan), or by three distinct points after the second symbol split (cyan to red). Note the discontinuity of $q_\lambda(X|T)$ at each symbol split (without the discontinuity, the trajectory around a symbol split would look like a branching). Right: corresponding SR pattern, i.e., corresponding output for the convex hull condition (blue: satisfied; red: not satisfied; dashed white lines: critical values $\lambda_c(i)$ of either λ_1 or λ_2). For instance, the critical value $\lambda_c(2) \approx 0.33$ corresponds, on the bottleneck trajectories (left), to the symbol split from two to three symbols (cyan crosses). Note that $\lambda_c(1) \approx 0$. The respective $p(Y|X)$ corresponding to this figure and to Figures 5 and 6 are plotted in Appendix E.

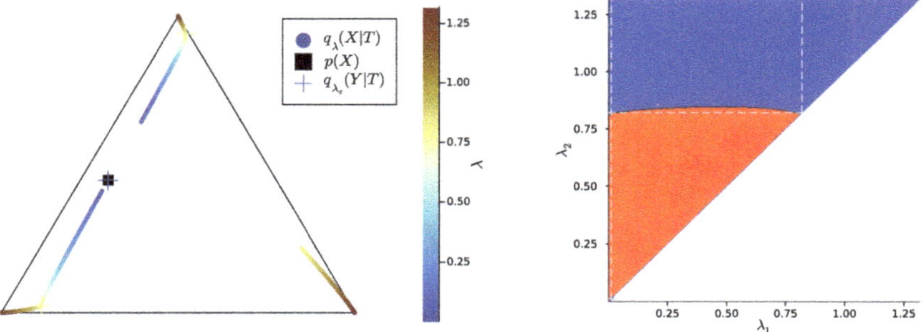

Figure 5. Same as Figure 4, with a different example distribution $p(X, Y)$ such that $|\mathcal{X}| = |\mathcal{Y}| = 3$.

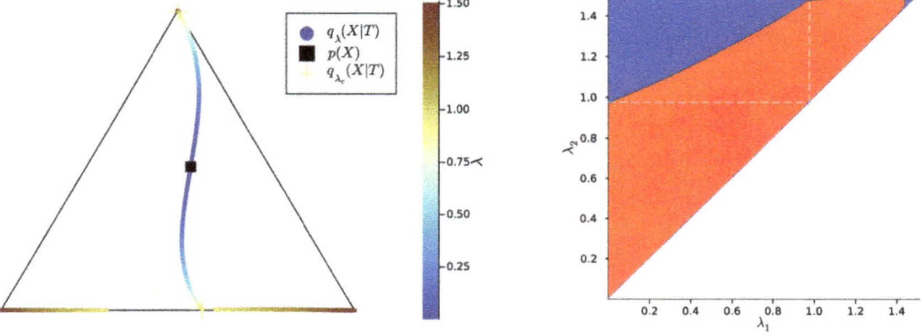

Figure 6. Same as Figure 4, with a different example distribution $p(X, Y)$ such that $|\mathcal{X}| = |\mathcal{Y}| = 3$.

Let us first give a general description of the bottleneck trajectories. For $\lambda \approx 0$, the $q_\lambda(X|t)$ all coincide with the source distribution $p(X)$. This should be the case, as, for $0 = \lambda = I(X; T)$, the bottleneck T is independent of X. Then, when λ increases, the trajectories seem piecewise continuous, where each discontinuity corresponds to a symbol split, i.e., a change in effective cardinality (see Section 1.3). We mark with a cross, for each $t \in \mathcal{T}$, the $q_\lambda(X|t) = q_{\lambda_c}(X|t)$ located just before such a change in effective cardinality.

In the examples of Figures 4–6, as $|\mathcal{X}| = 3$, there are two symbol splits, corresponding to that from one to two and two to three symbols, respectively. Eventually, for large λ, the last continuous segment of bottleneck trajectories corresponds to effective cardinality $k(T_\lambda) = |\mathcal{X}|$, and, for the maximal λ, each corner of the source simplex $\Delta_\mathcal{X}$ is reached by $q(X|t)$ for some $t \in \mathcal{T}$. This means that for maximum λ, there is a deterministic bijective relationship between T and X. The latter is expected: for maximum λ, bottlenecks are minimal sufficient statistics of X for Y [72]; where for $p(X, Y)$ sampled uniformly on the simplex, these minimal sufficient statistics are, with probability 1, just permutations of X.

Definition 6. *In the following, we refer to the piece of trajectory where the bottleneck's effective cardinality $k = k(T_\lambda)$ is equal to the integer i as the "segment $k = i$", i.e., it is the segment where $q_\lambda(X|T)$ corresponds to exactly i distinct points on the source simplex $\Delta_\mathcal{X}$; for instance, in Figure 4, the segment $k = 2$ corresponds to the first piece of trajectory spanning colors from dark blue to cyan.*

Notation 2. *We denote by $\lambda_c(i)$ the trade-off parameter's critical value corresponding to the i-th change in effective cardinality, i.e., the symbol split from i to $i + 1$ symbols. Here, we will only need to consider the critical values $\lambda_c(1) = 0$ and $\lambda_c(2)$, corresponding to the splits from one to two and two to three symbols, respectively.*

Let us now come back to the question of successive refinement: for which parameters $\lambda_1 < \lambda_2$ is the convex hull condition satisfied? The right-hand sides of Figures 4–6 provide the answers corresponding to trajectories on the respective left-hand sides—where blue and red mean that the condition is and is not satisfied, respectively. Moreover, we highlight with dashed white vertical and horizontal lines the critical parameter values $\lambda_1 = \lambda_c(i)$ and $\lambda_2 = \lambda_c(i)$, respectively, at which the symbol split occurs (see Appendix B.8 for details on the computation of these symbols splits). Note that we always have $\lambda_c(1) \approx 0$, which is expected, as a bottleneck T corresponding to some $\lambda = I(X; T) > 0$ must necessarily define at least two distinct $q_\lambda(X|t)$.

First, in these examples as in most non-reported examples, the convex hull condition (right) breaks as long as $\lambda_2 < \lambda_c(2)$, i.e., as long as the finer bottleneck's effective cardinality is at most $k = 2$. This can also be read from the bottleneck trajectories (left): if the condition was satisfied for all $\lambda_1 < \lambda_2 < \lambda_c(2)$, for instance, then the segment $k = 2$ would be a line segment. This is clearly not the case in Figures 4 and 6, and even though visually it virtually seems to be the case in Figure 5, the segment $k = 2$ happens to be very slightly curved, which is enough to break the convex hull condition. In other words, for $\lambda_1 < \lambda_2 < \lambda_c(i)$, several-stage processing seems to induce, in these examples, a nonzero loss of information optimality.

Then, for $\lambda_2 > \lambda_c(2)$, even though there is no single general pattern, the trajectory's structure at the bifurcation seems to impact successive refinement. Indeed, at the bifurcation at $\lambda_c(2)$, the set $Hull\{q_{\lambda_2}(X|t), t \in \mathcal{T}\}$ opens up along a new, third dimension, and keeps widening when λ_2 increases. This allows it to (gradually in Figures 4 and 6, or virtually straight away in Figure 5) encompass the segment $k = 2$ because it "overcomes" the curvature of this piece of trajectory. For instance, in Figure 4, because the segment $k = 2$ is strongly curved, the convex hull condition gets satisfied for all $\lambda_1 < \lambda_c(2)$ only if λ_2 is significantly larger than $\lambda_c(2)$. On the contrary, because in Figure 5, the segment $k = 2$ is virtually not curved, it is almost as soon as $\lambda_2 > \lambda_c(2)$ that the convex hull condition is satisfied for all $\lambda_1 < \lambda_c(2)$.

In Figure 6, the lack of successive refinement for $\lambda_2 > \lambda_c(2)$ does not seem to be due to the same phenomenon as the one just described. Generally speaking, we observed a whole variety of SR patterns (see Appendix F for more examples), and our aim here is not to try to interpret all of them. However, despite this diversity, the SR patterns that we studied typically shared a common qualitative feature: the bifurcation structure of the bottleneck trajectories seemingly participates in shaping these SR patterns. Mostly, it seems typically necessary, for SR to hold, that the larger parameter λ_2 has crossed the bifurcation value $\lambda_c(2)$, because the non-zero curvature of the segment $k = 2$ can only be "overcome" by opening the set $Hull\{q_{\lambda_2}(X|t), t \in \mathcal{T}\}$ along a new dimension, through the symbol split at $\lambda_2 = \lambda_c(2)$. This phenomenon will be explored in more details in Section 3.2.

Besides this relationship between SR and the structure of bottleneck bifurcations, this numerical study suggests a generalisation of the notion of successive refinement. Indeed, in Figure 5 for instance, even though the right-hand side asserts that successive refinement does not hold for $\lambda_1 < \lambda_2 < \lambda_c(2)$, the virtually linear piece of trajectory on the left-hand side suggests that this is "almost" the case. In the next section, we formalise this intuition.

3. Soft Successive Refinement of the IB

The minimal experiments from Section 2.3 suggest the intuition that even though successive refinement might not always hold exactly, when broken, it might still be "close" to being satisfied. More generally speaking, let us recall that we are trying here to understand the informationally optimal limits of several-stage information processing. As our numerical experiments suggest that the IB problem is not always successively refinable, it is desirable to *quantify* the lack of successive refinement—i.e., the lack of informational optimality induced by several-stage processing. These considerations lead to the notion of *soft successive refinement* [18], which we define and motivate in this section. As we will see, this generalisation of exact SR does not depend on the specific structure of the IB setting; rather, it can also be used as a generalisation of exact SR for *any* rate-distortion scenario.

3.1. Formalism

Let us first focus on the case $n = 2$: we thus want to quantify the amount of information captured by a coarse bottleneck T_1 and then discarded by a finer bottleneck T_2. Let us recall that, from Proposition 1, bottlenecks T_1 and T_2 achieve successive refinement if there exists an extension $q(X, T_1, T_2)$ of $q_1(X, T_1)$ and $q_2(X, T_2)$ such that, under q, we have the Markov chain $X - T_2 - T_1$, which is equivalent to $I_q(X; T_1|T_2) = 0$. It thus seems natural to quantify soft successive refinement with the conditional mutual information $I_q(X; T_1|T_2)$. However, the IB method does not entirely define the relationship between distinct bottlenecks; formally, there is a whole polytope $\Delta_{q_1,q_2} \subseteq \Delta_{\mathcal{X} \times T_1 \times T_2}$ of possible extensions $q(X, T_1, T_2)$ of $q_1(X, T_1)$ and $q_2(X, T_2)$ (see Section 1.3). Among these possible extensions, it seems natural to search for those that minimise the violation of the SR condition $I_q(X; T_1|T_2) = 0$. This leads us to use the *unique information* [40]

$$UI(X : T_1 \setminus T_2) := \min_{q \in \Delta_{q_1,q_2}} I_q(X; T_1|T_2). \tag{8}$$

This quantity was already defined in [40] in the context of partial information decomposition [61–64], and it happens to be relevant to us for several reasons.

First of all, it depends only on the distributions $q_1(X, T_1)$ and $q_2(X, T_2)$, which are indeed the only distributions provided by the IB framework. Second, from Proposition 1, there is successive refinement if and only if there are two bottlenecks T_1 and T_2 such that $UI_{q_1,q_2}(X : T_1 \setminus T_2) = 0$. Third, it is thoroughly argued in [40] that (8) is a good measure of the information that only T_1, and not T_2, has about X, which is an interpretation that coincides neatly with the intuition that we want to operationalise here. Eventually, Proposition 6 below, which first requires some definitions, provides an information-geometric justification.

Definition 7. *For Δ a probability simplex and $E_1, E_2 \subseteq \Delta$, we define*

$$D_{KL}(E_1||E_2) := \inf_{r_1 \in E_1, r_2 \in E_2} D_{KL}(r_1||r_2),$$

where D_{KL} is the Kullback–Leibler divergence: $D_{KL}(r_1||r_2) := \sum_{a \in \mathcal{A}} r_1(a) \log\left(\frac{r_1(a)}{r_2(a)}\right)$, if the probability distributions r_1 and r_2 are defined on the discrete alphabet $\mathcal{A}$.

Definition 8. *The* successive refinement set $\Delta_{SR,n} \subseteq \Delta_{\mathcal{X} \times \mathcal{T}_1 \times \cdots \times \mathcal{T}_n}$ *is the set of distributions r on $\mathcal{X} \times \mathcal{T}_1 \times \cdots \times \mathcal{T}_n$ such that, under r, the Markov chain $X - T_n - \cdots - T_1$ holds.*

Note that $\Delta_{SR,n}$ does not require its elements to be extensions of any fixed bottleneck distributions $q_i(X, T_i)$ but imposes the Markov chain that characterises SR (see Proposition 1). SR is achieved for bottlenecks $q_1(X, T_1), \ldots, q_n(X, T_n)$ if and only if the successive refinement set $\Delta_{SR,n}$ and the extension set $\Delta_{q_1,\ldots,q_n}$ share a common distribution $q \in \Delta_{SR,n} \cap \Delta_{q_1,\ldots,q_n}$. In general (for $n = 2$), the following proposition can easily be derived:

Proposition 6. *For fixed distributions $q_1 = q_1(X, T_1)$, $q_2 = q_2(X, T_2)$, we have*

$$UI(X : T_1 \setminus T_2) = D_{KL}(\Delta_{q_1,q_2}||\Delta_{SR,2}). \tag{9}$$

Proof. See Appendix C.1. $\square$

In this sense, $UI(X : T_1 \setminus T_2)$ quantifies "how far" the joint distributions extending the bottlenecks T_1 and T_2 are from making the successive refinement condition $X - T_2 - T_1$ hold true, where the "distance" is understood as a minimised Kullback–Leibler divergence. Our new measure of soft SR is continuous:

Proposition 7 ([73], Property P.7). *The unique information $UI(X : T_1 \setminus T_2)$ is a continuous function of the probabilities $q_1(X, T_1)$ and $q_2(X, T_2)$.*

Remark 3. *In particular, if $UI(X : T_1 \setminus T_2)$ has a discontinuity as a function of the parameter λ_1 or λ_2, which define the bottleneck distribution $q_{\lambda_1}(X, T_1)$ or $q_{\lambda_2}(X, T_2)$, respectively, then this can only be a consequence of a discontinuity of the probability $q_{\lambda_1}(X, T_1)$ as a function of λ_1 or $q_{\lambda_2}(X, T_2)$ as a function of λ_2, itself, respectively. This consideration will be useful for analysing our numerical experiments in Section 3.2.*

Moreover, the formulation (9) of unique information suggests a natural generalisation to an arbitrary number of processing stages:

Definition 9. *Let $T_1, \ldots, T_n$ be bottlenecks with respective parameters $\lambda_1 < \cdots < \lambda_n$, and $q_i(X, T_i)$ their respective individual distributions. One can quantify soft successive refinement, or, equivalently, the lack of successive refinement, through the divergence $D_{KL}(\Delta_{q_1,\ldots,q_n}||\Delta_{SR,n})$.*

While [74] proposes a provably convergent algorithm to compute $UI(X : T_1 \setminus T_2)$, to the best of our knowledge, there currently exists no provably convergent algorithm to compute $D_{KL}(\Delta_{q_1,\ldots,q_n}||\Delta_{SR,n})$ for $n > 2$. Our numerical investigations (see Section 3.2) will stick to the case $n = 2$, but this generalisation makes soft SR in particular, at least conceptually for now, more relevant to deep learning (see Section 4.2).

For the sake of completeness, let us point out that for each λ, there is a whole set of solutions $q_\lambda(T|X)$—or, equivalently, $q_\lambda(X, T)$—to the IB problem (1). Thus, the unique information, which is defined as a function of specific bottleneck distributions $q_1(X, T_1)$ and $q_2(X, T_2)$, could *a priori* not be uniquely defined by the corresponding trade-off parameters λ_1 and λ_2. This subtlety is further explained in Appendix D, where we also formulate a conjecture that would prove that, at least in the case of a strictly concave information curve, the trade-off parameters do uniquely define the unique information.

3.2. Numerical Results on Minimal Examples

A provably convergent algorithm that computes, in the discrete case, the unique information (8), was provided in [74]. In this section, we use the authors' implementation of this algorithm (https://github.com/infodeco/computeUI, accessed on 12 September 2023) to qualitatively investigate, on minimal examples, the landscapes of unique information (UI) and their relationship to the bottleneck trajectories on the simplex.

In Figures 7–9 (left), we plot again the same bottlenecks trajectories as in Figures 4–6 (left), but compare them this time with the unique information $UI(X : T_1 \setminus T_2)$, plotted as a function of λ_1 and λ_2 (right). We also plot, in Figures 10–12, some representative examples of the exact SR patterns (left) and UI landscapes (right) for slightly larger source and relevancy cardinalities, where $p(X, Y)$ is, as above, uniformly sampled — the explicit distributions $p(X, Y)$ corresponding to Figures 10–12 can be found at https://gitlab.com/uh-adapsys/successive-refinement-ib/. (see Appendix F for additional examples of comparison of the UI landscapes with bottleneck trajectories, and with the exact SR patterns.) Once again, we highlight with dashed white vertical and horizontal lines the critical parameter values $\lambda_1 = \lambda_c(i)$ and $\lambda_2 = \lambda_c(i)$, respectively, where, as expected, $\lambda_c(1) \approx 0$. We will first describe, for a fixed $p(X, Y)$, the relative variations in unique information as a function of λ_1 and λ_2. Then, we will compare the absolute values of unique information to the information globally processed by the system.

For all Figures from Figures 7–9, the UI landscape partly mirrors the respective exact SR pattern of Figures 4–6 (right). However, within the region where these latter figures answered a binary "no" to the question of exact SR, Figures 7–9 reveal a sharply uneven variation in the violation of SR, where, for important ranges of trade-off parameters, the unique information is negligible comparative to others. For instance, even though Figure 5 (right) seems to indicate that SR does not hold for $\lambda_1 < \lambda_2 < \lambda_c(2)$, the corresponding UI in Figure 8 (right) is virtually zero on a large part of this set of parameters, while still peaking for λ_2 close to $\lambda_c(2)$. This richer structure of the unique information landscape is further evidenced by Figures 10–12.

Moreover, the unique information landscapes seem shaped by the bottleneck trajectories. Most importantly, the influence of IB bifurcations on SR can be seen even more clearly with soft than with exact SR. In particular, in Figures 10–12, it seems that along the lines where one of the trade-off parameters crosses a critical value, the UI often goes through discontinuities, or at least sharp variations in either λ_1, λ_2, or both directions. In particular, even though patterns widely vary across different example distributions $p(X, Y)$, unique information can significantly *drop* when λ_2 crosses a critical value from below—a feature observed in both shown and non-shown examples. As we know that the unique information is continuous, the apparent discontinuity should be one of the bottleneck probability $q_{\lambda_2}(X, T_2)$ itself (see Proposition 7 and Remark 3). This is consistent with the observation from Section 2.3 that, at symbol splits, the trajectory of $q_\lambda(X|T)$ often seems to go through a discontinuity. Further, the fact that the sharp variation in UI is a *decrease* in this quantity (in increasing order of λ_2) is intuitively consistent with the fact that the bottleneck trajectory's discontinuity often induces a sudden "widening" (in increasing order of λ) of

$$\mathcal{H}_T := Hull\{q(X|t), t \in \mathcal{T}\}.$$

Indeed, for fixed λ_1, when λ_2 crosses a critical value from below, the corresponding symbol split means that $\mathcal{H}_{T_2}$ "widens" by opening up a new dimension, so it "more easily" encompasses $\mathcal{H}_{T_1}$, yielding as a consequence a drop in unique information. Recalling our intuition (see Section 2.2) that $\mathcal{H}_T$ describes the information content that a bottleneck T contains about the source X, the feature just described can be interpreted in the following way: the IB bifurcations seem to induce a sudden "expansion" (in increasing order of λ) of the information content carried by the bottleneck about the source, which makes the latter's content more easily contain the information content of coarser bottlenecks.

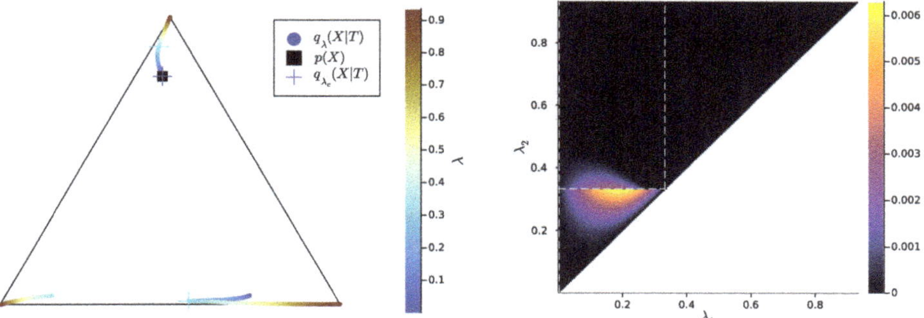

Figure 7. Left: example trajectory of $q_\lambda(X|T)$ as a function of $\lambda = I(X;T)$ (crosses: value of $q_{\lambda_c}(X|T)$ just before a symbol split at a critical parameter λ_c). Right: corresponding unique information, in bits (color), expressed as a function of the pair of trade-off parameters (white dashed lines indicate critical values $\lambda_c(i)$ of either λ_1 or λ_2.). For instance, the critical value $\lambda_c(2) \approx 0.33$ (right) corresponds, on the bottleneck trajectories (left), to the symbol split from two to three symbols (cyan crosses). The respective $p(Y|X)$ corresponding to this figure and to Figures 8 and 9 are plotted in Appendix E.

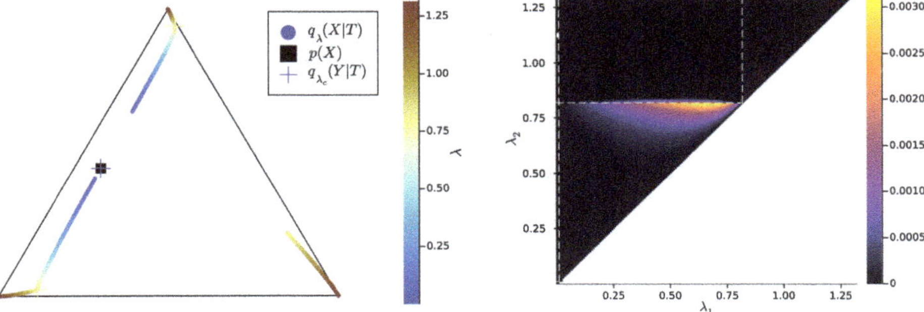

Figure 8. Same as Figure 7, where the example distribution $p(X, Y)$ is that of Figure 5.

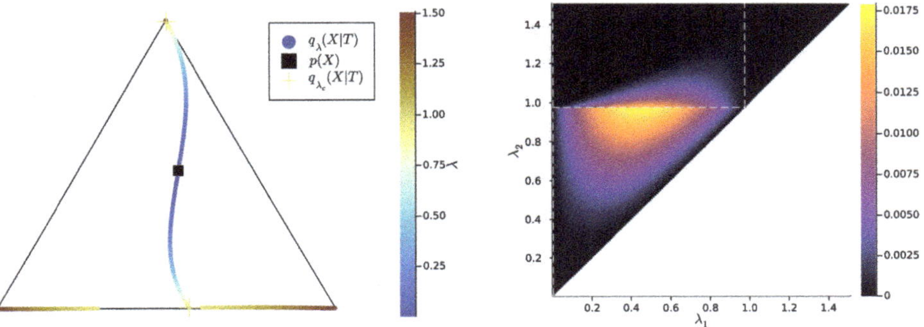

Figure 9. Same as Figure 7, where the example distribution $p(X, Y)$ is that of Figure 6.

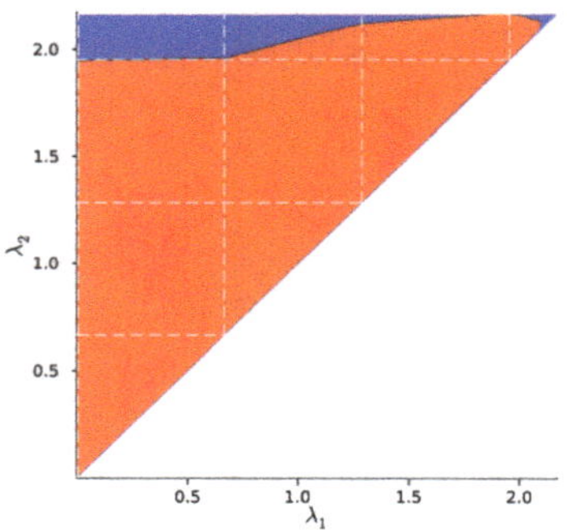 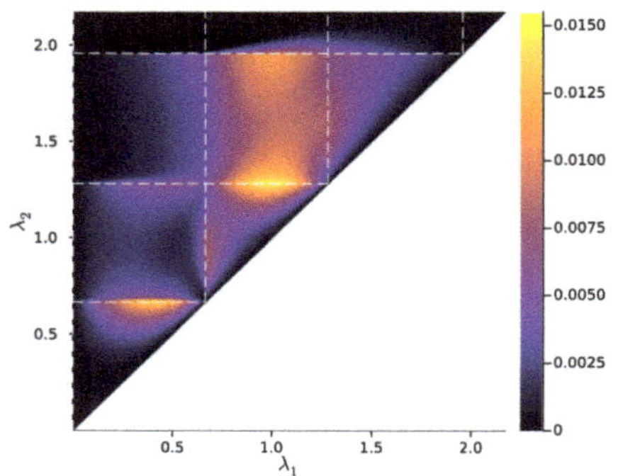

Figure 10. New example of an exact SR pattern and the corresponding UI landscape over trade-off parameters $\lambda_1 < \lambda_2$, where, here, $|\mathcal{X}| = 5$ and $|\mathcal{Y}| = 3$. Left: exact SR pattern, i.e., output for the convex hull condition (blue: satisfied, red: not satisfied). Right: corresponding UI landscape, in bits (color). White dashed lines indicate critical values $\lambda_c(i)$ of either λ_1 or λ_2. Note that (i) the binary notion of exact SR (left) filters out most of the structure unveiled by UI (right), (ii) the UI landscape seems highly impacted by IB bifurcations, and (iii) the UI is in any case always small, even though not entirely negligible. See main text for more details.

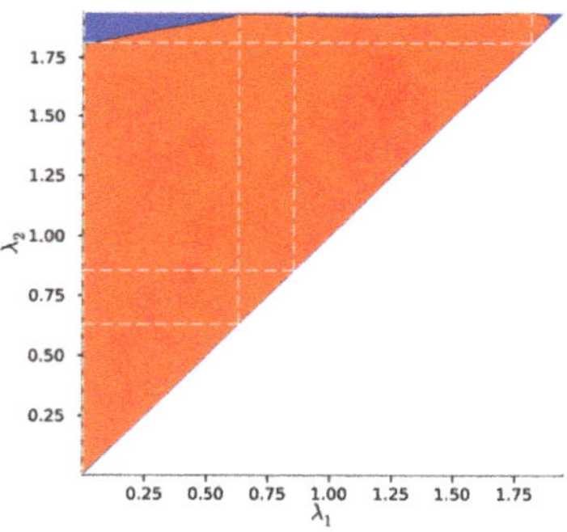 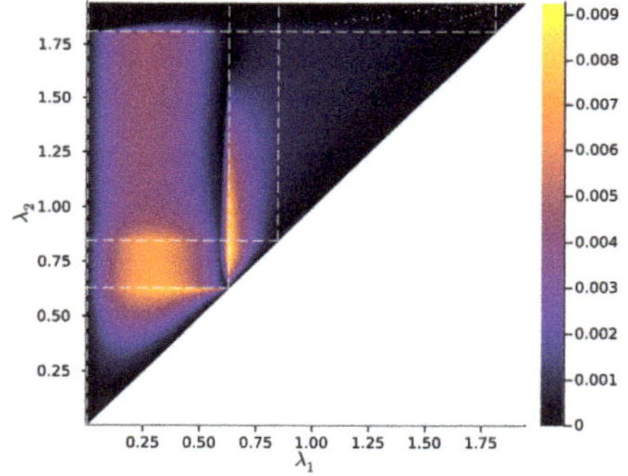

Figure 11. Same as Figure 10, with a new example distribution $p(X, Y)$, where, here, $|\mathcal{X}| = 5$ and $|\mathcal{Y}| = 3$. Besides the white orthogonal dashed lines, other white dots correspond to values of (λ_1, λ_2) for which the algorithm did not converge (see main text for a comment on this lack of convergence).

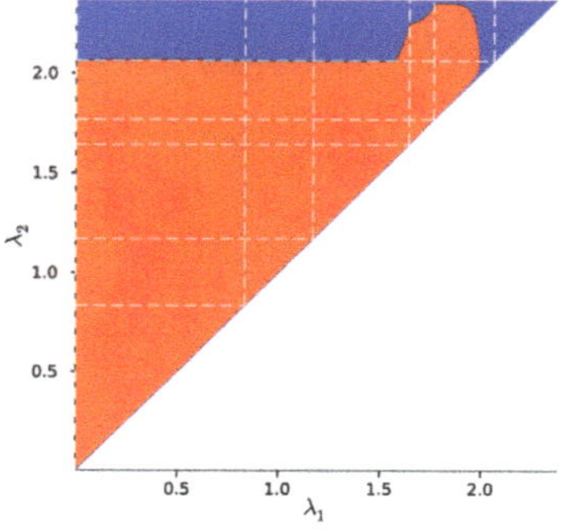 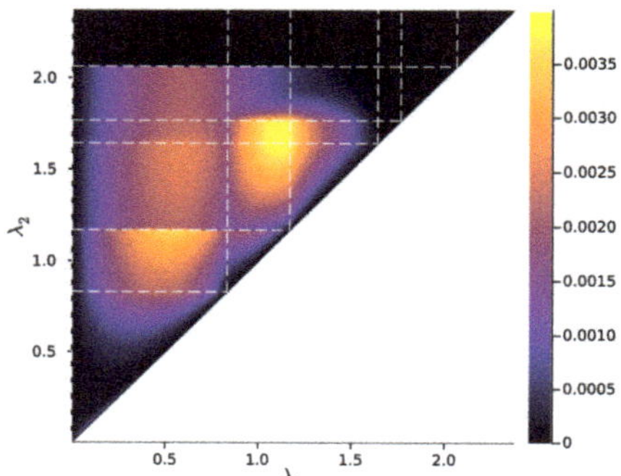

Figure 12. Same as Figure 10, with a new example distribution $p(X, Y)$, where, here, $|\mathcal{X}| = 7$ and $|\mathcal{Y}| = 5$.

Note, however, that these simple numerical results do not allow one to discriminate between the interpretation of the UI's sharp variations at bifurcations as a discontinuity with regard to trade-off parameters, or a discontinuity of the UI's *differential*. For instance, if the derivative with regard to λ_2 discontinuously takes a value close to $-\infty$ for λ_2 slightly larger than some λ_c, then the UI graph can seem discontinuous at finite numerical resolution, even if, formally, only the UI's differential is so. On the other hand, as an example, bifurcations can be characterised precisely as points of discontinuities of the derivatives, with regard to the trade-off parameter, of $I(T; X)$ and $I(T; Y)$ [43,75], even though the functions themselves are continuous [2,75]. A more involved analysis distinguishing discontinuities of UI from those of its differential is left to future work. In any case, the interpretation as a discontinuity of the differential rests on a weaker assumption, which is still sufficient for explaining the numerical results.

More generally, these results suggest that for a several-stage processing that is IB-optimal at each stage, to minimise the information discarded along stages, the trade-off parameters should rather lie close to well-chosen IB bifurcations. If this happens to be a general feature of the IB framework, it would have implications for incremental learning. Indeed, coming back to the modelling of embodied agents (see Section 1), for instance, it would mean that organisms that are poised close to information optimality by evolution should have a very specific structure of developmental learning, where the stages of learning should be discrete and determined by the right trade-off parameters.

Eventually, a last crucial feature was also satisfied on these minimal examples: whatever the structure of bottleneck trajectories, the maximal UI was significantly lower than the mutual information $I(X; T_1, T_2)$ between the external source X and the system's internal representations (T_1, T_2). More precisely, for an extension $q(X, T_1, T_2)$ of $q_{\lambda_1} := q_{\lambda_1}(X, T_1)$ and $q_{\lambda_2} := q_{\lambda_2}(X, T_2)$ that achieves the minimum in (8), let us define

$$\sigma(q_{\lambda_1}, q_{\lambda_2}) := \frac{UI_{q_{\lambda_1}, q_{\lambda_2}}(X : T_1 \setminus T_2)}{I_q(X; T_1, T_2)}.$$

Note that decomposing $I_q(X; T_1, T_2)$, where $q \in \Delta_{q_1, q_2}$, with the chain rule for mutual information shows that this quantity only depends on q_{λ_1} and q_{λ_2}: thus here, $\sigma(q_{\lambda_1}, q_{\lambda_2})$ is indeed well-defined by q_{λ_1} and q_{λ_2}. The maximum ratio over all trade-off parameters $\lambda_1 < \lambda_2$ was typically of the order of 1% in our minimal experiments; for instance, it was 1.89%, 0.39%, 1.82%, 2.03%, 1.34%, and 0.31% for the IB problems corresponding to Figures 7–12, respectively. Among all the (shown and non-shown) studied examples, it never exceeded 5.4%, and we did not notice an increase in this maximum ratio when the source or relevancy cardinalities were increased (the largest cardinalities that we experimented with were $|\mathcal{X}| = 20$, $|\mathcal{Y}| = 10$). In short, even though several-stage processing might incur a non-negligible loss of information optimality in the IB sense, these results suggest that this loss could often be significantly limited. Of course, here as in Section 2.3, on the one hand, the numerical results are purely phenomenological, and, on the other, it is at this stage far from being clear that the qualitative insights brought by these minimal experiments generalise well to more complex situations. However, they exhibit the potentially crucial qualitative features of exact and soft successive refinement in the IB framework, which can be targeted by further theoretical research.

4. Alternative Interpretations: Decision Problems and Deep Learning

The notion of successive refinement presented in this work builds on the intuition of the optimal incorporation of information. However, alternative interpretations can be given to the very same mathematical notion. First, thanks to the Sherman–Stein–Blackwell theorem [45,65], the rate-distortion-theoretic notion of SR can be shown to be equivalent to a specific order relation between the encoder of the finer bottleneck $q(T_2|X)$ and that of the coarser one $q(T_1|X)$, namely the *Blackwell order*. This point of view turns SR into an operational *decision-theoretic statement*; in short, there is SR when, for *any* task and *any* source distribution $p(X)$, the optimal performance is better (or at least as good) when decisions are based on the output of $q(T_2|X)$ than when they are based on the output of $q(T_1|X)$. Second, the Markov chain (4) characterising successive refinement makes it directly relevant [46] to the IB analysis of deep neural networks [49–56]. In the next two sections, we make these connections explicit and relate them to this paper's investigations.

4.1. Successive Refinement, Decision Problems, and Orders on Encoder Channels

Here, we show that exact and soft successive refinement can be, in the discrete case at least, understood in terms of optimally solving decision problems on arbitrary tasks, through orders on the encoder channels $q(T|X)$ (or more precisely, pre-orders: i.e., we will consider binary relations that are reflexive and transitive). We will rely on [45], where these orders were considered.

Let us first make clear what we mean here by a decision problem. Consider a state variable X over a finite set $\mathcal{X}$, another finite set $\mathcal{A}$ of possible actions, and a reward function $u = u(x, a)$ that depends on both the value x of the state X, and the chosen action $a \in \mathcal{A}$. The agent's observation is not the state X itself, but only the output T of X through some stochastic channel $\kappa := p(T|X)$ (where we assume here that the observation space $\mathcal{T}$ is finite). To each observation-dependent policy $\pi = \pi(A|T)$ corresponds an expected reward

$$\mathbb{E}_\pi(u(X, A)) := \sum_t p(t)\mathbb{E}_{(X,A)\sim p(X|t)\pi(A|t)}(u(X, A)),$$

where $p(X|t)$ is determined from $\kappa := p(T|X)$, $p(X)$ through the Bayes rule, and $p(X|t)$ $\pi(A|t)$ denotes the product measure of $p(X|t)$ and $\pi(A|t)$. Solving the decision problem $(p(X), \mathcal{A}, u)$ for the observation channel κ means choosing a policy that yields an optimal expected reward

$$\mathcal{R}(p(X), \kappa, u) := \max_\pi \mathbb{E}_\pi(u(X, A)).$$

For instance, any Markov decision process can be seen as a decision problem as defined above (for discrete time and finite state-space, number of possible actions at each state, and horizon). In this case, $\mathcal{X}$ and $\mathcal{T}$ are the spaces of state trajectories and observation trajectories, respectively, that an agent can go through along one episode; $\mathcal{A}$ is the space of action sequences that can be chosen along the episode; and u is the cumulative reward, i.e., the (potentially discounted) sum of rewards obtained at each time-step in the episode. (See, e.g., [76] for more details on the terminology used in this example.)

We can now define the following order [45]:

Definition 10. *For two channels κ and μ, we write $\kappa \sqsupseteq_{\mathcal{X}} \mu$, if, for any decision problem $(p(X), \mathcal{A}, u)$, we have*

$$\mathcal{R}(p(X), \kappa, u) \geq \mathcal{R}(p(X), \mu, u).$$

In short, $\kappa \sqsupseteq_{\mathcal{X}} \mu$ means that, for any conceivable task based on any data distribution $p(X)$ over the fixed data space $\mathcal{X}$, the observation channel κ can yield a performance at least as good as that of the observation channel μ—if combined with a well-chosen policy. The second order is the *Blackwell order* [65]:

Definition 11. *For two channels κ and μ, we write $\kappa \sqsupseteq'_{\mathcal{X}} \mu$ if there exists a channel η such that $\mu = \eta \circ \kappa$, where "$\circ$" denotes the composition of channels, i.e., such that $M_\mu = M_\eta M_\kappa$, where M_μ, M_η, and M_κ are the column transition matrices corresponding to μ, η, and κ.*

It turns out that successive refinement can be characterised by either of these two orders, thanks to the Sherman–Stein–Blackwell theorem [45,65]. In other words, SR, which is *a priori* not a decision-theoretic statement, turns into one through its equivalence with the Blackwell order:

Proposition 8. *Let $0 < \lambda_1 < \lambda_2$. The following are equivalent:*

(i)   There is successive refinement for parameters (λ_1, λ_2).
(ii) There are bottlenecks T_1, T_2 of respective parameters λ_1, λ_2 such that

$$q(T_2|X) \sqsupseteq_{\mathcal{X}} q(T_1|X).$$

(iii) There are bottlenecks T_1, T_2 of respective parameters λ_1, λ_2 such that

$$q(T_2|X) \sqsupseteq'_{\mathcal{X}} q(T_1|X).$$

Proof. Using the Markov chain characterisation (point (ii) in Proposition 1), the result is nothing more than a reformulation of Theorem 4 in [45] in the language of the present paper. Note that, to use this theorem, we need to assume that the source X is fully supported, but this is indeed an assumption that we are using along the whole paper because it does not incur any loss of generality (see Section 1.3). $\square$

Let us highlight the intuitive meaning of Proposition 8. Point (ii) means that there is SR when the coarse representation T_1 can be retrieved by post-processing the finer representation T_2—which has implications in terms of feed-forward processing (see Section 4.2).

Now, the equivalence of SR with point (iii) relies on the mathematically deep part of the Sherman–Stein–Blackwell theorem [45], and provides a new operational meaning to SR. Namely, there is SR when, for *any* distribution $p(X)$ on the source, and *any* reward function, the optimal performance is at least as good when the decisions are based on the output of $q(T_2|X)$, seen as an observation channel, than when they are based on the output of $q(T_1|X)$. Let us stress that the fact that $q(T_2|X)$ defines a finer bottleneck than $q(T_1|X)$ crucially depends on $p(X, Y)$, i.e., on the specific source distribution $p(X)$, and on how the latter relates to the specific relevancy variable through $p(Y|X)$. Proposition 8 shows that SR describes a much more "universal" relation between the channels $q(T_1|X)$ and $q(T_2|X)$.

For example, assume that evolution poises the sensors of a given biological organism at optimality in the IB sense [10,16], i.e., if X is the environment, S some sensor's output (e.g., a retina's ganglion cells activation), and Y a behaviourally relevant feature (e.g., the edibility of food), then S is a bottleneck for $p(X, Y)$. Successive refinement here means that if the sensor S_2 is finer than S_1 as a bottleneck for the fixed feature Y relevant to a particular task, then S_2 will afford to the organism—if combined with the right decision making—better performances than S_1 on any other task, for any other input distribution $p(X)$. In other words, S_2 is then "universally better" than S_1, which is a very strong (and somewhat unexpected) generalisation.

Eventually, the unique information that we chose as our measure of soft SR has initially been thought precisely as measuring the deviation from the order "$\sqsupseteq_{\mathcal{X}}$" (see arguments in [45]). Unique information can thus, for instance, be understood as quantifying the deviation from a finer IB-optimal sensor to be "universally better" than a coarser one.

4.2. Successive Refinement and Deep Learning

As suggested by Remark 1 and Proposition 8-(ii), successive refinement can be equally well understood in terms of feed-forward processing, an interpretation which is particularly relevant to deep neural networks. Indeed, while the information bottleneck theory of deep learning [49–51] is still under debate [52–56], our results can be connected to some of this theory's specific claims concerning the benefits of hidden and output layers' IB-optimality.

Let $L_1, \ldots, L_n$ denote the successive layers of a feed-forward deep neural network (DNN), which is fed with an input X and attempts to extract, within it, information about a target variable Y, thus satisfying the Markov chain [49]

$$Y - X - L_1 - \cdots - L_n. \tag{10}$$

One of the claims of the IB theory of DNNs [49–51] is that, once converged, a DNN's hidden and output layers lie close to the information curve of the IB problem defined by $p(X, Y)$, with each new layer corresponding to a coarser trade-off parameter. The performance and generalisation abilities of DNNs would rely on this IB-optimality of networks after training. While these claims have been challenged [52,77], the identified caveats have sparked a still ongoing line of research [54–56], which suggests that more nuanced versions of the initial claims might still hold. Most importantly for us here, numerical results suggest that layer-by-layer training with the IB Lagrangian as the loss function induces a performance on par with end-to-end training with cross-entropy loss [54], while recent theoretical work proved that the IB trade-off optimises a bound on the generalisation error [56]. In other words, the IB method seems to be relevant at least as a normative, if not descriptive, framework for DNNs. Thus, an interesting informationally optimal limit to compare a given DNN to is a sequence of variables $L_1, \ldots, L_n$ that

(i)   Satisfy the Markov chain (10); and

(ii)   Are each bottlenecks with source X and relevancy Y, for respective trade-off parameters $\lambda_1 > \cdots > \lambda_n$.

However, it is not clear that variables satisfying those conditions even exist; actually, it is the case if and only if the IB problem is $(\lambda_n, \ldots, \lambda_1)$-successively refinable. Indeed, points (i) and (ii) are exactly the conditions of point (iii) in Proposition 1, with $T_i := L_{n-i}$, and the order of trade-off parameters reversed as well. In this sense, the notion of exact successive refinement is relevant to deep learning; in particular—as suggested by the numerical results from Section 2.3—it might well be the case that there is successive refinement only for well-chosen combinations of trade-off parameters. In this case, an IB-optimal DNN should be designed and trained in such a way that its successive layers implement a compression corresponding to these well-chosen trade-off parameters.

Remark 4. *The single-letter formulation above mirrors, in large part, the asymptotic coding version of [46]. More precisely, Ref. [46] defines in asymptotic coding terms a feed-forward processing pipeline where each layer tries to extract, from the input coming from the previous layer, information about a potentially distinct relevancy Y_i. Theorem 2 in [46] shows that, for constant relevancy $Y_i := Y$, the notion of "successive refinement" defined there by the authors happens to be equivalent to points (i) and (ii) above, and thus to our notion of "successive refinement". In particular, the deep learning interpretation presented in this section also has an operational formulation in terms of asymptotic coding.*

Now, if exact SR describes the situation where each layer of a DNN can potentially reach the information curve, is our notion of soft SR also relevant to deep learning? Note that, here,

- We know that the variables $L_1, \ldots, L_n$ must satisfy $X - L_1 - \cdots - L_n$, i.e., we know that the joint distribution $q := q(X, L_n, \ldots, L_1)$ must be in Δ_{SR};
- And we want to know "how close" we can choose this joint distribution q to one whose marginals $q(X, L_1), \ldots, q(X, L_n)$ coincide with bottleneck distributions $q_1 := q_1(X, T_1), \ldots, q_n := q_n(X, T_n)$, respectively, of parameters $\lambda_1 > \cdots > \lambda_n$, respectively, i.e., we want to know how close we can choose q to the set $\Delta_{q_1, \ldots, q_n}$.

Thus, the quantity $D_{KL}(\Delta_{q_1, \ldots, q_n} || \Delta_{SR,n})$ can also be interpreted as a measure of the deficiency of a DNN's layers from all those simultaneously being bottlenecks. Note, however, that, in previous sections, we knew that any joint distribution $q(X, T_1, \ldots, T_n)$ had to be in the extension set $\Delta_{q_1, \ldots, q_n}$, and wanted to know "how close" to the successive refinement set $\Delta_{SR,n}$, in the KL sense, we could choose it. On the contrary, in the case of DNNs, we know that any $q(X, T_1, \ldots, T_n)$ must be in $\Delta_{SR,n}$—because the bottlenecks correspond to a DNN's layer—and want to know "how close" to $\Delta_{q_1, \ldots, q_n}$ we can choose it.

From this perspective, the numerical results of Section 3.2 suggest interesting properties, or at least desirable features, of DNNs. First, if the fact that the UI is typically low generalises well from our minimal investigation to the much richer deep learning setting, this would imply that even in situations where a DNN's successive layers cannot all lie exactly along the information curve, they might still be able to remain reasonably close to it. Second, the fact that UI (or its differential) seems to go through a discontinuity close to well-chosen bifurcations—such that the UI sharply drops when λ_2 crosses the bifurcation from below—suggests that, for each layer of the DNN to be individually as IB-optimal as possible, their corresponding trade-off parameters should each lie close to these IB bifurcations. This resonates with previous considerations suggesting that DNNs' hidden layers should [49] or might indeed do [50] lie at IB bifurcations.

5. Limitations and Future Work

Our convex hull characterisation intertwines the question of exact SR with the more fundamental question of the structure of decoder curves

$$\left\{ (\lambda \mapsto q_\lambda(X|t)), \, t \in \mathcal{T} \right\} \tag{11}$$

on the source simplex $\Delta_\mathcal{X}$, a question for which the convexity approach to the IB problem [35–39] seems promising. In short, this approach reformulates the IB problem to that of finding the lower convex envelope of a well-chosen function F_β, defined on the source simplex $\Delta_\mathcal{X}$, and parameterised by the information curve's inverse slope β (see Appendix B.7). More precisely, bottlenecks are essentially characterised by the fact that the lower convex envelope must be achieved by convex combinations of the points $F_\beta(q(X|t))$; this approach thus provides analytical tools for proving key properties of the set of trajectories (11), which would then have consequences for SR through the convex hull condition. Despite the limited scope of the result itself, the proof of Proposition 5 gives an example of such a fruitful interaction, thus suggesting a way forward for further theoretical progress. As a first step, one could try to use the convexity approach to the IB to prove our Conjecture 1 about the unicity, up to permutations and injectivity of $q(X|T)$, for canonical bottlenecks and the strictly concave information curve. This would both simplify our convex hull characterisation of SR for the case of the strictly concave information curve (see Appendix D) and provide in itself a crucial property of the curves (11). Generally speaking, leveraging, through our convex hull characterisation, the convexity approach to the IB problem might allow one to (i) identify new wholly refinable IB problems, but also (ii) produce general methods to identify, for a given distribution $p(X, Y)$, the combination of parameters for which exact SR holds.

It must be stressed that even though we motivate the successive refinement of the IB by diverse scientific questions in Sections 2 and 4, in this work, we do not model any concrete system. Rather, our minimal numerical experiments target the qualitative exploration of the formalised problem. Our results might in turn be relevant for future modelling work (see the last paragraph of this section), but the most pressing aspect is to first develop further the theoretical and computational framework. In particular, it seems important to describe formally the apparent discontinuity of UI (or its differential) as a function of the trade-off parameters λ_1 and λ_2 at IB bifurcations (through that of the $q_\lambda(X, T)$ as functions of λ); to describe more formally why the UI tends to peak and then drop close to IB bifurcations; to provide global bounds on UI in general or as functions of the source and relevancy distribution $p(X, Y)$; or to make formal the informal relationship between the "extent to which" the convex hull condition is broken, and variations in UI. Another interesting contribution would be to provide an asymptotic coding interpretation to unique information; indeed, the deviation from successive refinement is more classically quantified as a difference between asymptotic rates or distortions (see, e.g., [60]), and it is not clear whether or not this interpretation can be made for UI. Numerically speaking, one could design algorithms allowing for the computation of UI for continuous $p(X, Y)$ and/or more than two processing stages. Indeed, the algorithm from [74] only encompasses the case of discrete variables and two processing stages. One could, for instance, take inspiration from [74] to formulate the quantity $D_{KL}(\Delta_{q_1,\ldots,q_n}||\Delta_{SR,n})$ as a double minimisation problem over separate parameters, allowing for an alternating optimisation algorithm.

The deep learning interpretation of (exact and soft) SR depends crucially on some aspects of the ongoing debate on the IB theory of deep learning [49–52,54–56]. In this regard, it would be interesting to directly measure the unique information between different layers of a DNN or determine whether or not having the layers lying close to IB bifurcations does induce better performance or generalisation capabilities.

Let us point out that our framework considers that the source of information X and the target variable Y are the same along all processing stages. More general frameworks could allow for variations in either the source of information (as in the case in temporal series) or the target variable (as is the case in transfer learning). Frameworks for both these kinds of extensions have already been proposed [46,78], and it would be interesting to study if, in these cases as well, the specific nature of the IB problem imprints the informationally optimal limits of several-stage processing.

Eventually, we deem the interpretation in terms of the incorporation of information to be particularly relevant to modelling adaptive behaviour. For instance, for a given developmental or skill-learning problem on a given task, our framework could help in distinguishing situations where the choice of the successive representations' complexity along incrementally learning the task does not matter (i.e., when there is successive refinement) from situations where these complexities must be minutely weighed, so as to avoid as much as possible the "waste" of cognitive work along the way (i.e., when the unique information is not negligible and unevenly distributed). In the latter case, our framework, once mature, might precisely describe those sequences of representations' complexity that minimise the "waste" of cognitive work from one learning stage to another, thereby potentially identifying key stages of skill or developmental learning. Future work should keep in mind the horizon of identifying such qualitative features and producing measures capturing the relevant phenomena for experimental research in these areas.

6. Conclusions

Our approach in this paper is three-fold: to bring together in a common framework existing work on the exact successive refinement of the IB and related topics; to develop further this common framework, particularly through a geometric approach to the problem; and to then open up a line of research on the soft successive refinement of the IB.

The formal unity that we make explicit in this paper is mainly that between these three scientific questions: (*i*) that of informationally optimal incorporation of information—relevant in particular to developmental and skill learning; (*ii*) that of informationally optimal feed-forward processing—relevant in particular to describing and designing deep neural networks (DNNs); and (*iii*) that of channel order in statistical decision theory—which provides clear interpretations of distinct bottlenecks' comparison in terms of universal informativeness of an agent's sensor. Indeed, while we focused for most of the paper on the information incorporation interpretation, we saw in Section 4 that the two other ones are as legitimate as the first one.

Once the formal problem is motivated and set, we turn to the mathematical analysis of it. We first note that, for jointly Gaussian vectors (X,Y) or for deterministic $p(Y|X)$, successive refinability can be easily drawn from existing IB literature [33,34]. Then, we propose a new geometric characterisation of SR, which builds on the intuition that what is "known" by a bottleneck is the convex hull of its decoder conditional probabilities. This new point of view, associated with an active approach that reformulates the IB problem as that of finding the lower convex envelope of a well-chosen function [35–39], provides a new tool for theoretical research on this topic. We exemplify this potential fertility by proving, thanks to the combination of our convex hull characterisation with the convexity approach to the IB, the successive refinability of binary source X and binary relevancy Y (Proposition 5). This convex hull characterisation also allows one to numerically investigate SR with a linear program, which can be helpful for computational studies on this topic.

Our own minimal numerical experiments suggest that (i) successive refinement does not always hold for the IB, (ii) the successive refinement patterns are shaped by IB bifurcations, and (iii) even when successive refinement seems to break, sometimes it is "close" to being satisfied, in the sense of the convex hull condition being only "slightly" violated.

To formalise this latter intuition, we propose to soften the traditional notion of SR into a *quantification* of the loss of information optimality incurred by several-stage processing. For that purpose, we call on the measure of unique information (UI) used in [40]. Intuitively, this quantity measures the information that only the coarser bottleneck T_1, and not the finer one T_2, holds about the source X, and it can be generalised to an arbitrary number of processing stages. Our minimal experiments, in the case of two processing stages, unveil a rich structure of soft SR that was partially hidden by exact SR, which only makes the distinction between vanishing UI (if there is SR) and positive UI (if there is no SR). Even though the UI landscapes depend strongly on the distribution $p(X, Y)$ that defines the IB problem, some qualitative features seem to emerge: (i) the "more" the convex hull condition is broken, the higher the unique information; (ii) the IB bifurcations crucially shape the UI landscape, with sharp decreases in unique information in particular when the finer trade-off parameter λ_2 crosses a bifurcation critical value; and (iii) in any case, this violation of successive refinement seems to always be mild compared to the system's globally processed information.

The features exhibited by these numerical experiments offer a "first outlook" of potentially general properties of exact and soft successive refinement for the IB problem, thus providing a guide for future theoretical research. These potential properties might provide interesting perspectives on the scientific questions that motivate the formalism, particularly in terms of the incorporation of fresh information into already learned models, and deep learning. For instance, the apparently important role of bifurcations in exact and soft successive refinement suggests that informationally optimal several-stage learning or processing should ideally be organised along well-chosen "checkpoints" on the information plane. Moreover, if the loss of information optimality induced by this sequential processing is indeed typically low (even though not entirely negligible) for the IB framework, this could be taken as an indication that incremental learning might be made highly efficient. These potential features thus provide a strong incentive to bring the formal framework presented here closer to maturity—for instance, along the lines of research proposed in Section 5.

Author Contributions: Conceptualisation, H.C., N.C.V. and D.P.; methodology, H.C., N.C.V. and D.P.; software, H.C.; validation, H.C., N.C.V. and D.P.; formal analysis, H.C., N.C.V. and D.P.; investigation, H.C., N.C.V. and D.P.; resources, N.C.V. and D.P.; writing—original draft preparation, H.C.; writing—review and editing, H.C., N.C.V. and D.P.; visualisation, H.C.; supervision, N.C.V. and D.P.; project administration, D.P.; funding acquisition, D.P. All authors have read and agreed to the published version of the manuscript.

Funding: H.C. and D.P. were funded by the Pazy Foundation under grant ID 195.

Institutional Review Board Statement: Not applicable.

Data Availability Statement: The code that we used for this work can be found at https://gitlab.com/uh-adapsys/successive-refinement-ib/, along with the explicit values of the example distributions $p(X, Y)$ that we used to generate our figures.

Acknowledgments: Thanks to Johannes Rauh and Pradeep Banerjee for insightful comments on unique information [40] and the iterative algorithm proposed to compute it in [74].

Conflicts of Interest: The authors declare no conflict of interest. The funders had no role in the design of the study; in the collection, analyses, or interpretation of data; in the writing of the manuscript; or in the decision to publish the results.

Abbreviations

The following abbreviations are used in this manuscript:

IB	Information Bottleneck
SR	Successive Refinement
UI	Unique Information
DNN	Deep Neural Network

Appendix A. Section 1 Details

Appendix A.1. Effective Cardinality

In [43], the effective cardinality of a Lagrangian bottleneck T is defined as the number of distinct $q(Y|t)$, whereas it is defined as that of distinct $q(X|t)$ in our Definition 1. However, as mentioned in Section 1.3, both choices happen to be equivalent:

Proposition A1. *Let $q(T|X)$ be a fixed solution to the Lagragian IB problem (3) for some β, where $p(X,Y)$ is discrete. The number of distinct $q(X|t)$ and that of distinct $q(Y|t)$ are equal.*

Proof. It is proven in [1] that a solution to the Lagrangian IB must satisfy for all $t \in \mathcal{T}$, $x \in \mathcal{X}$ the self-consistent equation

$$q(t|x) \;=\; \frac{q(t)}{Z(x)} \, \exp\big(-\beta \, D_{\mathrm{KL}}(p(Y|x) \,||\, q(Y|t))\big), \tag{A1}$$

where $Z(x)$ is the normalisation factor, and

$$q(t) := \sum_x q(t|x)p(x), \tag{A2}$$

$$q(y|t) := \sum_x p(y|x)q(x|t), \tag{A3}$$

with

$$q(x|t) := \frac{q(t|x)p(x)}{q(t)}. \tag{A4}$$

If $q(Y|t_1) = q(Y|t_2)$, then (A1) implies that $q(t_1|X) = q(t_2|X)$, which, combined with (A2), implies that we also have $q(t_1) = q(t_2)$. These two new equalities, combined with (A4), then prove that $q(X|t_1) = q(X|t_2)$. Conversely, if $q(X|t_1) = q(X|t_2)$, then Equation (A3) proves that $q(Y|t_1) = q(Y|t_2)$. $\square$

Crucially, it is proven in [43] that a given Lagrangian bottleneck T can be reduced to effective cardinality while still being a bottleneck for the same trade-off parameter β by merging all bottleneck symbols $t_1 \ldots, t_r$ with equal decoder distributions $q(X|t_1) = \cdots = q(X|t_r)$ into a new symbol $[t]$ defined by

$$q([t]|x) := \sum_{i=1}^{r} q(t_i|x).$$

Moreover, this merging can also be carried out for primal bottlenecks (this is not a direct consequence of Proposition A1, as it is known that when the information curve is not strictly concave, the primal and Lagrangian problems might not be exactly equivalent [39,67]).

Proposition A2. *Let T be a primal bottleneck of parameter λ, i.e., a solution to (1), where $p(X,Y)$ is discrete. The bottleneck obtained from T by merging the symbols t with identical $q(X|t)$ is still a solution to (A38) for the same parameter λ.*

Proof. We will use the following reparametrisation of the IB problem (1) (see Section 2.2):

$$\underset{\substack{(q(T),q(X|T))\,:\\ \sum_t q(t)q(X|t)=p(X)\\ T-X-Y,\, I(X;T)\leq\lambda}}{\arg\max} \quad I(Y;T). \tag{A5}$$

We thus consider the bottleneck T from Proposition A2's statement as defined by a pair $(q(T),q(X|T))$ satisfying $\sum_t q(t)q(X|t) = p(X)$. Now, assume that there exist $t_1, t_2 \in \mathcal{T}$ such that $q(X|t_1) = q(X|t_2)$. Then,

$$\sum_x q(x|t_1) \log\left(\frac{q(x|t_1)}{p(x)}\right) = \sum_x q(x|t_2) \log\left(\frac{q(x|t_2)}{p(x)}\right),$$

so that

$$\begin{aligned}
I(X;T) &= \sum_{t,x} q(t)q(x|t) \log\left(\frac{q(x|t)}{p(x)}\right) \\
&= \alpha_{t_1,t_2} \sum_x q(x|t_1) \log\left(\frac{q(x|t_1)}{p(x)}\right) + \sum_{t\notin\{t_1,t_2\},\,x} q(t)q(x|t) \log\left(\frac{q(x|t)}{p(x)}\right),
\end{aligned} \tag{A6}$$

where $\alpha_{t_1,t_2} := q(t_1) + q(t_2)$. Moreover,

$$q(Y|t_1) = \sum_x q(x|t_1)p(y|x) = \sum_x q(x|t_2)p(y|x) = q(Y|t_2),$$

so that, similarly,

$$I(Y;T) = \alpha_{t_1,t_2} \sum_y q(y|t_1) \log\left(\frac{q(y|t_1)}{p(y)}\right) + \sum_{t\notin\{t_1,t_2\},\,y} q(t)q(y|t) \log\left(\frac{q(y|t)}{p(y)}\right). \tag{A7}$$

Eventually,

$$p(X) = \sum_t q(t)q(X|t) = \alpha_{t_1,t_2}q(X|t_1) + \sum_{t\notin\{t_1,t_2\}} q(t)q(X|t), \tag{A8}$$

where the first equality comes from the fact that $(q(T),q(X|T))$ is a solution to (A5) (so, in particular, it must satisfy the hard constraints required in the optimisation problem). Let us define the bottleneck $\tilde{T}$ on $\tilde{\mathcal{T}} := \mathcal{T} \setminus t_2$ by

$$\tilde{q}(t) := \begin{cases} \alpha_{t_1,t_2} & \text{if } t = t_1 \\ q(t) & \text{if } t \in \mathcal{T} \setminus \{t_1, t_2\}, \end{cases}$$

and, for all $t \in \mathcal{T} \setminus \{t_2\}$,

$$\tilde{q}(X|t) := q(X|t).$$

The last line of (A6) can then be rewritten as

$$\begin{aligned}
I(X;T) &= \tilde{q}(t_1) \sum_x \tilde{q}(x|t_1) \log\left(\frac{\tilde{q}(x|t_1)}{p(x)}\right) + \sum_{t\notin\{t_1,t_2\},\,x} \tilde{q}(t)\tilde{q}(x|t) \log\left(\frac{\tilde{q}(x|t)}{p(x)}\right) \\
&= \sum_{t\in\tilde{\mathcal{T}}\,x\in\mathcal{X}} \tilde{q}(t)\tilde{q}(x|t) \log\left(\frac{\tilde{q}(x|t)}{p(x)}\right) \\
&= I(X,\tilde{T}).
\end{aligned}$$

Similarly, from (A7), we obtain $I(Y;\tilde{T}) = I(Y;T)$, while from (A8), we have $\sum_t \tilde{q}(t)\tilde{q}(X|t) = p(X)$. In other words, $\tilde{T}$ is also a solution to the reparametrised primal bottleneck problem

(A5), for the same parameter λ. Moreover, it is clear that our definition of $(\tilde{q}(T), \tilde{q}(X|T))$ implies that $\tilde{q}(t_1|x) = q(t_1|x) + q(t_2|x)$ and $\tilde{q}(t|x) = q(t|x)$ for $t \in \mathcal{T} \setminus \{t_1, t_2\}$, so $\tilde{T}$ is the variable obtained from T by merging the symbols t_1 and t_2. The result follows by iterating this argument until all the $\tilde{q}(X|t)$ are distinct. $\square$

Appendix B. Section 2 Details

Appendix B.1. Proof of Proposition 1

$(i) \Rightarrow (ii)$: Suppose that there are variables T_1 and $T_i := (T_{i-1}, S_i)$ for $2 \leq i \leq n$ such that each T_i is a bottleneck with parameter λ_i. Unrolling the iterative definitions of the T_i, we obtain

$$T_i = (T_1, S_2, \ldots, S_i),$$

which implies that, if $j < i$, then T_j is a deterministic function of T_i; in other words, given T_i, the variable T_j is independent of any other variable. So, first, we have $X - T_n - T_{n-1}$. Now, assume that for a given i, we have

$$X - T_n - \cdots - T_i. \tag{A9}$$

Given T_i, the variable T_{i-1} is independent of any other variable, so, in particular,

$$(X, T_n, \ldots, T_{i+1}) - T_i - T_{i-1}. \tag{A10}$$

The Markov chains (A9) and (A10) together imply that

$$X - T_n - \cdots - T_{i-1}.$$

Thus, a recurrence from $i = n$ to $i = 1$ proves that we do have $X - T_n - \cdots - T_1$, where, by assumption, each T_i is indeed a bottleneck of parameter λ_i.

$(iii) \Rightarrow (i)$: For all i, the Markov chain (5) implies that

$$I(X; T_i) = I(X; T_i'),$$
$$I(Y; T_i) = I(Y; T_i'),$$

where $T_i' := (T_i, \ldots, T_1)$. The Markov chain (5) also implies that these T_i' satisfy $Y - X - T_i'$. Thus, the T_i' are also bottlenecks with respective trade-off parameters $\lambda_1, \ldots, \lambda_n$. But, by construction, they satisfy $T_i' = (T_{i-1}', S_i)$, where, here, $S_i := T_i$.

$(ii) \Rightarrow (iii)$. We merely define $q(X, T_1, \ldots, T_n, Y)$ through the density

$$q(x, t_1, \ldots, t_n, y) := q(x, t_1, \ldots, t_n) q(y|x).$$

From this construction and the fact that each individual bottleneck must by definition satisfy $Y - X - T_i$, it is clear that $q(X, T_1, \ldots, T_n, Y)$ is indeed an extension of the individual bottleneck probabilities $q(X, Y, T_i)$. Moreover, by construction, we have

$$Y - X - (T_n, \ldots, T_1).$$

This latter Markov chain, combined with the assumed Markov chain (4), together imply that the Markov chain (5) holds.

Appendix B.2. Operational Interpretation of Successive Refinement

This section describes the operational interpretation—for the case of discrete variables X, Y—of successive refinement, which was already proposed in [30,31], as well as, in a slightly more general fashion, in [32]. We will here rely on the content from the latter work (even though our notations will be different). We will denote, for a variable Z, by Z^l, the concatenation of l i.i.d. variables with the same law as Z.

Definition A1. *For $l \in \mathbb{N}$, an n-stage $(l, M_1, \ldots, M_n)$-code consists of n encoder functions*

$$\phi_i^l \ : \ \mathcal{X}^l \to \{1, \ldots, M_i\}$$

and n decoder functions

$$\psi_i^l \ : \ \{1, \ldots, M_1\} \times \cdots \times \{1, \ldots, M_i\} \to \mathcal{Y}^l.$$

For a given source X, the i-th output of the $(l, M_1, \ldots, M_n)$-code will be written

$$\hat{Y}_i^l := \psi_i^l(\phi_1^l(X^l), \ldots, \phi_i^l(X^l)).$$

Intuitively, each new encoder extracts additional information from the same source, and, crucially, each new decoder is allowed to rely on *all* the information encoded until the i-th stage. Note that the output space of the decoder is modelled on that of the relevancy variable because this is the one about which one wants to extract information.

Definition A2. *The* relevance-complexity region *is the set of tuples $(R_1, \ldots, R_n, \mu_1, \ldots, \mu_n)$ such that there exists a sequence of n-stage $(l, M_1, \ldots, M_n)$-codes for all $1 \leq i \leq n$,*

$$\forall l \in \mathbb{N}, \quad \frac{1}{l} \log M_i \leq R_i$$

and

$$\forall l \in \mathbb{N}, \quad \frac{1}{l} I(Y^l; \hat{Y}_i^l) \geq \mu_i.$$

Intuitively, for a tuple to be in the relevance-complexity region, there must be an n-stage code such that the i-th encoder adds information at a rate no larger than R_i, and the i-th decoder yields information about the target variable Y no lower than μ_i. In other words, the relevance-complexity region is made of all the tuples that are achievable by n-stage codes.

Now, let us give the operational definition of successive refinement. We will denote, for a parameter λ, by $I_Y(\lambda)$, the maximum value of $I(Y; T)$ in the primal IB problem (1).

Definition A3. *Let $0 \leq \lambda_1 < \cdots < \lambda_n$. An IB problem defined by $p(X, Y)$ is said to be* operationally successively refinable, *or O-SR, for rates $(\lambda_1, \ldots, \lambda_n)$, if the tuple*

$$(\lambda_1 , \lambda_2 - \lambda_1 , \ldots , \lambda_n - \lambda_{n-1}, I_Y(\lambda_1), \ldots, I_Y(\lambda_n))$$

is in the relevance-complexity region.

Intuitively, in the case $n = 2$, assume one is given a total rate λ_2 to "spend" on encoding a source X. One can choose to encode the source in a single processing stage, yielding at best, after decoding, asymptotic relevant information $I_Y(\lambda_2)$ (see [2]). Alternatively, one can choose to break up the total rate λ_2 into two rates $R_1 := \lambda_1 < \lambda_2$ and $R_2 := \lambda_2 - \lambda_1$, and successively encode potentially different aspects of the source at these rates. Operational SR means that even though this second alternative "spends" the total rate λ_2 along two distinct stages, it can still, after decoding, also yield asymptotic relevant information of $I_Y(\lambda_2)$. Naturally, in this case, the relevant information decodable from only the first stage must also be the optimal one, i.e., $I_Y(\lambda_1)$—otherwise, the "waste" in spending the rate λ_1 would prevent the second-stage decoder, which partially relies on the information encoded at the first stage, from ever achieving the optimal relevant information $I_Y(\lambda_2)$.

We then have the following single-letter characterisation:

Proposition A3. *The IB problem defined by $p(X, Y)$ is O-SR for rates $(\lambda_1, \ldots, \lambda_n)$ if and only if there exist variables $T_1, \ldots, T_n$ such that*

(i)   We have the Markov chain $Y - X - T_n - \cdots - T_1$;

(ii)   The variables $T_1, \ldots, T_n$ are each bottlenecks with respective parameters $\lambda_1, \ldots, \lambda_n$.

Proof. This single-letter characterisation is a consequence of Remark 1 in [32], which states the following: a tuple $(R_1, \ldots, R_n, \mu_1, \ldots, \mu_n)$ is in the relevance-complexity region if and only if there exist variables $T_1, \ldots, T_n$ such that the Markov chain $Y - X - T_n - \cdots - T_1$ holds, and such that, for all $i = 1, \ldots, n$,

$$\sum_{j=1}^{i} I(X; T_j | T_1, \ldots, T_{j-1}) \leq \sum_{j=1}^{i} R_j, \tag{A11}$$

$$I(Y; T_i) \geq \mu_i. \tag{A12}$$

By simplifying the left-hand side in (A11) through the chain rule for mutual information, defining $\lambda_i := \sum_{j=1}^{i} R_j$, and applying the statement with $\mu_i := I_Y(\lambda_i)$, we obtain that the IB problem is O-SR for rates $(\lambda_1, \ldots, \lambda_n)$ if and only if there exist variables $T_1, \ldots, T_n$ such that

1.   We have the Markov chain $Y - X - T_n - \cdots - T_1$; and
2.   We have, for all $i = 1, \ldots, n$,

$$I(X; T_i) \leq \lambda_i, \tag{A13}$$

$$I(Y; T_i) \geq I_Y(\lambda_i). \tag{A14}$$

However, if point 1 above holds, then, particularly for all $i = 1, \ldots, n$, we have the Markov chain $Y - X - T_i$. As a consequence, by definition of the primal IB problem (1), the inequality in (A14) can be replaced by an equality, and thus point 2 as a whole can be replaced by the condition that T_i is a bottleneck of parameter λ_i for the IB problem defined by $p(X, Y)$. Hence, we are left with points (i) and (ii) of Theorem A3's statement. $\square$

It is worth mentioning that our Proposition A3 is also essentially Theorem 7 in [30], which proves the same single-letter characterisation for the same operational problem—up to the difference that the result is limited to $n = 2$, and that the latter work does not consider any decoder functions ψ_i^j. Moreover, Proposition A3 is a consequence of Lemma 4 in [31].

It is clear that the conditions of Theorem A3 are exactly those of Proposition 4-(iii), so the operational Definition A3 and the single-letter Definition 5 are equivalent; in other words, the notion studied in our work does have an operational interpretation. Crucially, the operational construction of Definitions A1–A3 also goes clearly along the interpretation in terms of the successive incorporation of information.

Appendix B.3. Proof of Proposition 2

First of all, note that even though in the Definition 5 of successive refinement, the term "bottleneck" refers to a solution to the primal problem (1), the definition makes as much sense if now by "bottleneck" we mean a solution to the Lagrangian problem (3). This is, therefore, what we will be speaking about in this section. With this Lagrangian version, the Markov chain characterisation given by Proposition 1 still holds. More precisely:

Proposition A4. *Let (X, Y) be jointly Gaussian, and $1 \leq \beta_1 < \cdots < \beta_n$. The following are equivalent:*

(i)   There is successive refinement for Lagrangian parameters $(\beta_1, \ldots, \beta_n)$.

(ii) *There exist Lagrangian bottlenecks $T_1, \ldots, T_n$, of common source X and relevancy Y, with respective parameters $\beta_1, \ldots, \beta_n$, and an extension $q(Y, X, T_1, \ldots, T_n)$ of the $q_i := q_i(Y, X, T_i)$, such that, under q, we have the Markov chain*

$$Y - X - T_n - \cdots - T_1. \tag{A15}$$

Proof. One can directly verify that the proof given for Proposition 1 (see Appendix B.1) does not involve the explicit form of the IB problem, so the very same proof can be used for the Lagrangian formulation. $\square$

The statement of Proposition 2 is now fully explicit.

Proof of Proposition 2. For the case of the Lagrangian IB problem with jointly Gaussian source X and relevancy Y, an analytic solution was given in [75], which proves among other things that the functions $(\beta \mapsto I_\beta(X; T))$ and $(\beta \mapsto I_\beta(Y; T))$ are continuous and increasing, where $I_\beta(X; T)$ and $I_\beta(Y; T)$ are defined by bottlenecks T of Lagrangian trade-off parameter β. Let us define

$$\beta_{IB}(X, Y) := \sup \{\beta \in \mathbb{R} : I_\beta(X; T) = 0\},$$

where we must have $\beta_{IB}(X, Y) \geq 1$ (see Section 1.3). Moreover, from the continuity of the function $(\beta \mapsto I_\beta(X; T))$, this supremum is a maximum, and from the monotonicity of the latter function, $I_\beta(X; T) = 0$ for all $\beta \leq \beta_{IB}(X, Y)$, whereas, by definition of $\beta_{IB}(X, Y)$, we have $I_\beta(X; T) > 0$ for all $\beta > \beta_{IB}(X, Y)$. Thus, $\beta_{IB}(X, Y)$ delimits trivial from non-trivial solutions, and we can, without loss of generality, choose $\beta \geq \beta_{IB}(X, Y)$.

Let us now turn to the *semigroup structure* of the Gaussian IB problem, which was both defined and proved in [33]. In short, this structure means that one can *compose* two Gaussian bottlenecks, while still obtaining a Gaussian bottleneck for the original problem. More precisely, let $\beta_2 > \beta_{IB}(X, Y)$, and define T_2 as the analytical solution to the Lagrangian IB from [75]. This provides one with a joint distribution $q_2(Y, X, T_2)$, which, importantly for us here, happens to define a Gaussian vector as well. Then, we consider a new IB problem with still the same relevancy variable Y, but now with T_2 as the source, i.e.,

$$\underset{q(T_1|T_2)\,:\,T_1-T_2-Y,}{\arg\min} \quad I(T_2; T_1) - \beta_1' \, I(Y; T_1), \tag{A16}$$

where $\beta_1' \geq \beta_{IB}(T_2, Y)$. As T_2 and Y are jointly Gaussian, the problem above is again a Gaussian IB problem, so we can again analytically define a solution T_1 with the formulas from [75], yielding a distribution $q_1(Y, T_2, T_1)$. The semigroup structure proven in [33] refers to the following feature:

Proposition A5. *Assume that T_1 and T_2 are built as above, and define the extension $q(Y, X, T_1, T_2)$ of $q_1(Y, X, T_1)$ and $q_2(Y, X, T_2)$ through*

$$q(y, x, t_1, t_2) := q_2(y, x, t_2)q_1(t_1|t_2). \tag{A17}$$

Then, the marginal $q(Y, X, T_1)$ defines a Lagrangian bottleneck of source X and relevancy Y for some parameter β_1 uniquely defined, with $\beta_{IB}(X, Y) \leq \beta_1 < \beta_2$.

Thus, we can define a binary operator "$\circ$", which, for every $\beta_2 > \beta_{IB}(X, Y) \geq 1$ and $\beta_1' \geq \beta_{IB}(T_1, Y)$, provides the parameter $\beta_1 := \beta_2' \circ \beta_1$ defined by Proposition A5. Ref. [33] gives an explicit formula for this binary operator :

$$\beta_1' \circ \beta_2 = \frac{\beta_1' \beta_2}{\beta_1' + \beta_2 - 1}, \tag{A18}$$

which is well-defined for $\beta_2 > \beta_{IB}(X,Y)$ and $\beta_1' \geq \beta_{IB}(T_2,Y)$, because $\beta_{IB}(X,Y) \geq 1$ and $\beta_{IB}(T_2,Y) \geq 1 \geq 0$ imply that $\beta_1' + \beta_2 - 1 > 0$. This formula implies the following:

Proposition A6. *Let* $\beta_2 > \beta_{IB}(X,Y)$. *For any* β_1 *such that* $\beta_{IB}(X,Y) \leq \beta_1 < \beta_2$, *there exists a* β_1' *such that* $\beta_1 = \beta_1' \circ \beta_2$.

Proof. Let f denote the function $\beta_1' \mapsto \beta_1' \circ \beta_2$, which is well-defined and continuous on the interval $[\beta_{IB}(T_1,Y), +\infty[$. It is clear from formula (A18) that

$$\lim_{\beta_1' \to \infty} f(\beta_1') = \beta_2. \tag{A19}$$

On the other hand, note first that as $\beta_{IB}(T_2,Y)$ delimits trivial from non-trivial solutions, we have $I_{\beta_{IB}(T_2,Y)}(T_2;T_1) = 0$. But, by construction, under q given by Equation (A17), we have the Markov chain $Y - X - T_2 - T_1$. Thus, $I_{\beta_{IB}(T_2,Y)\circ\beta_2}(X;T_1) \leq I_{\beta_{IB}(T_2,Y)}(T_2,T_1)$, i.e., $I_{\beta_{IB}(T_2,Y)\circ\beta_2}(X;T_1) = 0$. So, by definition of $\beta_{IB}(X,Y)$, we have

$$\beta_{IB}(T_2,Y) \circ \beta_2 \leq \beta_{IB}(X,Y), \tag{A20}$$

i.e.,

$$f(\beta_{IB}(T_2,Y)) \leq \beta_{IB}(X,Y). \tag{A21}$$

Now, Equations (A19) and (A21), combined with the continuity of f, imply that

$$[\beta_{IB}(X,Y),\beta_2[\subseteq f([\beta_{IB}(T_2,Y),\infty[),$$

which yields the result. $\quad\square$

Now let us consider a family of parameters $\beta_{IB}(X,Y) \leq \beta_1 < \cdots < \beta_n$. By iterating Propositions A5 and A6 used together, we obtain that there exist bottlenecks $T_1, \ldots, T_n$ of common source X and relevancy Y, with respective parameters $\beta_1, \ldots, \beta_n$, and an extension $q(Y, X, T_1, \ldots, T_n)$ of these bottlenecks defined by

$$q(y, x, t_1, \ldots, t_n) := q(y, x, t_n)q(t_{n-1}|t_n)\ldots q(t_1|t_2).$$

By construction, under q, the Markov chain $Y - X - T_n - \cdots - T_1$ holds. In other words, condition (ii) from Proposition A4 is satisfied, which proves the successive refinability of jointly Gaussian vectors for the Lagrangian IB problem. $\quad\square$

Appendix B.4. Proof of Proposition 3

Here, for $\alpha \in [0,1]$, we denote by T_α the variable defined by

$$\begin{aligned}
q(T_\alpha = Y|X) &= \alpha \\
q(T_\alpha = e|X) &= 1 - \alpha,
\end{aligned} \tag{A22}$$

where e denotes a dummy symbol not pertaining to either $\mathcal{X}$ or $\mathcal{Y}$. It was proven in [67] that, for every primal parameter $\lambda \in [0, I(X;Y)]$, there exists an α such that T_α is a bottleneck of parameter λ. Note that we must have

$$\lambda = I(X;T_\alpha) = \alpha I(X;Y), \tag{A23}$$

where the first equality comes the general fact that a bottleneck must saturate the information constraint in (1) (see Section 1.3), and the second equality is a direct computation from (A22). Thus, α is a bijective and increasing function of λ, and it is sufficient, for proving successive refinement, to prove that, for $0 \leq \alpha_1 < \cdots < \alpha_n \leq 1$, we can design a joint distribution $q(X, T_{\alpha_1}, \ldots, T_{\alpha_n})$ such that we have the Markov chain

$$X - T_{\alpha_n} - \cdots - T_{\alpha_1}.$$

Let us first focus on the case $n = 2$. We define a bottleneck $T_2 := T_{\alpha_2}$, i.e, we set $q(X, T_2) := q(X, T_{\alpha_2})$ and then a distribution $q(T_1, T_2)$ through

$$q(T_1 = Y | T_2 = Y) := \frac{\alpha_1}{\alpha_2}$$
$$q(T_1 = e | T_2 = Y) := \frac{\alpha_2 - \alpha_1}{\alpha_2}$$
$$q(T_1 = Y | T_2 = e) := 0$$
$$q(T_1 = e | T_2 = e) := 1.$$

We then define an extension $q(X, T_1, T_2)$ of $q(X, T_2)$ and $q(T_1, T_2)$ through

$$q(x, t_1, t_2) := q(x, t_2) q(t_1 | t_2),$$

which implies by construction the Markov chain $X - T_2 - T_1$. But it also implies that

$$q(T_1 = Y | x) = q(T_1 = Y | T_2 = Y) q(T_2 = Y | x) + q(T_1 = Y | T_2 = e) q(T_2 = e | x)$$
$$= \frac{\alpha_1}{\alpha_2} \alpha_2 + 0 \times (1 - \alpha_2)$$
$$= \alpha_1,$$

and thus, necessarily, $q(T_1 = e | X) = 1 - \alpha_1$. So, $q(X, T_1) = q(X, T_{\alpha_1})$. Thus, we built a joint law $q(X, T_{\alpha_1}, T_{\alpha_2})$ such that $X - T_{\alpha_2} - T_{\alpha_1}$, which proves successive refinement for the case $n = 2$. The case of arbitrary n follows by direct iteration of the previous reasoning, where one starts from defining $q(X, T_n)$ through $T_n := T_{\alpha_n}$, and then iteratively defines $q(X, T_i, T_{i+1}, \ldots, T_n)$ through a well-chosen $q(T_i | T_{i+1})$ and the Markov chain condition $X - T_n - \cdots - T_{i+1} - T_i$.

Appendix B.5. Proof of Proposition 4

The result is a consequence of the following general fact, where we will eventually set $U := T_1$, $V := T_2$, and $W := X$.

Proposition A7. *Let $q(U, W)$ and $q(V, W)$ be full-support consistent distributions, defined on discrete alphabets $\mathcal{U} \times \mathcal{W}$ and $\mathcal{V} \times \mathcal{W}$, respectively. Consider the following properties:*

(i) *There exists an extension $\tilde{q}(U, V, W)$ of $q(U, W)$ and $q(V, W)$ under which the Markov chain $U - V - W$ holds.*

(ii) *For each $u \in \mathcal{U}$, there exists a family of convex combination coefficients $\{\alpha_{v,u} , v \in \mathcal{V}\}$ such that*

$$q(W | u) = \sum_v \alpha_{v,u} \, q(W | v).$$

Then, we always have $(i) \Rightarrow (ii)$ and, if, moreover, the channel $q(W | V)$ is injective, then we also have $(ii) \Rightarrow (i)$, and the extension $\tilde{q}$ is uniquely defined.

Note the abuse of notations in the statement of Proposition A7: we write q for both $q(U, V)$ and $q(V, W)$, which are distinct distributions on partially distinct alphabets, even though they are consistent; in addition, along the proof, context, if not explicit statements, will make clear which distribution we are referring to.

Proof. Along the proof, we will be using the fact that a probability distribution is equivalent to a family of convex combination coefficients several times; indeed, both notions define a family of non-negative numbers such that their sum equals one.

$(i) \Rightarrow (ii)$. For all u, w, assumption (i) provides a $\tilde{q}(U, V, W)$ such that

$$
\begin{aligned}
q(w|u) &= \tilde{q}(w|u) \\
&= \sum_v \tilde{q}(w, v|u) \\
&= \sum_v \tilde{q}(v|u)\, \tilde{q}(w|v) \\
&= \sum_v \tilde{q}(v|u)\, q(w|v),
\end{aligned}
$$

where the first and fourth equalities use the fact that $\tilde{q}(U, V, W)$ is an extension of $q(U, W)$ and $q(V, W)$, and the third equality uses the fact that, under $\tilde{q}(U, V, W)$, the Markov chain $U - V - W$ holds. Let us define $\alpha_{v,u} := q(v|u)$. For each $u \in \mathcal{U}$, the family $\{\alpha_{v,u},\ v \in \mathcal{V}\}$ is a probability distribution, and thus a family of convex combination coefficients.

$(ii) \Rightarrow (i)$. We want to design a distribution $\tilde{q}$ that is both consistent with $q(U, W)$ and $q(V, W)$, and satisfies $U - V - W$. Thus, such a distribution is wholly defined by $\tilde{q}(V|U)$, because it must satisfy

$$
\begin{aligned}
\tilde{q}(u, v, w) &= \tilde{q}(u)\tilde{q}(v|u)\tilde{q}(w|v) \\
&= q(u)\tilde{q}(v|u)q(w|v),
\end{aligned}
\tag{A24}
$$

where $q(U)$ is obtained by marginalising $q(U, W)$, whereas $q(W|V)$ is obtained from $q(V, W)$. Assumption (ii) provides a candidate: let us define $\tilde{q}(v|u) := \alpha_{v,u}$, which makes sense because, for each u, the family $(\alpha_{v,u})_v$ is made of convex combination coefficients. From assumption (ii), for all u, w,

$$
q(w|u) = \sum_v \tilde{q}(v|u)\, q(w|v),
\tag{A25}
$$

and the corresponding $\tilde{q}(U, V, W)$ defined through Equation (A24) satisfies the Markov chain $U - V - W$.

To prove that $\tilde{q}$ is an extension of $q(U, W)$ and $q(V, W)$, let us prove first that $\tilde{q}$ is consistent with $q(U, W)$. We have

$$
\begin{aligned}
\tilde{q}(u, w) &= \sum_v \tilde{q}(u, v, w) \\
&= \sum_v q(u)\tilde{q}(v|u)q(w|v) \\
&= q(u) \sum_v \tilde{q}(v|u)q(w|v) \\
&= q(u)q(w|u) \\
&= q(u, w),
\end{aligned}
$$

where the first equality is the definition of the marginal $\tilde{q}(u, w)$; the second equality uses Equation (A24); and the fourth equality uses (A25). Thus, $\tilde{q}(U, V, W)$ is consistent with $q(U, W)$.

Now, let us prove that $\tilde{q}(V, W) = q(V, W)$. This is equivalent to the channel $\tilde{q}(V|U)$ sending the marginal $q(U)$ on the marginal $q(V)$:

Lemma A1. *We have $\tilde{q}(V, W) = q(V, W)$ if and only if*

$$
\tilde{Q}_{vu} q_u = q_v,
\tag{A26}
$$

where q_u and q_v are the column vectors defined by $q(U)$ and $q(V)$, respectively, and $\tilde{Q}_{vu}$ is the column transition matrix defined by $\tilde{q}(V|U)$.

Proof. For all v, w,

$$\tilde{q}(v, w) = \sum_u \tilde{q}(u, v, w)$$
$$= \Big(\sum_u q(u)\tilde{q}(v|u) \Big) q(w|v),$$

where the first equality is the definition of the marginal $\tilde{q}(v, w)$, and the second one uses Equation (A24). Thus, for all v, w,

$$\tilde{q}(v, w) = q(v, w) \quad \Leftrightarrow \quad q(v, w) = \Big(\sum_u q(u)\tilde{q}(v|u) \Big) \tilde{q}(w|v)$$

$$\Leftrightarrow \quad q(v)q(w|v) = \Big(\sum_u q(u)\tilde{q}(v|u) \Big) q(w|v),$$

and, eventually, for all v, w,

$$\tilde{q}(v, w) = q(v, w) \quad \Leftrightarrow \quad q(w|v) = 0 \ \text{ or } \ q(v) = \sum_u q(u)\tilde{q}(v|u). \tag{A27}$$

Let us momentarily fix $v \in \mathcal{V}$. Since $q(W|v)$ is a probability, there must be some w_0 such that $q(w_0|v) > 0$. Choosing that w_0, we find that, for the given v, the vector equality $\tilde{q}(v, W) = q(v, W)$ implies, through Equation (A27), that the scalar equality $q(v) = \sum_u q(u)\tilde{q}(v|u)$. By now applying this reasoning to each $v \in \mathcal{V}$, we obtain that $\tilde{q}(V, W) = q(V, W)$ implies that

$$\forall v \in \mathcal{V}, \quad \sum_u q(u)\tilde{q}(v|u) = q(v), \tag{A28}$$

whose matrix formulation is precisely (A26). Conversely, if (A28) holds, then Equation (A27) shows that $\tilde{q}(V, W) = q(V, W)$. $\square$

We now prove that Equation (A26) indeed holds. Let us also write Q_{wv} and Q_{wu} for the column transition matrices defined by $q(W|V)$ and $q(W|U)$, respectively. Then, Equation (A25), which, here, is our assumption, can be rewritten as

$$Q_{wu} = Q_{wv}\tilde{Q}_{vu}. \tag{A29}$$

Thus,

$$Q_{wv}\tilde{Q}_{vu}q_u = Q_{wu}q_u = q_w = Q_{wv}q_v$$

where q_w is the column vector defined by $q(W)$, and the second and third equalities are the matrix versions of the decompositions $q(W) = \sum_u q(u)q(W|u)$ and $q(W) = \sum_v q(v)q(W|v)$, respectively. In other words,

$$Q_{wv}(\tilde{Q}_{vu}q_u - q_v) = 0. \tag{A30}$$

The injectivity of Q_{wv} implies that (A26) indeed holds, so, from Lemma A1, we have $\tilde{q}(V, W) = q(V, W)$. We have thus proven that $\tilde{q}$ extends both $q(U, W)$ and $q(V, W)$, so point (ii) holds.

Eventually, let us prove the uniqueness. Let $\tilde{q}' := \tilde{q}'(U, V, W)$ be another extension of $q(U, W)$ and $q(V, W)$ such that, under $\tilde{q}'$, the Markov chain $U - V - W$ holds. For the same reasons as above, $\tilde{q}'$ must satisfy Equation (A24) with $\tilde{q}$ replaced by $\tilde{q}'$, so $\tilde{q}'$ is wholly specified by $\tilde{q}'(V|U)$, and is enough to prove that $\tilde{q}'(V|U) = \tilde{q}(V|U)$. Now, using the assumptions of consistency and the Markov chain for $\tilde{q}'$, we obtain

$$
\begin{aligned}
q(w|u) &= \tilde{q}'(w|u) \\
&= \sum_v \tilde{q}'(v, w|u) \\
&= \sum_v \tilde{q}'(v|u)\tilde{q}'(w|v) \\
&= \sum_v \tilde{q}'(v|u)q(w|v),
\end{aligned}
\tag{A31}
$$

i.e., in matrix terms, if $\tilde{Q}'_{uv}$ is the column transition matrix representing $\tilde{q}'(V|U)$,

$$
Q_{wu} = Q_{wv}\tilde{Q}'_{vu}.
$$

Combining this with Equation (A29), we have $Q_{wv}(\tilde{Q}'_{vu} - \tilde{Q}_{vu}) = 0$. In other words, if c_i is the i-th column of $\tilde{Q}'_{vu} - \tilde{Q}_{vu}$, then $Q_{wv}c_i = 0$, which, by injectivity of Q_{wv}, means that $c_i = 0$. Thus, $\tilde{Q}'_{vu} - \tilde{Q}_{vu} = 0$, i.e., $\tilde{q}(U|V) = \tilde{q}'(U|V)$.

This ends the proof of Proposition A7. $\quad\square$

Now, first of all, note that if we set $U := T_1$, $V := T_2$ and $W := X$, then point (ii) in Proposition A7 is equivalent to the convex hull condition (7).

If there is successive refinement for parameters (λ_1, λ_2), then, from Proposition 1, there are bottlenecks T_1, T_2 of parameters λ_1, λ_2, respectively, such that $X - T_2 - T_1$; and the direction $(i) \Rightarrow (ii)$ of Proposition A7 implies that the convex hull condition (7) is satisfied.

Conversely, assume that the convex hull condition is satisfied for some bottlenecks T_1, T_2 of parameters λ_1, λ_2, respectively, such that $q_2(X|T_2)$ is injective. Then, the sense $(ii) \Rightarrow (i)$ of Proposition A7 shows that there exists a unique extension $\tilde{q}(X, T_1, T_2)$ of $q_1(X, T_1)$ and $q_2(X, T_2)$ such that we have $X - T_2 - T_1$. We then conclude with the Markov chain characterisation of successive refinement (Proposition 1).

Appendix B.6. Linear Program Used to Compute the Convex Hull Condition (7)

Consider, for points $u, v_1, \ldots, v_k \in \mathbb{R}^m$, the condition

$$
u \in Hull\{v_i, \ i = 1, \ldots, k\}.
\tag{A32}
$$

A linear program can be used to check whether this condition holds or not; in short, it consists of the first step of the simplex method (see, e.g., [68], Section 5.6), which asserts the existence or not of an initial feasible basis, and computes this basis if it exists. More precisely, let us first note V the $m \times k$ matrix whose columns are the points v_i, and define

$$
M := \begin{pmatrix} & V & \\ 1 & \cdots & 1 \end{pmatrix}, \qquad \tilde{u} := \begin{pmatrix} u \\ 1 \end{pmatrix}.
$$

Then, condition (A32) can be reformulated as

$$
\exists \alpha := (\alpha_1, \ldots, \alpha_k) \in \mathbb{R}^k : \quad \begin{cases} M\alpha = \tilde{u}, \\ \alpha_i \geq 0 \ \text{for} \ i = 1, \ldots, k. \end{cases}
\tag{A33}
$$

We now consider the linear program defined for the augmented variable

$$
\tilde{\alpha} := (\alpha_1, \ldots, \alpha_k, \alpha_{k+1}, \ldots, \alpha_{k+m+1}) \in \mathbb{R}^{k+m+1}
$$

as

$$\min_{\substack{\tilde{M}\tilde{\alpha}=\tilde{u} \\ \forall i=1,\ldots,k+m+1,\ \alpha_i\geq 0}} \alpha_{k+1}+\cdots+\alpha_{k+m+1}, \tag{A34}$$

where $\tilde{M} := (M\,|\,I_{m+1})$ is obtained by appending the $(m+1)\times(m+1)$ identity matrix to M to the right. It can be directly verified that (A33), and thus, equivalently, (A32), holds if and only if the minimum is 0 in the linear program (A34), and that if this is the case, then the first k coordinates $\alpha_1,\ldots,\alpha_k$ of any of the program's solutions provide coefficients for obtaining u as a convex combination of the v_i.

Now, consider two bottleneck distributions $q_1 := q_1(X, T_1)$ and $q_2 := q_2(X, T_2)$ such that $q(X|T_2)$ is injective. We want to check the convex hull condition (7), which holds if and only if for every $t_1 \in \mathcal{T}_1$, we have

$$q(X|t_1) \in Hull\{q(X|t_2),\ t_2 \in \mathcal{T}_2\}. \tag{A35}$$

This condition can be checked, for every fixed t_1, with the linear program described above, where if the condition holds, the algorithm also outputs a family of coefficients $(\alpha_{t_2,t_1})_{t_2}$ such that

$$q(X|t_1) = \sum_{t_2} \alpha_{t_2,t_1} q(X|t_2). \tag{A36}$$

Let us define $q(t_2|t_1) := \alpha_{t_2,t_1}$ and a joint distribution $q(X, T_1, T_2)$ through

$$q(x, t_1, t_2) := q_1(t_1)q(t_2|t_1)q_2(x|t_2). \tag{A37}$$

By construction, under q, we have the Markov chain $X - T_2 - T_1$. Moreover thanks to Equation (A36) and the injectivity of $q(X|T_2)$, Proposition A7 shows that q is indeed an extension of $q_1(X, T_1)$ and $q_2(X, T_2)$. Thus, the linear program above allows one both to check whether or not the convex hull condition holds and, when it does, to obtain Theorem 4's unique extension $q(X, T_1, T_2)$ such that $X - T_2 - T_1$.

Let us turn to considering the algorithm's complexity. For each $t_1 \in \mathcal{T}_1$, we want to know if the point $q(X|t_1)$, which is made of $m = |\mathcal{X}|$ coordinates, is in the convex hull of $k = |\mathcal{T}_2|$ points, where we can always choose $|\mathcal{T}_2| \leq |\mathcal{X}| + 1$ (see Section 1.3). One can directly verify that the linear program (A34) thus consists of at most $2|\mathcal{X}| + 2$ variables and $3|\mathcal{X}| + 2$ equality and inequality constraints. Moreover, we want to check condition (A35) for every $t_1 \in \mathcal{T}_1$, where we can always choose $|\mathcal{T}_1| \leq |\mathcal{X}| + 1$. As a consequence, the time complexity of checking the convex hull condition (7) can be bounded as $O((|\mathcal{X}| + 1)\,K)$, where K is the complexity bound of a linear program with $2|\mathcal{X}| + 2$ variables and $3|\mathcal{X}| + 2$ constraints. By changing the multiplicative constant in the definition of the $O(\cdot)$ notation, the bound $O((|\mathcal{X}| + 1)\,K)$ clearly simplifies to $O(|\mathcal{X}|\,K)$. Eventually, Ref. [69] shows that

$$K = \tilde{O}\left(|\mathcal{X}|^{\omega} \log\left(\frac{|\mathcal{X}|}{\delta}\right)\right)$$

where $\omega \approx 2.38$ corresponds to the complexity of matrix multiplication and δ is the relative accuracy. Here the notation $\tilde{O}(\cdot)$ hides polylogarithmic factors: i.e., for two functions f and g defined over positive integers, $f(n) = \tilde{O}(g(n))$ means that there exists some $r \in \mathbb{N}$ such that $f(n) = O(g(n)\log^r(g(n)))$. Overall, the convex hull condition (7) can thus be checked with an algorithm of time complexity no worse than $\tilde{O}(|\mathcal{X}|^{\omega+1} \log(\frac{|\mathcal{X}|}{\delta}))$.

Note that as the convex hull condition holds if and only if the linear program's output is 0 for all $t_1 \in \mathcal{T}_1$, in numerical computations, the threshold for rounding the program's output impacts the answer. In our numerical experiments, we chose the threshold 10^{-6}.

Appendix B.7. Proof of Proposition 5

We will first present the framework developed in [35,37] and the original content of this proof, which starts with Lemma A4 below. A full plan of this proof is presented in the main text.

We already noticed (in Section 2.2) that the primal IB problem (1) can be reformulated as an optimisation over the pairs $(q(T), q(X|T))$, i.e., Equation (6). Using the identity $I(U; V) = H(U) - H(U|V)$, and recalling that a bottleneck T must satisfy $I(X; T) = \lambda$ [37], we can further reformulate the problem (6) as

$$\underset{\substack{(q(T), q(X|T)) \\ \sum_t q(t)q(X|t)=p(X) \\ H(X|T)=\nu}}{\arg\min} \quad H(Y|T), \tag{A38}$$

where $\nu := H(X) - \lambda$. In particular, we can assume, without loss of generality, that $0 \leq \nu \leq H(X)$ (see Section 1.3), where $\nu = H(X)$ corresponds to $I(X; T) = 0$. Similarly as we denoted before by $I_Y(\lambda)$ the maximum in the classic IB problem (1), here, we denote by $H_Y(\nu)$ the minimum in (A38). Rather than considering the information curve, i.e., the graph of I_Y, and following [35] upon which we rely, here, we consider the graph of H_Y, which we will refer to as the conditional entropy (CE) curve. This curve is convex [35], and it is just an affine translation of the information curve. Let us now define, for $\beta \geq 1$, the function

$$F_\beta : \Delta_{\mathcal{X}} \to \mathbb{R}$$

$$p \mapsto H(\kappa p) - \beta^{-1} H(p),$$

where κ is the column transition matrix defined by the conditional probability $p(Y|X)$. Note that, for $p = p(X)$, we have $\kappa p = p(Y)$. (In this section, we choose notations close to those from [37], as long as they do not clash with the ones we already established; most notably, what we denote here by β would correspond to β^{-1} in [37].)

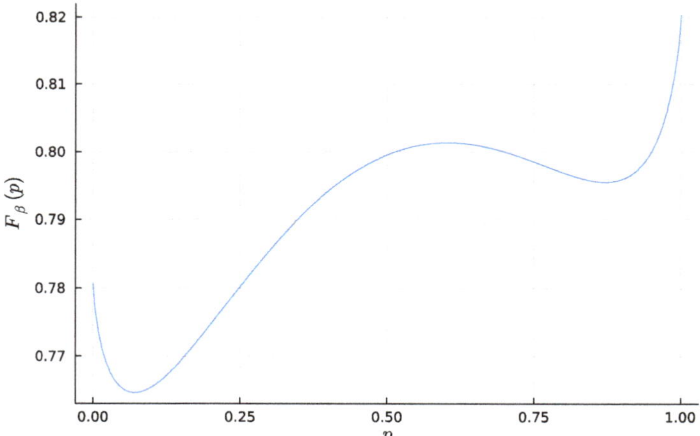

Figure A1. The function F_β for example values of β and $p(X, Y)$, where the source and relevancy are binary. Here, on the x-axis, p parameterises the binary distribution $[p, 1 - p]$.

The function F_β is plotted in Figure A1 for example values of β and $p(X, Y)$, where the source and relevancy are binary. As a difference in concave functions, the function is a priori neither concave nor convex, but we can define its *lower convex envelope*, i.e., the largest convex function, which is still inferior or equal to F_β everywhere: we will denote it by

$\mathcal{K}_{\cup}(F_{\beta})$. In Section IV in [35], through convex duality arguments, the following relationship between bottlenecks and F_{β} was proven:

Proposition A8. *If a pair $(q(T), q(X|T))$ solves the reformulated primal IB problem (A38), then*

$$\sum_{t} q(t) F_{\beta}(q(X|t)) = \mathcal{K}_{\cup}(F_{\beta})(p(X)), \tag{A39}$$

for some $\beta \geq 1$ such that β^{-1} is the slope of a tangent to the CE curve at the point $(v, H_Y(v))$.

Let us also define the set of points where F_{β} differs from its lower convex envelope:

$$\mathcal{P}(\beta) := \{ p \in \Delta_{\mathcal{X}} \; : \; F_{\beta}(p) \neq \mathcal{K}_{\cup}(F_{\beta})(p) \}, \tag{A40}$$

which will happen to be crucial for our considerations on successive refinement. As already noted (see [37], Section II.B), this set grows when β increases:

Lemma A2. *If $\beta_1 \leq \beta_2$, then $\mathcal{P}(\beta_1) \subseteq \mathcal{P}(\beta_2)$.*

Proof. For the sake of self-containedness, we reproduce the computation from [37]. Let $p \notin \mathcal{P}(\beta_2)$, which means that $\mathcal{K}_{\cup}\left(F_{\beta_2}\right)(p) = F_{\beta_2}(p)$. For all $\beta_1 \leq \beta_2$,

$$\begin{aligned}
F_{\beta_1}(p) &= H(\kappa p) - \beta_1^{-1} H(p) \\
&= F_{\beta_2}(p) - (\beta_1^{-1} - \beta_2^{-1}) H(p),
\end{aligned}$$

so

$$\begin{aligned}
\mathcal{K}_{\cup}\left(F_{\beta_1}\right)(p) &= \mathcal{K}_{\cup}\left(F_{\beta_2} - (\beta_1^{-1} - \beta_2^{-1})H\right)(p) \\
&\geq \mathcal{K}_{\cup}\left(F_{\beta_2}\right)(p) + \mathcal{K}_{\cup}\left(-(\beta_1^{-1} - \beta_2^{-1})H\right)(p) \\
&= \mathcal{K}_{\cup}\left(F_{\beta_2}\right)(p) - (\beta_1^{-1} - \beta_2^{-1})H(p),
\end{aligned}$$

where the last equality comes from the convexity of the function $p \mapsto -(\beta_1^{-1} - \beta_2^{-1})H(p)$. Thus,

$$\begin{aligned}
\mathcal{K}_{\cup}\left(F_{\beta_1}\right)(p) &\geq \mathcal{K}_{\cup}\left(F_{\beta_1}\right)(p) - (\beta_1^{-1} - \beta_2^{-1})H(p) \\
&= F_{\beta_2}(p) - (\beta_1^{-1} - \beta_2^{-1})H(p) \\
&= F_{\beta_1}(p).
\end{aligned}$$

But, by definition, we have $\mathcal{K}_{\cup}\left(F_{\beta_1}\right)(p) \leq F_{\beta_1}(p)$, so $\mathcal{K}_{\cup}\left(F_{\beta_1}\right)(p) = F_{\beta_1}(p)$; in other words, $p \notin \mathcal{P}(\beta_1)$. Thus, we have proved that $\mathcal{P}(\beta_2)^c \subseteq \mathcal{P}(\beta_1)^c$, which is equivalent to $\mathcal{P}(\beta_1) \subseteq \mathcal{P}(\beta_2)$. $\square$

Let us now assume that $|\mathcal{X}| = |\mathcal{Y}| = 2$. As we already proved successive refinability for deterministic $p(Y|X)$ in Proposition 3, we can assume that $p(Y|X)$ is not deterministic. But, the case of $|\mathcal{X}| = |\mathcal{Y}| = 2$ and non-deterministic $p(Y|X)$ is exhaustively studied in [35] (Section IV.A, IV.B and IV.D). The latter work implies that, in this case:

Lemma A3. *Let $0 \leq v < H(X)$, let $(q(T), q(X|T))$ be a a solution to (A38) with parameter v, and let β be given by Proposition A8. Then, the set $\mathcal{P}(\beta)$ is a non-empty open interval and, for a pair $(q(T), q(X|T))$ to satisfy (A39), the set of points*

$$\{ q(X|t), \ t \in \mathcal{T} \}$$

must coincide with the extreme points of the interval $\mathcal{P}(\beta)$.

Equipped with these previously established facts, we can leverage them to prove successive refinement when $|\mathcal{X}| = |\mathcal{Y}| = 2$ and $p(Y|X)$ is not deterministic. Note that the computations from [35] that yield Lemma A3 extract crucial information from the fact that the sign of F_β'' is given here by a quadratic polynomial. These computations are not straightforwardly generalisable to larger source and relevancy cardinalities—even though they might serve as inspiration for potential generalisations. Let us start with the following lemma.

Lemma A4. *Let $0 \leq v < H(X)$. Then, we can assume, without loss of generality, that $|\mathcal{T}| = 2$. Moreover, in this case, a solution $(q(T), q(X|T))$ to the reformulated IB problem (A38) is such that $q(X|T)$, seen as a probability transition matrix, is injective.*

Proof. Let $(q(T), q(X|T))$ be a solution to (A38) for parameter v, and let β be given by Proposition A8. From Lemma A3, each $q(X|t)$ must correspond to one of the two extreme points of the interval $\mathcal{P}(\beta)$. Moreover, Proposition A2 ensures that, for any primal bottleneck (or equivalently, any solution to (A38)), we still obtain a bottleneck for the same parameter if we merge symbols t with identical $q(X|t)$. Thus, we can assume, without loss of generality, that $|\mathcal{T}| = 2$, and, in this case, the decoder $q(X|T)$ is, up to permutation of bottleneck symbols, uniquely defined by β.

Moreover, as $\mathcal{P}(\beta)$ is open and non-empty, these extreme points are distinct; in other words, the column transition matrix Q defined by $q(X|T)$ has its columns made of two distinct points on the simplex $\Delta_\mathcal{X}$. These points must thus be linearly independent as vectors in $\mathbb{R}^2$, so the rank of Q is 2. By the null rank theorem and as $|\mathcal{T}| = 2$, this implies that Q is injective. $\square$

Let us now first consider SR for the case of $n = 2$ processing stages. Let $0 < \lambda_1 < \lambda_2 \leq H(X)$, and let T_1, T_2 be solutions to the primal IB problem (1) of respective parameters λ_1, λ_2. Equivalently, T_1 and T_2 are solutions to the reformulated IB problem (A38) with resp. parameters v_1, v_2, where $0 \leq v_2 < v_1 < H(X)$. From Lemma A4, we can assume that $q(X|T_2)$ is injective. Moreover, from Proposition A8, the bottleneck pairs $(q(T_1), q(X|T_1))$ and $(q(T_2), q(X|T_2))$ are solutions to (A39) for parameters β_1, β_2, respectively, which correspond to inverse slopes of the CE curve at $(v_1, H_Y(v_1))$ and $(v_2, H_Y(v_2))$, respectively. By convexity of the CE curve [35], we have $\beta_1 \leq \beta_2$. Thus, from Lemma A2,

$$\mathcal{P}(\beta_1) \subseteq \mathcal{P}(\beta_2).$$

This is equivalent to

$$Hull\big(\overline{\mathcal{P}(\beta_1)}\big) = \overline{\mathcal{P}(\beta_1)} \subseteq \overline{\mathcal{P}(\beta_2)} = Hull\big(\overline{\mathcal{P}(\beta_2)}\big),$$

where $\overline{E}$ denotes the closure of a set E, so, here, $\mathcal{P}(\beta_i)$ and $\overline{\mathcal{P}(\beta_i)}$ only differ by taking or not taking the segment's extreme points, and the equalities come from the convexity of this segment. From Lemma A3, this can be rewritten as

$$Hull\big\{q(X|t_1), \, t_1 \in \mathcal{T}_1\big\} \subseteq Hull\big\{q(X|t_2), \, t_2 \in \mathcal{T}_2\big\}.$$

But this is exactly the convex hull condition (7). As we chose an injective $q(X|T_2)$, we can use the convex hull characterisation (Theorem 4) to conclude that T_1 and T_2 achieve successive refinement. Thus, we have proved SR for $n = 2$ stages.

Appendix B.8. Computation of Bifurcations Values

In this work, we compute the bottlenecks' bifurcation parameters as the values where the effective cardinality changes [43]: i.e., a bifurcation is a trade-off parameter value λ for which the number of distinct $q_\lambda(X|t)$ changes in a neighborhood of λ (see Section 1.3). With this naive method, the threshold chosen to numerically equate points $q(X|t)$ impacts the computed critical values, which could be avoided by using more sophisticated methods for computing these bifurcation values [42,43,71]. However, the bifurcation values computed by our naive method did correspond, on our minimal examples, to parameters where the smoothness of the functions $I_X(\beta) := I_\beta(X; T)$ and $I_Y(\beta) := I_\beta(Y; T)$ breaks. Thus, our method seemingly identifies discontinuities of the first-order derivative of I_X and I_Y, which are those of second-order derivatives of the Lagrangian in (3) (see Corollary 1 in [43]). In this sense, our naive method still identifies the IB bifurcations, if defined as second-order bifurcations of the IB Lagragian as in, e.g., [42,43].

Appendix C. Section 3 Details

Appendix C.1. Proof of Proposition 6

We recall that Δ_{q_1,q_2} is the space of extensions $q(X, T_1, T_2)$ of $q_1(X, T_1)$ and $q_2(X, T_2)$, and that $\Delta_{SR,2}$ is the space of all distributions $r(X, T_1, T_2)$ (not necessarily consistent with q_1 and q_2) under which the Markov chain $X - T_2 - T_1$ holds. We write the proof for discrete variables for ease of presentation, but the very same proof works for continuous variables if we replace sums by integrals. For $q(X, T_1, T_2) \in \Delta_{q_1,q_2}$ and $r(X, T_1, T_2) \in \Delta_{SR,2}$,

$$
\begin{aligned}
D_{KL}(q||r) &= \sum q(x, t_1, t_2) \log\left(\frac{q(x, t_1, t_2)}{r(x, t_1, t_2)}\right) \\
&= \sum q(x, t_1, t_2) \log\left(\frac{q(x, t_2)q(t_1|x, t_2)}{r(x, t_2)r(t_1|t_2)}\right) \\
&= \sum q(x, t_1, t_2) \log\left(\frac{q(t_1|x, t_2)}{r(t_1|t_2)}\right) + D_{KL}(q(X, T_2)||r(X, T_2)) \\
&\geq \sum q(x, t_1, t_2) \log\left(\frac{q(t_1|x, t_2)}{r(t_1|t_2)}\right) \\
&= \sum q(x, t_1, t_2) \log\left(\frac{q(t_1|x, t_2)}{q(t_1|t_2)}\right) + \sum q(t_2) D_{KL}(q(T_1|t_2)||r(T_1|t_2)) \\
&\geq \sum q(x, t_1, t_2) \log\left(\frac{q(t_1|x, t_2)}{q(t_1|t_2)}\right)
\end{aligned}
\tag{A41}
$$

The last term is $D_{KL}(q||r_0)$, with

$$
r_0(X, T_1, T_2) := q(X)q(T_2|X)q(T_1|T_2) \in \Delta_{SR,2},
$$

because, under r_0, the Markov chain $X - T_2 - T_1$ holds. So, from the last inequality in (A41),

$$
\inf_{r \in \Delta_{SR,2}} D_{KL}(q||r) = D_{KL}(q||r_0).
$$

But, the last term of (A41) is also $I_q(X; T_1|T_2)$. Thus,

$$
\begin{aligned}
D_{KL}(\Delta_{q_1,q_2}||\Delta_{SR}) &= \inf_{q \in \Delta_{q_1,q_2}} \inf_{r \in \Delta_{SR}} D_{KL}(q, r) \\
&= \inf_{q \in \Delta_{q_1,q_2}} D_{KL}(q||r_0) \\
&= \inf_{q \in \Delta_{q_1,q_2}} I_q(X; T_1|T_2) \\
&= UI(X : T_1 \setminus T_2).
\end{aligned}
$$

Appendix D. The Unicity and Injectivity Conjecture, and Technical Subtleties It Would Solve

In this section, we describe in more details some technical subtleties encountered in the main text, and present a conjecture that, if true, would make them fade away in cases where the information curve is strictly concave. Let us start by stating the conjecture, which is also interesting in itself. We recall that a bottleneck T is in canonical form when all the pointwise conditional probabilities $q(X|t)$ are distinct (see Section 1.3), and that every primal bottleneck can be reduced to canonical form (see Proposition A2).

Conjecture 1. *Let $p(X, Y)$ be such that the information curve is strictly concave. Then, the set of solutions $(q(T), q(X|T))$ to the primal IB problem (6) that are expressed in canonical form is such that*

(i)   The pair $(q(T), q(X|T))$ is, up to permuting bottleneck symbols, uniquely determined.
(ii)   The channel $q(X|T)$, seen as a linear operator on probability distributions, is injective.

Note that point (ii) in the conjecture was always numerically satisfied in our minimal numerical experiments, where we also always observed a strictly concave information curve. The strict concavity assumption is necessary for this conjecture to be possibly true, because it has been shown that for a non-strictly concave information curve, the channel $q(X|T)$ can be non-injective [39].

The convex hull characterisation of exact SR, i.e., Theorem 4, would, with Conjecture 1, be made more complete for the strictly concave case. Indeed, one can prove that the conjecture would imply the following one (Conjecture 2 can be obtained by combining Theorem 4 and Conjecture 1; as the latter is in any case not a statement for now, we omit the details):

Conjecture 2. *Let X and Y be discrete variables, let $\lambda_1 < \lambda_2$, and assume that the information curve is strictly concave. Then, there is successive refinement for parameters (λ_1, λ_2) if and only if, equivalently:*

(i)   There exist bottlenecks T_1, T_2 of parameters λ_1, λ_2, respectively, such that the convex hull condition (7) holds;
(ii)   For any bottlenecks T_1, T_2 of parameters λ_1, λ_2, respectively, the convex hull condition (7) holds.

In particular, assuming that the information curve is strictly concave and that Conjecture 1 is true, then if the convex hull condition breaks for some bottlenecks T_1 and T_2 of parameters λ_1 and λ_2, respectively, this is enough to conclude that there is not SR for parameters (λ_1, λ_2). Recalling that, in our numerical experiments, we observed strictly concave information curves, this would make the exact SR patterns in Figures 4–6 (right), Figures 10–12 (left), and Figure A3 (middle) exact characterisations of successive refinement. On the contrary, with Theorem 4 in its current state, in the latter figures, we are indeed guaranteed that SR holds in the blue areas where the convex hull condition is satisfied, but we are not formally guaranteed that SR does not hold in the red areas where the convex hull condition breaks. This is the reason for why, in the main text, we refer to these figures as mere numerical proxies for successive refinement.

Conjecture 1 being true would also, in the strictly concave case, solve a potential ambiguity in the definition of soft SR. Indeed, we do not provide, in Section 3.1, any formal guarantee that the quantity $UI(X : T_1 \setminus T_2)$ does not depend on the choice of the bottlenecks T_1 and T_2, among all those that solve the IB problems with respective trade-off parameters λ_1 and λ_2. To make sure that there is no such dependency, we should rather consider

$$\delta(\lambda_1, \lambda_2) := \inf_{q(T_1|X) \in IB(\lambda_1),\, q(T_2|X) \in IB(\lambda_2)} UI(X : T_1 \setminus T_2),$$

where, here, $IB(\lambda)$ denotes the set of distributions $q(T|X)$ that solve the IB problem (1) with trade-off parameter λ. In practice, there currently exists, to the best of our knowledge, no algorithm to compute, for a given bottleneck problem, *all* the solutions in $IB(\lambda)$. This is the reason for why, in this paper, we stick to computing $UI(X : T_1 \setminus T_2)$ for fixed bottlenecks T_1 and T_2. Careful readers should take this number to be, a priori, only an upper bound on the true measure of soft successive refinement $\delta(\lambda_1, \lambda_2)$.

However, one can directly verify that either permuting bottleneck symbols t or merging those with identical $q(X|t)$—so as to obtain a canonical bottleneck—leaves the unique information invariant. Thus, if Conjecture 1-(i) is true, it proves that, for a strictly concave information curve, $UI(X : T_1 \setminus T_2)$ is actually uniquely defined by the trade-off parameters λ_1, λ_2, because any pair of corresponding bottlenecks T_1 and T_2 results in the same unique information.

Eventually, Conjecture 1 seems interesting in itself. Indeed, it would provide crucial information on the trajectory of the bottlenecks' pointwise decoders $q_\lambda(X|t)$ over λ, which could then help for theoretical advances on the successive refinement of the IB.

Appendix E. Sample $p(Y|X)$ Used in Sections 2.3 and 3.2

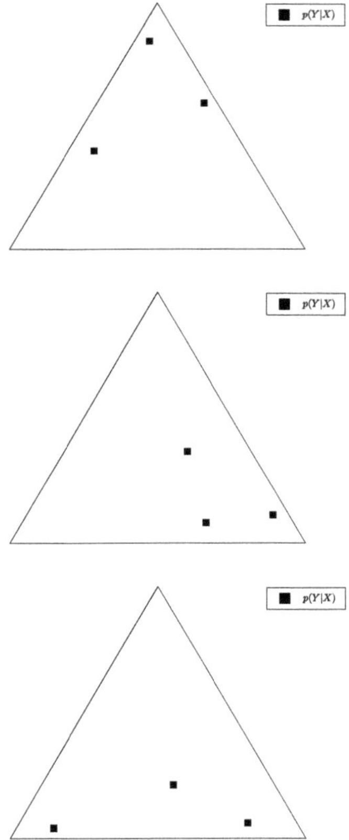

Figure A2. Plot of the sample distributions $p(Y|X)$ used in, respectively, from top to bottom: (i) Figures 4 and 7; (ii) Figures 5 and 8; (iii) Figures 6 and 9. The simplex depicted here is $\Delta_{\mathcal{Y}}$, where $|\mathcal{Y}| = 3$, and each black square corresponds to a symbol-wise conditional probability $p(Y|x) \in \Delta_{\mathcal{Y}}$. Note that the corresponding $p(X) \in \Delta_{\mathcal{X}}$ is shown in the left parts of Figures 4–9, which depict the simplex $\Delta_{\mathcal{X}}$, where, here, we also have $|\mathcal{X}| = 3$. The explicit values of the corresponding $p(X, Y)$ can be found at: https://gitlab.com/uh-adapsys/successive-refinement-ib/.

Appendix F. Additional Plots for Exact and Soft Successive Refinement

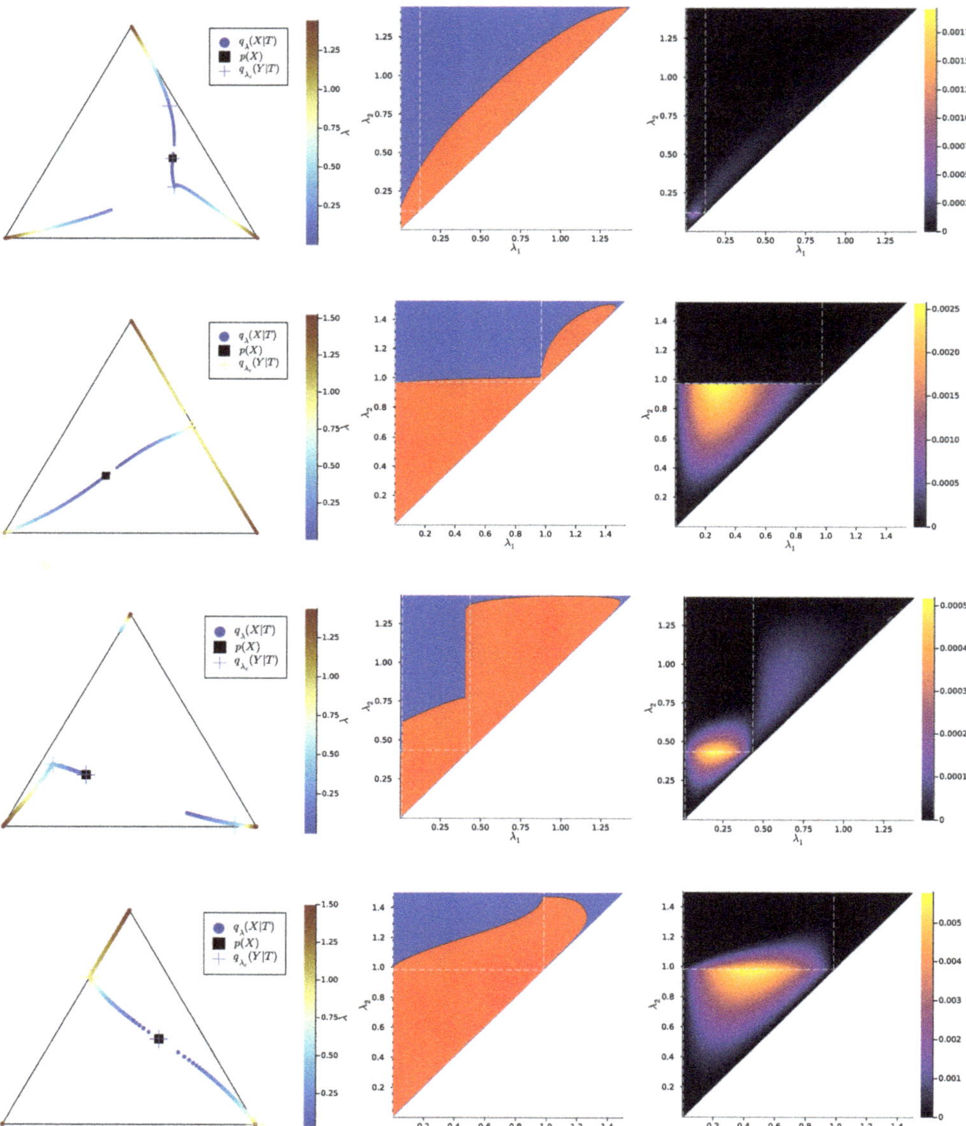

Figure A3. Additional examples for $|\mathcal{X}| = |\mathcal{Y}| = 3$: comparison of bottleneck trajectories (left) with exact SR patterns (center) and unique information landscapes (right). See Figures 4 and 7 for more details on the legends. The conditional distributions $p(Y|X)$ corresponding to each row in this figure are plotted in Figure A4. The explicit values of the corresponding $p(X, Y)$ can be found at: https://gitlab.com/uh-adapsys/successive-refinement-ib/.

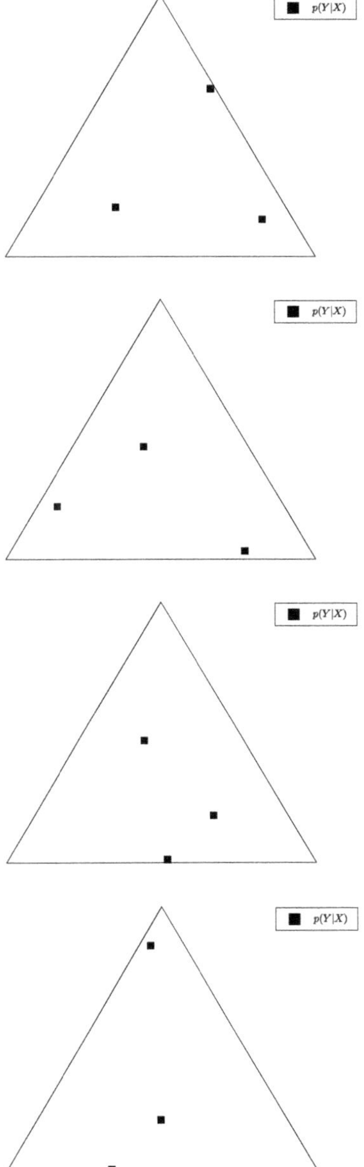

Figure A4. Sample distributions $p(Y|X)$ used in Figure A3, where the vertical order here corresponds to that of Figure A3. The simplex depicted here is $\Delta_{\mathcal{Y}}$, where $|\mathcal{Y}| = 3$, and each black square corresponds to a symbol-wise conditional probability $p(Y|x) \in \Delta_{\mathcal{Y}}$. Note that the corresponding $p(X) \in \Delta_{\mathcal{X}}$ is shown in the left parts of each row in Figure A3, which depict the simplex $\Delta_{\mathcal{X}}$, where, here, we also have $|\mathcal{X}| = 3$. The explicit values of the corresponding $p(X,Y)$ can be found at: https://gitlab.com/uh-adapsys/successive-refinement-ib/.

References

1. Tishby, N.; Pereira, F.; Bialek, W. The Information Bottleneck Method. In Proceedings of the 37th Allerton Conference on Communication, Control and Computation, 22–24 September 1999; Volume 49.
2. Gilad-Bachrach, R.; Navot, A.; Tishby, N. An Information Theoretic Tradeoff between Complexity and Accuracy. In *Learning Theory and Kernel Machines*; Lecture Notes in Computer Science; Springer: Berlin/Heidelberg, Germany, 2003. [CrossRef]
3. Bialek, W.; De Ruyter Van Steveninck, R.R.; Tishby, N. Efficient representation as a design principle for neural coding and computation. In Proceedings of the 2006 IEEE International Symposium on Information Theory, Seattle, WA, USA, 9–14 July 2006; pp. 659–663. [CrossRef]
4. Creutzig, F.; Globerson, A.; Tishby, N. Past-future information bottleneck in dynamical systems. *Phys. Rev. E* **2009**, *79*, 041925. [CrossRef]
5. Amir, N.; Tiomkin, S.; Tishby, N. Past-future Information Bottleneck for linear feedback systems. In Proceedings of the 2015 54th IEEE Conference on Decision and Control (CDC), Osaka, Japan, 15–18 December 2015; pp. 5737–5742. [CrossRef]
6. Sachdeva, V.; Mora, T.; Walczak, A.M.; Palmer, S.E. Optimal prediction with resource constraints using the information bottleneck. *PLoS Comput. Biol.* **2021**, *17*, e1008743. [CrossRef] [PubMed]
7. Klampfl, S.; Legenstein, R.; Maass, W. Spiking Neurons Can Learn to Solve Information Bottleneck Problems and Extract Independent Components. *Neural Comput.* **2009**, *21*, 911–959. [CrossRef] [PubMed]
8. Buesing, L.; Maass, W. A Spiking Neuron as Information Bottleneck. *Neural Comput.* **2010**, *22*, 1961–1992. [CrossRef]
9. Chalk, M.; Marre, O.; Tkačik, G. Toward a unified theory of efficient, predictive, and sparse coding. *Proc. Natl. Acad. Sci. USA* **2018**, *115*, 186–191. [CrossRef]
10. Palmer, S.E.; Marre, O.; Berry, M.J.; Bialek, W. Predictive information in a sensory population. *Proc. Natl. Acad. Sci. USA* **2015**, *112*, 6908–6913. [CrossRef]
11. Wang, S.; Segev, I.; Borst, A.; Palmer, S. Maximally efficient prediction in the early fly visual system may support evasive flight maneuvers. *PLoS Comput. Biol.* **2021**, *17*, e1008965. [CrossRef] [PubMed]
12. Buddha, S.K.; So, K.; Carmena, J.M.; Gastpar, M.C. Function Identification in Neuron Populations via Information Bottleneck. *Entropy* **2013**, *15*, 1587–1608. [CrossRef]
13. Kleinman, M.; Wang, T.; Xiao, D.; Feghhi, E.; Lee, K.; Carr, N.; Li, Y.; Hadidi, N.; Chandrasekaran, C.; Kao, J.C. A cortical information bottleneck during decision-making. *bioRxiv* **2023**. [CrossRef]
14. Nehaniv, C.L.; Polani, D.; Dautenhahn, K.; te Beokhorst, R.; Cañamero, L. Meaningful Information, Sensor Evolution, and the Temporal Horizon of Embodied Organisms. In *Artificial life VIII*; ICAL 2003; MIT Press: Cambridge, MA, USA, 2002; pp. 345–349.
15. Klyubin, A.; Polani, D.; Nehaniv, C. Organization of the information flow in the perception-action loop of evolved agents. In Proceedings of the 2004 NASA/DoD Conference on Evolvable Hardware, Seattle, WA, USA, 24–26 June 2004; pp. 177–180. [CrossRef]
16. van Dijk, S.G.; Polani, D. Informational Drives for Sensor Evolution. Vol. ALIFE 2012: The Thirteenth International Conference on the Synthesis and Simulation of Living Systems, ALIFE 2022: The 2022 Conference on Artificial Life. 2012. Available online: https://direct.mit.edu/isal/proceedings-pdf/alife2012/24/333/1901044/978-0-262-31050-5-ch044.pdf (accessed on 12 September 2023).
17. Möller, M.; Polani, D. Emergence of common concepts, symmetries and conformity in agent groups—An information-theoretic model. *Interface Focus* **2023**, *13*, 20230006. [CrossRef]
18. Catenacci Volpi, N.; Polani, D. Space Emerges from What We Know-Spatial Categorisations Induced by Information Constraints. *Entropy* **2020**, *20*, 1179. [CrossRef]
19. Zaslavsky, N.; Kemp, C.; Regier, T.; Tishby, N. Efficient compression in color naming and its evolution. *Proc. Natl. Acad. Sci. USA* **2018**, *115*, 201800521. [CrossRef]
20. Zaslavsky, N.; Garvin, K.; Kemp, C.; Tishby, N.; Regier, T. The evolution of color naming reflects pressure for efficiency: Evidence from the recent past. *bioRxiv* **2022**. [CrossRef]
21. Tucker, M.; Levy, R.P.; Shah, J.; Zaslavsky, N. Trading off Utility, Informativeness, and Complexity in Emergent Communication. *Adv. Neural Inf. Process. Syst.* **2022**, *35*, 22214–22228.
22. Pacelli, V.; Majumdar, A. Task-Driven Estimation and Control via Information Bottlenecks. *arXiv* **2018**, arXiv:1809.07874. [CrossRef]
23. Lamb, A.; Islam, R.; Efroni, Y.; Didolkar, A.; Misra, D.; Foster, D.; Molu, L.; Chari, R.; Krishnamurthy, A.; Langford, J. Guaranteed Discovery of Control-Endogenous Latent States with Multi-Step Inverse Models. *arXiv* **2022**, arXiv:2207.08229. [CrossRef]
24. Goyal, A.; Islam, R.; Strouse, D.; Ahmed, Z.; Larochelle, H.; Botvinick, M.; Levine, S.; Bengio, Y. Transfer and Exploration via the Information Bottleneck. In Proceedings of the International Conference on Learning Representations, New Orleans, LA, USA, 6–9 May 2019.
25. Koshelev, V. Hierarchical Coding of Discrete Sources. *Probl. Peredachi Inf.* **1980**, *16*, 31–49.
26. Equitz, W.; Cover, T. Successive refinement of information. *IEEE Trans. Inf. Theory* **1991**, *37*, 269–275. [CrossRef]
27. Rimoldi, B. Successive refinement of information: Characterization of the achievable rates. *IEEE Trans. Inf. Theory* **1994**, *40*, 253–259. [CrossRef]
28. Tuncel, E.; Rose, K. Computation and analysis of the N-Layer scalable rate-distortion function. *IEEE Trans. Inf. Theory* **2003**, *49*, 1218–1230. [CrossRef]

29. Kostina, V.; Tuncel, E. Successive Refinement of Abstract Sources. *IEEE Trans. Inf. Theory* **2019**, *65*, 6385–6398. [CrossRef]
30. Tian, C.; Chen, J. Successive Refinement for Hypothesis Testing and Lossless One-Helper Problem. *IEEE Trans. Inf. Theory* **2008**, *54*, 4666–4681. [CrossRef]
31. Tuncel, E. Capacity/Storage Tradeoff in High-Dimensional Identification Systems. In Proceedings of the 2006 IEEE International Symposium on Information Theory, Seattle, WA, USA, 9–14 July 2006; pp. 1929–1933. [CrossRef]
32. Mahvari, M.M.; Kobayashi, M.; Zaidi, A. On the Relevance-Complexity Region of Scalable Information Bottleneck. *arXiv* **2020**, arXiv:2011.01352. [CrossRef]
33. Kline, A.G.; Palmer, S.E. Gaussian information bottleneck and the non-perturbative renormalization group. *New J. Phys.* **2022**, *24*, 033007. [CrossRef]
34. Kolchinsky, A.; Tracey, B.D.; Van Kuyk, S. Caveats for information bottleneck in deterministic scenarios. *arXiv* **2018**, arXiv:1808.07593. [CrossRef]
35. Witsenhausen, H.; Wyner, A. A conditional entropy bound for a pair of discrete random variables. *IEEE Trans. Inf. Theory* **1975**, *21*, 493–501. [CrossRef]
36. Hsu, H.; Asoodeh, S.; Salamatian, S.; Calmon, F.P. Generalizing Bottleneck Problems. In Proceedings of the 2018 IEEE International Symposium on Information Theory (ISIT), Vail, CO, USA, 17–22 June 2018; pp. 531–535. [CrossRef]
37. Asoodeh, S.; Calmon, F. Bottleneck Problems: An Information and Estimation-Theoretic View. *Entropy* **2020**, *22*, 1325. [CrossRef]
38. Dikshtein, M.; Shamai, S. A Class of Nonbinary Symmetric Information Bottleneck Problems. *arXiv* **2021**, arXiv:cs.IT/2110.00985.
39. Benger, E.; Asoodeh, S.; Chen, J. The Cardinality Bound on the Information Bottleneck Representations is Tight. *arXiv* **2023**, arXiv:cs.IT/2305.07000.
40. Bertschinger, N.; Rauh, J.; Olbrich, E.; Ay, N. Quantifying Unique Information. *Entropy* **2013**, *16*, 2161–2183. [CrossRef]
41. Parker, A.E.; Gedeon, T.; Dimitrov, A. The Lack of Convexity of the Relevance-Compression Function. *arXiv* **2022**, arXiv:2204.10957. [CrossRef]
42. Wu, T.; Fischer, I. Phase Transitions for the Information Bottleneck in Representation Learning. *arXiv* **2020**, arXiv:2001.01878.
43. Zaslavsky, N.; Tishby, N. Deterministic Annealing and the Evolution of Information Bottleneck Representations. 2019. Available online: https://www.nogsky.com/publication/2019-evo-ib/2019-evo-IB.pdf (accessed on 12 September 2023).
44. Ngampruetikorn, V.; Schwab, D.J. Perturbation Theory for the Information Bottleneck. *Adv. Neural Inf. Process. Syst.* **2021**, *34*, 21008–21018.
45. Bertschinger, N.; Rauh, J. The Blackwell relation defines no lattice. In Proceedings of the 2014 IEEE International Symposium on Information Theory, Honolulu, HI, USA, 29 June–4 July 2014; pp. 2479–2483. [CrossRef]
46. Yang, Q.; Piantanida, P.; Gündüz, D. The Multi-layer Information Bottleneck Problem. *arXiv* **2017**, arXiv:1711.05102. [CrossRef]
47. Cover, T.; Thomas, J. *Elements of Information Theory*; Wiley-Interscience: Hoboken, NJ, USA, 2006.
48. Zaidi, A.; Estella-Aguerri, I.; Shamai (Shitz), S. On the Information Bottleneck Problems: Models, Connections, Applications and Information Theoretic Views. *Entropy* **2020**, *22*, 151. [CrossRef] [PubMed]
49. Tishby, N.; Zaslavsky, N. Deep Learning and the Information Bottleneck Principle. In Proceedings of the 2015 IEEE Information Theory Workshop, ITW 2015, Jerusalem, Israel, 26 April–1 May 2015. [CrossRef]
50. Shwartz-Ziv, R.; Tishby, N. Opening the Black Box of Deep Neural Networks via Information. 2017. Available online: http://xxx.lanl.gov/abs/1703.00810 (accessed on 12 September 2023).
51. Shwartz-Ziv, R.; Painsky, A.; Tishby, N. Representation Compression and Generalization in Deep Neural Networks, 2019. Available online: https://openreview.net/pdf?id=SkeL6sCqK7 (accessed on 12 September 2023).
52. Saxe, A.M.; Bansal, Y.; Dapello, J.; Advani, M.; Kolchinsky, A.; Tracey, B.D.; Cox, D.D. On the information bottleneck theory of deep learning. *J. Stat. Mech. Theory Exp.* **2019**, *2019*, 124020. [CrossRef]
53. Achille, A.; Soatto, S. Emergence of Invariance and Disentanglement in Deep Representations. In Proceedings of the 2018 Information Theory and Applications Workshop (ITA), San Diego, CA, USA, 11–16 February 2018; pp. 1–9. [CrossRef]
54. Elad, A.; Haviv, D.; Blau, Y.; Michaeli, T. Direct Validation of the Information Bottleneck Principle for Deep Nets. In Proceedings of the 2019 IEEE/CVF International Conference on Computer Vision Workshop (ICCVW), Seoul, Korea, 27–28 October 2019; pp. 758–762. [CrossRef]
55. Lorenzen, S.S.; Igel, C.; Nielsen, M. Information Bottleneck: Exact Analysis of (Quantized) Neural Networks. In Proceedings of the International Conference on Learning Representations, Virtual Event, 25–29 April 2022.
56. Kawaguchi, K.; Deng, Z.; Ji, X.; Huang, J. How Does Information Bottleneck Help Deep Learning? 2023. Available online: https://proceedings.mlr.press/v202/kawaguchi23a/kawaguchi23a.pdf (accessed on 12 September 2023).
57. Yousfi, Y.; Akyol, E. Successive Information Bottleneck and Applications in Deep Learning. In Proceedings of the 2020 54th Asilomar Conference on Signals, Systems, and Computers, Pacific Grove, CA, USA, 1–4 November 2020; pp. 1210–1213. [CrossRef]
58. No, A. Universality of Logarithmic Loss in Successive Refinement. *Entropy* **2019**, *21*, 158. [CrossRef]
59. Nasser, R. On the input-degradedness and input-equivalence between channels. In Proceedings of the 2017 IEEE International Symposium on Information Theory (ISIT), Aachen, Germany, 25–30 June 2017; pp. 2453–2457. [CrossRef]
60. Lastras, L.; Berger, T. All sources are nearly successively refinable. *IEEE Trans. Inf. Theory* **2001**, *47*, 918–926. [CrossRef]
61. Williams, P.L.; Beer, R.D. Nonnegative Decomposition of Multivariate Information. 2010. Available online: https://arxiv.org/pdf/1004.2515 (accessed on 12 September 2023).

62. Bertschinger, N.; Rauh, J.; Olbrich, E.; Jost, J. Shared Information—New Insights and Problems in Decomposing Information in Complex Systems. In Proceedings of the European Conference on Complex Systems, 2012; Springer International Publishing: Berlin/Heidelberg, Germany, 2013; pp. 251–269. [CrossRef]
63. Griffith, V.; Koch, C. Quantifying Synergistic Mutual Information. In *Guided Self-Organization: Inception*; Prokopenko, M., Ed.; Springer: Berlin/Heidelberg, Germany, 2014; pp. 159–190. [CrossRef]
64. Harder, M.; Salge, C.; Polani, D. Bivariate measure of redundant information. *Phys. Rev. E* **2013**, *87*, 012130. [CrossRef] [PubMed]
65. Blackwell, D. Equivalent Comparisons of Experiments. *Ann. Math. Stat.* **1953**, *24*, 265–272. [CrossRef]
66. Lemaréchal, C. Lagrangian Relaxation. In *Computational Combinatorial Optimization: Optimal or Provably Near-Optimal Solutions*; Jünger, M.; Naddef, D., Eds.; Springer: Berlin/Heidelberg, Germany, 2001; pp. 112–156. [CrossRef]
67. Kolchinsky, A.; Tracey, B.; Wolpert, D. Nonlinear Information Bottleneck. *Entropy* **2017**, *21*, 1181. [CrossRef]
68. Matousek, J.; Gärtner, B. *Understanding and Using Linear Programming*, 1st ed.; Springer: Berlin/Heidelberg, Germany, 2007.
69. van den Brand, J. A Deterministic Linear Program Solver in Current Matrix Multiplication Time. In Proceedings of the Thirty-First Annual ACM-SIAM Symposium on Discrete Algorithms; Society for Industrial and Applied Mathematics (SODA'20), Salt Lake City, UT, USA, 5–8 January 2020; pp. 259–278.
70. Rose, K. Deterministic annealing for clustering, compression, classification, regression, and related optimization problems. *Proc. IEEE* **1998**, *86*, 2210–2239. [CrossRef]
71. Gedeon, T.; Parker, A.E.; Dimitrov, A.G. The Mathematical Structure of Information Bottleneck Methods. *Entropy* **2012**, *14*, 456–479. [CrossRef]
72. Shamir, O.; Sabato, S.; Tishby, N. Learning and generalization with the information bottleneck. *Theor. Comput. Sci.* **2010**, *411*, 2696–2711. [CrossRef]
73. Rauh, J.; Banerjee, P.K.; Olbrich, E.; Jost, J. Unique Information and Secret Key Decompositions. In Proceedings of the 2019 IEEE International Symposium on Information Theory (ISIT), Paris, France, 7–12 July 2019. [CrossRef]
74. Banerjee, P.; Rauh, J.; Montufar, G. Computing the Unique Information. In Proceedings of the 2018 IEEE International Symposium on Information Theory (ISIT), Vail, CO, USA, 17–22 June 2018; pp. 141–145. [CrossRef]
75. Chechik, G.; Globerson, A.; Tishby, N.; Weiss, Y. Information bottleneck for Gaussian variables. *J. Mach. Learn. Res.* **2005**, *6*, 165–188.
76. Sutton, R.S.; Barto, A.G. *Reinforcement Learning: An Introduction*, 2nd ed.; The MIT Press: Cambridge, MA, USA, 2018.
77. Goldfeld, Z.; Polyanskiy, Y. The Information Bottleneck Problem and its Applications in Machine Learning. *IEEE J. Sel. Areas Inf. Theory* **2020**, *1*, 19–38. [CrossRef]
78. Mahvari, M.M.; Kobayashi, M.; Zaidi, A. Scalable Vector Gaussian Information Bottleneck. In Proceedings of the 2021 IEEE International Symposium on Information Theory (ISIT), Melbourne, Australia, 12–20 July 2021; pp. 37–42. [CrossRef]

Article

The Double-Sided Information Bottleneck Function [†]

Michael Dikshtein [1,*], **Or Ordentlich [2]** and **Shlomo Shamai (Shitz) [1]**

[1]   Department of Electrical and Computer Engineering, Technion, Haifa 3200003, Israel
[2]   School of Computer Science and Engineering, The Hebrew University of Jerusalem, Jerusalem 9190401, Israel
*   Correspondence: michaeldic@campus.technion.ac.il
†   This paper is an extended version of our paper published in 2021 IEEE International Symposium on Information Theory.

Abstract: A double-sided variant of the information bottleneck method is considered. Let (X, Y) be a bivariate source characterized by a joint pmf P_{XY}. The problem is to find two independent channels $P_{U|X}$ and $P_{V|Y}$ (setting the Markovian structure $U \to X \to Y \to V$), that maximize $I(U; V)$ subject to constraints on the relevant mutual information expressions: $I(U; X)$ and $I(V; Y)$. For jointly Gaussian X and Y, we show that Gaussian channels are optimal in the low-SNR regime but not for general SNR. Similarly, it is shown that for a doubly symmetric binary source, binary symmetric channels are optimal when the correlation is low and are suboptimal for high correlations. We conjecture that Z and S channels are optimal when the correlation is 1 (i.e., $X = Y$) and provide supporting numerical evidence. Furthermore, we present a Blahut–Arimoto type alternating maximization algorithm and demonstrate its performance for a representative setting. This problem is closely related to the domain of biclustering.

Keywords: information bottleneck; lossy compression; remote source coding; biclustering

Citation: Dikshtein, M.; Ordentlich, O.; Shamai, S. The Double-Sided Information Bottleneck Function. *Entropy* **2022**, *24*, 1321. https://doi.org/10.3390/e24091321

Academic Editors: Gerhard Bauch and Jan Lewandowsky

Received: 25 July 2022
Accepted: 13 September 2022
Published: 19 September 2022

Publisher's Note: MDPI stays neutral with regard to jurisdictional claims in published maps and institutional affiliations.

1. Introduction

The *information bottleneck* (IB) method [1] plays a central role in advanced lossy source compression. The analysis of classical source coding algorithms is mainly approached via the rate-distortion theory, where a fidelity measure must be defined. However, specifying an appropriate distortion measure in many real-world applications is challenging and sometimes infeasible. The IB framework introduces an essentially different concept, where another variable is provided, which carries the relevant information in the data to be compressed. The quality of the reconstructed sequence is measured via the mutual information metric between the reconstructed data and the relevance variables. Thus, the IB method provides a universal fidelity measure.

In this work, we extend and generalize the IB method by imposing an additional bottleneck constraint on the relevant variable and considering noisy observation of the source. In particular, let (X, Y) be a bivariate source characterized by a fixed joint probability law P_{XY} and consider all Markov chains $U \to X \to Y \to V$. The Double-Sided Information Bottleneck (DSIB) function is defined as [2]:

$$R_{P_{XY}}(C_u, C_v) \triangleq \max I(U; V), \qquad (1)$$

where the maximization is over all $P_{U|X}$ and $P_{V|Y}$ satisfying $I(U; X) \leq C_u$ and $I(V; Y) \leq C_v$. This problem is illustrated in Figure 1. In our study, we aim to determine the maximum value and the achieving conditional distributions $(P_{U|X}, P_{V|Y})$ (test channels) of (1) for various fixed sources P_{XY} and constraints C_u and C_v.

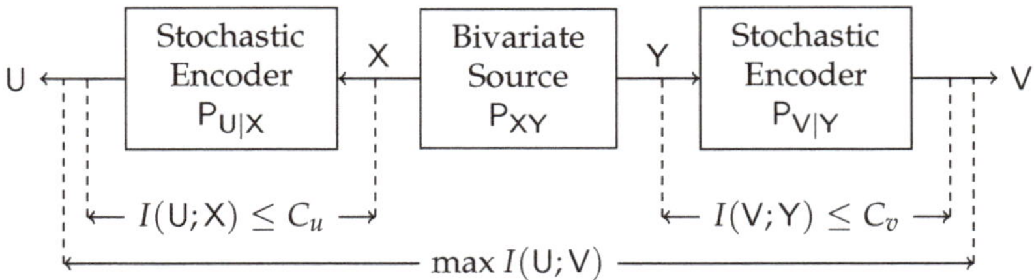

Figure 1. Block diagram of the Double-Sided Information Bottleneck function.

The problem we consider originates from the domain of clustering. Clustering is applied to organize similar entities in unsupervised learning [3]. It has numerous practical applications in data science, such as: joint word-document clustering, gene expression [4], and pattern recognition. The data in those applications are arranged as a contingency table. Usually, clustering is performed on one dimension of the table, but sometimes it is helpful to apply clustering on both dimensions of the contingency table [5], for example, when there is a strong correlation between the rows and the columns of the table or when high-dimensional sparse structures are handled. The input and output of a typical biclustering algorithm are illustrated in Figure 2. Consider an $S \times T$ data matrix (a_{st}). Find partitions $\mathcal{B}_k \subseteq \{1, \ldots, S\}$ and $\mathcal{C}_l \subseteq \{1, \ldots, T\}, k = 1, \ldots, K, l = 1, \ldots, L$ such that all elements of the "biclusters" [6] $(a_{st})_{s \in \mathcal{B}_k, t \in \mathcal{C}_l}$ are homogeneous. The measure of homogeneity depends on the application.

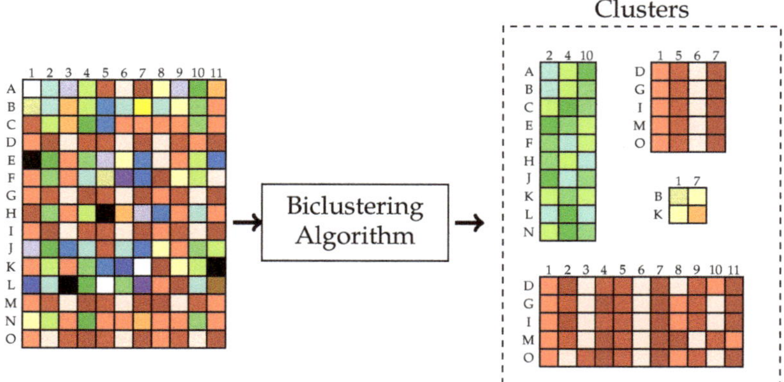

Figure 2. Illustration of a typical biclustering algorithm.

This problem can also be motivated by a remote source coding setting. Consider a latent random variable W, which satisfies $U \leftarrow X \leftarrow W \rightarrow Y \rightarrow V$ and represents a source of information. We have two users that observe noisy versions of W, i.e., X and Y. Those users try to compress the observed noisy data so that their reconstructed versions, U and V, will be comparable under the maximum mutual information metric. The problem we consider also bears practical applications. Imagine a distributed sensor network where the different edges measure a noisy version of a particular signal but are not allowed to communicate with each other. Each of the nodes performs compression of the received signal. Under the DSIB framework, we can find the optimal compression schemes that preserve the reconstructed symbols' proximity subject to the mutual information measure.

Dhillon et al. [7] initiated an information-theoretic approach to biclustering. They have regarded the normalized non-negative contingency table as a joint probability distribution matrix of two random variables. Mutual information was proposed as a measure for optimal co-clustering. An optimization algorithm was presented that intertwines both row and column clustering at all stages. Distributed clustering from a proper information-theoretic perspective was first explicitly considered by Pichler et al. [2]. Consider the model illustrated in Figure 3. A bivariate memory-less source with joint law P_{XY} generates n i.i.d. copies (X^n, Y^n) of (X, Y). Each sequence is observed at two different encoders, and each encoder generates a description of the observed sequence, $f_n(X^n)$ and $g_n(Y^n)$. The objective is to construct the mappings f_n and g_n such that the normalized mutual information between the descriptions would be maximal while the description coding has bounded rate constraints. Single-letter inner and outer bounds for a general P_{XY} were derived. An example of a *doubly symmetric binary source* (DSBS) was given, and several converse results were established. Furthermore, connections were made to the standard IB [1] and the *multiple description* CEO problems [8]. In addition, the equivalence of information-theoretic biclustering problem to hypothesis testing against independence with multiterminal data compression and a pattern recognition problem was established in [9,10], respectively.

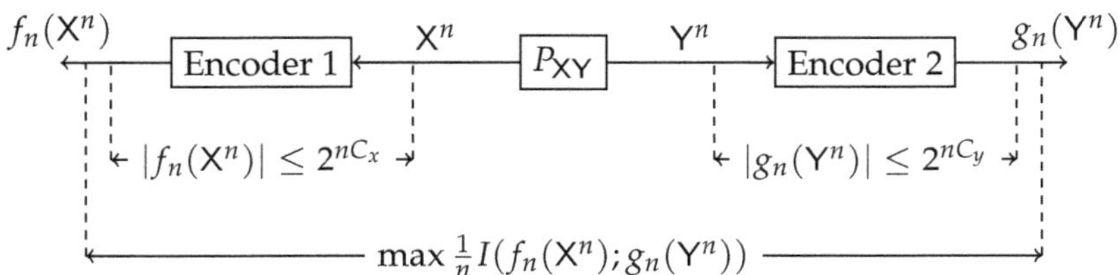

Figure 3. Block diagram of the information-theoretic biclustering problem.

The DSIB problem addressed in our paper is, in fact, a single-letter version of the *distributed clustering* setup [2]. The inner bound in [2] coincides with our problem definition. Moreover, if the Markov condition $U \to X \to Y \to Z$ is imposed on the multi-letter variant, then those problems are equivalent. A similar setting, but with a maximal correlation criterion between the reconstructed random variables, has been considered in [11,12]. Furthermore, it is sometimes the case that the optimal biclustering problem is more straightforward to solve than its standard, single-sided, clustering counterpart. For example, the Courtade–Kumar conjecture [13] for the standard single-sided clustering setting was ultimately proven for the biclustering setting [14]. A particular case, where (X, Y) are drawn from DSBS distribution and the mappings f_n and g_n are restricted to be Boolean functions, was addressed in [14]. The bound $I(f_n(X^n); g_n(Y^n)) \leq I(X; Y)$ was established, which is tight if and only if f_n and g_n are dictator functions.

1.1. Related Work

Our work extends the celebrated standard (single-sided) IB (SSIB) method introduced by Tishby et al. [1]. Indeed, consider the problem illustrated in Figure 4. This single-sided counterpart of our work is essentially a remote source coding problem [15–17], choosing the distortion measure as the logarithmic loss. The random variable U represents the noisy version (X) of the source (Y) with a constrained number of bits ($I(U; X) \leq C$), and the goal is to maximize the relevant information in U regarding Y (measured by the mutual information between Y and U). In the standard IB setup, $I(U; X)$ is referred to as the complexity of U, and $I(Y; U)$ is referred to as the relevance of U.

For the particular case where (U, X, Y) are discrete, an optimal $P_{U|X}$ can be found by iteratively solving a set of self-consistent equations. A generalized Blahut–Arimoto algorithm [18–21] was proposed to solve those equations. The optimal test-channel $P_{U|X}$ was characterized using a variation principle in [1]. A particular case of deterministic mappings from X to U was considered in [22], and algorithms that find those mappings were described.

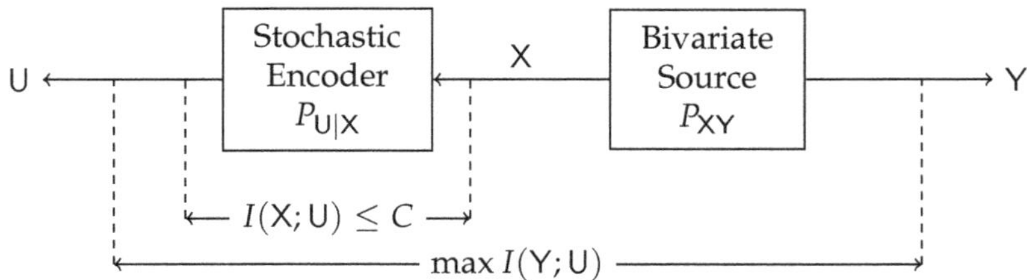

Figure 4. Block diagram of the Single-Sided Information Bottleneck function.

Several representing scenarios have been considered for the SSIB problem. The setting where the pair (X, Y) is a *doubly symmetric binary source* (DSBS) with transition probability p was addressed from various perspectives in [17,23,24]. Utilizing Mrs. Gerber's Lemma (MGL) [25], one can show that the optimal test-channel for the DSBS setting is a BSC. The case where $(\mathbf{X}, \mathbf{Y})$ are jointly multivariate Gaussians in the SSIB framework was first considered in [26]. It was shown that the optimal distribution of $(\mathbf{U}, \mathbf{X}, \mathbf{Y})$ is also jointly Gaussian. The optimality of the Gaussian test channel can be proven using EPI [27], or exploiting I-MMSE and Single Crossing Property [28]. Moreover, the proof can be easily extended to jointly Gaussian random vectors $(\mathbf{X}, \mathbf{Y})$ under the I-MMSE framework [29].

In a more general scenario where $X = Y + Z$ and only Z is fixed to be Gaussian, it was shown that discrete signaling with deterministic quantizers as test-channel sometimes outperforms Gaussian P_X [30]. This exciting observation leads to a conjecture that discrete inputs are optimal for this general setting and may have a connection to the input amplitude constrained AWGN channels where it was already established that discrete input distributions are optimal [31–33]. One reason for the optimality of discrete distributions stems from the observation that constraining the compression rate limits the usable input amplitude. However, as far as we know, it remains an open problem.

There are various related problems considered in the literature that are equivalent to the SSIB; namely, they share a similar single-letter optimization problem. In the *conditional entropy bound* (CEB) function, studied in [17], given a fixed bivariate source (X, Y) and an equality constraint on the conditional entropy of X given U, the goal is to minimize the conditional entropy of Y given U over the set of U such that $U \to X \to Y$ constitute a Markov chain. One can show that CEB is equivalent to SSIB. The *common reconstruction* CR setting [34] is a source coding with a side-information problem, also known as Wyner–Ziv coding, as depicted in Figure 5; with an additional constraint, the encoder can reconstruct the same sequence as the decoder. Additional assumption of log-loss fidelity results in a single-letter rate-distortion region equivalent to the SSIB. In the problem of *information combining* (IC) [23,35], motivated by message combining in LDPC decoders, a source of information, P_Y, is observed through two test-channels $P_{X|Y}$ and $P_{Z|Y}$. The IC framework aims to design those channels in two extreme approaches. For the first, IC asks what those channels should be to make the output pair (X, Z) maximally informative regarding Y. On the contrary, IC also considers how to design $P_{X|Y}$ and $P_{Z|Y}$ to minimize the information in (X, Z) regarding Y. The problem of minimizing IC can be shown to be equivalent to the SSIB. In fact, if (X, Y) is a DSBS, then by [23], $P_{Z|Y}$ is a *binary symmetric channel* (BSC), recovering similar results from (Section IV.A of [17]).

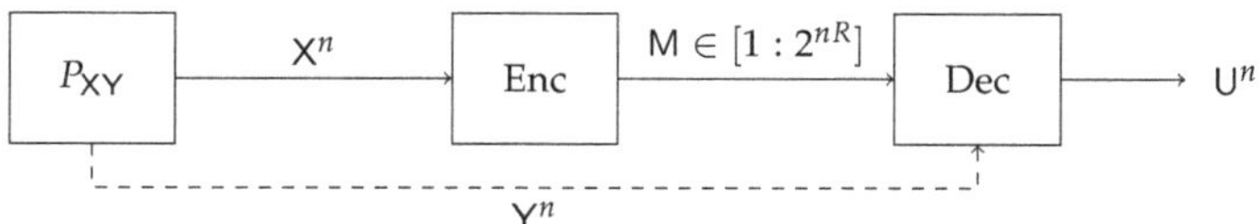

Figure 5. Block diagram of Source Coding with Side Information.

The IB method has been extended to various network topologies. A multilayer extension of the IB method is depicted in Figure 6. This model was first considered in [36]. A multivariate source $(X, Y_1, \ldots, Y_L)$ generates a sequence of n i.i.d. copies $(X^n, Y_1^n, \ldots, Y_L^n)$. The receiver has access only to the sequence X^n while $(Y_1^n, \ldots, Y_L^n)$ are hidden. The decoder performs a consecutive L-stage compression of the observed sequence. The representation at step k must be maximally informative about the respective hidden sequence Y_k, $k \in \{1, 2, \ldots, L\}$. This setup is highly motivated by the structure of deep neural networks. Specific results were established for the binary and Gaussian sources.

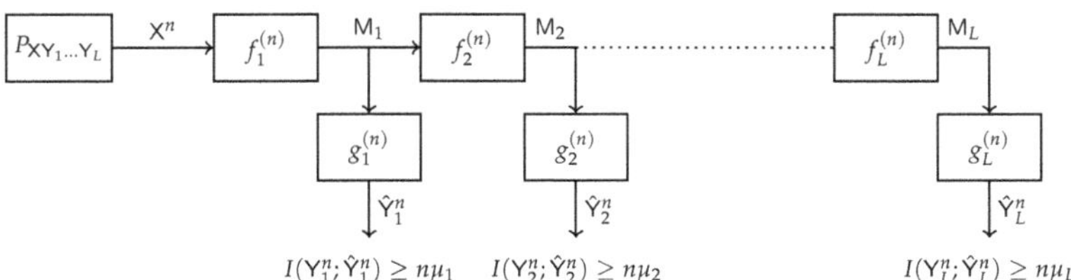

Figure 6. Block diagram of the Multi-Layer IB.

The model depicted in Figure 7 represents a multiterminal extension of the standard IB. A set of receivers observe noisy versions $(X_1, X_2, \ldots, X_K)$ of some source of information Y. The channel outputs $(X_1, X_2, \ldots, X_K)$ are conditionally independent given Y. The receivers are connected to the central processing unit through noiseless but limited-capacity backhaul links. The central processor aims to attain a good prediction $\hat{Y}$ of the source Y based on compressed representations of the noisy version of Y obtained from the receivers. The quality of prediction is measured via the mutual information merit between Y and $\hat{Y}$. The Distributive IB setting is essentially a CEO source coding problem under logarithmic loss (log-loss) distortion measure [37]. The case where $(X, Y_1, \ldots, Y_K)$ are jointly Gaussian random variables was addressed in [20], and a Blahut–Arimoto-type algorithm was proposed. An optimized algorithm to design quantizers was proposed in [38].

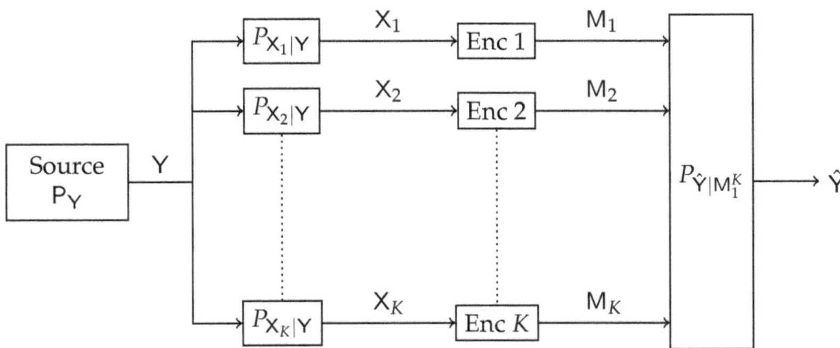

Figure 7. Block diagram of the Distributive IB.

A cooperative multiterminal extension of the IB method was proposed in [39]. Let (X_1^n, X_2^n, Y^n) be n i.i.d. copies of the multivariate source (X_1, X_2, Y). The sequences X_1^n and X_2^n are observed at encoders 1 and 2, respectively. Each encoder sends a representation of the observed sequence through a noiseless yet rate-limited link to the other encoder and the mutual decoder. The decoder attempts to reconstruct the latent representation sequence Y^n based on the received descriptions. As shown in Figure 8, this setup differs from the CEO setup [40] since the encoders can cooperate during the transmission. The set of all feasible rates of complexity and relevance were characterized, and specific regions for the binary and Gaussian sources were established. There are many additional variations of multi-user IB in the literature [20,26,35–37,39–44].

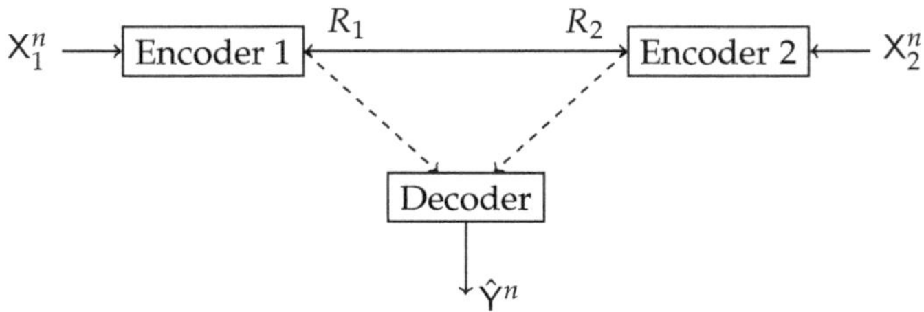

Figure 8. Block diagram of the Collaborative IB.

The IB problem connects to many timely aspects, such as *capital investment* [43], *distributed learning* [45], *deep learning* [46–52], and *convolutional neural networks* [53,54]. Moreover, it has been recently shown that the IB method can be used to reduce the data transfer rate and computational complexity in 5G LDPC decoders [55,56]. The IB method has also been connected with constructing good polar codes [57]. Due to the exponential output-alphabet growth of polarized channels, it becomes demanding to compute their capacities to identify the location of "frozen bits". Quantization is employed in order to reduce the computation complexity. The quality of the quantization scheme is assessed via mutual information preservation. It can be shown that the corresponding IB problem upper bounds the quantization technique. Quantization algorithms based upon the IB method were considered in [58–60]. Furthermore, a relationship between the KL means algorithm and the IB method has been discovered in [61].

A recent comprehensive tutorial on the IB method and related problems is given in [24]. Applications of IB problem in *machine learning* are detailed in [26,45–47,51,52,62].

1.2. Notations

Throughout the paper, random variables are denoted using a sans-serif font, e.g., X, their realizations are denoted by the respective lower-case letters, e.g., x, and their alphabets are denoted by the respective calligraphic letters, e.g., $\mathcal{X}$. Let $\mathcal{X}^n$ stand for the set of all n-tuples of elements from $\mathcal{X}$. An element from $\mathcal{X}^n$ is denoted by $x^n = (x_1, x_2, \ldots, x_n)$ and substrings are denoted by $x_i^j = (x_i, x_{i+1}, \ldots, x_j)$. The cardinality of a finite set, say $\mathcal{X}$, is denoted by $|\mathcal{X}|$. The probability mass function (pmf) of X, the joint pmf of X and Y, and the conditional pmf of X given Y are denoted by P_X, P_{XY}, and $P_{X|Y}$, respectively. The expectation of X is denoted by $\mathbb{E}[X]$. The probability of an event $\mathcal{E}$ is denoted as $P(\mathcal{E})$.

Let X and Y be an n-ary and m-ary random variables, respectively. The marginal probability vector is denoted by a lowercase boldface letter, i.e.,

$$\mathbf{q} \triangleq (P(X = 1), P(X = 2), \ldots, P(X = n))^T. \tag{2}$$

The probability vector of an n-ary uniform random variable is denoted by $\mathbf{u}_n$. We denote by T the transition matrix from X to Y, i.e.,

$$T_{ij} \triangleq P(Y = i | X = j), \qquad 1 \leq i \leq m, 1 \leq j \leq n. \tag{3}$$

The entropy of n-ary probability vector $\mathbf{q}$ is given by $h(\mathbf{q})$, where

$$h(\mathbf{q}) \triangleq - \sum_{i=1}^{n} q_i \log q_i. \tag{4}$$

Throughout this paper all logarithms are taken to base 2 unless stated otherwise. We denote the ones complemented with a bar, i.e., $\bar{x} = 1 - x$. The binary convolution of $x, y \in [0, 1]$ is defined as $x * y \triangleq x\bar{y} + \bar{x}y$. The binary entropy function is defined by $h_b(p) : [0, 1] \rightarrow [0, 1]$, i.e., $h_b(p) \triangleq -p \log p - \bar{p} \log \bar{p}$, and $h_b^{-1}(\cdot)$ its inverse, restricted to $[0, 1/2]$.

Let X and Y be a pair of random variables with joint pmf P_{XY} and marginal pmfs $P_X = \mathbf{q}_x$ and $P_Y = \mathbf{q}_y$. Furthermore, let T ($\bar{T}$) be the transition matrix from X (Y) to Y (X). The mutual information between X and Y is defined as:

$$I(X;Y) = I(\mathbf{q}_x, T) = I(\mathbf{q}_y, \bar{T}) = \sum_{x \in \mathcal{X}} \sum_{y \in \mathcal{Y}} P_{XY}(x,y) \log \frac{P_{XY}(x,y)}{P_X(x)P_Y(y)}. \tag{5}$$

1.3. Paper Outline

Section 2 gives a proper definition of the DSIB optimization problem, mentions various results directly related to this work, and provides some general preliminary results. The spotlight of Section 3 is on the binary (X, Y), where we derive bounds on the respective DSIB function and show a complete characterization for extreme scenarios. The jointly Gaussian (X, Y) is considered in Section 4, where an elegant representation of an objective function is presented, and complete characterization in the low-SNR regime is established. A Blahut–Arimoto-type alternating maximization algorithm will be presented in Section 5. Representative numerical evaluation of the bounds and the proposed algorithm will be provided in Section 6. Finally, a summary and possible future directions will be described in Section 7. The prolonged proofs are postponed to the Appendix A.

2. Problem Formulation and Basic Properties

The DSIB function is a multi-terminal extension of the standard IB [1]. First, we briefly remind the latter's definition and give related results that will be utilized for its double-sided counterpart. Then, we provide a proper definition of the DSIB optimization problem and present some general preliminaries.

2.1. The Single-Sided Information Bottleneck (SSIB) Function

Definition 1 (SSIB). *Let* (X, V) *be a pair of random variables with* $|\mathcal{X}| = n$, $|\mathcal{V}| = m$, *and fixed* P_{XV}. *Denote by* $\mathbf{q}$ *the marginal probability vector of* X, *and let* T *be the transition matrix from* X *to* V, *i.e.,*

$$T_{ij} \triangleq P(V = i | X = j), \qquad 1 \le i \le m, \quad 1 \le j \le n.$$

Consider all random variables U *satisfying the Markov chain* $U \to X \to V$. *The SSIB function is defined as:*

$$\hat{R}_T(\mathbf{q}, C) \triangleq \underset{P_{U|X}}{maximize} \quad I(U;V)$$
$$subject\ to \quad I(X;U) \le C. \tag{6}$$

Remark 1. *The SSIB problem defined in (6) is equivalent (has similar solution) to the CEB problem considered in [17].*

Although the optimization problem in (6) is well defined, the auxiliary random variable U may have an unbounded alphabet. The following lemma provides an upper bound on the cardinality of $\mathcal{U}$, thus relaxing the optimization domain.

Lemma 1 (Lemma 2.2 of [17]). *The optimization over* U *in (6) can be restricted to* $|\mathcal{U}| \le n + 1$.

Remark 2. *A tighter bound, namely* $|\mathcal{U}| \le n$, *was previously proved in [63] for the corresponding dual problem, namely, the IB Lagrangian. However, since* $\hat{R}_T(\mathbf{q}, C)$ *is generally not a strictly convex function of* C, *it cannot be directly applied for the primal problem (6).*

Note that the SSIB optimization problem (6) is basically a convex function maximization over a convex set; thus, the maximum is attained on the boundary of the set.

Lemma 2 (Theorem 2.5 of [17]). *The inequality constraint in (6) can be replaced by equality constraint, i.e.,* $I(X;U) = C$.

2.2. The Double-Sided Information Bottleneck (DSIB) Function

Definition 2 (DSIB). *Let (X, Y) be a pair of random variables with $|\mathcal{X}| = n$, $|\mathcal{Y}| = m$ and fixed P_{XY}. Consider all the random variables U and V satisfying the Markov chain $U \to X \to Y \to V$. The DSIB function $R : [0, H(X)] \times [0, H(Y)] \to \mathbb{R}_+$ is defined as:*

$$R_{P_{XY}}(C_u, C_v) \triangleq \underset{P_{U|X}, P_{V|Y}}{\text{maximize}} \quad I(U; V)$$
$$\text{subject to} \quad I(X; U) \leq C_u \text{ and } I(Y; V) \leq C_v. \tag{7}$$

The achieving conditional distributions $P_{U|X}$ and $P_{V|Y}$ will be termed as the optimal test-channels. Occasionally, we will drop the subscript denoting the particular choice of the bivariate source P_{XY}.

Note that (7) can be expressed in the following equivalent form:

$$R(C_u, C_v) \triangleq \underset{P_{V|Y}}{\text{maximize}} \quad \underset{P_{U|X}}{\text{maximize}} \quad I(U; V).$$
$$\text{subject to} \quad \text{subject to}$$
$$I(Y; V) \leq C_v \quad I(X; U) \leq C_u \tag{8}$$

Evidently, we can define (8) using (6). Indeed, fix $P_{V|Y}$ so that it satisfies $I(Y; V) \leq C_v$. Denote by $T_{V|Y}$ the transition matrix from Y to V and by $T_{Y|X}$ the transition matrix from X to Y, respectively, i.e.,

$$(T_{V|Y})_{ik} \triangleq P(V = i | Y = k), \quad 1 \leq i \leq |\mathcal{V}|, 1 \leq k \leq m,$$
$$(T_{Y|X})_{kj} \triangleq P(Y = k | X = j), \quad 1 \leq k \leq m, 1 \leq j \leq n.$$

Denote by $\mathbf{q}_x$ and $\mathbf{q}_y$ the marginal probability vectors of X and Y, respectively, and consider the inner maximization term in (8). Since $P_{V|Y}$ and P_{XY} are fixed, then $P_{XV} = \sum_y P_{V|Y}(\cdot|y) P_{XY}(\cdot, y)$ is also fixed. Denote by $T_{V|X} \triangleq T_{V|Y} T_{Y|X}$ the transition matrix from X to V. Therefore, the inner maximization term in (8) is just the SSIB function with parameters $T_{V|X}$ and C_u, namely, $\hat{R}_{T_{V|X}}(\mathbf{q}_x, C_u)$. Hence, our problem can also be interpreted in the following two equivalent ways:

$$R(C_u, C_v) \triangleq \underset{T_{V|Y}}{\text{maximize}} \quad \hat{R}_{T_{V|Y} T_{Y|X}}(\mathbf{q}_x, C_u)$$
$$\text{subject to} \quad I(\mathbf{q}_y, T_{V|Y}) \leq C_v; \tag{9}$$

or, similarly, by interchanging the order of maximization in (8), it can be expressed as follows:

$$R(C_u, C_v) \triangleq \underset{T_{U|X}}{\text{maximize}} \quad \hat{R}_{T_{U|X} T_{X|Y}}(\mathbf{q}_y, C_v)$$
$$\text{subject to} \quad I(\mathbf{q}_x, T_{U|X}) \leq C_u, \tag{10}$$

where $T_{U|X}$ is the transition matrix from X to U, and $T_{X|Y}$ is the transition matrix from Y to X. This representation gives us a different perspective on our problem as an optimal compressed representation of the relevance random variable for the IB framework.

Remark 3. *Taking $C_v = \infty$ in (9) results in an deterministic channel from Y to V, i.e., $V = Y$. Thus, the DSIB problem defined in (7) reduces to the SSIB problem (6).*

The bound from Lemma 1 can be utilized to give cardinality bounding for the double-sided problem.

Proposition 1. *For the DSIB optimization problem defined in (7), it suffices to consider random variables* U *and* V *with cardinalities* $|\mathcal{U}| \leq n + 1$ *and* $|\mathcal{V}| \leq m + 1$.

Proof. Let $T_{\mathsf{U}|\mathsf{X}}$ and $T_{\mathsf{V}|\mathsf{Y}}$ be two arbitrary transition matrices. By Lemma 1, there exists $T_{\tilde{\mathsf{U}}|\mathsf{X}}$ with $|\tilde{\mathcal{U}}| \leq n + 1$ such that $I(\tilde{\mathsf{U}};\mathsf{V}) \geq I(\mathsf{U};\mathsf{V})$ and $I(\mathsf{X};\tilde{\mathsf{U}}) \leq C_u$. Similarly, $T_{\mathsf{V}|\mathsf{Y}}$ can be replaced with $T_{\tilde{\mathsf{V}}|\mathsf{Y}}$, $|\tilde{\mathcal{V}}| \leq m + 1$ such that $I(\tilde{\mathsf{U}};\tilde{\mathsf{V}}) \geq I(\tilde{\mathsf{U}},\mathsf{V}) \geq I(\mathsf{U};\mathsf{V})$, and $I(\mathsf{Y};\tilde{\mathsf{V}}) \leq C_v$. Therefore, there exists an optimal solution with $|\mathcal{U}| \leq n + 1$ and $|\mathcal{V}| \leq m + 1$. $\square$

In the following two sections, we will present the primary analytical outcomes of our study. First, we consider the scenario where our bivariate source is binary, specifically DSBS. Then, we handle the case where X and Y are jointly Gaussian.

3. Binary (X,Y)

Let (X,Y) be a DSBS with parameter p, i.e.,

$$\mathsf{P}_{\mathsf{XY}}(x,y) = \frac{1}{2}(p \cdot \mathbb{1}(x \neq y) + (1 - p)\mathbb{1}(x = y)). \tag{11}$$

We entitle the respective optimization problem (7) as the *binary double-sided information bottleneck* (BDSIB) and emphasize its dependence on the parameter p as $R(C_u, C_v, p)$.

The following proposition states that the cardinality bound from Lemma 1 can be tightened in the binary case.

Proposition 2. *Considering the optimization problem in (6) with* $\mathsf{X} = \mathrm{Ber}(q)$ *and* $|\mathcal{Y}| = 3$, *binary* U *is optimal.*

The proof of this proposition is postponed to Appendix A. Using similar justification for Proposition 1 combined with Proposition 2, we have the following strict cardinality formula for the BDSIB setting.

Proposition 3. *For the respective DSBS setting of (7), it suffices to consider random variables* U *and* V *with cardinalities* $|\mathcal{U}| = |\mathcal{V}| = 2$.

Note that the above statement is not required for the results in the rest of this section to hold and will be mainly applied to justify our conjectures via numerical simulations.

We next show that the specific objective function for the binary setting of (7), i.e, the mutual information between U and V, has an elegant representation which will be useful in deriving lower and upper bounds.

Lemma 3. *The mutual information between* U *and* V *can be expressed as follows:*

$$I(\mathsf{U};\mathsf{V}) = \mathbb{E}_{\mathsf{P}_{\mathsf{U}} \times \mathsf{P}_{\mathsf{V}}}[K(\mathsf{U},\mathsf{V},p) \log K(\mathsf{U},\mathsf{V},p)], \tag{12}$$

where the expectation is taken over the product measure $\mathsf{P}_{\mathsf{U}} \times \mathsf{P}_{\mathsf{V}}$, U *and* V *are binary random variables satisfying:*

$$\mathsf{P}(\mathsf{U} = 0) = \frac{\alpha_1 - \frac{1}{2}}{\alpha_1 - \alpha_0}, \qquad \mathsf{P}(\mathsf{V} = 0) = \frac{\beta_1 - \frac{1}{2}}{\beta_1 - \beta_0}, \tag{13}$$

the kernel $K(u,v,p)$ *is given by:*

$$K(u,v,p) = 2\alpha_u * \beta_v * p = 1 - (1 - 2p)(1 - 2\alpha_u)(1 - 2\beta_v), \tag{14}$$

and the reverse test-channels are defined by: $\alpha_u \triangleq \mathsf{P}(\mathsf{X} = 1|\mathsf{U} = u)$, $\beta_v \triangleq \mathsf{P}(\mathsf{Y} = 0|\mathsf{V} = v)$. *Furthermore, since* $|(1 - 2p)(1 - 2\alpha_u)(1 - 2\beta_v)| < 1$, *utilizing Taylor's expansion of* $\log(1 - x)$, *we obtain:*

$$I(\mathsf{U};\mathsf{V}) = \sum_{n=2}^{\infty} \frac{(1 - 2p)^n \mathbb{E}[(1 - 2\alpha_\mathsf{U})^n]\mathbb{E}[(1 - 2\beta_\mathsf{V})^n]}{n(n - 1)}. \tag{15}$$

The general cascade of test-channels and the DSBS, defined by $\{\alpha_u\}_{u=0}^{1}$, $\{\beta_v\}_{v=0}^{1}$ and p, is illustrated in Figure 9. The proof of Lemma 3 is postponed to Appendix B.

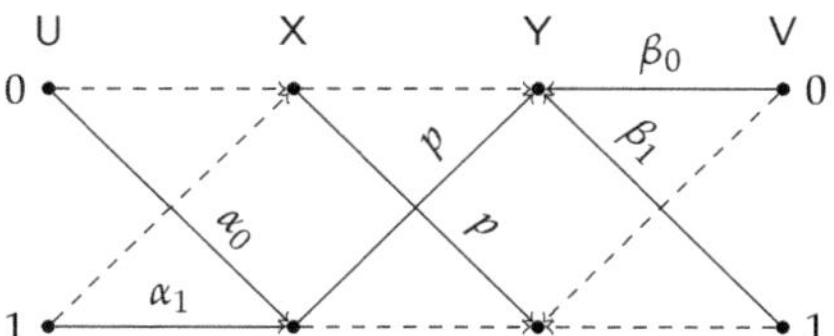

Figure 9. General test-channel construction of the BDSIB function.

We next examine some corner cases for which $R(C_u, C_v, p)$ is fully characterized.

3.1. Special Cases

A particular case where we have a complete analytical solution is when p tends to $1/2$.

Theorem 1. *Suppose* $p = \frac{1}{2} - \epsilon$, *and consider* $\epsilon \to 0$. *Then*

$$R(C_u, C_v, \epsilon) = 2\epsilon^2 \log e \cdot (1 - 2h_b^{-1}(1 - C_u))^2(1 - 2h_b^{-1}(1 - C_v))^2 + o(\epsilon^2), \tag{16}$$

and it is achieved by taking $\mathsf{P}_{\mathsf{U}|\mathsf{X}}$ *and* $\mathsf{P}_{\mathsf{V}|\mathsf{Y}}$ *as BSC test-channels satisfying the constraints with equality.*

This theorem follows by considering the low SNR regime in Lemma 3 and is proved in Appendix D. For the lower bound we take $\mathsf{P}_{\mathsf{U}|\mathsf{X}}$ and $\mathsf{P}_{\mathsf{V}|\mathsf{Y}}$ to be BSCs.

In Section 6 we will give a numerical evidence that BSC test-channels are in fact optimal provided that p is sufficiently large. However, for small p this is no longer the case and we believe the following holds.

Conjecture 1. *Let* $\mathsf{X} = \mathsf{Y}$, *i.e.,* $p = 0$. *The optimal test-channels* $\mathsf{P}_{\mathsf{U}|\mathsf{X}}$ *and* $\mathsf{P}_{\mathsf{V}|\mathsf{X}}$ *that achieve* $R(C_u, C_v, 0)$ *are Z-channel and S-channel respectively.*

Remark 4. *Our results in the numerical section strongly support this conjecture. In fact they prove it within the resolution of the experiments, i.e., for optimizing over a dense set of test-channels rather then all test-channels. Nevertheless, we were not able to find an analytical proof for this result.*

Remark 5. *Suppose* $\mathsf{X} = \mathsf{Y}$, $I(\mathsf{X};\mathsf{U}) = C_u$, *and* $I(\mathsf{X};\mathsf{V}) = C_v$. *Since* $I(\mathsf{U};\mathsf{V}) = I(\mathsf{U};\mathsf{X}) + I(\mathsf{V};\mathsf{X}) - I(\mathsf{X};\mathsf{U},\mathsf{V})$ *(as* $\mathsf{U} \to \mathsf{X} \to \mathsf{Y} \to \mathsf{V}$ *form a Markov chain in this order) then maximizing* $I(\mathsf{U};\mathsf{V})$ *is equivalent to minimizing* $I(\mathsf{X};\mathsf{U},\mathsf{V})$, *namely, minimizing information combining as in [23,35]. Therefore, Conjecture 1 is equivalent to the conjecture that among all channels with* $I(\mathsf{X};\mathsf{U}) \geq C_u$ *and* $I(\mathsf{Y};\mathsf{V}) \geq C_v$, *Z and S are the worst channels for information combining.*

This observation leads us the following additional conjecture.

Conjecture 2. *The test-channels* $\mathsf{P}_{\mathsf{U}|\mathsf{X}}$ *and* $\mathsf{P}_{\mathsf{V}|\mathsf{X}}$ *that maximize* $I(\mathsf{X};\mathsf{U},\mathsf{V})$ *are both Z channels.*

Remark 6. *Suppose now that p is arbitrary and assume that one of the channels $\mathsf{P}_{\mathsf{U}|\mathsf{X}}$ or $\mathsf{P}_{\mathsf{V}|\mathsf{Y}}$ is restricted to be a binary memoryless symmetric (BMS) channel (Chapter 4 of [64]), then the maximal $I(\mathsf{U};\mathsf{V})$ is attained by BSC channels, as those are the symmetric channels minimizing $I(\mathsf{X};\mathsf{U},\mathsf{V})$ [23]. It is not surprising that once the BMS constraint is removed, symmetric channels are no longer optimal (see the discussion in (Section VI.C of [23])).*

Consider now the case $\mathsf{X} = \mathsf{Y}$ ($p = 0$) with an additional symmetry assumption $C_u = C_v$. The most reasonable apriori guess is that the optimal test-channels $\mathsf{P}_{\mathsf{U}|\mathsf{X}}$ and $\mathsf{P}_{\mathsf{V}|\mathsf{X}}$ are the same up to some permutation of inputs and outputs. Surprisingly, this is not the case, unless they are BSC or Z channels, as the following negative result states.

Proposition 4. *Suppose $C_u = C_v$ and the transition matrix from X to V, given by*

$$T_{\mathsf{V}|\mathsf{X}} = \begin{pmatrix} a & b \\ 1-a & 1-b \end{pmatrix},\tag{17}$$

satisfies $I(\mathbf{u}_2, T_{\mathsf{V}|\mathsf{X}}) = C_v$. Consider the respective SSIB optimization problem

$$\hat{R}_{T_{\mathsf{V}|\mathsf{X}}}(\mathbf{u}_2, C_u) = \max_{\mathsf{P}_{\mathsf{U}|\mathsf{X}}\,:\, I(\mathsf{U};\mathsf{X})\leq C_u} I(\mathsf{U};\mathsf{V}).\tag{18}$$

The optimal $\mathsf{P}_{\mathsf{U}|\mathsf{X}}$ that attains (6) with $\mathbf{q}_{\mathsf{X}} = \mathbf{u}_2$ and $C = C_u$ does not equal to $\mathsf{P}_{\mathsf{V}|\mathsf{X}}$ or any permutation of $\mathsf{P}_{\mathsf{V}|\mathsf{X}}$, unless $\mathsf{P}_{\mathsf{V}|\mathsf{X}}$ is a BSC or a Z channel.

The proof is based on [17] and is postponed to Appendix E.
As for the case of $\mathsf{X} \neq \mathsf{Y}$, i.e., $p \neq 0$, we have the following conjecture.

Conjecture 3. *For every $(C_u, C_v) \in [0,1] \times [0,1]$, there exists $\theta(C_u, C_v)$, such that for every $p > \theta(C_u, C_v)$ the achieving test-channels $\mathsf{P}_{\mathsf{U}|\mathsf{X}}$ and $\mathsf{P}_{\mathsf{V}|\mathsf{Y}}$ are BSC with parameters $\alpha = h_b^{-1}(1 - C_u)$ and $\beta = h_b^{-1}(1 - C_v)$ respectively.*

We will provide supporting arguments for this conjecture via numerical simulations in Section 6.

3.2. Bounds

In this section we present our lower and upper bounds on the BDSIB function, then we compare them for various channel parameters. The proofs are postponed to Appendix F. For the simplicity of the following presentation we define

$$g_b(x) \triangleq \frac{1}{2(1-x)} h_b(x), \quad x \in [0, 1/2],\tag{19}$$

denote $g_b^{-1}(\cdot)$ as its inverse restricted to $[0,1]$, and $\hbar(x) \triangleq -x \log x$.

Proposition 5. *The BDSIB function is bounded from below by*

$$R(C_u, C_v, p) \geq$$
$$\max \begin{cases} 1 - h_b(\alpha * \beta * p), \\ 1 - \dfrac{1}{2\bar{\delta}\bar{\zeta}}\left[\hbar(\delta * \zeta * p) + (1-2\zeta)\cdot\hbar(\bar{\delta} * p) + (1-2\delta)\cdot\hbar(\bar{\zeta} * p) + (1-2\delta)(1-2\zeta)\cdot\hbar(p)\right], \end{cases}\tag{20}$$

where $\alpha = h_b^{-1}(1 - C_u)$, $\beta = h_b^{-1}(1 - C_v)$, $\delta = g_b^{-1}(1 - C_u)$, and $\zeta = g_b^{-1}(1 - C_v)$.

All terms in the RSH of (20) are attained by taking test-channels that match the constraints with equality and plugging them in Lemma 3. In particular: the first term is achieved by BSC test channels with transition probabilities α and β; the second term is achieved by taking $P_{U|X}$ be a $Z(\delta)$ channel and $P_{V|Y}$ be an $S(\zeta)$ channel. The aforementioned test-channel configurations are illustrated in Figure 10.

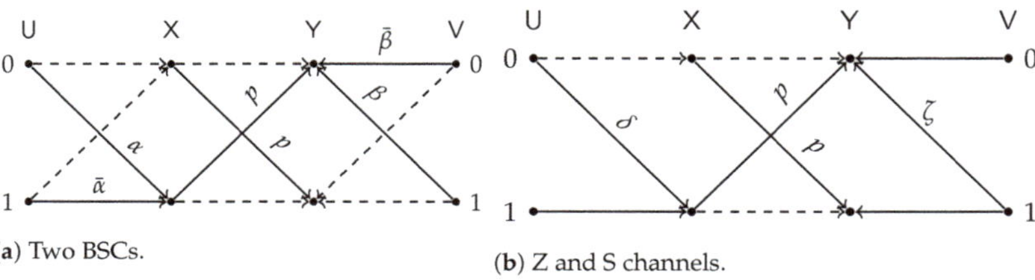

(**a**) Two BSCs.

(**b**) Z and S channels.

Figure 10. Test-channel that achieve the lower bound of Proposition 5.

We compare the different lower bounds derived in Proposition 5 for various values of constraints. The achievable rate vs channel transition probability p is shown in Figure 11. Our first observation is that BSC test-channels outperform all other choices for almost all values of p. However, Figure 12 gives a closer look on small values of p. It is evident that the combination of Z and S test-channels outperforms any other schemes for small values of p. We have used this observation as one supporting evidence to Conjecture 1.

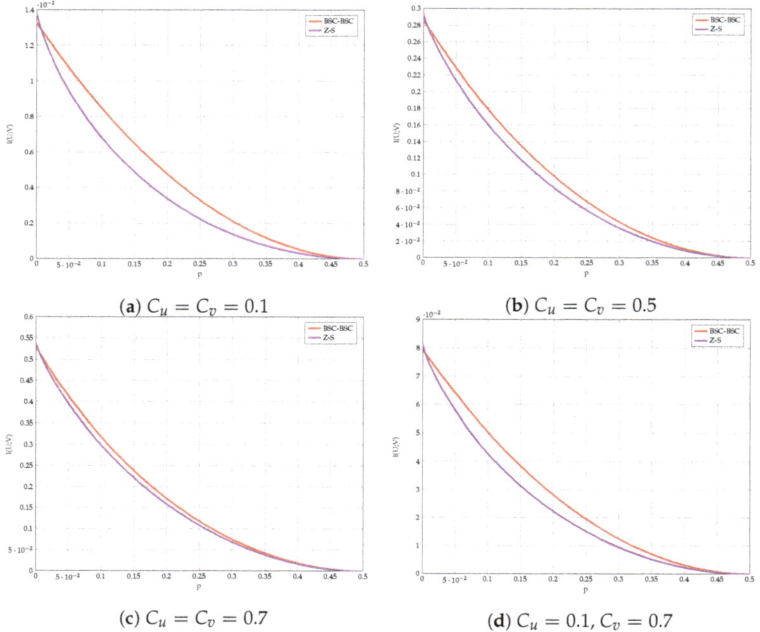

(**a**) $C_u = C_v = 0.1$

(**b**) $C_u = C_v = 0.5$

(**c**) $C_u = C_v = 0.7$

(**d**) $C_u = 0.1, C_v = 0.7$

Figure 11. Comparison of the lower bounds.

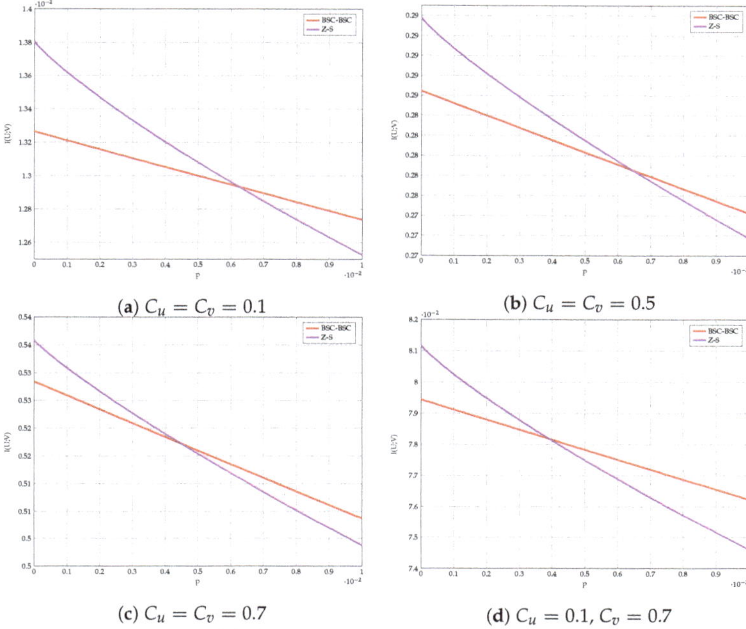

(**a**) $C_u = C_v = 0.1$

(**b**) $C_u = C_v = 0.5$

(**c**) $C_u = C_v = 0.7$

(**d**) $C_u = 0.1, C_v = 0.7$

Figure 12. Comparison of the lower bounds in high SNR regime.

We proceed to give an upper bound.

Proposition 6. *A general upper bound on BDSIB is given by*

$$R(C_u, C_v, p) \leq \min \begin{cases} (1 - 2p)^2 (1 - 2h_b^{-1}(1 - C_u)^2 (1 - 2h_b^{-1}(1 - C_v)^2, \\ \min\{1 - h_b(h_b^{-1}(1 - C_u) * p), 1 - h_b(h_b^{-1}(1 - C_v) * p)\}. \end{cases} \tag{21}$$

Note that the first term can be derived by applying Jensen's inequality on (12), and the second term is a combination of the standard IB and the cut-set bound. We postpone the proof of Proposition 6 to Appendix F.

Remark 7. *Since* $p = \frac{1}{2} - \epsilon$, *we have a factor 2 loss in the first term compared to the precise behavior we have found for* $p \approx \frac{1}{2}$ *in Theorem 1. This loss comes from the fact that the bound in (21) actually upper bounds the χ-squared mutual information between* U *and* V. *It is well-known that for very small* $I(\mathsf{X}; \mathsf{Y})$ *we have that* $I(\mathsf{X}; \mathsf{Y}) \approx 1/2 I_{\chi^2}(\mathsf{X}; \mathsf{Y})$, *see [65].*

We compare the different upper bounds from Proposition 6 in Figure 13 for various bottleneck constraints, and in Figure 14 for various values of channel transition probabilities p. We observe that there are regions of C and p for which Jensen's based bound outperforms the standard IB bound.

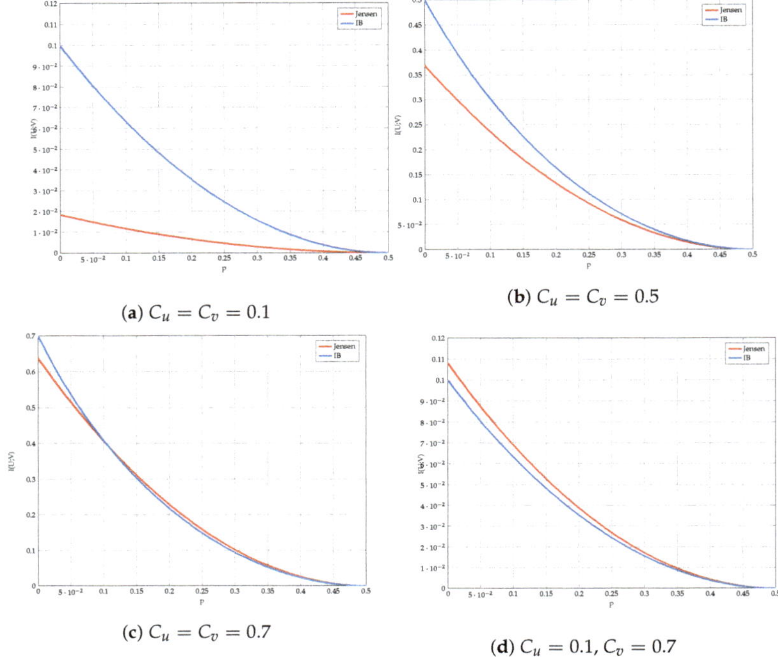

(**a**) $C_u = C_v = 0.1$

(**b**) $C_u = C_v = 0.5$

(**c**) $C_u = C_v = 0.7$

(**d**) $C_u = 0.1, C_v = 0.7$

Figure 13. Comparison of the upper bounds for various values of (C_u, C_v).

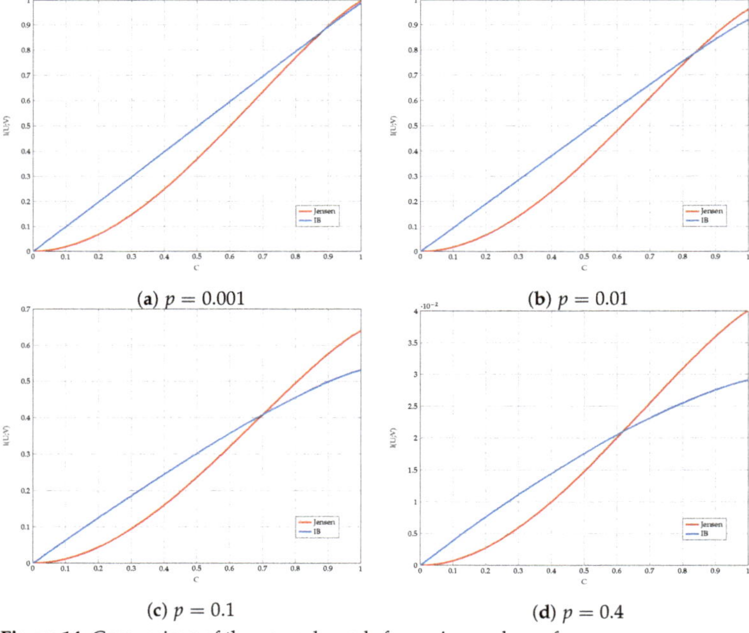

(**a**) $p = 0.001$

(**b**) $p = 0.01$

(**c**) $p = 0.1$

(**d**) $p = 0.4$

Figure 14. Comparison of the upper bounds for various values of p.

Finally, we compare the best lower and upper bounds from Propositions 5 and 6 for various values of channel parameters in Figure 15. We observe that the bounds are tighter for asymmetric constraints and high transition probabilities.

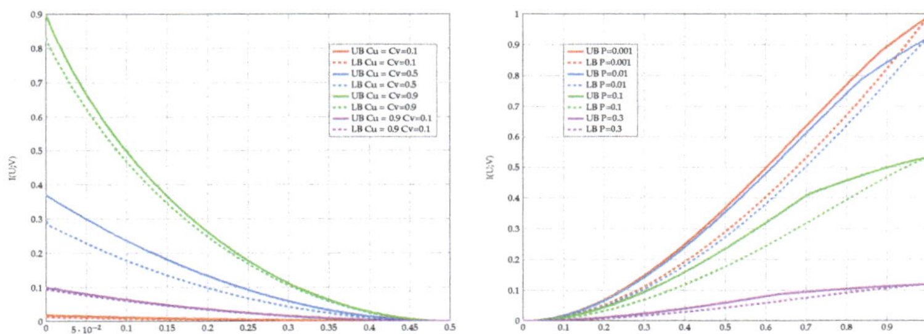

Figure 15. Capacity bounds for various values of p and $C = C_u = C_v$.

4. Gaussian (X, Y)

In this section we consider a specific setting where (X, Y) is the normalized zero mean Gaussian bivariate source, namely,

$$\begin{pmatrix} \mathsf{X} \\ \mathsf{Y} \end{pmatrix} \sim \mathcal{N}\left(\begin{pmatrix} 0 \\ 0 \end{pmatrix}, \begin{pmatrix} 1 & \rho \\ \rho & 1 \end{pmatrix} \right). \tag{22}$$

We establish achievability schemes and show that Gaussian test-channels $\mathsf{P}_{\mathsf{U}|\mathsf{X}}$ and $\mathsf{P}_{\mathsf{V}|\mathsf{Y}}$ are optimal for vanishing SNR. Furthermore we show an elegant representation of the problem through *probabilistic Hermite polynomials* which are defined by

$$H_n(x) \triangleq (-1)^n e^{\frac{x^2}{2}} \frac{\mathrm{d}^n}{\mathrm{d}x^n} e^{-\frac{x^2}{2}}, \quad n \in \mathbb{N}_0. \tag{23}$$

We denote the Gaussian DSIB function with explicit dependency on ρ as $R(C_u, C_v, \rho)$.

Proposition 7. *Let $H_n(\cdot)$ be the nth order probabilistic Hermite polynomial, then the objective function of (7) for the Gaussian setting is given by*

$$I(\mathsf{U};\mathsf{V}) = \mathbb{E}_{\mathsf{U}\mathsf{V}}\left[\log\left(\sum_{n=0}^{\infty} \frac{\rho^n}{n!} \mathbb{E}\left[H_n(\mathsf{X}) | \mathsf{U} \right] \mathbb{E}\left[H_n(\mathsf{Y}) | \mathsf{V} \right] \right) \right]. \tag{24}$$

This representation follows by considering $I(\mathsf{U};\mathsf{V}) = D(\mathsf{P}_{\mathsf{U}\mathsf{V}} || \mathsf{P}_\mathsf{U} \cdot \mathsf{P}_\mathsf{V})$ and expressing $\frac{\mathsf{P}_{\mathsf{U}\mathsf{V}}}{\mathsf{P}_\mathsf{U} \cdot \mathsf{P}_\mathsf{V}}$ using Mehler Kernel [66]. Mehler Kernel decomposition is a special case of a much richer family of Lancaster distributions [67]. The proof of Proposition 7 is relegated to Appendix G.

Now we give two lower bounds on $R(C_u, C_v, \rho)$. Our first lower bound is established by choosing $\mathsf{P}_{\mathsf{U}|\mathsf{X}}$ and $\mathsf{P}_{\mathsf{V}|\mathsf{Y}}$ to be additive Gaussian channels, satisfying the mutual information constraints with equality.

Proposition 8. *A lower bound on $R(C_u, C_v, \rho)$ is given by*

$$R(C_u, C_v, \rho) \geq -\frac{1}{2} \log\left(1 - \rho^2 \left(1 - 2^{-2C_u} \right) \left(1 - 2^{-2C_v} \right) \right). \tag{25}$$

The proof of this bound is developed in Appendix H.

Although it was shown in [26] that choosing the test-channel to be Gaussian is optimal for the single-sided variant, it is not the case for its double-sided extension. We will show this by examining a specific set of values for the rate constraints, $(C_u, C_v) = (1,1)$. Furthermore, we choose the test channels $P_{U|X}$ and $P_{V|Y}$ to be deterministic quantizers.

Proposition 9. *Let* $(C_u, C_v) = (1,1)$, *then*

$$R(1,1,\rho) \geq 1 - h_2\left(\frac{\arccos\rho}{\pi}\right). \tag{26}$$

The proof of this bound is developed in Appendix I.

We compare the bounds from Propositions 8 and 9 with $(C_u, C_v) = (1,1)$ in Figure 16. The most unexpected observation here is that the deterministic quantizers lower bound outperform the Gaussian test-channels for high values of ρ. The crossing point of those bounds is given by

$$\rho_{\mathrm{cros}} = \frac{e}{\sqrt{1 + e^2}} \to \sqrt{SNR_{\mathrm{cros}}} = \frac{\rho_{\mathrm{cros}}}{\sqrt{1 - \rho_{\mathrm{cros}}^2}} = e. \tag{27}$$

We proceed to present our upper bound on $R(C_u, C_v, \rho)$. This bound is a combination of the cutset bound and the single-sided Gaussian IB.

Proposition 10. *An upper bound on* (7) *with Gaussian* (X, Y) *setting* (22) *is given by*

$$R(C_u, C_v, \rho) \leq \min\left\{-\frac{1}{2}\log(1 - \rho^2(1 - 2^{-2C_u})), -\frac{1}{2}\log(1 - \rho^2(1 - 2^{-2C_v}))\right\}. \tag{28}$$

We compare the best lower and upper bounds from Propositions 8–10 in Figure 17. We observe that the bounds become tighter as the constraint increases and in the low-SNR regime.

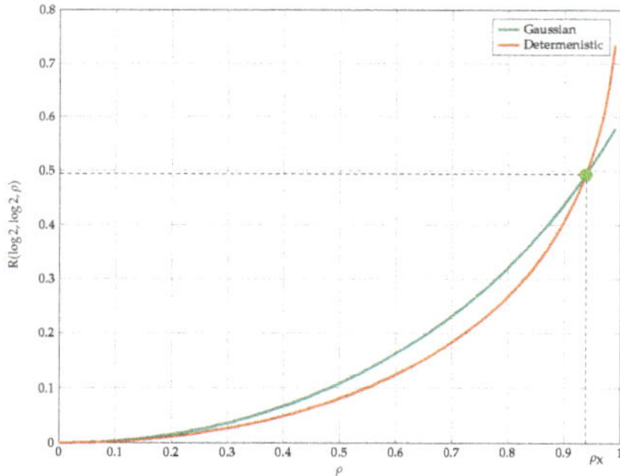

Figure 16. Comparison of the lower bounds from Propositions 8 and 9.

4.1. Low-SNR Regime

For $\rho \to 0$, the exact asymptotic behavior of the Gaussian (Proposition 8) and deterministic (Proposition 9) test-channels, respectively, for $C_u = C_v = 1$ bit is given by:

$$\lim_{\rho \to 0} -\frac{1}{2}\log\left(1 - \rho^2(1 - 2^{-2C_u})(1 - 2^{-2C_v})\right) = \frac{9\log e}{32}\rho^2 + o(\rho^2),$$

$$\lim_{\rho \to 0} 1 - h_2\left(\frac{\arccos\rho}{\pi}\right) = \frac{2\log e}{\pi^2}\rho^2 + o(\rho^2).$$

Hence, the Gaussian choice outperforms the second lower bound for vanishing SNR. The following theorem establishes that Gaussian test-channels are optimal for low-SNR.

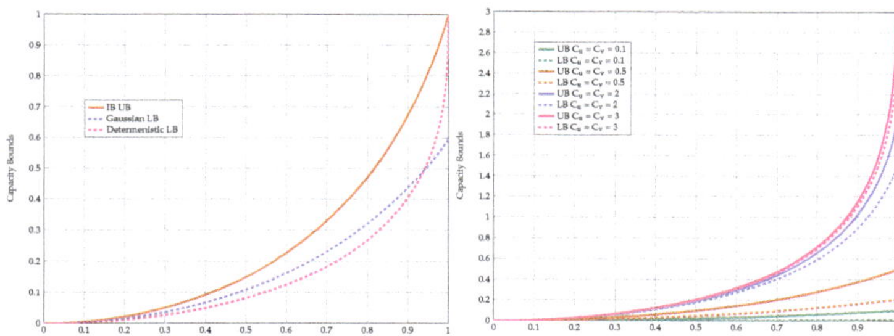

Figure 17. Capacity bounds for various values of p and $C = C_u = C_v = 1$.

Theorem 2. *For small ρ, the GDSIB function is given by:*

$$R(C_u, C_v, \rho) = \frac{\rho^2 \log e}{2}(1 - 2^{-2C_u})(1 - 2^{-2C_v}) + o(\rho^2).$$ (29)

The lower bound follows from Proposition 8. The upper bound is established by considering the kernel representation from Proposition 7 in the limit of vanishing ρ. The detailed proof is given in Appendix J.

4.2. Optimality of Symbol-by-Symbol Quantization When $X = Y$

Consider an extreme scenario for which $X = Y \sim \mathcal{N}(0, 1)$. Taking the encoders $\mathsf{P}_{\mathsf{U}|\mathsf{X}}$ and $\mathsf{P}_{\mathsf{V}|\mathsf{X}}$ as a symbol-by-symbol deterministic quantizers satisfying:

$$H(\mathsf{U}) = H(\mathsf{V}) = \min\{C_u, C_v\},$$

we achieve the optimum

$$I(\mathsf{U}; \mathsf{V}) = \min\{C_u, C_v\}.$$

5. Alternating Maximization Algorithm

Consider the DSIB problem for DSBS with parameter p analyzed in Section 3. The respective optimization problem involves simultaneous search of the maximum over the sets $\{\mathsf{P}_{\mathsf{U}|\mathsf{X}}\}$ and $\{\mathsf{P}_{\mathsf{V}|\mathsf{Y}}\}$. An alternating maximization, namely, fixing $\mathsf{P}_{\mathsf{U}|\mathsf{X}}$, then finding the respective optimal $\mathsf{P}_{\mathsf{V}|\mathsf{Y}}$ and vice versa, is sub-optimal in general and may result in convergent to a saddle point. However, for the case $p = 0$ with symmetric bottleneck constraints, Proposition 4 implies that such point exists only for the BSC and Z/S channels. This motivates us to believe that performing an alternating maximization procedure on (9) will not result in sub-optimal saddle point, but rather converge to the optimal solution also for the general discrete (X, Y).

Thus, we propose an alternating maximization algorithm. The main idea is to fix $\mathsf{P}_{\mathsf{V}|\mathsf{Y}}$ and then compute $\mathsf{P}^*_{\mathsf{U}|\mathsf{X}}$ that attains the inner term in (9). Then, using $\mathsf{P}^*_{\mathsf{U}|\mathsf{X}}$, we find the optimal $\mathsf{P}^*_{\mathsf{V}|\mathsf{Y}}$ that attains the inner term in (10). Then, we repeat the procedure in alternating manner until convergence.

Note that inner terms of (9) and (10) are just the standard IB problem defined in (6). For completeness, we state here the main result from [1] and adjust it for our problem. Consider the respective Lagrangian of (6) given by:

$$L(\mathsf{P}_{\mathsf{U}|\mathsf{X}}, \lambda) = I(\mathsf{U}; \mathsf{V}) + \lambda(C - I(\mathsf{X}; \mathsf{U})).$$ (30)

Lemma 4 (Theorem 4 of [1]). *The optimal test-channel that maximizes (30) satisfies the equation:*

$$P_{U|X}(u|x) = \frac{P_U(u)}{Z(x,\beta)} e^{-\beta D(P_{V|X=x} \| P_{V|U=u})}, \tag{31}$$

where $\beta \triangleq 1/\lambda$ and $P_{V|U}$ is given via Bayes' rule, as follows:

$$P_{V|U}(v|u) = \frac{1}{P_U(u)} \sum_x P_{V|X}(v|x) P_{U|X}(u|x) P_X(x). \tag{32}$$

In a very similar manner to the Blahut–Arimoto algorithm [18], the self-consistent equations can be adapted into converging, alternating iterations over the convex sets $\{P_{U|X}\} = \Delta_n^{\otimes n}$, $\{P_U\} = \Delta_n$, and $\{P_{V|U}\} = \Delta_n^{\otimes n}$, as stated in the following lemma.

Lemma 5 (Theorem 5 of [1]). *The self-consistent equations are satisfied simultaneously at the minima of the functional:*

$$F(P_{U|X}, P_U, P_{Y|U}) = I(U;X) + \beta \mathbb{E}\left[D(P_{V|X} \| P_{V|U}) \right], \tag{33}$$

where the minimization is performed independently over the convex sets of $\{P_{U|X}\} = \Delta_n^{\otimes n}$, $\{P_U\} = \Delta_n$, and $\{P_{V|U}\} = \Delta_n^{\otimes n}$. The minimization is performed by the converging alternation iterations as described in Algorithm 1.

Algorithm 1: IB iterative algorithm IBAM(args)

Input: $P_{U|X}^{(0)}, P_{XY}, \beta, \epsilon$

$R^{(0)} = 0$

$t \leftarrow 0$

while $\Delta R \geq \epsilon$ **do**

$\quad P_U(u) \leftarrow \sum_x P_X(x) P_{U|X}^{(t)}(u|x)$

$\quad P_{X|U} \leftarrow \frac{P_{U|X}^{(t)} P_X}{P_U}$

$\quad P_{Y|U}(y|u) \leftarrow \sum_x P_{Y|X}(y|x) P_{X|U}(x|u)$

$\quad P_{U|X}^{(t+1)}(u|x) \leftarrow \frac{P_U(u)\exp\left(-\beta D(P_{Y|X=x}\|P_{Y|U=u})\right)}{\sum_u P_U(u)\exp\left(-\beta D(P_{Y|X=x}\|P_{Y|U=u})\right)}$

$\quad R^{(t+1)} \leftarrow I(P_U^{(t+1)}, P_{Y|U}^{(t+1)})$

$\quad \Delta R = |R^{(t+1)} - R^{(t)}|$

$\quad t \leftarrow t+1$

Output: $P_{U|X}^{(t)}(u|x)$

Next, we propose a combined algorithm to solve the optimization problem from (7). The main idea is to fix one of the test-channels, i.e., $P_{V|Y}$, and then find the corresponding optimal opposite test-channel, i.e., $P_{U|X}$, using Algorithm 1. Then, we apply again Algorithm 1 by switching roles, i.e., fixing the opposite test-channel, i.e., $P_{U|X}$, and then identifying the optimal $P_{V|Y}$. We repeat this procedure until convergence of the objective function $I(U;V)$. We summarize the proposed composite method in Algorithm 2.

Remark 8. *Note that every alternating step of the algorithm involves finding an optimal (β^*, η^*) that corresponds to the respective problem constraints (C_u, C_v). We have chosen to implement this exploration step using a bisection-type method. This may result that the actual pair (C_u, C_v) is ϵ-far away from the desired constraint.*

Algorithm 2: DSIB iterative algorithm DSIBAM(args)

Input: $P_{U|X}^{(0)}, P_{V|Y}^{(0)}, P_{XY}, C_u, C_v, \epsilon$

$R^{(0)} = 0$

$s \leftarrow 0$

while $\Delta R \geq \epsilon$ **do**

    $P_{XV}(x,v) \leftarrow \sum_y P_{V|Y}^{(s)}(v|y)P_{XY}(x,y)$

    $P_{U|X}(\beta) \leftarrow IBAM(P_{U|X}^{(s)}, P_{XV}, \beta, \epsilon)$

    $\beta^* \leftarrow \arg\min_\beta |I(P_X, P_{U|X}(\beta)) - C_u|$

    $P_{U|X}^{(s+1)} \leftarrow P_{U|X}(\beta^*)$

    $P_{YU}(y,u) \leftarrow \sum_x P_{U|X}^{(s+1)}(u|x)P_{XY}(x,y)$

    $P_{V|Y}(\eta) \leftarrow IBAM(P_{V|Y}^{(s)}, P_{YU}, \eta, \epsilon)$

    $\eta^* \leftarrow \arg\min_\eta |I(P_Y, P_{V|Y}(\eta)) - C_v|$

    $P_{V|Y}^{(s+1)} \leftarrow P_{V|Y}(\eta^*)$

    $P_{UV}(u,v) = \sum_y P_{V|Y}^{(s+1)}(v,y)P_{YU}(y,u)$

    $R^{(s+1)} \leftarrow I(P_{UV})$

    $\Delta R \leftarrow |R^{(s+1)} - R^{(s)}|$

    $s \leftarrow s + 1$

$C_u^{(s)} \leftarrow I(P_X, P_{U|X}^{(s)})$

$C_v^{(s)} \leftarrow I(P_Y, P_{V|Y}^{(s)})$

Output: $P_{U|X}^{(s)}, P_{V|Y}^{(s)}, R^{(s)}, C_u^{(s)}, C_v^{(s)}$

---

6. Numerical Results

In this section, we focus on the DSBS setting of Section 3. In the first part of this section, we will show using a brute-force method the existence of a sharp, phase-transition phenomena in the optimal test-channels $P_{U|X}$ and $P_{V|Y}$ vs. DSBS parameter p. In the second part of this section, we will evaluate the alternating maximization algorithm proposed in Section 5; then, we compare its performance to the brute-force method.

6.1. Exhaustive Search

In this set of simulations, we again fix the transition matrix from Y to V characterized by the parameters:

$$T = \begin{pmatrix} a & b \\ 1-a & 1-b \end{pmatrix}, \tag{34}$$

chosen such that $I(Y; V) = C_v$. This choice defines a path $b = f(a)$ in the (a, b) plain. Then, for every such T we optimize $I(U; V)$ for different values of the DSBS parameter p. The results for a specific choice of $(C_u, C_v) = (0.4, 0.6)$ vs. a for different values of p are plotted in Figure 18. Note that the region of a corresponds to the continuous conversion from a Z channel ($a = 0$) to a BSC ($a = a_{\max}$). We observe here a very sharp transition from the optimality of Z-S channels to BSC channels configuration for a small change in p. This kind of behavior continues to hold with a different choice of ($C_u = 0.1, C_v = 0.9$), as can be seen in Figure 19.

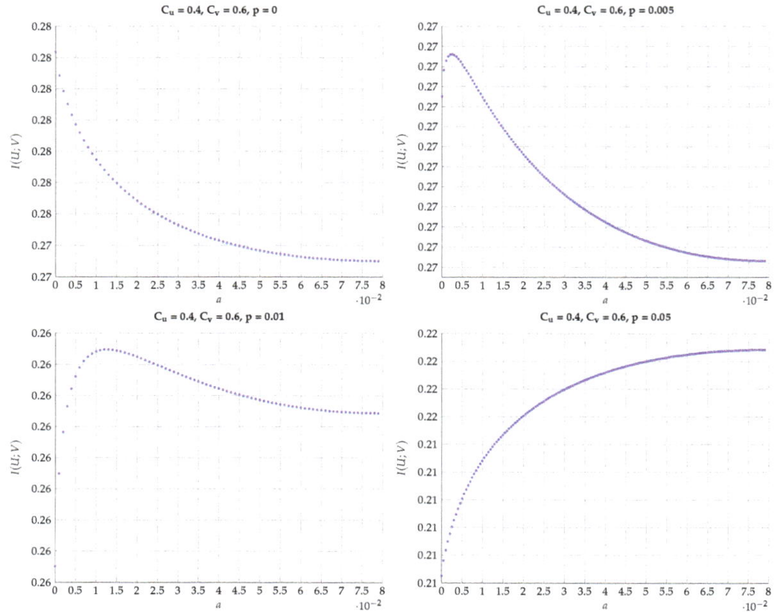

Figure 18. Maximal $I(\mathsf{U};\mathsf{V})$ for fixed values $(C_u, C_v) = (0.4, 0.6)$ and different values of p.

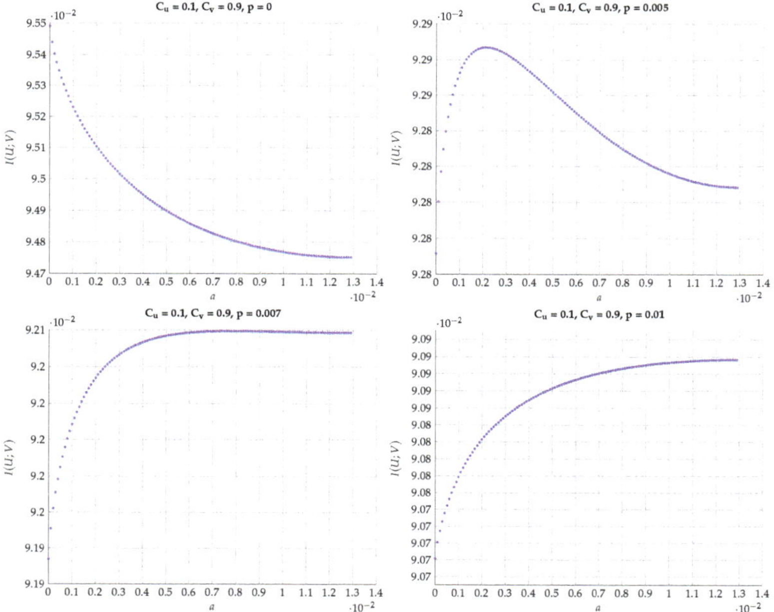

Figure 19. Maximal $I(\mathsf{U};\mathsf{V})$ for fixed values $(C_u, C_v) = (0.1, 0.9)$ and different values of p.

Next, we would like to emphasize this sharp phase transition phenomena by plotting the optimal a that achieves the maximal $I(U;V)$ vs the DSBS parameter p. The results for various combinations of C_u and C_v are presented in Figures 20 and 21. We observe that the curves are convex for $p \in [0, p_{th})$ and constant for $p > p_{th}$ with $a = a_{bsc}$. Furthermore, the derivative of $a(p)$ for $p \to p_{th}$ tends to ∞.

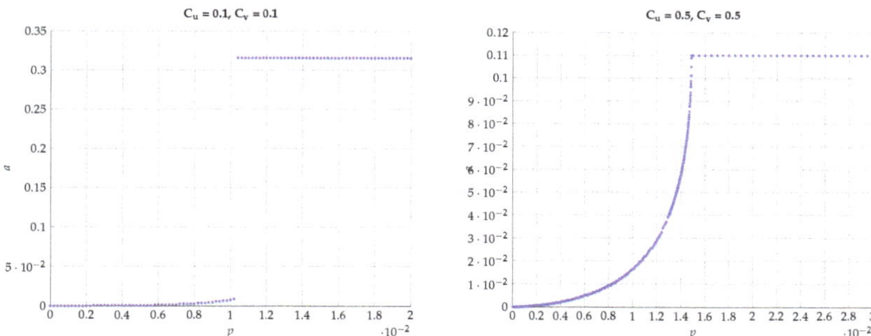

Figure 20. Optimal value of a for various values of C_u and C_v.

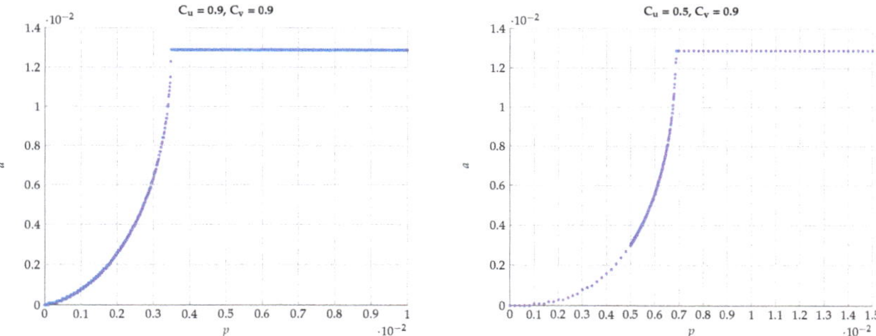

Figure 21. Optimal value of a for various values of C_u and $C_v = 0.9$.

One may further claim that there is no sharp transition to the BSC test-channels $P_{U|X}$ and $P_{V|Y}$ as p grows away from zero, but rather only approaches BSC. To convince the reader that the optimal test channels are exactly BSC, we performed an alternating maximization experiment. We fixed $p > 0$, C_u and C_v. Then we have chosen $P_{V|Y}$ as an almost BSC channel satisfying $I(Y;V) \le C_v$ and found the channel $P_{X|U}$ that maximizes $I(U;V)$ subject to $I(X;U) \le C_u$. Then, we fixed the channel $P_{X|U}$ and found the $P_{Y|V}$ that maximizes $I(U;V)$ subject to $I(Y;V) \le C_v$. We have repeated this alternating maximization procedure until it converges. The transition matrices were parameterized as follows:

$$T_{Y|V} = \begin{pmatrix} q_0 & q_1 \\ 1 - q_0 & 1 - q_1 \end{pmatrix}, \qquad T_{X|U} = \begin{pmatrix} p_0 & p_1 \\ 1 - p_0 & 1 - p_1 \end{pmatrix}. \tag{35}$$

The results for different values of p, C_u, and C_v are shown in Figures 22–24. We observe that p_0 and q_0 rapidly converge to their respective BSC values satisfying the mutual information constraints. Note that the last procedure is still an exhaustive search, but it is performed in alternating fashion between the sets $\{P_{U|X}\}$ and $\{P_{V|Y}\}$.

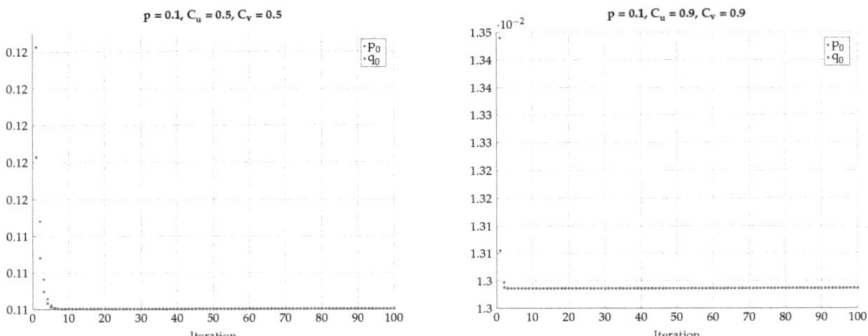

Figure 22. Alternating maximization with exhaustive search for various p, C_u, C_v.

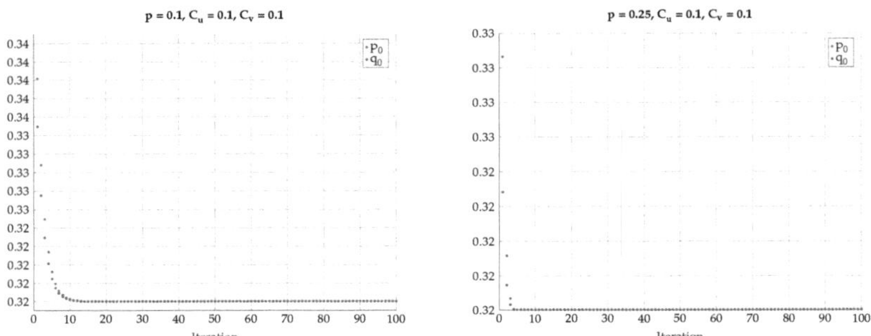

Figure 23. Alternating maximization with exhaustive search for various p, C_u, C_v.

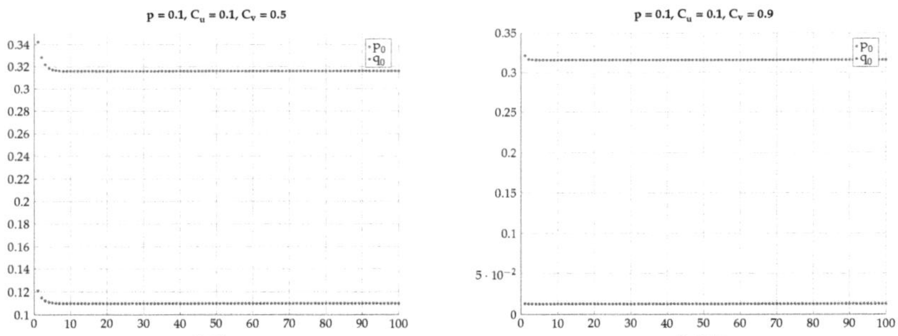

Figure 24. Alternating maximization with exhaustive search for various p, C_u, C_v.

6.2. Alternating Maximization

In this section, we will evaluate the algorithm proposed in Section 5. We focus on the DSBS setting of Section 3 with various values of problem parameters.

First, we explore the convergence behavior of the proposed algorithm. Figure 25 shows the objective function $I(\mathsf{U};\mathsf{V})$ on every iteration step for representative fixed-channel transition parameters p and the constraints C_u and C_v. We observe a slow convergence to a final value for $p = 0$ and $C_u = C_v = 0.2$, but once the constraints and the transition probability are increased, the algorithm converges much more rapidly. The non-monotonic behavior in some regimes is justified with the help of Remark 8. In Figure 26, we see the respective test-channel probabilities $\alpha_0 + \alpha_1$, $1 - \alpha_0$, $\beta_0 + \beta_1$, and $1 - \beta_1$. First, note that if $\alpha_0 + \alpha_1 = 1$, then $\mathsf{P}_{\mathsf{X}|\mathsf{U}}$ is a BSC. Similarly, if $\beta_0 + \beta_1 = 1$, then $\mathsf{P}_{\mathsf{Y}|\mathsf{V}}$ is a BSC. Second, if $1 - \alpha_0 = 1$, then $\mathsf{P}_{\mathsf{X}|\mathsf{U}}$ is a Z-channel. Similarly, if $1 - \beta_1 = 1$, then $\mathsf{P}_{\mathsf{Y}|\mathsf{V}}$ is an S-channel. We observe that for $p = 0$, the test-channels $\mathsf{P}_{\mathsf{X}|\mathsf{U}}$ and $\mathsf{P}_{\mathsf{Y}|\mathsf{V}}$ converge to Z- and S-channels, respectively. As for all other settings, the test-channels converge to BSC channels.

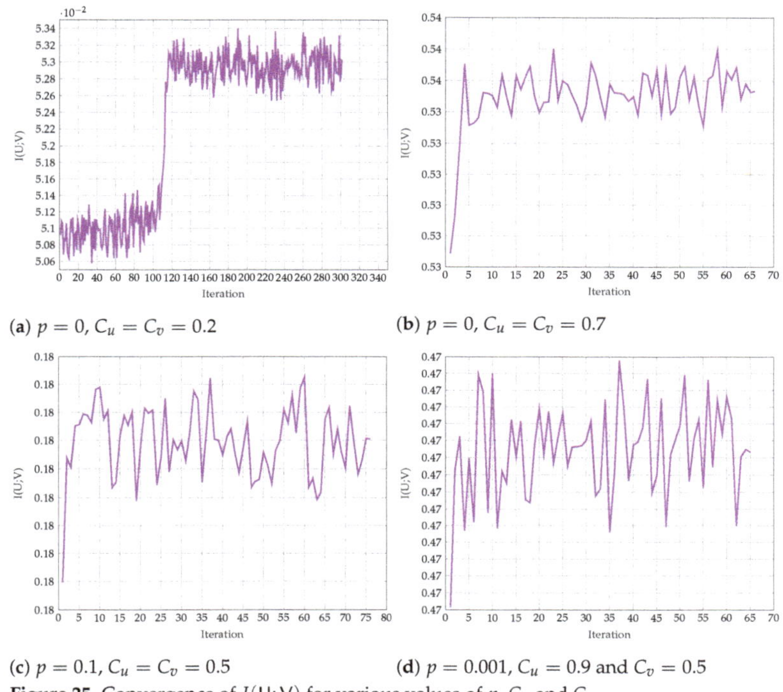

(a) $p = 0$, $C_u = C_v = 0.2$

(b) $p = 0$, $C_u = C_v = 0.7$

(c) $p = 0.1$, $C_u = C_v = 0.5$

(d) $p = 0.001$, $C_u = 0.9$ and $C_v = 0.5$

Figure 25. Convergence of $I(\mathsf{U};\mathsf{V})$ for various values of p, C_u and C_v.

Finally, we compare the outcome of Algorithm 2 to the optimal solution achieved by the brute-force method, namely, evaluating (12) for every $\mathsf{P}_{\mathsf{U}|\mathsf{X}}$ and $\mathsf{P}_{\mathsf{V}|\mathsf{Y}}$ that satisfy the problem constraints. The results for various values of channel parameters are shown in Figure 27. We observe that the proposed algorithm achieves the optimum for any DSBS parameter p and some representative constraints C_u and C_v.

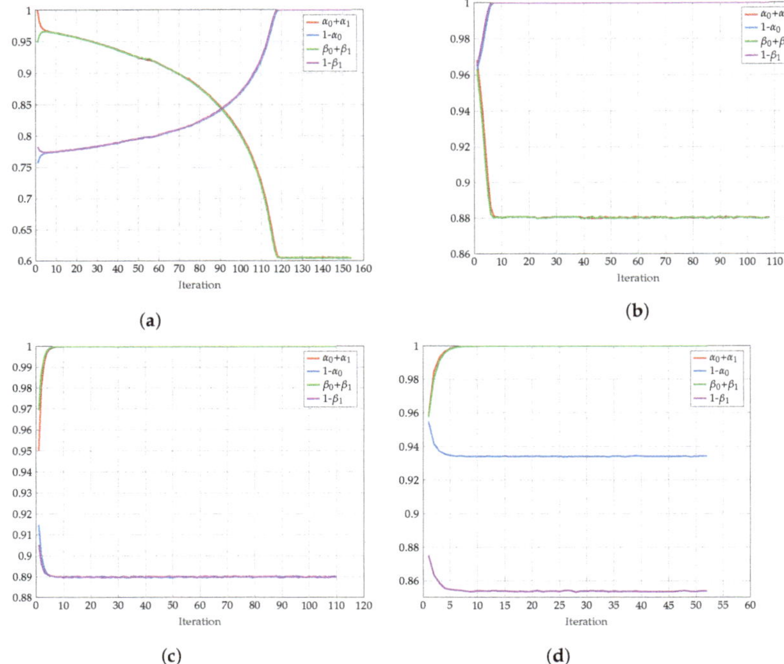

Figure 26. Convergence of $I(U;V)$ p with: (**a**) $C_u = C_v = 0.2$, $p = 0$; (**b**) $C_u = C_v = 0.7$, $p = 0$; (**c**) $C_u = C_v = 0.5$, $p = 0.1$; (**d**) $C_u = 0.65, C_v = 0.4$, $p = 0.1$.

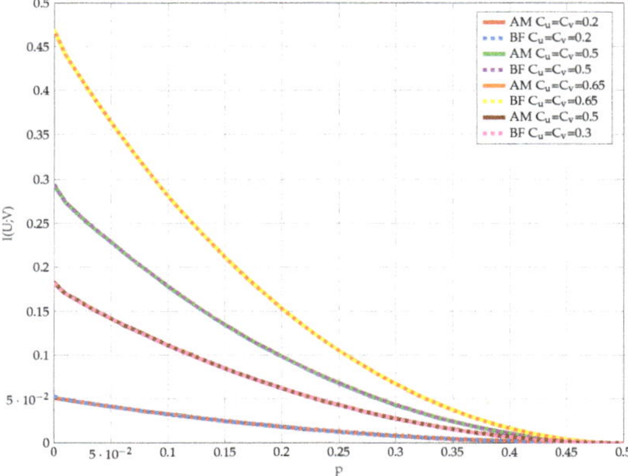

Figure 27. Comparison of the proposed alternating maximization algorithm and the brute-force search method for various problem parameters.

7. Concluding Remarks

In this paper, we have considered the Double-Sided Information Bottleneck problem. Cardinality bounds on the representation's alphabets were obtained for an arbitrary discrete bivariate source. When X and Y are binary, we have shown that taking binary auxiliary random variables is optimal. For DSBS, we have shown that BSC test-channels are optimal when $p \to 0.5$. Furthermore, numerical simulations for arbitrary p indicate that Z -and S-channels are optimal for $p = 0$. As for the Gaussian bivariate source, representation of $I(\mathsf{U};\mathsf{V})$ utilizing Hermite polynomials was given. In addition, the optimality of the Gaussian test-channels was demonstrated for vanishing SNR. Moreover, we have constructed a lower bound attained by deterministic quantizers that outperforms the jointly Gaussian choice at high SNR. Note that the solution for the n-letter problem $\max \frac{1}{n} I(\mathsf{U};\mathsf{V})$ for $\mathsf{U} \to \mathsf{X}^n \to \mathsf{Y}^n \to \mathsf{V}$ under constraints $I(\mathsf{U};\mathsf{X}^n) \leq nC_u$ and $I(\mathsf{V};\mathsf{Y}^n) \leq nC_v$ does not tensorize in general. For $\mathsf{X}^n = \mathsf{Y}^n \sim \mathrm{Ber}^{\otimes n}(0.5)$, we can easily achieve the cut-set bound $I(\mathsf{U};\mathsf{V})/n = \min\{C_u, C_v\}$. In addition, if time-sharing is allowed, the results change drastically.

Finally, we have proposed an alternating maximization algorithm based on the standard IB [1]. For the DSBS, it was shown that the algorithm converges to the global optimal solution.

Author Contributions: Conceptualization, O.O. and S.S.; methodology, M.D., O.O., and S.S.; software, M.D.; formal analysis, M.D.; writing—original draft preparation, M.D.; writing—review and editing, M.D.; supervision, S.S. All authors have read and agreed to the published version of the manuscript.

Funding: The work has been supported by the European Union's Horizon 2020 Research And Innovation Programme, grant agreement no. 694630, by the ISF under Grant 1641/21, and by the WIN consortium via the Israel minister of economy and science.

Conflicts of Interest: The authors declare no conflict of interest.

Appendix A. Proof of Proposition 2

Before proceeding to proof Proposition 2, we need the following auxiliary results.

Lemma A1. *Let $\mathsf{P}_{\mathsf{Y}|\mathsf{X}}$ be an arbitrary binary-input, ternary-output channel, parameterized using the following transition matrix:*

$$T \triangleq \begin{pmatrix} a & b \\ c & d \\ 1-a-c & 1-b-d \end{pmatrix}. \tag{A1}$$

Consider the function $p \mapsto \phi(p, \lambda) = h(T\mathbf{p}) - \lambda h_b(p)$ defined on $[0,1]$. This function has the following properties:

1. *If $\phi(p, \lambda)$ is linear on a sub-interval of $[0,1]$, then it is linear for every $p \in [0,1]$.*
2. *Otherwise, it is strictly convex over $[0,1]$ or there are points p_l and p_u such that $0 < p_l < p_u < 1$ where*

$$\phi(p, \lambda) = \begin{cases} \text{strictly convex} & 0 < p < p_l = \mathcal{I}_1, \\ \text{strictly concave} & p_l < p < p_u = \mathcal{I}_2, \\ \text{strictly convex} & p_u < p < 1 = \mathcal{I}_3. \end{cases} \tag{A2}$$

We postpone the proof of this lemma to Appendix K.

Lemma A2. *The convex envelope of $\phi(\cdot)$ at any point $q \in [0,1]$ can be obtained as a convex combination of only points in $\mathcal{I}_1$ and $\mathcal{I}_3$.*

We postpone the proof of this lemma to Appendix L and proceed to proof Proposition 2. Note that if $F_T(x)$ is strictly convex in $[0, h_b(q)]$, then by the paragraph following (Theorem 2.3 of [17]) $|\mathcal{U}| = 2$, we are done.

From now on, we consider the case where $F_T(x)$ is not strictly convex. Then, there is an interval $\mathcal{L} \subset [0, h(q)]$ and $a \in \mathbb{R}_+$ such that

$$F_T(x) = a + \lambda_{\mathcal{L}} \cdot x \quad \forall x \in \mathcal{L}. \tag{A3}$$

Let $\mathbf{t}_0$ and $\mathbf{t}_1$ represent the columns of T corresponding to $\mathsf{X} = 0$ and $\mathsf{X} = 1$, respectively. Moreover let $q \triangleq \mathsf{P}(\mathsf{X} = 0)$ and $\mathbf{p} \triangleq (p, \bar{p})^T$ be the probability vector of an arbitrary binary random variable, where $\bar{p} \triangleq 1 - p$.

Assume $x_1, x_2 \in \mathcal{L}$ and $x_1 \neq x_2$. Then, there must be $\{\alpha_{1i}, p_{1i}\}_{i=1,2,3}$ and $\{\alpha_{2i}, p_{2i}\}_{i=1,2,3}$ such that

$$\sum_{i=1}^{3} \alpha_{1i} p_{1i} = q, \qquad \sum_{i=1}^{3} \alpha_{1i} h_b(p_{1i}) = x_1, \qquad \sum_{i=1}^{3} \alpha_{1i} h(T\mathbf{p}_{1i}) = a + \lambda_{\mathcal{L}} x_1, \qquad \text{(A4)}$$

$$\sum_{i=1}^{3} \alpha_{2i} p_{2i} = q, \qquad \sum_{i=1}^{3} \alpha_{2i} h_b(p_{2i}) = x_2, \qquad \sum_{i=1}^{3} \alpha_{2i} h(T\mathbf{p}_{2i}) = a + \lambda_{\mathcal{L}} x_2. \qquad \text{(A5)}$$

Lemma A3. *The set $\{p_{11}, p_{12}, p_{13}, p_{21}, p_{22}, p_{23}\}$ must contain at least three distinct points.*

We postpone the proof of this lemma to Appendix M.

Consider the function $p \mapsto \phi(p) = \phi(p, \lambda_{\mathcal{L}}) = h(T\mathbf{p}) - \lambda_{\mathcal{L}} h_b(p)$ defined on $[0, 1]$. We have that

$$\sum_{i=1}^{3} \alpha_{1i} \phi(p_{1i}) = \sum_{i=1}^{3} \alpha_{2i} \phi(p_{2i}) = a. \qquad \text{(A6)}$$

In addition, if we define $\psi(\cdot)$ to be the lower convex envelope of $\phi(\cdot)$, then $\psi(q) = a$. Thus, the lower convex envelope of $\phi(\cdot)$ at q is attained by two linear combinations.

By Lemma Lemma A3, the set $\{p_{11}, p_{12}, p_{13}, p_{21}, p_{22}, p_{23}\}$ must contain at least three distinct points, say $\{p_{11}, p_{21}, p_{22}\}$. Due to Lemma A2, they are all in $\mathcal{I}_1 \cup \mathcal{I}_3$. Furthermore, by the pigeonhole principle, we must have that one of the intervals contains at least two points. Assume WLOG that $\{p_{11}, p_{21}\} \in \mathcal{I}_1$. For any $\gamma \in [0, 1]$, let $S = \bar{\gamma}\alpha_{11} + \gamma\alpha_{21}$ and consider the following set of weights/probabilities:

$$\left\{ \left(S, \frac{\bar{\gamma}\alpha_{11}}{S} \cdot p_{11} + \frac{\gamma\alpha_{21}}{S} \cdot p_{21} \right), (\bar{\gamma}\alpha_{12}, p_{12}), (\bar{\gamma}\alpha_{13}, p_{13}), (\gamma\alpha_{22}, p_{22}), (\gamma\alpha_{23}, p_{23}) \right\}. \qquad \text{(A7)}$$

Note that

$$S + \bar{\gamma}\alpha_{12} + \bar{\gamma}\alpha_{13} + \gamma\alpha_{22} + \gamma\alpha_{23} = 1, \qquad \text{(A8)}$$

and

$$\bar{\gamma}\alpha_{11} \cdot p_{11} + \gamma\alpha_{21} \cdot p_{21}\bar{\gamma}\alpha_{12} \cdot p_{12} + \bar{\gamma}\alpha_{13}, p_{13} + \gamma\alpha_{22} \cdot p_{22} + \gamma\alpha_{23} \cdot p_{23} = q, \qquad \text{(A9)}$$

but since $\{p_{11}, p_{21}\} \in \mathcal{I}_1$

$$S \cdot \phi\left(\frac{\bar{\gamma}\alpha_{11}}{S} \cdot p_{11} + \frac{\gamma\alpha_{21}}{S} \cdot p_{21} \right) + \bar{\gamma}\alpha_{12}\phi(p_{12}) + \bar{\gamma}\alpha_{13}\phi(p_{13}) + \gamma\alpha_{22}\phi(p_{22}) + \gamma\alpha_{23}\phi(p_{23}) \qquad \text{(A10)}$$

$$< S \cdot \left(\frac{\bar{\gamma}\alpha_{11}}{S} \cdot \phi(p_{11}) + \frac{\gamma\alpha_{21}}{S} \cdot \phi(p_{21}) \right) + \bar{\gamma}\alpha_{12}\phi(p_{12}) + \bar{\gamma}\alpha_{13}\phi(p_{13}) + \gamma\alpha_{22}\phi(p_{22}) + \gamma\alpha_{23}\phi(p_{23}) = a, \qquad \text{(A11)}$$

thus, it attains a smaller value than a, provided that ϕ is strictly convex on $\mathcal{I}_1$. This contradicts our assumption that the convex envelope at q equals a, and thus $\phi(\cdot)$ must contain a linear segment in $\mathcal{I}_1$.

By Lemma A1, this can happen only if p is linear for every $p \in [0, 1]$. In particular:

$$h(T\mathbf{p}) - \lambda_{\mathcal{L}} h_b(p) = \phi(p) = (1 - p)\phi(0) + p\phi(1) = (1 - p)h(\mathbf{t}_0) + ph(\mathbf{t}_1). \qquad \text{(A12)}$$

Note that for any choice of $\mathsf{P}_{\mathsf{X}|\mathsf{U}=u}$

$$H(\mathsf{Y}|\mathsf{U} = u) = h(T\mathbf{p}_u) \qquad \text{(A13)}$$

$$= \phi(p_u) + \lambda_{\mathcal{L}} h_b(p_u) \qquad \text{(A14)}$$

$$= (1 - p_u)h(\mathbf{t}_0) + p_u h(\mathbf{t}_1) + \lambda_{\mathcal{L}} h_b(p_u). \qquad \text{(A15)}$$

Taking the expectation we obtain:

$$H(\mathsf{Y}|\mathsf{U}) = (1-q)h(\mathsf{t_0}) + qh(\mathsf{t_1}) + \lambda_{\mathcal{L}}x. \tag{A16}$$

This implies that

$$F_T(x) = (1-q)h(\mathsf{t_0}) + qh(\mathsf{t_1}) + \lambda_{\mathcal{L}}x, \tag{A17}$$

and this is attained by any choice of $\mathsf{P_{U|X}}$ satisfying $H(\mathsf{X}|\mathsf{U}) = x$. In particular the choice $\mathsf{U} = \mathsf{X} \oplus \mathsf{Z}$, where $\mathsf{Z} \sim \mathrm{Ber}(\delta)$ is statistically independent of X and is chosen such that $H(\mathsf{X}|\mathsf{U}) = x$, attains $F_T(x)$. Thus, $|\mathcal{U}| = 2$ suffices even if $F_T(x)$ is not strictly convex.

Appendix B. Proof of Lemma 3

Let $\mathsf{P_{U|X}}$ and $\mathsf{P_{V|Y}}$ be the test-channels from X to U and from Y to V, respectively. The joint probability function of U and V can be expressed via Bayes' rule and the Markov chain condition $\mathsf{U} \to \mathsf{X} \to \mathsf{Y} \to \mathsf{V}$ as:

$$\mathsf{P_{UV}}(u,v) = 4 \cdot \mathsf{P_U}(u)\mathsf{P_V}(v) \sum_{x,y} \mathsf{P_{X|U}}(x|u)\mathsf{P_{XY}}(x,y)\mathsf{P_{Y|V}}(y|v). \tag{A18}$$

Since $I(\mathsf{U};\mathsf{V}) = \mathbb{E}[\log(\mathsf{P_{UV}}/\mathsf{P_U} \times \mathsf{P_V})]$, we define $K(u,v,p)$ as the ratio between the joint distribution of U and V relative to the respective product measure. Note that:

$$K(u,v,p) \triangleq \frac{\mathsf{P_{UV}}(u,v)}{\mathsf{P_U}(u)\mathsf{P_V}(v)} \tag{A19}$$

$$= 4 \sum_{x,y} \mathsf{P_{X|U}}(x|u)\mathsf{P_{XY}}(x,y)\mathsf{P_{Y|V}}(y|v) \tag{A20}$$

$$= 2(\mathsf{P_{X|U}}(0|u) \cdot \bar{p} \cdot \mathsf{P_{Y|V}}(0|u) + \mathsf{P_{X|U}}(1|u) \cdot p \cdot \mathsf{P_{Y|V}}(0|u)) \tag{A21}$$

$$+ 2(\mathsf{P_{X|U}}(0|u) \cdot p \cdot \mathsf{P_{Y|V}}(1|u) + \mathsf{P_{X|U}}(1|u) \cdot \bar{p} \cdot \mathsf{P_{Y|V}}(1|u)). \tag{A22}$$

Denoting $\alpha_u \triangleq \mathsf{P_{X|U}}(1|u)$ and $\beta_v \triangleq \mathsf{P_{Y|V}}(0|v)$, we obtain:

$$K(u,v,p) = 2(\bar{\alpha}_u \bar{p}\beta_v + \alpha_u p\beta_v + \bar{\alpha}_u p\bar{\beta}_v + \alpha_u \bar{p}\bar{\beta}_v) = 2\alpha_u * \beta_v * p. \tag{A23}$$

The last expression can also be represented as follows:

$$2\alpha_u * \beta_v * p = 2(1-p)(\alpha_u + \beta_v - 2\alpha_u\beta_v) + 2p(1 - \alpha_u - \beta_v + 2\alpha_u\beta_v) \tag{A24}$$

$$= 2\alpha_u + 2\beta_v - 4\alpha_u\beta_v + 2p(1 - 2\alpha_u - 2\beta_v + 4\alpha_u\beta_v) \tag{A25}$$

$$= 1 - (1-2p)(1 - 2\alpha_u - 2\beta_v + 4\alpha_u\beta_v) \tag{A26}$$

$$= 1 - (1-2p)(1 - 2\alpha_u)(1 - 2\beta_v). \tag{A27}$$

Thus,

$$I(\mathsf{U};\mathsf{V}) = \sum_{u,v} \mathsf{P_{UV}}(u,v) \log \frac{\mathsf{P_{UV}}(u,v)}{\mathsf{P_U}(u)\mathsf{P_V}(v)} \tag{A28}$$

$$= \sum_{u,v} \mathsf{P_U}(u)\mathsf{P_V}(v)K(u,v,p) \log K(u,v,p). \tag{A29}$$

Furthermore, note that since $|(1-2p)(1-2\alpha_u)(1-2\beta_v)| < 1$, we can utilize Taylor's expansion of $\log(1-x)$ to obtain:

$$\log K(u,v,p) = -\sum_{n=1}^{\infty} \frac{(1-2p)^n(1-2\alpha_u)^n(1-2\beta_v)^n}{n}, \tag{A30}$$

and

$$K(u, v, p) \log K(u, v, p) = - \sum_{n=1}^{\infty} \frac{(1-2p)^n (1-2\alpha_u)^n (1-2\beta_v)^n}{n}$$
$$+ \sum_{n=1}^{\infty} \frac{(1-2p)^{n+1} (1-2\alpha_u)^{n+1} (1-2\beta_v)^{n+1}}{n}. \tag{A31}$$

Therefore:

$$I(\mathsf{U}; \mathsf{V}) = - \sum_{n=1}^{\infty} \frac{(1-2p)^n \mathbb{E}[(1-2\alpha_\mathsf{U})^n] \mathbb{E}[(1-2\beta_\mathsf{V})^n]}{n} \tag{A32}$$

$$+ \sum_{n=1}^{\infty} \frac{(1-2p)^{n+1} \mathbb{E}[(1-2\alpha_\mathsf{V})^{n+1}] \mathbb{E}[(1-2\beta_\mathsf{V})^{n+1}]}{n} \tag{A33}$$

$$\overset{(a)}{=} - \sum_{n=2}^{\infty} \frac{(1-2p)^n \mathbb{E}[(1-2\alpha_\mathsf{U})^n] \mathbb{E}[(1-2\beta_\mathsf{V})^n]}{n} \tag{A34}$$

$$+ \sum_{n=1}^{\infty} \frac{(1-2p)^{n+1} \mathbb{E}[(1-2\alpha_\mathsf{V})^{n+1}] \mathbb{E}[(1-2\beta_\mathsf{V})^{n+1}]}{n} \tag{A35}$$

$$= - \sum_{n=1}^{\infty} \frac{(1-2p)^{n+1} \mathbb{E}[(1-2\alpha_\mathsf{U})^{n+1}] \mathbb{E}[(1-2\beta_\mathsf{V})^{n+1}]}{n+1} \tag{A36}$$

$$+ \sum_{n=1}^{\infty} \frac{(1-2p)^{n+1} \mathbb{E}[(1-2\alpha_\mathsf{V})^{n+1}] \mathbb{E}[(1-2\beta_\mathsf{V})^{n+1}]}{n} \tag{A37}$$

$$= \sum_{n=1}^{\infty} (1-2p)^{n+1} \mathbb{E}\left[(1-2\alpha_\mathsf{U})^{n+1}\right] \mathbb{E}\left[(1-2\beta_\mathsf{V})^{n+1}\right] \cdot \frac{1}{n(n+1)}, \tag{A38}$$

where (a) follows since $\mathbb{E}[\alpha_\mathsf{U}] = \mathbb{E}[\beta_\mathsf{V}] = \frac{1}{2}$. This completes the proof.

Appendix C. Auxiliary Concavity Lemma

As a preliminary step to proving Theorem 1, we will need the following auxiliary lemma.

Lemma A4. *The function $f(x) = (1 - 2h_b^{-1}(x))^2$ is concave.*

Proof. Denoting $g(x) \triangleq h_b^{-1}(x)$, we have $f(x) = (1 - 2g(x))^2$. Since $f(x)$ is twice differentiable, it is sufficient to show that $f'(x)$ is decreasing. The first derivative is given by:

$$f'(x) = -4(1 - 2g(x))g'(x). \tag{A39}$$

Since

$$h_b(x) = -x \log x - (1-x) \log(1-x), \tag{A40}$$

$$h_b'(x) = \log \frac{1-x}{x}, \tag{A41}$$

$$h_b''(x) = -\frac{1}{(1-x)\ln 2} - \frac{1}{x \ln 2} = -\frac{1}{x(1-x)\ln 2}, \tag{A42}$$

utilizing the inverse function derivative property, we obtain:

$$g(x) = h_b^{-1}(x), \tag{A43}$$

$$g'(x) = \frac{1}{h'(g(x))} = \frac{1}{\log \frac{1-g(x)}{g(x)}}. \tag{A44}$$

In addition, the second order derivative is given by:

$$g''(x) = \frac{\log e}{\log^3 \frac{1-g(x)}{g(x)}} \frac{g(x)}{1-g(x)} \frac{1}{(g(x))^2} \tag{A45}$$

$$= \frac{\log e}{\log^3 \frac{1-g(x)}{g(x)}} \frac{1}{(1-g(x))g(x)} \tag{A46}$$

$$= \frac{(g'(x))^3}{\ln 2 \, g(x)(1-g(x))}. \tag{A47}$$

Define

$$r(t) \triangleq \frac{-4(1-2t)}{\log \frac{1-t}{t}}. \tag{A48}$$

Note that $f'(x) = r(g(x))$. Since $g(x)$ is increasing, in order to show that $f'(x)$ decreasing, it suffices to show that $r(t)$ decreasing. The first order derivative of $r(t)$ is given by:

$$r'(t) = \frac{8}{\log^2 \frac{1-t}{t}} \frac{1}{t(1-t)} \left(t(1-t) \log \frac{1-t}{t} - \frac{1-2t}{\ln 4} \right).$$

Define $\alpha \triangleq 1 - 2t$ such that $t = \frac{1}{2}(1-\alpha)$. Note that $\alpha \in [0,1]$. We obtain:

$$r'(t) = \frac{32}{\log^2 \frac{1+\alpha}{1-\alpha}} \frac{1}{1-\alpha^2} \left(\frac{1}{4}(1-\alpha^2) \log \frac{1+\alpha}{1-\alpha} - \frac{\alpha}{\ln 4} \right).$$

Now, making use of the expansion $\log(1+x) = \sum_{k=1}^{\infty} (-1)^{k+1} \frac{x^k}{k}$, we have:

$$\log \frac{1+\alpha}{1-\alpha} = \sum_{k=1}^{\infty} (-1)^{k+1} \frac{\alpha^k}{k} - \sum_{k=1}^{\infty} (-1)^{k+1} \frac{(-\alpha)^k}{k} = 2 \sum_{k \text{ odd}} \frac{\alpha^k}{k}.$$

Thus,

$$\frac{1}{4}(1-\alpha^2) \log \frac{1+\alpha}{1-\alpha} - \frac{\alpha}{\ln 4}$$

$$= \frac{1}{2} \sum_{k \text{ odd}} \frac{\alpha^k}{k} - \frac{1}{2} \sum_{k \text{ odd}} \frac{\alpha^{k+2}}{k} - \frac{\alpha}{\ln 4}$$

$$= \alpha \left(\frac{1}{2} - \frac{1}{\ln 4} \right) + \frac{1}{2} \sum_{\substack{k \text{ odd} \\ k \geq 3}} \alpha^k \left(\frac{1}{k} - \frac{1}{k-2} \right)$$

$$\stackrel{(a)}{<} \alpha \left(\frac{1}{2} - \frac{1}{\ln e^2} \right) - \sum_{\substack{k \text{ odd} \\ k \geq 3}} \frac{\alpha^k}{k(k-2)} < 0,$$

where (a) follows since $\alpha > 0$. Thus, $r'(t) < 0$ and $f(x)$ is concave. $\quad\square$

Appendix D. Proof of Theorem 1

Plugging $p \leftarrow \frac{1}{2} - \epsilon$ ($\epsilon \triangleq \frac{1}{2} - p$) in (14), we obtain:

$$K(u,v,\epsilon) = 1 + 2\epsilon(1 - 2\alpha_u)(1 - 2\beta_v). \tag{A49}$$

Now, we rewrite $I(U;V)$ with explicit dependency on ϵ as:

$$I(\epsilon) = \sum_{u,v} P_U(u) P_V(v) K(u,v,\epsilon) \log K(u,v,\epsilon). \tag{A50}$$

We would like to expand $I(\epsilon)$ with Taylor series around $\epsilon = 0$. Note that $I(0) = 0 = I'(\epsilon)|_{\epsilon=0}$. Furthermore, the second derivative is given by:

$$I''(\epsilon)|_{\epsilon=0} = 4\log e \cdot \left(\sum_u P_U(u)(1 - 2\alpha_u)^2 \right)\left(\sum_v P_V(v)(1 - 2\beta_v)^2 \right).$$

Hence,

$$I(\epsilon) = 2\epsilon^2 \log e \cdot \left(\sum_u P_U(u)(1 - 2\alpha_u)^2 \right)\left(\sum_v P_V(v)(1 - 2\beta_v)^2 \right) + o(\epsilon^2).$$

Now, note that

$$\alpha_u = \begin{cases} h_2^{-1}(H(X|U = u)), & \alpha_u \le \frac{1}{2} \\ 1 - h_2^{-1}(H(X|U = u)), & \alpha_u > \frac{1}{2} \end{cases} \tag{A51}$$

with similar relation for β_v. Therefore,

$$\begin{aligned} I(\epsilon) &= \frac{2\epsilon^2}{\ln 2} \cdot \mathbb{E}_u\left[(1 - 2h_2^{-1}(H(X|U = u)))^2 \right] \cdot \mathbb{E}_v\left[(1 - 2h_2^{-1}(H(Y|V = v)))^2 \right] + o(\epsilon^2) \\ &\le 2\epsilon^2 \log e \cdot (1 - 2h_2^{-1}(H(X|U)))^2(1 - 2h_2^{-1}(H(Y|V)))^2 + o(\epsilon^2) \\ &\le 2\epsilon^2 \log e \cdot (1 - 2h_2^{-1}(1 - C_x))^2(1 - 2h_2^{-1}(1 - C_y))^2 + o(\epsilon^2), \end{aligned}$$

where the first inequality follows since the function $f : x \mapsto (1 - 2h_2^{-1}(x))^2$ is concave by Lemma A4 and applying Jensen's inequality, and the second inequality follows from rate constraints.

Appendix E. Proof of Proposition 4

Suppose that the optimal test-channel $P_{V|X}$ is given by the following transition matrix:

$$T_{V|X} = \begin{pmatrix} a & b \\ 1 - a & 1 - b \end{pmatrix}. \tag{A52}$$

Assume in contradiction that the opposite optimal test-channel $P_{U|X}$ is symmetric to $P_{V|X}$ and is given by:

$$T_{U|X} = \begin{pmatrix} 1 - b & 1 - a \\ b & a \end{pmatrix}. \tag{A53}$$

Applying Bayes' rule on (A53), we obtain:

$$T_{X|U} = \begin{pmatrix} 1 - \alpha_0 & 1 - \alpha_1 \\ \alpha_0 & \alpha_1 \end{pmatrix} = \begin{pmatrix} \frac{\bar{b}}{\bar{a}+\bar{b}} & \frac{b}{a+b} \\ \frac{\bar{a}}{\bar{a}+\bar{b}} & \frac{a}{a+b} \end{pmatrix}. \tag{A54}$$

It was shown in (Section IV.D of [17]) that for fixed $P_{V|X}$ given by (A52), the optimal $P_{X|U}$ must satisfy the following equation:

$$\begin{aligned} (a - b)(h_b(\alpha_1) - h_b(\alpha_0))(h'_b(\hat{\alpha}_0) - h'_b(\hat{\alpha}_1)) &+ (h'_b(\alpha_1) - h'_b(\alpha_0))(h_b(\hat{\alpha}_1) - h_b(\hat{\alpha}_0)) \\ + (a - b)(\alpha_1 - \alpha_0)(h'_b(\alpha_0)h'_b(\hat{\alpha}_1) - h'_b(\alpha_1)h'_b(\hat{\alpha}_0)) &= 0, \end{aligned} \tag{A55}$$

where $\hat{\alpha}_0 \triangleq a\alpha_0 + b\bar{\alpha}_0$ and $\hat{\alpha}_1 \triangleq a\alpha_1 + b\bar{\alpha}_1$. Plugging α_0 and α_1 from (A54) in (A55) results in a contradiction, thus establishing the proof of Proposition 4.

Appendix F. Proof of Proposition 6

By Lemma 3, the objective function of (7) for a DSBS setting, denoted here by $I(p)$, is given by:

$$I(p) = \mathbb{E}_{P_U \times P_V}[K(U, V, p) \log K(U, V, p)], \tag{A56}$$

where $K(u, v, p)$ can be expressed as:

$$K(u, v, p) = 1 + (1 - 2p)(1 - 2\alpha_u * \beta_v) = 1 + (1 - 2p)(1 - 2\alpha_u)(1 - 2\beta_v). \tag{A57}$$

Since $\log(1 + x) \le x$, we have the following upper bound on $I(p)$:

$$I(p) = \sum_{u,v} P_U(u) P_V(v) K(u, v, p) \log K(u, v, p) \tag{A58}$$

$$\le \sum_{u,v} P_U(u) P_V(v)(1 + (1 - 2p)(1 - 2\alpha_u)(1 - 2\beta_v))(1 - 2p)(1 - 2\alpha_u)(1 - 2\beta_v) \tag{A59}$$

$$= (1 - 2p)(1 - 2\sum_u P_U(u)\alpha_u)(1 - 2\sum_v P_V(v)\beta_v) \tag{A60}$$

$$+ (1 - 2p)^2 \sum_u P_U(u)(1 - 2\alpha_u)^2 \sum_v P_V(v)(1 - 2\beta_v)^2 \tag{A61}$$

$$= (1 - 2p)(1 - 2\sum_u P_U(u)P(X = 1|U = u))(1 - 2\sum_v P_V(v)P(Y = 1|V = v)) \tag{A62}$$

$$+ (1-2p)^2 \sum_u P_U(u)(1-2P(X = 1|U = u))^2 \sum_v P_V(v)(1-2P(Y = 1|V = v))^2$$

$$= (1 - 2p)(1 - 2P(X = 1))(1 - 2P(Y = 1)) \tag{A63}$$

$$+ (1-2p)^2 \sum_u P_U(u)(1-2h_2^{-1}(H(X|U = u)))^2 \sum_v P_V(v)(1-2h_2^{-1}(H(Y = |V = v)))^2$$

$$\overset{(a)}{\le} (1 - 2p)^2(1 - 2h_2^{-1}(H(X|U)))^2(1 - 2h_2^{-1}(H(Y = |V)))^2 \tag{A64}$$

$$\overset{(b)}{\le} (1 - 2p)^2(1 - 2h_2^{-1}(1 - C_x)^2(1 - 2h_2^{-1}(1 - C_y)^2, \tag{A65}$$

where the inequality in (a) follows from Lemma A4 and inequality in (b) follows from the problem constraints.

Appendix G. Proof of Proposition 7

We assume U and V are continuous RVs. The proof for the discrete case is identical. The joint density $f_{UV}(u, v)$ can be expressed with explicit dependency on ρ as follows:

$$f(u, v; \rho) \triangleq f_U(u) f_V(v) \iint_{\mathbb{R}^2} f_{X|U}(x|u) M(x, y; \rho) f_{Y|V}(y|v) dx dy,$$

where $M(x, y; \rho) = \sum_{n=0}^{\infty} \frac{\rho^n}{n!} H_n(x) H_n(y)$ [66]. Similarly, $I(U; V)$ can also be written with explicit dependency on ρ

$$I(\rho) \triangleq I_\rho(U; V) = \int \int f(u, v; \rho) \log \frac{f(u, v; \rho)}{f_U(u) f_V(v)} du dv.$$

Appendix H. Proof of Proposition 8

Let (U, X, Y, V) be jointly Gaussian Random variables, such that

$$X = \sigma_{UX} U + \sqrt{1 - \sigma_{UX}^2} Z_u, \qquad Y = \sigma_{YV} V + \sqrt{1 - \sigma_{YV}^2} Z_v,$$

where $Z_u \sim \mathcal{N}(0,1)$, $Z_v \sim \mathcal{N}(0,1)$, $Z_u \perp U$, $Z_v \perp V$. Due to Proposition 7, the mutual information for jointly Gaussian (U, X, Y, V) is given by

$$
\begin{aligned}
I(U;V) &= \mathbb{E}_{UV}\left[\log\left(\sum_{n=0}^{\infty}\frac{\rho^n}{n!}\mathbb{E}[H_n(X)|U]\mathbb{E}[H_n(Y)|V]\right)\right] \\
&\overset{(a)}{=} \mathbb{E}_{UV}\left[\log\left(\sum_{n=0}^{\infty}\frac{(\rho\sigma_{UX}\sigma_{YV})^n}{n!}H_n(U)H_n(V)\right)\right] \\
&\overset{(b)}{=} \mathbb{E}_{UV}\left[\log\left(\frac{1}{\sqrt{1-\rho^2\sigma_{UX}^2\sigma_{YV}^2}}\exp\left(\frac{2\rho\sigma_{UX}\sigma_{YV}UV - \rho^2\sigma_{UX}^2\sigma_{YV}^2(U^2+V^2)}{2(1-\rho^2\sigma_{UX}^2\sigma_{YV}^2)}\right)\right)\right] \\
&= -\frac{1}{2}\log(1-\rho^2\sigma_{UX}^2\sigma_{YV}^2) + \frac{\rho\sigma_{UX}\sigma_{YV}}{1-\rho^2\sigma_{UX}^2\sigma_{YV}^2}\mathbb{E}[UV] - \frac{\rho^2\sigma_{UX}^2\sigma_{YV}^2}{2(1-\rho^2\sigma_{UX}^2\sigma_{YV}^2)}\left(\mathbb{E}\left[U^2\right]+\mathbb{E}\left[V^2\right]\right) \\
&= -\frac{1}{2}\log(1-\rho^2\sigma_{UX}^2\sigma_{YV}^2),
\end{aligned}
$$

where (a) and (b) follow from the properties of Mehler Kernel [66].

By the Mutual Information constraints we have:

$$
\sigma_{UX}^2 = 1 - e^{-2C_u} \qquad \sigma_{YV}^2 = 1 - e^{-2C_v}. \tag{A66}
$$

Hence,

$$
I(U;V) = -\frac{1}{2}\log(1-\rho^2(1-e^{-2C_u})(1-e^{-2C_v})). \tag{A67}
$$

Appendix I. Proof of Proposition 9

We choose U and V to be deterministic functions of X and Y, respectively, i.e., $U = \mathrm{sign}(X)$ and $V = \mathrm{sign}(Y)$. In such case, the rate constraints are met with equality, namely, $I(U;X) = 1 = I(Y;V)$. We proceed to evaluate the achievable rate:

$$
\begin{aligned}
I(U;V) &= 1 - P(U=0)h_2(P(V=1|U=0)) - P(U=1)h_2(P(V=0|U=1)) \\
&\overset{(a)}{=} 1 - h_2(P(U \neq V)),
\end{aligned}
$$

where equality in (a) follows since $P(V=1|U=0) = P(V=0|U=1)$ by symmetry. We therefore obtain the following formula for the "error probability":

$$
P(V \neq U) = 1 - P(X<0, Y<0) - P(X>0, Y>0) \overset{(a)}{=} 1 - 2P(X<0, Y<0),
$$

where (a) also follows from symmetry. Utilizing Sheppard's Formula (Chapter 5, p.107 of [68]), we have $1 - 2P(X<0, Y<0) = \frac{\arccos\rho}{\pi}$. This completes the proof of the proposition.

Appendix J. Proof of Theorem 2

We would like to approximate $I(\rho)$ in the limit $\rho \to 0$ using a Taylor series up to a second order in ρ. As a first step, we evaluate the first two derivatives of $f(u,v;\rho)$ at $\rho = 0$. Note that $M(x,y;0) = 1$ and

$$
\frac{dM}{d\rho}\Big|_{\rho=0} = xy, \qquad \frac{d^2M}{d\rho^2}\Big|_{\rho=0} = (x^2-1)(y^2-1). \tag{A68}
$$

Thus, $f(u,v;0) = f_U(u)f_V(v)$,

$$
\frac{df}{d\rho}\Big|_{\rho=0} = f_U(u)f_V(v)\mathbb{E}[X|U=u]\mathbb{E}[Y|V=v],
$$

and

$$\frac{\mathrm{d}^2 f}{\mathrm{d}\rho^2}\Big|_{\rho=0} = f_U(u)f_V(v)\int_{-\infty}^{\infty}\int_{-\infty}^{\infty} f_{X|U}(x|u)\frac{\mathrm{d}^2 M(x,y;\rho)}{\mathrm{d}\rho^2}\Big|_{\rho=0}f_{Y|V}(y|v)\mathrm{d}x\mathrm{d}y \tag{A69}$$

$$= f_U(u)f_V(v)\left(\int_{-\infty}^{\infty}(x^2-1)f_{X|U}(x|u)\mathrm{d}x\right)\left(\int_{-\infty}^{\infty}(y^2-1)f_{Y|V}(y|v)\mathrm{d}y\right) \tag{A70}$$

$$= f_U(u)f_V(v)\left(\mathbb{E}[X^2|U=u]-1\right)\left(\mathbb{E}[Y^2|V=v]-1\right). \tag{A71}$$

Expanding $I(\rho)$ in Taylor series around $\rho=0$ gives us $I(0)=0=\frac{\mathrm{d}I(\rho)}{\mathrm{d}\rho}\big|_{\rho=0}$ and

$$\frac{\mathrm{d}^2 I(\rho)}{\mathrm{d}\rho^2}\Big|_{\rho=0} = \log e \cdot \mathbb{E}\left[(\mathbb{E}[X|U])^2\right]\mathbb{E}\left[(\mathbb{E}[Y|V])^2\right].$$

Thus,

$$I(\rho) = \frac{\rho^2 \log e}{2}\mathbb{E}\left[(\mathbb{E}[X|U])^2\right]\mathbb{E}\left[(\mathbb{E}[Y|V])^2\right] + o(\rho^2). \tag{A72}$$

Note that $\mathbb{E}[X]=\mathbb{E}[\mathbb{E}[X|U]]$ and

$$1 = \mathbb{E}[X^2] = \mathbb{E}\left[\mathbb{E}[X^2|U]\right] = \mathbb{E}[\mathrm{var}[X|U]] + \mathbb{E}\left[(\mathbb{E}[X|U])^2\right]. \tag{A73}$$

In addition, by (Corollary to Theorem 8.6.6 of [69]), $\mathbb{E}[\mathrm{var}[X|U]] \geq \frac{1}{2\pi e}e^{2h(X|U)}$. Moreover, from MI constraint, we have

$$I(X;U) = h(X) - h(X|U) = \frac{1}{2}\log(2\pi e) - h(X|U) \leq C_u,$$

and therefore $h(X|U) \geq \log(2\pi e) - C_u$. Thus, we obtain:

$$-C_u \leq \frac{1}{2}\log(\mathbb{E}[\mathrm{var}[X|U]]) \rightarrow \mathbb{E}[\mathrm{var}[X|U]] \geq 2^{-2C_u}. \tag{A74}$$

Combining (A73) and (A74), we obtain $\mathbb{E}\left[(\mathbb{E}[X|U])^2\right] \leq 1 - 2^{-2C_u}$.
In a very similar method, one can show that $\mathbb{E}\left[(\mathbb{E}[Y|V])^2\right] \leq 1 - 2^{-2C_v}$.
Thus, for $\rho \to 0$

$$I(\rho) \leq \frac{\rho^2 \log e}{2}(1 - 2^{-2C_u})(1 - 2^{-2C_v}) + o(\rho^2). \tag{A75}$$

Appendix K. Proof of Lemma A1

The function $\phi(p,\lambda)$ is a twice differentiable continuous function with respective second derivative given by

$$\frac{\partial^2 \phi(p,\lambda)}{\partial p^2} = \phi_{pp}(p,\lambda) = -\frac{(a-b)^2}{ap+b\bar{p}} - \frac{(c-d)^2}{cp+d\bar{p}} - \frac{(a-b+c-d)^2}{1-(a+c)p-(b+d)\bar{p}} + \frac{\lambda}{p\bar{p}}. \tag{A76}$$

The former can also be written as a proper rational function [70], i.e., $\phi_{pp}(p,\lambda) = \frac{N(p)}{D(p)}$, where

$$N(p) = \lambda(ap+b\bar{p})(cp+d\bar{p})(1-(a+c)p-(b+d)\bar{p}) - (a-b)^2(cp+d\bar{p})(1-(a+c)p$$
$$-(b+d)\bar{p})p\bar{p} - (c-d)^2(ap+b\bar{p})(1-(a+c)p-(b+d)\bar{p})p\bar{p}$$
$$-(a-b+c-d)^2(ap+b\bar{p})(cp+d\bar{p})p\bar{p}, \tag{A77}$$

and

$$D(p) = p\bar{p}(ap+b\bar{p})(cp+d\bar{p})(1-(a+c)p-(b+d)\bar{p}). \tag{A78}$$

Note that $\phi_{pp}(p,\lambda)$ equals $+\infty$ for $p \in \{0,1\}$ and hence is positive for this set of points.

1. Suppose $\phi(p,\lambda)$ is linear over some interval $\mathcal{I} \subset [a,b]$. In such case, its second derivative must be zero over this interval, which implies that $N(p)$ is zero over this interval. Since $N(p)$ is a degree 3 polynomial, it can be zero over some interval if and only if it is zero everywhere. Thus, if $\phi(p,\lambda)$ is linear over some interval $\mathcal{I}$, then it is non-linear for every $p \in [0,1]$.

2. For $p \in (0,1)$, $D(p) > 0$ and $N(p)$ is a degree 3 polynomial in p. Since $N(0^+) > 0$ and $N(1^-) > 0$, this polynomial has no sign changes or has exactly two sign changes in $(0,1)$. Therefore, either $\phi(p,\lambda)$ is convex or there are two points p_1 and p_2, $0 < p_1 < p_2 < 1$, such that $\phi(p,\lambda)$ is convex in $p \in [0,p_1] \cup [p_2,1]$ and concave in $p \in [p_1,p_2]$.

Appendix L. Proof of Lemma A2

Let $\mathcal{I}_2 = [c,d] \subset [0,1]$ and assume in contradiction that $\{\alpha_i, p_i\}_{i=1,2,3}$ attains the lower convex envelope at point q, and that $p_2 \in \mathcal{I}_2$. By assumption, we have that

$$\alpha_1 p_1 + \alpha_2 p_2 + \alpha_3 p_3 = q. \tag{A79}$$

We can write $p_2 = \bar{\gamma} c + \gamma d$ for some $\gamma \in (0,1)$ and still

$$\alpha_1 p_1 + \alpha_2 \bar{\gamma} c + \alpha_2 \gamma d + \alpha_3 p_3 = q. \tag{A80}$$

However, due to concavity of $\phi(\cdot)$ in $\mathcal{I}_2$, we must have

$$\alpha_1 \phi(p_1) + \alpha_2 \bar{\gamma} \phi(c) + \alpha_2 \gamma \phi(d) + \alpha_3 \phi(p_3) \le \alpha_1 \phi(p_1) + \alpha_2 \phi(\bar{\gamma} c + \gamma d) + \alpha_3 \phi(p_3)$$
$$= \alpha_1 \phi(p_1) + \alpha_2 \phi(p_2) + \alpha_3 \phi(p_3). \tag{A81}$$

This implies that there is a linear combination of point from $\mathcal{I}_1 \cup \mathcal{I}_3$ that attains a lower value than $\phi(q)$, contradicting the assumption that $\phi(q)$ is the lower convex envelope at point q. Since p_2 was arbitrary, the lemma holds.

Appendix M. Proof of Lemma A3

Assume in contradiction that there are no distinct points, i.e., it has $p_{11} = p_{12} = p_{13} = p_{21} = p_{22} = p_{23} = p$, then $p = q$ and $x_1 = x_2$, which contradicts the initial assumption that $x_1 \ne x_2$. Assume WOLG that $p_{11} = p_{12} = p_{13} = p_{21} = p_{22} = p$ but $p_{23} \ne p$. Since $p_{11} = p_{12} = p_{13} = p$ implies $p = q$, then p_{23} must be q as well in contradiction to the initial assumption.

Consider the following cases:

- $p_{11} = p_{12} = p_{13} = p_{21} = p_1, p_{22} = p_{23} = p_2$, $p_1 \ne p_2$: This implies $p_1 = q$. Furthermore,

$$\alpha_{21} q + \alpha_{22} p_2 + (1 - \alpha_{21} - \alpha_{22}) p_2 = q \to (1 - \alpha_{21}) p_2 = (1 - \alpha_{21}) q, \tag{A82}$$

which holds only if $p_2 = q$ in contradiction to our initial assumption.

- $p_{11} = p_{12} = p_{21} = p_{22} = p_1, p_{13} = p_{23} = p_2$, $p_1 \ne p_2$: This implies

$$(\alpha_{11} + \alpha_{12}) p_1 + (1 - \alpha_{21} - \alpha_{22}) p_2 = q = (\alpha_{21} + \alpha_{22}) p_1 + (1 - \alpha_{21} - \alpha_{22}) p_2, \tag{A83}$$

which holds only if $\alpha_{11} + \alpha_{12} = \alpha_{21} + \alpha_{22}$. In such case

$$x_1 = (\alpha_{11} + \alpha_{12}) h(p_1) + (1 - \alpha_{11} - \alpha_{12}) h(p_2)$$
$$= (\alpha_{21} + \alpha_{22}) h(p_1) + (1 - \alpha_{21} - \alpha_{22}) h(p_2) = x_2, \tag{A84}$$

in contradiction to the assumption $x_1 \ne x_2$.

Thus, the lemma holds.

References

1. Tishby, N.; Pereira, F.C.N.; Bialek, W. The information bottleneck method. In Proceedings of the 37th Annual Allerton Conference on Communication, Control and Computing, Monticello, IL, USA, 22–24 September 1999; pp. 368–377.
2. Pichler, G.; Piantanida, P.; Matz, G. Distributed information-theoretic clustering. *Inf. Inference J. Ima* **2021**, *11*, 137–166. [CrossRef]
3. Jain, A.K.; Dubes, R.C. *Algorithms for Clustering Data*; Prentice-Hall: Hoboken, NJ, USA, 1988.
4. Gupta, N.; Aggarwal, S. Modeling Biclustering as an optimization problem using Mutual Information. In Proceedings of the International Conference on Methods and Models in Computer Science (ICM2CS), Delhi, India, 14–15 December 2009 ; pp. 1–5.
5. Hartigan, J. Direct Clustering of a Data Matrix. *J. Am. Stat. Assoc.* **1972**, *67*, 123–129. [CrossRef]
6. Madeira, S.; Oliveira, A. Biclustering algorithms for biological data analysis: A survey. *IEEE/ACM Trans. Comput. Biol. Bioinform.* **2004**, *1*, 24–45. [CrossRef] [PubMed]
7. Dhillon, I.S.; Mallela, S.; Modha, D.S. Information-Theoretic Co-Clustering. In Proceedings of the Ninth ACM SIGKDD International Conference on Knowledge Discovery and Data Mining, 2003, (KDD '03), Washington, DC, USA, 24–27 August 2003; pp. 89–98.
8. Courtade, T.A.; Kumar, G.R. Which Boolean Functions Maximize Mutual Information on Noisy Inputs? *IEEE Trans. Inf. Theory* **2014**, *60*, 4515–4525. [CrossRef]
9. Han, T.S. Hypothesis Testing with Multiterminal Data Compression. *IEEE Trans. Inf. Theory* **1987**, *33*, 759–772. [CrossRef]
10. Westover, M.B.; O'Sullivan, J.A. Achievable Rates for Pattern Recognition. *IEEE Trans. Inf. Theory* **2008**, *54*, 299–320. [CrossRef]
11. Painsky, A.; Feder, M.; Tishby, N. An Information-Theoretic Framework for Non-linear Canonical Correlation Analysis. *arXiv* **2018**, arXiv:1810.13259.
12. Williamson, A.R. The Impacts of Additive Noise and 1-bit Quantization on the Correlation Coefficient in the Low-SNR Regime. In Proceedings of the 57th Annual Allerton Conference on Communication, Control, and Computing, Monticello, IL, USA, 24–27 September 2019; pp. 631–638.
13. Courtade, T.A.; Weissman, T. Multiterminal Source Coding Under Logarithmic Loss. *IEEE Trans. Inf. Theory* **2014**, *60*, 740–761. [CrossRef]
14. Pichler, G.; Piantanida, P.; Matz, G. Dictator Functions Maximize Mutual Information. *Ann. Appl. Prob.* **2018**, *28*, 3094–3101. [CrossRef]
15. Dobrushin, R.; Tsybakov, B. Information transmission with additional noise. *IRE Trans. Inf. Theory* **1962**, *8*, 293–304. [CrossRef]
16. Wolf, J.; Ziv, J. Transmission of noisy information to a noisy receiver with minimum distortion. *IEEE Trans. Inf. Theory* **1970**, *16*, 406–411. [CrossRef]
17. Witsenhausen, H.S.; Wyner, A.D. A Conditional Entropy Bound for a Pair of Discrete Random Variables. *IEEE Trans. Inf. Theory* **1975**, *21*, 493–501. [CrossRef]
18. Arimoto, S. An algorithm for computing the capacity of arbitrary discrete memoryless channels. *IEEE Trans. Inf. Theory* **1972**, *18*, 14–20. [CrossRef]
19. Blahut, R. Computation of channel capacity and rate-distortion functions. *IEEE Trans. Inf. Theory* **1972**, *18*, 460–473. [CrossRef]
20. Aguerri, I.E.; Zaidi, A. Distributed Variational Representation Learning. *IEEE Trans. Pattern Anal.* **2021**, *43*, 120–138. [CrossRef]
21. Hassanpour, S.; Wuebben, D.; Dekorsy, A. Overview and Investigation of Algorithms for the Information Bottleneck Method. In Proceedings of the SCC 2017: 11th International ITG Conference on Systems, Communications and Coding, Hamburg, Germany, 6–9 February 2017; pp. 1–6.
22. Slonim, N. The Information Bottleneck: Theory and Applications. Ph.D. Thesis, Hebrew University of Jerusalem, Jerusalem, Israel, 2002.
23. Sutskover, I.; Shamai, S.; Ziv, J. Extremes of information combining. *IEEE Trans. Inf. Theory* **2005**, *51*, 1313–1325. [CrossRef]
24. Zaidi, A.; Aguerri, I.E.; Shamai, S. On the Information Bottleneck Problems: Models, Connections, Applications and Information Theoretic Views. *Entropy* **2020**, *22*, 151. [CrossRef]
25. Wyner, A.; Ziv, J. A theorem on the entropy of certain binary sequences and applications–I. *IEEE Trans. Inf. Theory* **1973**, *19*, 769–772. [CrossRef]
26. Chechik, G.; Globerson, A.; Tishby, N.; Weiss, Y. Information Bottleneck for Gaussian Variables. *J. Mach. Learn. Res.* **2005**, *6*, 165–188.
27. Blachman, N. The convolution inequality for entropy powers. *IEEE Trans. Inf. Theory* **1965**, *11*, 267–271. [CrossRef]
28. Guo, D.; Shamai, S.; Verdú, S. The interplay between information and estimation measures. *Found. Trends Signal Process.* **2013**, *6*, 243–429. [CrossRef]
29. Bustin, R.; Payaro, M.; Palomar, D.P.; Shamai, S. On MMSE Crossing Properties and Implications in Parallel Vector Gaussian Channels. *IEEE Trans. Inf. Theory* **2013**, *59*, 818–844. [CrossRef]
30. Sanderovich, A.; Shamai, S.; Steinberg, Y.; Kramer, G. Communication Via Decentralized Processing. *IEEE Trans. Inf. Theory* **2008**, *54*, 3008–3023. [CrossRef]
31. Smith, J.G. The information capacity of amplitude-and variance-constrained scalar Gaussian channels. *Inf. Control.* **1971**, *18*, 203–219. [CrossRef]
32. Sharma, N.; Shamai, S. Transition points in the capacity-achieving distribution for the peak-power limited AWGN and free-space optical intensity channels. *Probl. Inf. Transm.* **2010**, *46*, 283–299. [CrossRef]

33. Dytso, A.; Yagli, S.; Poor, H.V.; Shamai, S. The Capacity Achieving Distribution for the Amplitude Constrained Additive Gaussian Channel: An Upper Bound on the Number of Mass Points. *IEEE Trans. Inf. Theory* **2019**, *66*, 2006–2022. [CrossRef]
34. Steinberg, Y. Coding and Common Reconstruction. *IEEE Trans. Inf. Theory* **2009**, *55*, 4995–5010. [CrossRef]
35. Land, I.; Huber, J. Information Combining. *Found. Trends Commun. Inf. Theory* **2006**, *3*, 227–330. [CrossRef]
36. Yang, Q.; Piantanida, P.; Gündüz, D. The Multi-layer Information Bottleneck Problem. In Proceedings of the IEEE Information Theory Workshop (ITW), Kaohsiung, Taiwan, 6–10 November 2017; pp. 404–408.
37. Berger, T.; Zhang, Z.; Viswanathan, H. The CEO Problem. *IEEE Trans. Inf. Theory* **1996**, *42*, 887–902. [CrossRef]
38. Steiner, S.; Kuehn, V. Optimization Of Distributed Quantizers Using An Alternating Information Bottleneck Approach. In Proceedings of the WSA 2019: 23rd International ITG Workshop on Smart Antennas, Vienna, Austria, 24–26 April 2019; pp. 1–6.
39. Vera, M.; Rey Vega, L.; Piantanida, P. Collaborative Information Bottleneck. *IEEE Trans. Inf. Theory* **2019**, *65*, 787–815. [CrossRef]
40. Ugur, Y.; Aguerri, I.E.; Zaidi, A. Vector Gaussian CEO Problem Under Logarithmic Loss and Applications. *IEEE Trans. Inf. Theory* **2020**, *66*, 4183–4202. [CrossRef]
41. Estella, I.; Zaidi, A. Distributed Information Bottleneck Method for Discrete and Gaussian Sources. In Proceedings of the International Zurich Seminar on Information and Communication (IZS), Zurich, Switzerland, 21–23 February 2018; pp. 35–39.
42. Courtade, T.A.; Jiao, J. An Extremal Inequality for Long Markov Chains. In Proceedings of the 52nd Annual Allerton Conference on Communication, Control, and Computing, Monticello, IL, USA, 1–3 October 2014; pp. 763–770.
43. Erkip, E.; Cover, T.M. The Efficiency of Investment Information. *IEEE Trans. Inf. Theory* **1998**, *44*, 1026–1040. [CrossRef]
44. Gács, P.; Körner, J. Common information is far less than mutual information. *Probl. Contr. Inform. Theory* **1973**, *2*, 149–162.
45. Farajiparvar, P.; Beirami, A.; Nokleby, M. Information Bottleneck Methods for Distributed Learning. In Proceedings of the 56th Annual Allerton Conference on Communication, Control, and Computing, Monticello, IL, USA, 2–5 October 2018; pp. 24–31.
46. Tishby, N.; Zaslavsky, N. Deep Learning and the Information Bottleneck Principle. In Proceedings of the Information Theory Workshop (ITW), Jeju Island, Korea, 11–15 October 2015; pp. 1–5.
47. Alemi, A.; Fischer, I.; Dillon, J.; Murphy, K. Deep Variational Information Bottleneck. In Proceedings of the International Conference on Learning Representations (ICLR), Toulon, France, 24–26 April 2017.
48. Shwartz-Ziv, R.; Tishby, N. Opening the black box of deep neural networks via information. *arXiv* **2017**, arXiv:1703.00810.
49. Gabrié, M.; Manoel, A.; Luneau, C.; Barbier, j.; Macris, N.; Krzakala, F.; Zdeborová, L. Entropy and mutual information in models of deep neural networks. In *Advances in NIPS*; Curran Associates, Inc.: Red Hook, NY, USA, 2018; Volume 31.
50. Goldfeld, Z.; van den Berg, E.; Greenewald, K.H.; Melnyk, I.; Nguyen, N.; Kingsbury, B.; Polyanskiy, Y. Estimating Information Flow in Neural Networks. *arXiv* **2018**, arXiv:1810.05728.
51. Amjad, R.A.; Geiger, B.C. Learning Representations for Neural Network-Based Classification Using the Information Bottleneck Principle. *IEEE Trans. Pattern Anal.* **2020**, *42*, 2225–2239. [CrossRef]
52. Saxe, A.M.; Bansal, Y.; Dapello, J.; Advani, M.; Kolchinsky, A.; Tracey, B.D.; Cox, D.D. On the information bottleneck theory of deep learning. *J. Stat. Mech. Theory Exp.* **2019**, *2019*, 1–34. [CrossRef]
53. Cheng, H.; Lian, D.; Gao, S.; Geng, Y. Evaluating Capability of Deep Neural Networks for Image Classification via Information Plane. In *Lecture Notes in Computer Science, Proceedings of the Computer Vision-ECCV 2018-15th European Conference, Munich, Germany, 8–14 September 2018, Proceedings, Part XI*; Ferrari, V., Hebert, M., Sminchisescu, C., Weiss, Y., Eds.; Springer: Berlin/Heidelberg, Germany, 2018; Volume 11215, pp. 181–195.
54. Yu, S.; Wickstrøm, K.; Jenssen, R.; Príncipe, J.C. Understanding Convolutional Neural Networks with Information Theory: An Initial Exploration. *IEEE Trans. Neural Netw. Learn. Syst.* **2021**, *32*, 435–442. [CrossRef]
55. Lewandowsky, J.; Stark, M.; Bauch, G. Information bottleneck graphs for receiver design. In Proceedings of the IEEE International Symposium on Information Theory, Barcelona, Spain, 10–15 July 2016; pp. 2888–2892.
56. Stark, M.; Wang, L.; Bauch, G.; Wesel, R.D. Decoding rate-compatible 5G-LDPC codes with coarse quantization using the information bottleneck method. *IEEE Open J. Commun. Soc.* **2020**, *1*, 646–660. [CrossRef]
57. Bhatt, A.; Nazer, B.; Ordentlich, O.; Polyanskiy, Y. Information-distilling quantizers. *IEEE Trans. Inf. Theory* **2021**, *67*, 2472–2487. [CrossRef]
58. Stark, M.; Shah, A.; Bauch, G. Polar code construction using the information bottleneck method. In Proceedings of the 2018 IEEE Wireless Communications and Networking Conference Workshops (WCNCW), Barcelona, Spain, 15–18 April 2018; pp. 7–12.
59. Shah, S.A.A.; Stark, M.; Bauch, G. Design of Quantized Decoders for Polar Codes using the Information Bottleneck Method. In Proceedings of the SCC 2019: 12th International ITG Conference on Systems, Communications and Coding, Rostock, Germany, 11–14 February 2019; pp. 1–6.
60. Shah, S.A.A.; Stark, M.; Bauch, G. Coarsely Quantized Decoding and Construction of Polar Codes Using the Information Bottleneck Method. *Algorithms* **2019**, *12*, 192. [CrossRef]
61. Kurkoski, B.M. On the Relationship Between the KL Means Algorithm and the Information Bottleneck Method. In Proceedings of the 11th International ITG Conference on Systems, Communications and Coding (SCC), Hamburg, Germany, 6–9 February 2017; pp. 1–6.
62. Goldfeld, Z.; Polyanskiy, Y. The Information Bottleneck Problem and its Applications in Machine Learning. *IEEE J. Sel. Areas Inf. Theory* **2020**, *1*, 19–38. [CrossRef]
63. Harremoes, P.; Tishby, N. The Information Bottleneck Revisited or How to Choose a Good Distortion Measure. In Proceedings of the 2007 IEEE International Symposium on Information Theory, Nice, France, 24–29 June 2007; pp. 566–570.

64. Richardson, T.; Urbanke, R. *Modern Coding Theory*; Cambridge University Press: Cambridge, UK, 2008.
65. Sason, I. On f-divergences: Integral representations, local behavior, and inequalities. *Entropy* **2018**, *20*, 383. [CrossRef] [PubMed]
66. Mehler, F.G. Ueber die Entwicklung einer Function von beliebig vielen Variablen nach Laplaceschen Functionen höherer Ordnung. *J. Reine Angew. Math.* **1866**, *66*, 161–176.
67. Lancaster, H.O. The Structure of Bivariate Distributions. *Ann. Math. Statist.* **1958**, *29*, 719–736. [CrossRef]
68. O'Donnell, R. *Analysis of Boolean Functions*, 1st ed.; Cambridge University Press: New York, NY, USA, 2014.
69. Cover, T.M.; Thomas, J.A. *Elements of Information Theory*; Wiley: Hoboken, NJ, USA, 2006.
70. Corless, M.J. *Linear Systems and Control : An Operator Perspective*; Monographs and Textbooks in Pure and Applied Mathematics; Marcel Dekker: New York, NY, USA, 2003; Volume 254.

Article

Counterfactual Supervision-Based Information Bottleneck for Out-of-Distribution Generalization

Bin Deng * and **Kui Jia** *

School of Electronic and Information Engineering, South China University of Technology, Guangzhou 510641, China

* Correspondence: eebindeng@mail.scut.edu.cn (B.D.); kuijia@scut.edu.cn (K.J.)

Abstract: Learning invariant (causal) features for out-of-distribution (OOD) generalization have attracted extensive attention recently, and among the proposals, invariant risk minimization (IRM) is a notable solution. In spite of its theoretical promise for linear regression, the challenges of using IRM in linear classification problems remain. By introducing the information bottleneck (IB) principle into the learning of IRM, the IB-IRM approach has demonstrated its power to solve these challenges. In this paper, we further improve IB-IRM from two aspects. First, we show that the key assumption of support overlap of invariant features used in IB-IRM guarantees OOD generalization, and it is still possible to achieve the optimal solution without this assumption. Second, we illustrate two failure modes where IB-IRM (and IRM) could fail in learning the invariant features, and to address such failures, we propose a *Counterfactual Supervision-based Information Bottleneck (CSIB)* learning algorithm that recovers the invariant features. By requiring counterfactual inference, CSIB works even when accessing data from a single environment. Empirical experiments on several datasets verify our theoretical results.

Keywords: out-of-distribution generalization; information bottleneck; causal learning

Citation: Deng, B.; Jia, K. Counterfactual Supervision-Based Information Bottleneck for Out-of-Distribution Generalization. *Entropy* **2023**, *25*, 193. https://doi.org/10.3390/e25020193

Academic Editors: Sotiris Kotsiantis, Gerhard Bauch and Jan Lewandowsky

Received: 10 October 2022
Revised: 14 December 2022
Accepted: 16 January 2023
Published: 18 January 2023

1. Introduction

Modern machine learning models are prone to catastrophic performance loss during deployment when the test distribution is different from the training distribution. This phenomenon has been repeatedly witnessed and intentionally exposed in many examples [1–5]. Among the explanations, shortcut learning [6] is considered as a main factor causing this phenomenon. A good example is the classification of images of cows and camels—a trained convolutional network tends to recognize cows or camels by learning spurious features from image backgrounds (e.g., green pastures for cows and deserts for camels), rather than learning the causal shape features of the animals [7]; decisions based on the spurious features would make the learned models fail when cows or camels appear in unusual or different environments. Machine learning models are expected to have the capability of out-of-distribution (OOD) generalization and avoid shortcut learning.

To achieve OOD generalization, recent theories [8–12] are motivated by causality literature [13,14] and resort to extraction of the invariant, causal features and establishing the relevant conditions under which machine learning models have guaranteed generalization. Among these works, invariant risk minimization (IRM) [8] is a notable learning paradigm that incorporates the invariance principle [15] into practice. In spite of the theoretical promise of IRM, it is only applicable to problems of linear regression. For other problems, such as linear classification, Ahuja et al. [12] first show that for OOD generalization, linear classification is more difficult (see Theorem 1) and propose a new learning method of information bottleneck-based invariant risk minimization (IB-IRM) based on the support overlap assumption (Assumption 7). In this work, we closely investigate the conditions identified in [12] and propose improved results for OOD generalization of linear classification.

Our technical contributions are as follows. In [12], a notion of support overlap of invariant features is assumed in order to make the OOD generalization of linear classification successful. In this work, we first show that this assumption is strong, but it is still possible to achieve such goal without this assumption. Then, we examine whether the IB-IRM proposed in [12] is sufficient to learn invariant features for linear classification and find that IB-IRM (and IRM) could fail in two modes. We then analyze two failure modes of IB-IRM and IRM, in particular when the spurious features in training environments capture sufficient information for the task of interest but have less information than the invariant features. Based on the above analyses, we propose a new method, termed counterfactual supervision-based information bottleneck (CSIB), to address such failures. We prove that without the need of the support overlap assumption, CSIB is theoretically guaranteed for the success of OOD generalization in linear classification. Notably, CSIB works even when accessing data from a single environment. Finally, we design three synthetic datasets and a colored MINST dataset based on our examples; experiments demonstrate the effectiveness of CSIB empirically.

The rest of this article is organized as follows. The learning problem of out-of-distribution (OOD) generalization is formulated in Section 2. In Section 3, we study the learnability of the OOD generalization with different assumptions to the training and test environments. Using these assumptions, two failure modes of previous methods (IRM and IB-IRM) are analysed in Section 4. Based on the above analysis, our method is then proposed in Section 5. The experiments are reported in Section 6. Finally, we discuss the related works in Section 7 and provide some conclusions and limitations of our work in Section 8. All the proofs and details of experiments are given in the Appendices A and B.

2. OOD Generalization: Background and Formulations

2.1. Background on Structural Equation Models

Before introducing our formulations of OOD generalization, we provide a detailed background on structural equation models (SEMs) [8,13].

Definition 1 (Structural Equation Model (SEM)). *A structural equation model (SEM) $\mathcal{C} := (\mathcal{S}, N)$ governing the random vector $X = (X_1, \ldots, X_d)$ is a set of structural equations:*

$$\mathcal{S}_i : X_i \leftarrow f_i(Pa(X_i), N_i),$$

where $Pa(X_i) \subseteq \{X_1, \ldots, X_d\} \setminus \{X_i\}$ are called the parents of X_i, and N_i are independent noise random variables. For every SEM, we yield a directed acyclic graph (DAG) $\mathcal{G}$ by adding one vertex for each X_i and directed edges from each parent in $Pa(X_i)$ (the causes) to child X_i (the effect).

Definition 2 (Intervention). *Consider an SEM $\mathcal{C} = (\mathcal{S}, N)$. An intervention e on $\mathcal{C}$ consists of replacing one or several of its structural equations to obtain an intervened SEM $\mathcal{C}^e = (\mathcal{S}^e, N^e)$, with structural equations:*

$$\mathcal{S}_i^e : X_i^e \leftarrow f_i^e(Pa^e(X_i^e), N_i^e),$$

The variable X^e is intervened if $\mathcal{S}_i \neq \mathcal{S}_i^e$ or $N_i \neq N_i^e$.

In an SEM $\mathcal{C}$, we can draw samples from the observational distribution $\mathbb{P}(X)$ according to the topological ordering of its DAG $\mathcal{G}$. We can also manipulate (intervene) a unique SEM $\mathcal{C}$ in different ways, indexed by e, to different but related SEMs $\mathcal{C}^e$, which results in different interventional distributions $\mathbb{P}(X^e)$. Such family of interventions are used to model the environments.

2.2. Formulations of OOD Generalization

In this paper, we study the OOD generalization problem by following the linear classification structural equation model below [12].

Assumption 1 (Linear classification SEM $\mathcal{C}_{ood}$).

$$Y \leftarrow \mathbf{1}(w_{inv}^* \cdot Z_{inv}) \oplus N, \quad N \sim Bernoulli(q), \ q < \frac{1}{2};$$
$$X \leftarrow S(Z_{inv}, Z_{spu}),$$

(1)

where $w_{inv}^* \in \mathbb{R}^m$ is the labeling hyperplane, $Z_{inv} \in \mathbb{R}^m$, $Z_{spu} \in \mathbb{R}^o$, $X \in \mathbb{R}^d$, $\oplus$ is the XOR operator, $S \in \mathbb{R}^{d \times (m+o)}$ is invertible ($d = m + o$), $\cdot$ is the dot product function, and $\mathbf{1}(a) = 1$ if $a \geq 0$ otherwise 0.

The SEM $\mathcal{C}_{ood}$ governs four random variables $\{X, Y, Z_{inv}, Z_{spu}\}$, and its directed acyclic graph (DAG) is illustrated in Figure 1a, where the exogenous noise variable N is omitted. Following Definition 2, each intervention e generates a new environment e with interventional distribution $\mathbb{P}(X^e, Y^e, Z_{inv}^e, Z_{spu}^e)$. We assume only the variables of X^e and Y^e are observable. In OOD generalization, we are interested in a set of environments $\mathcal{E}_{all}$ defined as below.

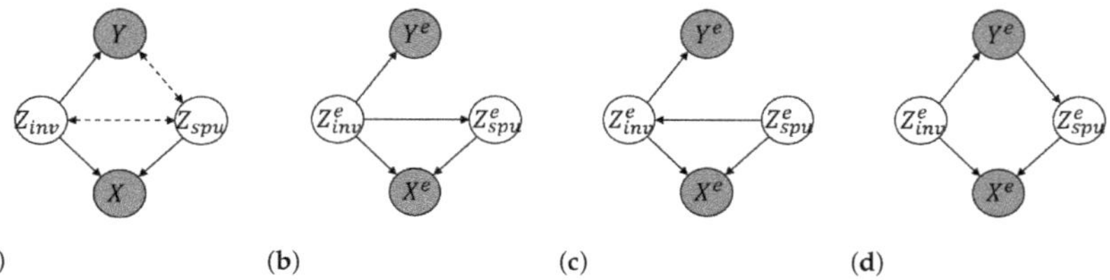

(a) (b) (c) (d)

Figure 1. (a) DAG of the SEM $\mathcal{C}_{ood}$ (Assumption 1); (b–d) DAGs of the interventional SEM $\mathcal{C}_{ood}^e$ in the training environments $\mathcal{E}_{tr}$ with respect to different correlations between Z_{inv} and Z_{spu}. Grey nodes denote observed variables, and white nodes represent unobserved variables. Dashed lines denote the edges which might vary across the interventional environments and even be absent in some scenarios, whilst solid lines indicate that they are invariant across all the environments. All exogenous noise variables are omitted in the DAGs.

Definition 3 ($\mathcal{E}_{all}$). *Consider the SEM $\mathcal{C}_{ood}$ (Assumption 1) and the learning goal of predicting Y from X. Then, the set of all environments $\mathcal{E}_{all}(\mathcal{C}_{ood})$ indexes all the interventional distributions $\mathbb{P}(X^e, Y^e)$ obtainable by valid interventions e. An intervention $e \in \mathcal{E}_{all}(\mathcal{C}_{ood})$ is valid as long as (i) the DAG remains acyclic, (ii) $\mathbb{P}(Y^e | Z_{inv}^e) = \mathbb{P}(Y | Z_{inv})$, and (iii) $\mathbb{P}(X^e | Z_{inv}^e, Z_{spu}^e) = \mathbb{P}(X | Z_{inv}, Z_{spu})$.*

Assumption 1 shows that Z_{inv} is the cause of the response Y. We name Z_{inv} the invariant features or causal features because $\mathbb{P}(Y^e | Z_{inv}^e) = \mathbb{P}(Y | Z_{inv})$ always holds among all valid interventional SEMs $\mathcal{C}_{ood}^e$, as defined in Definition 3. The Z_{spu} is called spurious features because $\mathbb{P}(Y^e | Z_{spu}^e)$ may vary in different environments of $\mathcal{E}_{all}$.

Let $D = \{D^e\}_{e \in \mathcal{E}_{tr}}$ be the training data gathered from a set of training environments $\mathcal{E}_{tr} \subset \mathcal{E}_{all}$, where $D^e = \{(x_i^e, y_i^e)\}_{i=1}^{n_e}$ is the dataset from environment e with each instance (x_i^e, y_i^e) i.i.d. drawn from $\mathbb{P}(X^e, Y^e)$. Let $\mathcal{X}^e \subseteq \mathbb{R}^d$ and $\mathcal{Y} \subseteq \{0, 1\}$ be the support sets of X^e and Y, respectively. Given observed data D, the goal of OOD generalization is to find a predictor $f : \mathbb{R}^d \rightarrow \mathcal{Y}$ such that it can perform well across a set of OOD environments (test environments) $\mathcal{E}_{ood}$ of interest, where $\mathcal{E}_{ood} \subseteq \mathcal{E}_{all}$. Formally, it is expected to minimize

$$\max_{e \in \mathcal{E}_{ood}} R^e(f),$$

(2)

where $R^e(f) := \mathbb{E}_{X^e, Y^e}[l(f(X^e), Y^e)]$ is the risk under the environment e with $l(\cdot, \cdot)$ the 0-1 loss function. Since $\mathcal{E}_{ood}$ may be different from $\mathcal{E}_{tr}$, this learning problem is called OOD generalization. We assume the predictor $f = w \circ \Phi$ includes a feature extractor $\Phi : \mathcal{X} \rightarrow \mathcal{H}$ and a classifier $w : \mathcal{H} \rightarrow \mathcal{Y}$. With a slight abuse of notation, we also let the classifier w

and feature extractor Φ be parameteried by themselves, respectively, as $w \in \mathbb{R}^{c+1}$ and $\Phi \in \mathbb{R}^{c \times d}$ with c the number of feature dimension.

2.3. Background on IRM and IB-IRM

To minimize Equation (2), two notable solutions of IRM [8] and IB-IRM [12] are listed as follows:

$$\text{IRM:} \quad \min_{w,\Phi} \frac{1}{|\mathcal{E}_{tr}|} \sum_{e \in \mathcal{E}_{tr}} R^e(w \circ \Phi), \text{ s.t. } w \in \arg\min_{\tilde{w}} R^e(\tilde{w} \circ \Phi), \forall e \in \mathcal{E}_{tr}, \tag{3}$$

$$\text{IB-IRM:} \quad \min_{w,\Phi} \sum_{e \in \mathcal{E}_{tr}} h^e(\Phi), \text{ s.t. } \frac{1}{|\mathcal{E}_{tr}|} \sum_{e \in \mathcal{E}_{tr}} R^e(w \circ \Phi) \leq rth, w \in \arg\min_{\tilde{w}} R^e(\tilde{w} \circ \Phi), \forall e \in \mathcal{E}_{tr}, \tag{4}$$

where $R^e(w \circ \Phi) = \mathbb{E}_{X^e, Y^e}[l(w \circ \Phi(X^e), Y^e)]$, and $h^e(\Phi) = H(\Phi(X^e))$ with H the Shannon entropy (or a lower bounded differential entropy), and rth is the threshold on the average risk. If we drop the invariance constraint from IRM and IB-IRM, we obtain standard empirical risk minimization (ERM) and information bottleneck-based empirical risk minimization (IB-ERM), respectively. The use of an entropy constraint in IB-IRM is inspired from the information bottleneck principle [16] where mutual information $I(X; \Phi(X))$ is used for information compression. Since the representation $\Phi(X)$ is a deterministic mapping of X, we have

$$I(X; \Phi(X)) = H(\Phi(X)) - H(\Phi(X)|X) = H(\Phi(X)), \tag{5}$$

thus minimizing the entropy of $\Phi(X)$ is equivalent to minimizing the mutual information $I(X; \Phi(X))$. In brief, the optimization goal of IB-IRM is to select the one that has the least entropy among all highly predictive invariant predictors.

3. OOD Generalization: Assumptions and Learnability

To study the learnability of OOD generalization, we make following definition.

Definition 4. *Given $\mathcal{E}_{tr} \subset \mathcal{E}_{all}$ and $\mathcal{E}_{ood} \subseteq \mathcal{E}_{all}$. We say an algorithm succeeds to solve OOD generalization with respect to $(\mathcal{E}_{tr}, \mathcal{E}_{ood})$ if the predictor $f^* \in \mathcal{F}$ returned by this algorithm satisfies the following equation:*

$$\max_{e \in \mathcal{E}_{ood}} R^e(f^*) = \min_{f \in \mathcal{F}} \max_{e \in \mathcal{E}_{ood}} R^e(f), \tag{6}$$

where $\mathcal{F}$ is the learning hypothesis (a function set including all possible linear classifier). Otherwise we say it fails to solve OOD generalization.

So far, we have omitted how different environments of $\mathcal{E}_{tr}$ and $\mathcal{E}_{ood}$ exactly are to enable OOD generalization. Different assumptions about $\mathcal{E}_{tr}$ and $\mathcal{E}_{ood}$ make the OOD generalization problem different.

3.1. Assumptions about the Training Environments $\mathcal{E}_{tr}$

Define the support set of the invariant (*resp.*, spurious) features Z_{inv}^e (*resp.*, Z_{spu}^e) in environment e as $\mathcal{Z}_{inv}^e$ (*resp.*, $\mathcal{Z}_{spu}^e$). In general, we make following assumptions to the invariant features $\mathcal{Z}_{inv}^e$ in the training environments $\mathcal{E}_{tr}$.

Assumption 2 (Bounded invariant features). *$\cup_{e \in \mathcal{E}_{tr}} \mathcal{Z}_{inv}^e$ is a bounded set. (A set $\mathcal{Z}$ is bounded if $\exists M < \infty$ such that $\forall z \in \mathcal{Z}, \|z\| \leq M$).*

Assumption 3 (Strictly separable invariant features). *$\forall z \in \cup_{e \in \mathcal{E}_{tr}} \mathcal{Z}_{inv}^e, w_{inv}^* \cdot z \neq 0$.*

The difficulties of OOD generalization are due to the spurious correlations between Z_{inv} and Z_{spu} in the training environments $\mathcal{E}_{tr}$. In this paper, we consider three modes induced by different correlations between Z_{inv} and Z_{spu} as shown below.

Assumption 4 (Spurious correlation 1). *Assume each $e \in \mathcal{E}_{tr}$,*

$$Z_{spu}^e \leftarrow AZ_{inv}^e + W^e; \tag{7}$$

where $A \in \mathbb{R}^{o \times m}$, and $W^e \in \mathbb{R}^o$ is a continuous (or discrete with each component supported on at least two distinct values), bounded, and zero mean noise variable.

Assumption 5 (Spurious correlation 2). *Assume each $e \in \mathcal{E}_{tr}$,*

$$Z_{inv}^e \leftarrow AZ_{spu}^e + W^e; \tag{8}$$

where $A \in \mathbb{R}^{m \times o}$, and $W^e \in \mathbb{R}^m$ is a continuous (or discrete with each component supported on at least two distinct values), bounded, and zero mean noise variable.

Assumption 6 (Spurious correlation 3). *Assume each $e \in \mathcal{E}_{tr}$,*

$$Z_{spu}^e \leftarrow W_1^e Y^e + W_0^e (1 - Y^e); \tag{9}$$

where $W_0^e \in \mathbb{R}^o$ and $W_1^e \in \mathbb{R}^o$ are independent noise variables.

For each $e \in \mathcal{E}_{tr}$, the DAGs of its corresponding interventional SEMs $\mathcal{C}_{ood}^e$ with respect to Assumptions 4–6 are illustrated in Figure 1b–d, respectively. It is worth noting that although the DAGs are identical across all training environments in each mode of Assumptions 4–6, the interventional SEMs $\mathcal{C}_{ood}^e$ among different training environments are different due to the interventions on the exogenous noise variables.

3.2. Assumptions about the OOD Environments $\mathcal{E}_{ood}$

Theorem 1 (Impossibility of guaranteed OOD generalization for linear classification [12]). *Suppose $\mathcal{E}_{ood} = \mathcal{E}_{all}$. If for all the training environments $\mathcal{E}_{tr}$, the latent invariant features are bounded and strictly separable, i.e., Assumptions 2 and 3 hold, then every deterministic algorithm fails to solve the OOD generalization.*

The above theorem shows that it is impossible to solve OOD generalization if $\mathcal{E}_{ood} = \mathcal{E}_{all}$. To make it learnable, Ahuja et al. [12] propose the support overlap assumption (Assumption 7) to the invariant features.

Assumption 7 (Invariant feature support overlap). $\forall e \in \mathcal{E}_{ood}, \mathcal{Z}_{inv}^e \subseteq \cup_{e' \in \mathcal{E}_{tr}} \mathcal{Z}_{inv}^{e'}.$

However, Assumption 7 is strong, and we would show that it is still possible to solve OOD generalization without this assumption. For better illustration, consider an OOD generalization task from $\mathbb{P}(X^{e_1}, Y^{e_2})$ to $\mathbb{P}(X^{e_2}, Y^{e_2})$ with $\mathcal{E}_{tr} = \{e_1\}$ and $\mathcal{E}_{ood} = \{e_2\}$, and the support sets of the corresponding invariant features $Z_{inv}^{e_1}$ and $Z_{inv}^{e_2}$ are intuitively illustrated in Figure 2c (assume $dim(Z_{inv}) = 2$ in this example). From Figure 2c, it is clear that although the support sets of invariant features between the two environments are different, it is still possible to solve OOD generalization if the learned feature extractor Φ only captures the invariant features, e.g., $\Phi(X) = Z_{inv}$.

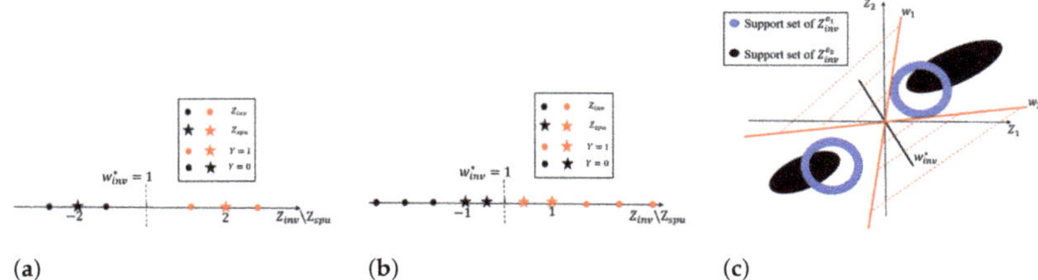

Figure 2. (a) Example 1; (b) example 2; (c) example illustration. Here, $dim(Z_{inv}) = 2$ and $Z_{inv} = (Z_1, Z_2)$. The blue and black regions represent the support sets of $Z^{e_1}_{inv}$ and $Z^{e_2}_{inv}$, corresponding to the environments e_1 and e_2, respectively. $\mathcal{E}_{tr} = \{e_1\}$ is the training environment and $\mathcal{E}_{ood} = \{e_2\}$ is the OOD environment. Although Assumption 7 does not hold in this example, any zero-error classifier with $\Phi(X) = Z_{inv}$ on the e_1 environment data would clearly make the classification error zero in e_2, thus succeeding to solve OOD generalization.

To make Assumption 7 weaker, we propose the following assumption.

Assumption 8. *Let* $\mathbb{P}(Z^{tr}_{inv}, Y^{tr}) = \frac{1}{|\mathcal{E}_{tr}|} \sum_{e \in \mathcal{E}_{tr}} \mathbb{P}(Z^e_{inv}, Y^e)$ *be the mixture distribution of invariant features in the training environments. Denote $\mathcal{A}$ be a hypothesis set including all linear classifiers mapping from $\mathbb{R}^m$ to $\mathcal{Y}$. $\forall e \in \mathcal{E}_{ood}$, assume $F_l(\mathbb{P}(Z^{tr}_{inv}, Y^{tr})) \subseteq F_l(\mathbb{P}(Z^e_{inv}, Y^e))$, where l is the 0-1 loss function and $F_l(\mathbb{P}(Z, Y)) = \arg\min_{f \in \mathcal{A}} \mathbb{E}_{Z,Y}[l(f(Z), Y)]$.*

Clearly, under the assumption of separable invariant features (Assumption 3), for any $e \in \mathcal{E}_{ood}$, Assumption 7 holds $\Rightarrow Z^e_{inv} \subseteq Z^{tr}_{inv} \Rightarrow F_l(\mathbb{P}(Z^{tr}_{inv}, Y^{tr})) \subseteq F_l(\mathbb{P}(Z^e_{inv}, Y^e)) \Rightarrow$. Assumption 8 holds, but not vice versa. Therefore, Assumption 8 is weaker than Assumption 7. We show that Assumption 8 could be substituted for Assumption 7 for the success of OOD generalization in our proposed method in Section 5.

4. Failures of IRM and IB-IRM

Under Spurious Correlation 1 (Assumption 4), the IB-IRM algorithm has been shown to enable OOD generalization, while IRM fails [12]. In this section, we would show that both IRM and IB-IRM could fail under Spurious Correlations 2 and 3 (Assumptions 5 and 6).

4.1. Failure under Spurious Correlation 2

Example 1 (Counter-Example 1). *Under Assumption 5, let $Z^e_{inv} \leftarrow Z^e_{spu} + W^e$ with $dim(Z^e_{inv}) = dim(Z^e_{spu}) = dim(W^e) = 1$ and $w^*_{inv} = 1$ be the generated classifier in Assumption 1. We assume two training environments and a OOD environment as:*

$$\mathcal{E}_{tr} = \{e_1, e_2\}; \quad \mathcal{E}_{ood} = \{e_3\};$$
$$e_1 : \mathbb{P}(Z^{e_1}_{spu} = -2) = 1, \mathbb{P}(W^{e_1} = -1) = 0.5, \mathbb{P}(W^{e_1} = 1) = 0.5;$$
$$e_2 : \mathbb{P}(Z^{e_2}_{spu} = 2) = 1, \mathbb{P}(W^{e_2} = -1) = 0.5, \mathbb{P}(W^{e_2} = 1) = 0.5;$$
$$e_3 : \mathbb{P}(Z^{e_3}_{spu} = 1) = 1, \mathbb{P}(W^{e_3} = -2) = 0.5, \mathbb{P}(W^{e_3} = 2) = 0.5.$$

Figure 2a shows the support points of these features in the training environments. Then, by applying any algorithm to solve the above example with $rth = q$, we would obtain a predictor of $f^* = w^* \circ \Phi^*$. Consider the prediction made by this model as (we ignore the classifier bias for convenience)

$$f^*(X^e) = f^*(S(Z^e_{inv}, Z^e_{spu})) = \mathbf{1}(\Phi^*_{inv} Z^e_{inv} + \Phi^*_{spu} Z^e_{spu}). \tag{10}$$

It is trivial to show that the f^* of $\Phi^*_{inv} = 0$ and $\Phi^*_{spu} = 1$ is an invariant predictor across training environments with classification error $R^{e_1} = R^{e_2} = q$, and it achieves the least entropy of $h^e(\Phi^*) = 0$ for each training environment e. Therefore, it is a solution of IB-IRM and IRM. However, the predictor of f^* relies on spurious features and has the test error $R^{e_3} = 0.5$; thus, it fails to solve the OOD generalization.

4.2. Failure under Spurious Correlation 3

Example 2 (Counter-Example 2). *Under Assumption 6, let $Z^e_{spu} \leftarrow W^e_1 Y^e + W^e_0 (1 - Y^e)$ with $dim(Z_{inv}) = dim(Z_{spu}) = dim(W^e_0) = dim(W^e_1) = 1$, Z^e_{inv} be a discrete variable supported uniformly on six points $\{-4, -3, -2, 2, 3, 4\}$ among all environments, and $w^*_{inv} = 1$ be the generated classifier in Assumption 1. We assume two training environments and a OOD environment as:*

$$\mathcal{E}_{tr} = \{e_1, e_2\}; \quad \mathcal{E}_{ood} = \{e_3\}$$
$$e_1 : \mathbb{P}(W^{e_1}_0 = -1) = 1, \mathbb{P}(W^{e_1}_1 = 1) = 1;$$
$$e_2 : \mathbb{P}(W^{e_2}_0 = -0.5) = 1, \mathbb{P}(W^{e_2}_1 = 0.5) = 1;$$
$$e_3 : \mathbb{P}(W^{e_3}_0 = 1) = 1, \mathbb{P}(W^{e_3}_1 = -1) = 1;$$

Figure 2b shows the support points of these features in the training environments. Then, by applying any algorithm to solve the above example with $rth = q$, we would obtain a predictor of $f^* = w^* \circ \Phi^*$. Consider the prediction made by this model as (we ignore the classifier bias for convenience):

$$f^*(X^e) = f^*(S(Z^e_{inv}, Z^e_{spu})) = \mathbf{1}(\Phi^*_{inv} Z^e_{inv} + \Phi^*_{spu} Z^e_{spu}). \tag{11}$$

It is trivial to show that the f^* of $\Phi^*_{inv} = 0$ and $\Phi^*_{spu} = 1$ is an invariant predictor across training environments with classification error $R^{e_1} = R^{e_2} = 0$, and it achieves the least entropy of $h^e(\Phi^*) = 1$ among all highly predictive predictors for each training environment e. and Therefore, it is a solution of IB-IRM and IRM. However, the predictor of f^* relies on spurious features and has the test error $R^{e_3} = 1$; thus, it fails to solve the OOD generalization.

4.3. Understanding the Failures

From the illustrations of the above simple examples, we can conclude that the failure of the invariance constraint for removing the spurious features is because the spurious features among all training environments are strictly linearly separable by their corresponding labels. This would make the predictor rely only on spurious features to achieve minimum training error and also be the invariant predictor across training environments. Since the label set is finite (with only two values in binary classification) in classification problems, such a phenomenon may exist. We state such failure mode formally as below.

Theorem 2. *Given any $\mathcal{E}_{tr} \subset \mathcal{E}_{all}$ and $\mathcal{E}_{ood} \subseteq \mathcal{E}_{all}$ satisfying Assumptions 2, 3, and 7, if two sets $\cup_{e \in \mathcal{E}_{tr}} Z^e_{spu}(Y^e = 1)$ and $\cup_{e \in \mathcal{E}_{tr}} Z^e_{spu}(Y^e = 0)$ are linearly separable and $H(Z^e_{inv}) > H(Z^e_{spu})$ on each training environment e, then IB-IRM (and IRM, ERM, or IB-ERM) with any $rth \in \mathbb{R}$ fails to solve the OOD generalization.*

The understanding of Theorem 2 is intuitive since when the spurious features in the training environments with respect to different labels are linearly separable, there is no algorithm that can distinguish spurious features from invariant features. Although the assumption of linear separation of the spurious features seems strong for this failure, it is easy to hold in high-dimensional space when $dim(Z_{spu})$ is large (common cases in practice such as image data). We show one case in Appendix A.3 that if the number of environments is $|\mathcal{E}_{tr}| < dim(Z_{spu})/2$ under Assumption 6, the spurious features in the training environments are probably separable by their labels. This is because in o-dimensional space there is a high probability that o randomly drawn distinct points are linearly separable for any two subsets.

5. Counterfactual Supervision-Based Information Bottleneck

In the above analyses, we have shown two failure modes of IB-IRM and IRM for OOD generalization in the linear classification problem. The key reason for the failure is due to the learned features $\Phi(X)$ that rely on spurious features. To prevent such failure, we present the counterfactual supervision-based information bottleneck (CSIB) learning algorithm for removing the spurious features progressively.

In general, the IB-ERM method is applied to extract features from the beginning of each iteration:

$$\min_{w,\Phi} \sum_{e \in \mathcal{E}_{tr}} h^e(\Phi) \quad \text{s.t.} \quad \frac{1}{|\mathcal{E}_{tr}|} \sum_{e \in \mathcal{E}_{tr}} R^e(w \circ \Phi) \leq rth \tag{12}$$

Due to the information bottleneck, only a part of the information of the input X are exploited in $\Phi(X)$. If the information of spurious features Z_{spu} exists in the learned features $\Phi(X)$, the idea of CSIB is to drop such information and meanwhile maintain the causal information (represented by invariant features Z_{inv}) as well. However, achieving such a goal faces two challenges: (1) how to determine whether $\Phi(X)$ contains spurious information of Z_{spu}? and (2) how to remove the information of Z_{spu}?

Fortunately, due to the orthogonality in the linear space, it is possible to disentangle the features that are exploited by $\Phi(X)$ (denoted as X_1) and the features that are not exploited by $\Phi(X)$ (denoted as X_2) via Singular Value Decomposition (SVD). Based on that, we could construct an SEM $\mathcal{C}_{new}$ governing three variables of X_1, X_2, and X. Therefore, by conducting counterfactual interventions on X_1 and X_2 in $\mathcal{C}_{new}$, we could solve the first challenge by requiring a single supervision on the counterfactual examples X'. For example, if we intervene on X_1 and find that the causal information remains in the resulting X', then the extracted features $\Phi(X)$ are definitely the spurious features. To address the second challenge, we replace the input by X_2 by filtering out the information of X_1 and conduct the same learning procedure from the beginning.

The learning algorithm of CSIB is illustrated in Algorithm 1, and Figure 3 shows the framework of CSIB. We show in Theorem 3 that CSIB is theoretically guaranteed to succeed to solve OOD generalization.

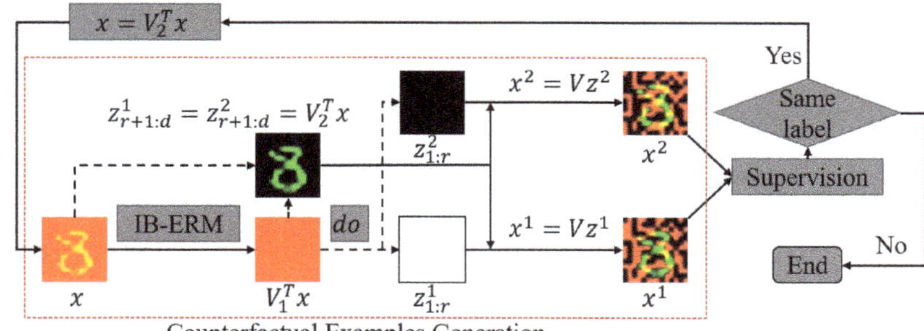

Figure 3. A simplified framework for the illustration of the proposed CSIB method.

Theorem 3 (Guarantee of CSIB). *Given any $\mathcal{E}_{tr} \subset \mathcal{E}_{all}$ and $\mathcal{E}_{ood} \subseteq \mathcal{E}_{all}$ satisfying Assumptions 2, 3, and 8, then for every spurious correlation of Assumptions 4, 5, and 6 (in this correlation mode, assume the spurious features are linearly separable in the training environments), the CSIB algorithm with $rth = q$ succeeds in solving the OOD generalization.*

Algorithm 1 Counterfactual Supervision-based Information Bottleneck (CSIB)

Input: $\mathbb{P}(X^e, Y^e)$, $e \in \mathcal{E}_{tr}$, $rth > 0$, $c \geq dim(Z_{inv})$, $M \gg 0$, and (x, y) is an example randomly drawn from $\mathbb{P}(X^e, Y^e)$.
Output: classifier $w \in \mathbb{R}^{c+1}$, feature extractor $\Phi = \mathbb{R}^{c \times d}$.
Begin:

  1: $Lv \leftarrow []; Lr \leftarrow []; \Phi' \leftarrow \mathbb{I}^{d \times d}$
  2: $d' \leftarrow dim(X^e)$
  3: Apply IB-ERM method (Equation (12)) to $\mathbb{P}(X^e, Y^e)$ and obtain $w^* \in \mathbb{R}^{c+1}$ and $\Phi^* \in \mathbb{R}^{c \times d'}$
  4: Apply SVD to Φ^* as $\Phi^* = U \Lambda V^T = [U_1, U_2][\Lambda_1, \mathbf{0}; \mathbf{0}, \mathbf{0}][V_1^T; V_2^T]$
  5: $r \leftarrow rank(\Phi^*)$
  6: $z_{1:r}^1 \leftarrow [-M, ..., -M]; z_{r+1:d'}^1 \leftarrow V_2^T \Phi' x$
  7: $z_{1:r}^2 \leftarrow [M, ..., M]; z_{r+1:d'}^2 \leftarrow V_2^T \Phi' x$
  8: $x^1 \leftarrow V z^1; x^2 \leftarrow V z^2$
  9: **if** Lv is not empty **then**
10:     $z_{old} \leftarrow []; i \leftarrow 0; x' \leftarrow x$
11:     **while** $i < len(Lv)$ **do**
12:         $z \leftarrow Lv[i]x'$
13:         $z_{old}.\text{append}(z)$
14:         $x' \leftarrow z_{Lr[i]:}$
15:         $i \leftarrow i + 1$
16:     **end while**
17:     $i \leftarrow 0$
18:     **while** $i < len(Lv)$ **do**
19:         $j \leftarrow len(Lv) - i$
20:         $z^1 \leftarrow z_{old}[j]; z^2 \leftarrow z_{old}[j]$
21:         $z_{Lr[j]:}^1 \leftarrow x^1; z_{Lr[j]:}^2 \leftarrow x^2$
22:         $x^1 \leftarrow Lv[j]^T z^1; x^2 \leftarrow Lv[j]^T z^2$
23:         $i \leftarrow i + 1$
24:     **end while**
25: **end if**
26: **if** $\text{label}(x^1) = \text{label}(x^2)$ **then**
27:     $Lr.\text{append}(r); Lv.\text{append}(V^T)$
28:     $X^e \leftarrow V_2^T X^e; \Phi' \leftarrow V_2^T \Phi'$
29:     Goto Step 2
30: **end if**
31: $w \leftarrow w^*; \Phi \leftarrow \Phi^*$

End

Remark 1. *CSIB succeeds to solve OOD generalization without assuming the support overlap to invariant features and could apply to multiple spurious modes where IB-IRM (as well as ERM, IRM, and IB-ERM) may fail. By introducing counterfactual inference and further supervision (usually conducted by a human) with several steps, CSIB works even when accessing data from a single environment, which is significant especially in the cases where multiple environments' data are not available.*

6. Experiments

6.1. Toy Experiments on Synthetic Datasets

We perform experiments on three synthetic datasets from different spurious correlations modes to verify our method—counterfactual, supervision-based, and information bottleneck (CSIB)—and compare them to ERM, IB-ERM, IRM, and IB-IRM. We follow the same protocol for tuning hyperparameters from [8,12,17] and report the classification error for all experiments. In the following, we first briefly describe the designed datasets and then report the main results. More experimental details can be found in the Appendix.

6.1.1. Datasets

Example 1/1S. The example is a modified one from the linear unit tests introduced in [17], which generalizes the cow/camel classification task with relevant backgrounds.

$$\theta_{cow} = \mathbf{1}_m, \quad \theta_{camel} = -\theta_{cow}, \quad v_{animal} = 10^{-2}$$
$$\theta_{grass} = \mathbf{1}_o, \quad \theta_{sand} = -\theta_{grass}, \quad v_{background} = 1.$$

The dataset D_e of each environment $e \in \mathcal{E}_{tr}$ is sampled from the following distribution

$$U^e \sim \text{Categorical}(p^e s^e, (1 - p^e)s^e, p^e(1 - s^e), (1 - p^e)(1 - s^e)),$$

$$Z^e_{inv} \sim \begin{cases} (\mathcal{N}_m(0, 0.1) + \theta_{cow})v_{animal} & \text{if } U^e \in \{1, 2\}, \\ (\mathcal{N}_m(0, 0.1) + \theta_{camel})v_{animal} & \text{if } U^e \in \{3, 4\}, \end{cases}$$

$$Z^e_{spu} \sim \begin{cases} (\mathcal{N}_o(0, 0.1) + \theta_{grass})v_{background} & \text{if } U^e \in \{1, 4\}, \\ (\mathcal{N}_o(0, 0.1) + \theta_{sand})v_{background} & \text{if } U^e \in \{2, 3\}, \end{cases}$$

$$Z^e \leftarrow (Z^e_{inv}, Z^e_{spu}), \quad X^e \leftarrow S(Z^e), \quad N \sim \text{Bernoulli}(q), \quad q < 0.5, \quad Y^e \leftarrow \mathbf{1}(\mathbf{1}_m^T Z^e_{inv}) \oplus N$$

We set $s^{e_0} = 0.5, s^{e_1} = 0.7$, and $s^{e_2} = 0.3$ for the first three environments, and $s^{e_j} \sim \text{Uniform}$ $(0.3, 0.7)$ for $j > 3$. The scrambling matrix S is an identical matrix in Example 1 and a random unitary matrix in Example 1S. Here, we set $p^e = 1$ and $q = 0$ for all environments to make the spurious features and the invariant features both linearly separable to confuse each other. The experiments on different values of q and p^e are presented in the Appendix, where we have found very interesting observations related to the inductive bias of neural networks.

Example 2/2S. This example is extended from Example 1 to show one of the failure modes of IB-IRM (as well as ERM, IRM, and IB-ERM) and how our method can be improved by intervention (counterfactual supervision). Given $w^e \in \mathbb{R}$, each instance in the environment data D^e is sampled by

$$\theta_{spu} = 5 \cdot \mathbf{1}_o, \quad \theta_w = w^e \cdot \mathbf{1}_m, \quad v_{spu} = 10^{-2}, \quad v_w = 1, \quad p, q \sim \text{Bernoulli}(0.5),$$
$$Z^e_{spu} = \mathcal{N}_o(0, 1)v_{spu} + (2p - 1) \cdot \theta_{spu}, \quad W^e = \mathcal{N}_m(0, 1)v_w + (2q - 1) \cdot \theta_w$$
$$Z^e_{inv} = AZ^e_{spu} + W^e, \quad Z^e \leftarrow (Z^e_{inv}, Z^e_{spu}), \quad X^e \leftarrow S(Z^e), \quad Y^e = \mathbf{1}(\mathbf{1}_m^T Z^e_{inv}),$$

where we set $m = o = 5$, and $A \in \mathbb{R}^{m \times o}$ is the identical matrix in our experiments. We set $w^{e_0} = 3$, $w^{e_1} = 2$, $w^{e_2} = 1$, and $w^{e_j} = \text{Uniform}(0, 3)$ if $j > 3$ for different training environments. This example shows clearer smaller entropy of spurious features than that of invariant features, which is opposite Example 1/1S.

Example 3/3S. This example extends from Example 2 and is similar to the construction of Example 2/2S. Let $w^e \sim \text{Uniform}(0, 1)$ for different training environments. Each instance in the environments e is sampled by

$$\theta_{inv} = \cdot 10 \cdot \mathbf{1}_m, \quad v_{inv} = 10, \quad v_{spu} = 1, \quad p, q \sim \text{Bernoulli}(0.5),$$
$$Z^e_{inv} = \mathcal{N}_m(0, 1)v_{inv} + (2p - 1) \cdot \theta_{inv}, \quad Y^e = \mathbf{1}(\mathbf{1}_m^T Z^e_{inv}),$$
$$Z^e_{spu} = 2(Y^e - 1) \cdot v_{spu} + (2q - 1) \cdot w^e \cdot \mathbf{1}_o, \quad Z^e \leftarrow (Z^e_{inv}, Z^e_{spu}), \quad X^e \leftarrow S(Z^e),$$

where we set $m = o = 5$ in our experiments. The spurious features have smaller entropy than the invariant features in this example, which is similar to Example 2/2S, but the invariant features significantly enjoy much larger margin than the spurious features, which is very different from the above two examples. We show a summary of the properties of these three datasets in Table 1 for a general view.

Table 1. Summary of three synthetic datasets. Note that for linearly separable features, their margin levels significantly influence the final learning classifier due to the implicit bias of the gradient descent [18]. Such bias would push the standard learning (such as cross-entropy loss) to focus more on the large-margin features. The margin with respect to a dataset (or features) $\mathcal{Z}$ (each instance has a label 0 or 1) is the minimum distance between a point in $\mathcal{Z}$ and the max-margin hyperplane, which separates $\mathcal{Z}$ by its labels.

Datasets	Margin Relationship	Entropy Relationship	Dim_{inv}	Dim_{spu}
Example 1/1S	$\mathrm{Margin}_{inv} \ll \mathrm{Margin}_{spu}$	$\mathrm{Entropy}_{inv} < \mathrm{Entropy}_{spu}$	5	5
Example 2/2S	$\mathrm{Margin}_{inv} \approx \mathrm{Margin}_{spu}$	$\mathrm{Entropy}_{inv} > \mathrm{Entropy}_{spu}$	5	5
Example 3/3S	$\mathrm{Margin}_{inv} \gg \mathrm{Margin}_{spu}$	$\mathrm{Entropy}_{inv} > \mathrm{Entropy}_{spu}$	5	5

6.1.2. Summary of Results

Table 2 shows the classification errors of different methods when training data comes from single, three, and six environments. We can see that ERM and IRM fail to recognize the invariant features in the experiment of Example 1/1S, where invariant features have smaller margin than spurious features do, while information bottleneck-based methods (IB-ERM, IB-IRM, and CSIB) show improved results due to the smaller entropy of the invariant features. Our method CSIB shows results consistent with IB-IRM in Example 1/1S when invariant features are extracted in the first run, which verifies the effectiveness of using the information bottleneck for OOD generalization. In another dataset of Example 2/2S, where the invariant features have larger entropy than spurious features do, we can see that only CSIB can remove the spurious features compared with the other method, although the information bottleneck-based method IB-ERM would degrade the performance of ERM by focusing more on the spurious features. In the third experiment of Example 3/3S, we can see that although ERM shows not-bad results due to the significantly larger margin of invariant features, our CSIB method still shows improvements by removing more spurious features. Notably, comparing the IB-ERM and IB-IRM when only spurious features are extracted (Example 2/2S, Example 3/3S), our CSIB method could effectively remove them by counterfactual supervision and then refocus on the invariant features. Note that the reason of non-zero average error and the fluctuant results of CSIB in some experiments is that the entropy minimization in the training process is less accurate, where entropy is substituted by variance for the ease of the optimization. Nevertheless, there always exists a case where the entropy is indeed truly minimized, and the error reaches zero (see (min) in the table) in Example 2/2S and Example 3/3S. In summary, CSIB consistently performs better in different spurious correlations modes and is especially more effective than IB-ERM and IB-IRM when the spurious features enjoy much smaller entropy than the invariant features do.

6.2. Experiments on Color MNIST Dataset

In this experiment, we set up a binary classification task for digit recognition and identify whether the digit is less than five or more than five. We use real-world dataset, the MNIST database of handwritten digits (http://yann.lecun.com/exdb/mnist/), for the construction. Following our learning setting, we use color information as the spurious features that correlates strongly with the class label. By construction, the label is more strongly correlated with the color than with the digit in the training environments, but this correlation is broken in the test environment. Specifically, the three designed environments (two training environments and one test environment containing 10,000 points each) of the color MNIST are as follows: first, we define a preliminary binary label $\hat{y}$ to the image base on the digit: $\hat{y} = 0$ for digits 0–4 and $\hat{y} = 1$ for 5–9. Second, we obtain the final label y by flipping $\hat{y}$ with probability 0.25. Then, we flip the final labels to obtain the color id, where the flipping probabilities with respect to two training environments and one test environment are 0.2 and 0.1, and 0.9. For better understanding, we randomly draw 20 examples for each label from each environment and visualize them in Figure 4.

Table 2. Main results: #Envs means the number of training environments, and (min) reports the minimal test classification error across different running seeds.

	#Envs	ERM (min)	IRM (min)	IB-ERM (min)	IB-IRM (min)	CSIB (min)
Example 1	1	0.50 ± 0.01 (0.49)	0.50 ± 0.01 (0.49)	$\mathbf{0.23 \pm 0.02}$ **(0.22)**	0.31 ± 0.10 (0.25)	$\mathbf{0.23 \pm 0.02}$ **(0.22)**
Example 1S	1	0.50 ± 0.00 (0.49)	0.50 ± 0.00 (0.50)	0.46 ± 0.04 (0.39)	$\mathbf{0.30 \pm 0.10}$ **(0.25)**	0.46 ± 0.04 (0.39)
Example 2	1	0.40 ± 0.20 (0.00)	0.50 ± 0.00 (0.49)	0.50 ± 0.00 (0.49)	0.46 ± 0.02 (0.45)	$\mathbf{0.00 \pm 0.00}$ **(0.00)**
Example 2S	1	0.50 ± 0.00 (0.50)	0.31 ± 0.23 (0.00)	0.50 ± 0.00 (0.50)	0.45 ± 0.01 (0.43)	$\mathbf{0.10 \pm 0.20}$ **(0.00)**
Example 3	1	0.16 ± 0.06 (0.09)	0.18 ± 0.03 (0.14)	0.50 ± 0.01 (0.49)	0.40 ± 0.20 (0.01)	$\mathbf{0.11 \pm 0.20}$ **(0.00)**
Example 3S	1	0.17 ± 0.07 (0.10)	$\mathbf{0.09 \pm 0.02}$ **(0.07)**	0.50 ± 0.00 (0.50)	0.50 ± 0.00 (0.50)	0.21 ± 0.24 (0.00)
Example 1	3	0.45 ± 0.01 (0.45)	0.45 ± 0.01 (0.45)	$\mathbf{0.22 \pm 0.01}$ **(0.21)**	0.23 ± 0.13 (0.02)	$\mathbf{0.22 \pm 0.01}$ **(0.21)**
Example 1S	3	0.45 ± 0.00 (0.45)	0.45 ± 0.00 (0.45)	0.41 ± 0.04 (0.34)	$\mathbf{0.27 \pm 0.11}$ **(0.11)**	0.41 ± 0.04 (0.34)
Example 2	3	0.40 ± 0.20 (0.00)	0.50 ± 0.00 (0.50)	0.50 ± 0.00 (0.50)	0.33 ± 0.04 (0.25)	$\mathbf{0.00 \pm 0.00}$ **(0.00)**
Example 2S	3	0.50 ± 0.00 (0.50)	0.37 ± 0.15 (0.15)	0.50 ± 0.00 (0.50)	0.34 ± 0.01 (0.33)	$\mathbf{0.10 \pm 0.20}$ **(0.00)**
Example 3	3	0.18 ± 0.04 (0.15)	0.21 ± 0.02 (0.20)	0.50 ± 0.01 (0.49)	0.50 ± 0.01 (0.49)	$\mathbf{0.11 \pm 0.20}$ **(0.00)**
Example 3S	3	0.18 ± 0.04 (0.15)	0.08 ± 0.03 (0.03)	0.50 ± 0.00 (0.50)	0.43 ± 0.09 (0.31)	$\mathbf{0.01 \pm 0.00}$ **(0.00)**
Example 1	6	0.46 ± 0.01 (0.44)	0.46 ± 0.09 (0.41)	$\mathbf{0.22 \pm 0.01}$ **(0.20)**	0.37 ± 0.14 (0.17)	$\mathbf{0.22 \pm 0.01}$ **(0.20)**
Example 1S	6	0.46 ± 0.02 (0.44)	0.46 ± 0.02 (0.44)	$\mathbf{0.35 \pm 0.10}$ **(0.23)**	0.42 ± 0.12 (0.28)	$\mathbf{0.35 \pm 0.10}$ **(0.23)**
Example 2	6	0.49 ± 0.01 (0.48)	0.50 ± 0.01 (0.48)	0.50 ± 0.00 (0.50)	0.30 ± 0.01 (0.28)	$\mathbf{0.00 \pm 0.00}$ **(0.00)**
Example 2S	6	0.50 ± 0.00 (0.50)	0.35 ± 0.12 (0.25)	0.50 ± 0.00 (0.50)	0.30 ± 0.01 (0.29)	$\mathbf{0.20 \pm 0.24}$ **(0.00)**
Example 3	6	0.18 ± 0.04 (0.15)	0.20 ± 0.01 (0.19)	0.50 ± 0.00 (0.49)	0.37 ± 0.16 (0.16)	$\mathbf{0.01 \pm 0.01}$ **(0.00)**
Example 3S	6	0.18 ± 0.04 (0.14)	$\mathbf{0.05 \pm 0.04}$ **(0.01)**	0.50 ± 0.00 (0.50)	0.50 ± 0.00 (0.50)	0.11 ± 0.20 (0.00)

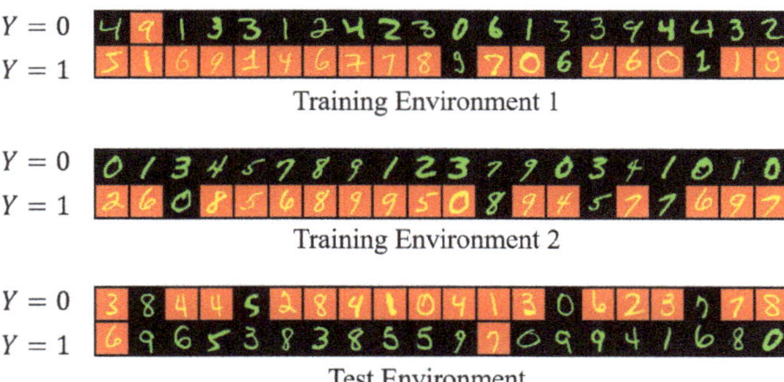

Figure 4. Visualization of the color mnist dataset.

The classification results on the color MNIST dataset are shown in Table 3. From the results, we can see that both ERM and IB-ERM methods almost surely use the color features to achieve the task. Although IRM and IB-IRM methods have shown some improvements over ERM, only our method can perform better than a random prediction, which demonstrates the effectiveness of CSIB.

Table 3. Classification accuracy (%) on color MNIST dataset. "Oracle" in the table means that the training and test data are in the same environment.

Methods	ERM	IRM	IB-ERM	IB-IRM	CSIB	Oracle
Accuracy	9.94 ± 0.28	20.39 ± 2.76	9.94 ± 0.28	43.84 ± 12.48	$\mathbf{60.03 \pm 1.28}$	84.72 ± 0.65

7. Related Works

We divide the works related to OOD generalization into two categories: theory and methods, though some of them belong to both.

7.1. Theory of OOD Generalization

Based on different definitions to the distributional changes, we review the corresponding theory by the following three categories.

Based on causality. Due to the close connection between the distributional changes and the interventions discussed in the theory of causality [13,14], the problem of OOD generalization is usually built in the framework of causal learning. The theory states that a response Y is directly caused only by its parents variables $X_{Pa(Y)}$, and all interventions other that those on Y do not change the conditional distribution of $\mathbb{P}(Y|X_{Pa(Y)})$. Such theory inspires a popular learning principle—the invariance principle—that aims to discover a set of variables such that they remain invariant to the response Y in all observed environments [15,19,20]. Invariant risk minimization (IRM) [8] is then proposed to learn a feature extractor Φ in an end-to-end way such that the optimal classifier based on the extracted features $\Phi(X)$ remains unchanged in each environment. The theory in [8] shows the guarantee of IRM for OOD generalization under some general assumptions but only focuses on the linear regression tasks. Different from the failure analyses of IRM for the classification tasks in [21,22], where the response Y is the cause of the spurious feature, Ahuja et al. [12] analyse another scenario when the invariant feature is the cause of the spurious feature and show that in this case, linear classification is more difficult than linear regression, where the invariance principle itself is insufficient to ensure the success of OOD generalization. They also claim that the assumption of support overlap of invariant features is necessarily needed. They then propose a learning principle of information bottleneck-based invariant risk minimization (IB-IRM) for linear classification, which shows how to address the failures of IRM by adding information bottleneck [16] into the learning. In this work, we closely investigate the conditions identified in [12] and first show that support overlap of invariant features is not necessarily needed for the success of OOD generalization. We further show several failure cases of IB-IRM and propose improved results for it.

Recently, some works tackle the challenge of OOD generalization in the nonlinear regime [23,24]. Commonly, both of them use variational autoencoder (VAE)-based models [25,26] to identify the latent variables from observations in the first stage. Then, these inferring latent variables are separated into two distinct parts of invariant (causal) and spurious (non-causal) features based on different assumptions. Specifically, Lu et al. [23,27] assume that the latent variables conditioned on some accessible side information such as the environment index or class label follow the exponential family distributions, and Liu et al. [24] directly disentangle the latent variables to two different parts during the inferring stage and assume that the marginal distributions of them are independent of each other. These assumptions, however, are rather strong in general. Nevertheless, these solutions aim to capture the latent variables such that the response given these variables is invariant for different environments, which could still fail because the invariance principle itself is insufficient for OOD generalization in the classification tasks, as shown in [12]. In this work, we focus on the linear classification only and show a new theory of a new method that addresses several OOD generalization failures in the linear settings. Our method could extend to the nonlinear regime by combining with the disentangled representation learning [28] or causal representation learning [29]. Specifically, once the latent representations are well disentangled, i.e., the latent features are represented by a linear transform of the causal features and spurious features, we then could apply our method to filter out the spurious features in the latent space such that only causal features remain.

Based on robustness. Different from those based on the causality, where different distributions are generated by intervention on a same SEM and the goal is to discover causal features, the robustness-based methods aim to protect the model against the potential distributional shifts within the uncertainty set, which is usually constrained by f-divergence [30] or Wasserstein distance [31]. This series of works is theoretically addressed by distributionally robust optimization (DRO) under a minimax framework [32,33]. Recently, some works tend to discover the connections between causality and robustness [34]. Although these works show less relevance to us, it is possible that a well-defined measure of distribution divergence could help to effectively extract causal features under the robustness framework. This would be an interesting avenue for future research.

Others. Some other works assume that the distributions (domains) are generated from a hyper-distribution and aim to minimize the average risk estimation error bound [35–37]. These works are often built based on the generalization theory under the independent and identically distributed (IID) assumption. The authors in [38] do not make any assumption on the distributional changes and only study the learnability of OOD generalization in a general way. All of these theories do not cover the OOD generalization problem under a single training environment or domain.

7.2. Methods of OOD Generalization

Based on the invariance principle. Inspired from the invariance principle [15,19], many methods are proposed by designing various loss to extract features to better satisfy the principle itself. IRMv1 [8] is the first objective to address this in an end-to-end way by adding a gradient penalty to the classifier. Following this work, Krueger et al. [9] suggest penalizing the variance of the risks, while Xie et al. [39] give the same objective but take the square root of the variance, and many other alternatives can also be found [40–42]. It is clear that all of these methods aim to find an invariant predictor. Recently, Ahuja et al. [12] found that for the classification problem, finding the invariant predictor is not enough to extract causal features since the features could include spurious information to make the predictor invariant across training environments, and they propose IB-IRM to address such a failure. Similar ideas to IB-IRM can also be found in the work [43,44], where different loss functions are proposed to achieve the same purpose. Specifically, Alesiani et al. [44] also use the information bottleneck (IB) for the help in dropping spurious correlations, but their analyses only focus on the scenario when spurious features are independent from the causal features, which could be considered as a special case of ours. More recently, Wang et al. [45] propose similar ideas to ours but only tackle the situation when the invariant features have the same distribution among all environments. In this work, we further show that IB-IRM could still fail in two cases due to the model only relying on spurious features to meet the task of interest. We then propose a counterfactual supervision-based information bottleneck (CSIB) method to address such failures and show improving results to prior works.

Based on distribution matching. It is worth noting that there are many works focused on learning domain invariant features representations [46–48]. Most of these works are inspired by the seminal theory of domain adaptation [49,50]. The goal of these methods is to learn a feature extractor Φ such that the marginal distribution of $\mathbb{P}(\Phi(X))$ or the conditional distribution of $\mathbb{P}(\Phi(X)|Y)$ is invariant across different domains. This is different from the invariance principle, where the goal is to make $\mathbb{P}(Y|\Phi(X))$ (or $\mathbb{E}(Y|\Phi(X))$) invariant. We refer readers to the papers of [8,51] for better understanding the details of why these distribution-matching-based methods often fail to address OOD generalization.

Others. Other related methods are varied, including by using data augmentation in both image level [52] or feature level [53], by removing spurious correlations through stable learning [54], and by utilizing the inductive bias of neural networks [3,55], etc. Most of these methods are empirically inspired from experiments and are verified on some specific datasets. Recently, empirical studies in [56,57] notice that the real effects of many OOD generalization (domain generalization) methods are weak, which indicates that the benchmark-based evaluation criteria may be inadequate to validate the OOD generalization algorithms.

8. Conclusions, Limitations and Future Work

In this paper, we focus on the OOD generalization problem of linear classification. We first revisit the fundamental assumptions and results of prior works and show that the condition of invariant features supporting overlap is not necessarily needed for the success of OOD generalization and thus propose a weaker counterpart. Then, we show two failure cases of IB-IRM (as well as ERM, IB-ERM, and IRM) and illustrate its intrinsic causes by theoretical analysis. We further propose a new method—counterfactual supervision-based information bottleneck (CSIB)—and theoretically prove its effectiveness under some weaker assumptions. CSIB works even when accessing data from a single environment and can easily extend to the multi-class problems. Finally, we design several synthetic datasets with our examples for experimental verification. Empirical observations among all comparing methods illustrate the effectiveness of the CSIB.

Since we only take the linear problem into account, including linear representation and linear classifier, any nonlinear case would not be guaranteed by our theoretical results, and thus CSIB may fail. Therefore, the same as prior works (IRM [8] and IB-IRM [12]), the nonlinear challenge is still an unsolved problem [21,22]. We believe this is of great value for investigating in future work since widely used data in the wild are nonlinearly generated. Another fruitful direction is to design a powerful algorithm for entropy minimization during the learning process of CSIB. Currently, we use the variance of features to replace the entropy of the features during optimization. However, variance and entropy are essentially different. A truly effective entropy minimization is the key to the success of CSIB. Another limitation of our method is that we have to require further supervision to the counterfactual examples during the learning process, although it only takes one time for a single step.

Author Contributions: Conceptualization, B.D. and K.J.; methodology, B.D.; software, B.D.; validation, B.D. and K.J.; formal analysis, B.D.; investigation, B.D.; resources, B.D.; data curation, B.D.; writing—original draft preparation, B.D.; writing—review and editing, B.D. and K.J.; visualization, B.D.; supervision, K.J. All authors have read and agreed to the published version of the manuscript.

Funding: This research (including the APC) was funded by Guangdong R&D key project of China (No.: 2019B010155001).

Institutional Review Board Statement: Not applicable.

Informed Consent Statement: Not applicable.

Data Availability Statement: Data is contained within the article or supplementary material.

Conflicts of Interest: The authors declare no conflict of interest.

Abbreviations

The following abbreviations are used in this manuscript:

OOD	Out-of-distribution
ERM	Empirical risk minimization
IRM	Invariant risk minimization
IB-ERM	Information bottleneck-based empirical risk minimization
IB-IRM	Information bottleneck-based invariant risk minimization
CSIB	Counterfactual supervision-based information bottleneck
DAG	Directed acyclic graph
SEM	Structure equation model
SVD	Singular value decomposition

Appendix A. Experiments Details

In this section, we provide more details on the experiments. The code to reproduce the experiments can be found at https://github.com/szubing/CSIB.

Appendix A.1. Optimization Loss of IB-ERM

The objective function of IB-ERM is as follows:

$$\min_{w,\Phi} \sum_{e \in \mathcal{E}_{tr}} h^e(\Phi) \quad \text{s.t.} \quad \frac{1}{|\mathcal{E}_{tr}|} \sum_{e \in \mathcal{E}_{tr}} R^e(w \circ \Phi) \leq rth. \tag{A1}$$

Since the entropy of $h^e(\Phi) = H(\Phi(X^e))$ is hard to estimate by a differential variable that can be optimized by using gradient descent, we follow [12] by using the variance instead of the entropy for optimization. The total loss function is given by

$$loss(w, \Phi) = \sum_{e \in \mathcal{E}_{tr}} \left(R^e(w \circ \Phi) + \lambda \mathrm{Var}(\Phi) \right) \tag{A2}$$

with a hyperparameter λ onto it.

Appendix A.2. Experiments Setup

Model, hyperparameters, loss, and evaluation. In all experiments, we follow the same protocol as prescribed by [12,17] for the model / hyperparameter selection, training, and evaluation. Except those specified, for all experiments across three examples and five comparing methods, the model is the same with a linear feature extractor $\Phi \in \mathbb{R}^{d \times d}$ followed by a linear classifier $w \in \mathbb{R}^{d+1}$. We use binary cross-entropy loss for classification. All hyperparameters, including the learning rate, the penalty term in IRM, or the λ associated with the $\mathrm{Var}(\Phi)$ in Equation (A2), etc., are randomly searched and selected by using 20 test samples for validation. The results reported in the main manuscript use three hyperparameter queries of each and average over five data seeds. The results when searching over more hyperparameter values are reported in the supplementary experiments. The search spaces of all the hyperparameters follow the same as in [12,17]. The classification test errors between 0 and 1 are reported.

Compute description. Our computing resource is one GPU of NVIDIA GeForce GTX 1080 Ti with 6 CPU cores of Intel(R) Core(TM) i7-8700 CPU @ 3.20GHz.

Existing codes and datasets used. In our experiments, we mainly rely on the following two github repositories: InvarianceUnitTests (https://github.com/facebookresearch/InvarianceUnitTests) and IB-IRM (https://github.com/ahujak/IB-IRM).

Appendix A.3. Supplementary Experiments

The purpose of the first supplementary experiment is to illustrate what the result would be when we increase the number of running seeds in the hyperparameters selection. These results are shown in Table A1, where we increase the number of hyperparameter queries to 10 of each. It is clear that overall, the results of the CSIB in Table A1 are much better and have less fluctuations than those in Table 2, and the conclusions remain almost the same as we have summarized in Section 6.1.2. This further verifies the effectiveness of the CSIB method.

Observation on different settings in Example 1/1S. In our main experiments of Example 1/1S, we set $p^e = 1$ and $q = 0$ to make the spurious features and the invariant features both linearly separable to confuse each other. Here, we analyse what the result would be if we vary their values. Following [17], we set $p^{e_0} = 0.95$, $p^{e_1} = 0.97$, $p^{e_2} = 0.99$, and $p^{e_j} \sim \mathrm{Uniform}(0.9, 1)$ to make spurious features linearly inseparable, and q is set to $0/0.05$ to make invariant features linearly separable/inseparable. Table A2 shows the corresponding results. Interestingly, we find that all methods except for IB-IRM have an ideal error rate (the same as the Oracle) when the spurious features are linearly inseparable ($p^e \neq 1$), even when the invariant features are linearly inseparable too ($q = 0.05$). Why would this happen? We then remove the linear embedding Φ. The results are presented in Table A3. Comparing the results between Tables A2 and A3, we found there is a significant inductive bias of the neural network, though the model is linear. Further analysis to such observation is out of the scope of this paper, but this would be an interesting avenue for future research.

Table A1. Supplementary results when using 10 hyperparameter queries. #Envs means the number of training environments, and (min) reports the minimal test classification error across different running data seeds.

	#Envs	ERM (min)	IRM (min)	IB-ERM (min)	IB-IRM (min)	CSIB (min)	Oracle (min)
Example 1	1	0.50 ± 0.01 (0.49)	0.50 ± 0.01 (0.49)	0.23 ± 0.02 (0.22)	0.31 ± 0.10 (0.25)	0.23 ± 0.02 (0.22)	0.00 ± 0.00 (0.00)
Example 1S	1	0.50 ± 0.00 (0.49)	0.50 ± 0.00 (0.49)	0.09 ± 0.04 (0.04)	0.30 ± 0.10 (0.25)	0.08 ± 0.04 (0.04)	0.00 ± 0.00 (0.00)
Example 2	1	0.40 ± 0.20 (0.00)	0.00 ± 0.00 (0.00)	0.50 ± 0.00 (0.49)	0.48 ± 0.03 (0.43)	0.00 ± 0.00 (0.00)	0.00 ± 0.00 (0.00)
Example 2S	1	0.50 ± 0.00 (0.50)	0.30 ± 0.25 (0.00)	0.50 ± 0.00 (0.50)	0.50 ± 0.01 (0.48)	0.00 ± 0.00 (0.00)	0.00 ± 0.00 (0.00)
Example 3	1	0.16 ± 0.06 (0.09)	0.03 ± 0.00 (0.03)	0.50 ± 0.01 (0.49)	0.41 ± 0.09 (0.25)	0.02 ± 0.01 (0.00)	0.00 ± 0.00 (0.00)
Example 3S	1	0.16 ± 0.06 (0.10)	0.04 ± 0.01 (0.02)	0.50 ± 0.00 (0.50)	0.41 ± 0.12 (0.26)	0.01 ± 0.01 (0.00)	0.00 ± 0.00 (0.00)
Example 1	3	0.44 ± 0.01 (0.44)	0.44 ± 0.01 (0.44)	0.21 ± 0.00 (0.21)	0.21 ± 0.10 (0.06)	0.21 ± 0.00 (0.21)	0.00 ± 0.00 (0.00)
Example 1S	3	0.45 ± 0.00 (0.44)	0.45 ± 0.00 (0.44)	0.09 ± 0.03 (0.05)	0.23 ± 0.13 (0.01)	0.09 ± 0.03 (0.05)	0.00 ± 0.00 (0.00)
Example 2	3	0.13 ± 0.07 (0.00)	0.00 ± 0.00 (0.00)	0.50 ± 0.00 (0.50)	0.33 ± 0.04 (0.25)	0.00 ± 0.00 (0.00)	0.00 ± 0.00 (0.00)
Example 2S	3	0.50 ± 0.00 (0.50)	0.14 ± 0.20 (0.00)	0.50 ± 0.00 (0.50)	0.34 ± 0.01 (0.33)	0.00 ± 0.00 (0.00)	0.00 ± 0.00 (0.00)
Example 3	3	0.17 ± 0.04 (0.14)	0.02 ± 0.00 (0.02)	0.50 ± 0.01 (0.49)	0.43 ± 0.08 (0.29)	0.01 ± 0.00 (0.00)	0.00 ± 0.00 (0.00)
Example 3S	3	0.17 ± 0.04 (0.13)	0.02 ± 0.00 (0.02)	0.50 ± 0.00 (0.50)	0.36 ± 0.18 (0.07)	0.01 ± 0.00 (0.00)	0.00 ± 0.00 (0.00)
Example 1	6	0.46 ± 0.01 (0.44)	0.46 ± 0.09 (0.41)	0.22 ± 0.01 (0.21)	0.41 ± 0.11 (0.26)	0.22 ± 0.01 (0.21)	0.00 ± 0.00 (0.00)
Example 1S	6	0.46 ± 0.02 (0.44)	0.46 ± 0.02 (0.44)	0.06 ± 0.04 (0.02)	0.45 ± 0.07 (0.41)	0.06 ± 0.04 (0.02)	0.00 ± 0.00 (0.00)
Example 2	6	0.21 ± 0.03 (0.17)	0.00 ± 0.00 (0.00)	0.50 ± 0.00 (0.50)	0.36 ± 0.03 (0.31)	0.00 ± 0.00 (0.00)	0.00 ± 0.00 (0.00)
Example 2S	6	0.50 ± 0.00 (0.50)	0.10 ± 0.20 (0.00)	0.50 ± 0.00 (0.50)	0.19 ± 0.16 (0.01)	0.00 ± 0.00 (0.00)	0.00 ± 0.00 (0.00)
Example 3	6	0.17 ± 0.03 (0.14)	0.02 ± 0.00 (0.02)	0.50 ± 0.00 (0.49)	0.37 ± 0.16 (0.16)	0.01 ± 0.00 (0.00)	0.00 ± 0.00 (0.00)
Example 3S	6	0.17 ± 0.03 (0.14)	0.02 ± 0.00 (0.02)	0.50 ± 0.00 (0.50)	0.46 ± 0.09 (0.28)	0.01 ± 0.00 (0.00)	0.00 ± 0.00 (0.00)

Observation on linearly separable properties of high-dimensional data. Here, we empirically show that for o-dimensional data, we have high probability that o randomly drawn points are linearly separable for any two subsets. To verify that, we design a random experiment as follows: (1) Let $o \in [100, 10{,}000]$, and we randomly draw o points from $[-1, 1]^o$, and give random labels to these o points of 0 or 1. (2) We train a linear classifier to fit these o points and report the final training error. (3) We perform (1) and (2) 100 times for different seeds. Our results show that for 100 runs, all training errors reach 0 for every o, which proves our conjecture.

Table A2. Results in Example 1/1S, where the learning model is a linear embedding $\Phi \in \mathbb{R}^{d \times d}$ followed by a linear classifier $w \in \mathbb{R}^{d+1}$.

	#Envs	$p^e = 1$?	q	ERM	IB-ERM	IB-IRM	CSIB	IRM	Oracle
Example 1	1	Yes	0	0.50 ± 0.01	0.23 ± 0.02	0.31 ± 0.10	0.23 ± 0.02	0.50 ± 0.01	0.00 ± 0.00
Example 1S	1	Yes	0	0.50 ± 0.00	0.46 ± 0.04	0.30 ± 0.10	0.46 ± 0.04	0.50 ± 0.00	0.00 ± 0.00
Example 1	3	Yes	0	0.45 ± 0.01	0.22 ± 0.01	0.23 ± 0.13	0.22 ± 0.01	0.45 ± 0.01	0.00 ± 0.00
Example 1S	3	Yes	0	0.45 ± 0.00	0.41 ± 0.04	0.27 ± 0.11	0.41 ± 0.04	0.45 ± 0.00	0.00 ± 0.00
Example 1	6	Yes	0	0.46 ± 0.01	0.22 ± 0.01	0.37 ± 0.14	0.22 ± 0.01	0.46 ± 0.09	0.00 ± 0.00
Example 1S	6	Yes	0	0.46 ± 0.02	0.35 ± 0.10	0.42 ± 0.12	0.35 ± 0.10	0.46 ± 0.02	0.00 ± 0.00
Example 1	1	No	0	0.00 ± 0.00	0.00 ± 0.00	0.15 ± 0.20	0.00 ± 0.00	0.00 ± 0.00	0.00 ± 0.00
Example 1S	1	No	0	0.00 ± 0.00	0.00 ± 0.00	0.12 ± 0.19	0.00 ± 0.00	0.00 ± 0.00	0.00 ± 0.00
Example 1	3	No	0	0.00 ± 0.00	0.00 ± 0.00	0.00 ± 0.00	0.00 ± 0.00	0.00 ± 0.00	0.00 ± 0.00
Example 1S	3	No	0	0.00 ± 0.00	0.00 ± 0.00	0.00 ± 0.01	0.00 ± 0.00	0.00 ± 0.00	0.00 ± 0.00
Example 1	6	No	0	0.00 ± 0.00	0.00 ± 0.00	0.30 ± 0.20	0.00 ± 0.00	0.00 ± 0.00	0.00 ± 0.00
Example 1S	6	No	0	0.00 ± 0.00	0.00 ± 0.00	0.31 ± 0.20	0.00 ± 0.00	0.04 ± 0.06	0.00 ± 0.00
Example 1	1	No	0.05	0.05 ± 0.00	0.05 ± 0.00	0.32 ± 0.22	0.05 ± 0.00	0.05 ± 0.00	0.05 ± 0.00
Example 1S	1	No	0.05	0.05 ± 0.00	0.05 ± 0.00	0.19 ± 0.17	0.05 ± 0.00	0.05 ± 0.00	0.05 ± 0.00
Example 1	3	No	0.05	0.05 ± 0.00	0.05 ± 0.00	0.07 ± 0.03	0.05 ± 0.00	0.05 ± 0.00	0.05 ± 0.00
Example 1S	3	No	0.05	0.05 ± 0.00	0.05 ± 0.00	0.05 ± 0.00	0.05 ± 0.00	0.05 ± 0.00	0.05 ± 0.00
Example 1	6	No	0.05	0.05 ± 0.00	0.05 ± 0.00	0.30 ± 0.21	0.05 ± 0.00	0.05 ± 0.00	0.05 ± 0.00
Example 1S	6	No	0.05	0.05 ± 0.00	0.05 ± 0.00	0.32 ± 0.19	0.05 ± 0.00	0.05 ± 0.00	0.05 ± 0.00

Table A3. Results in Example 1/1S, where the learning model is a linear classifier $w \in \mathbb{R}^{d+1}$ without linear embedding Φ. The CSIB must require a feature extractor, so there are not results related to the CSIB.

	#Envs	$p^e = 1$?	q	ERM	IB-ERM	IB-IRM	IRM	Oracle
Example 1	1	Yes	0	0.50 ± 0.01	0.25 ± 0.01	0.31 ± 0.10	0.50 ± 0.01	0.00 ± 0.00
Example 1S	1	Yes	0	0.50 ± 0.00	0.49 ± 0.01	0.30 ± 0.10	0.50 ± 0.00	0.00 ± 0.00
Example 1	3	Yes	0	0.44 ± 0.01	0.23 ± 0.01	0.21 ± 0.10	0.44 ± 0.01	0.00 ± 0.00
Example 1S	3	Yes	0	0.45 ± 0.00	0.44 ± 0.01	0.42 ± 0.04	0.45 ± 0.00	0.00 ± 0.00
Example 1	6	Yes	0	0.46 ± 0.01	0.27 ± 0.07	0.41 ± 0.11	0.46 ± 0.01	0.01 ± 0.01
Example 1S	6	Yes	0	0.46 ± 0.02	0.42 ± 0.08	0.46 ± 0.09	0.46 ± 0.02	0.01 ± 0.02
Example 1	1	No	0	0.50 ± 0.01	0.00 ± 0.00	0.15 ± 0.20	0.50 ± 0.01	0.00 ± 0.00
Example 1S	1	No	0	0.50 ± 0.00	0.00 ± 0.00	0.13 ± 0.19	0.50 ± 0.00	0.00 ± 0.00
Example 1	3	No	0	0.45 ± 0.01	0.00 ± 0.00	0.00 ± 0.00	0.45 ± 0.01	0.00 ± 0.00
Example 1S	3	No	0	0.45 ± 0.00	0.01 ± 0.02	0.08 ± 0.14	0.46 ± 0.02	0.00 ± 0.00
Example 1	6	No	0	0.46 ± 0.01	0.10 ± 0.16	0.30 ± 0.20	0.46 ± 0.01	0.01 ± 0.01
Example 1S	6	No	0	0.46 ± 0.01	0.24 ± 0.19	0.41 ± 0.12	0.47 ± 0.03	0.01 ± 0.02
Example 1	1	No	0.05	0.50 ± 0.01	0.05 ± 0.00	0.32 ± 0.22	0.50 ± 0.01	0.05 ± 0.00
Example 1S	1	No	0.05	0.50 ± 0.01	0.05 ± 0.01	0.20 ± 0.17	0.50 ± 0.00	0.05 ± 0.00
Example 1	3	No	0.05	0.45 ± 0.01	0.05 ± 0.00	0.07 ± 0.03	0.47 ± 0.01	0.05 ± 0.00
Example 1S	3	No	0.05	0.45 ± 0.01	0.07 ± 0.03	0.11 ± 0.11	0.46 ± 0.01	0.05 ± 0.00
Example 1	6	No	0.05	0.47 ± 0.01	0.14 ± 0.14	0.30 ± 0.21	0.47 ± 0.01	0.05 ± 0.00
Example 1S	6	No	0.05	0.47 ± 0.01	0.27 ± 0.18	0.42 ± 0.11	0.47 ± 0.01	0.05 ± 0.01

Then, we look back to Theorem 2. For real data, such as an image, the dimension of spurious features o is often high. Assume different environments enjoy different spurious points randomly; then, from the above observation, there is a high probability that the following events will occur: For any labeling data in the n training environments with $n < o/2$ (2 is due to the binary label), models could achieve zero training error by relying on spurious features only. This illustrates why prior methods easily fail to address OOD generalization under Assumption 6.

Appendix B. Proofs

Appendix B.1. Preliminary

Before our proofs, we first review some useful properties related to the entropy [12,58].

Entropy. For discrete random variable $X \sim \mathbb{P}_X$ with support $\mathcal{X}$, its entropy (Shannon entropy) is defined as

$$H(X) = - \sum_{x \in \mathcal{X}} \mathbb{P}_X(X = x) \log(\mathbb{P}_X(X = x)) \tag{A3}$$

The differential entropy of the continuous random variable $X \sim \mathbb{P}_X$ with support $\mathcal{X}$ is given by

$$h(X) = - \int_{x \in \mathcal{X}} p_X(x) \log(p_X(x)) dx, \tag{A4}$$

where $p_X(x)$ is the probability density function of the distribution $\mathbb{P}_X$. Sometimes, we may confuse using $H(X)$ or $h(X)$ to represent its entropy no matter whether X is discrete or continuous.

Lemma A1. *If X and Y are discrete random variables that are independent, then*

$$H(X + Y) \geq \max\{H(X), H(Y)\}. \tag{A5}$$

Proof. Define $Z = X + Y$. Since $X \perp Y$, we have

$$
\begin{aligned}
H(Z|X) &= -\sum_{x \in \mathcal{X}} \mathbb{P}_X(x) \sum_{z \in \mathcal{Z}} \mathbb{P}_{Z|X}(Z = z | X = x) \log(\mathbb{P}_{Z|X}(Z = z | X = x)) \\
&= -\sum_{x \in \mathcal{X}} \mathbb{P}_X(x) \sum_{z \in \mathcal{Z}} \mathbb{P}_{Y|X}(Y = z - x | X = x) \log(\mathbb{P}_{Y|X}(Y = z - x | X = x)) \\
&= -\sum_{x \in \mathcal{X}} \mathbb{P}_X(x) \sum_{z \in \mathcal{Z}} \mathbb{P}_Y(Y = z - x) \log(\mathbb{P}_Y(Y = z - x)) \\
&= -\sum_{x \in \mathcal{X}} \mathbb{P}_X(x) \sum_{y \in \mathcal{Y}} \mathbb{P}_Y(Y = y) \log(\mathbb{P}_Y(Y = y)) \\
&= H(Y),
\end{aligned}
$$

and similar we have $H(Z|Y) = H(X)$. Therefore,

$$H(X + Y) = I(Z, X) + H(Z|X) = I(Z, X) + H(Y) \geq H(Y) \tag{A6}$$
$$H(X + Y) = I(Z, Y) + H(Z|Y) = I(Z, Y) + H(X) \geq H(X). \tag{A7}$$

This completes the proof. $\square$

Lemma A2. *If X and Y are continuous random variables that are independent, then*

$$h(X + Y) \geq \max\{h(X), h(Y)\}. \tag{A8}$$

Proof. Define $Z = X + Y$. Since $X \perp Y$, we have

$$
\begin{aligned}
h(Z|X) &= -\int_{x \in \mathcal{X}} p_X(x) \int_{z \in \mathcal{Z}} p_{Z|X}(Z = z | X = x) \log(p_{Z|X}(Z = z | X = x)) dx dz \\
&= -\int_{x \in \mathcal{X}} p_X(x) \int_{z \in \mathcal{Z}} p_{Y|X}(Y = z - x | X = x) \log(p_{Y|X}(Y = z - x | X = x)) dx dz \\
&= -\int_{x \in \mathcal{X}} p_X(x) \int_{z \in \mathcal{Z}} p_Y(Y = z - x) \log(p_Y(Y = z - x)) dx dz \\
&= -\int_{x \in \mathcal{X}} p_X(x) dx \int_{y \in \mathcal{Y}} p_Y(Y = y) \log(p_Y(Y = y)) dy \\
&= h(Y),
\end{aligned}
$$

and similar, we have $h(Z|Y) = h(X)$. Therefore,

$$h(X + Y) = I(Z, X) + h(Z|X) = I(Z, X) + h(Y) \geq h(Y) \tag{A9}$$
$$h(X + Y) = I(Z, Y) + h(Z|Y) = I(Z, Y) + h(X) \geq h(X). \tag{A10}$$

This completes the proof. $\square$

Lemma A3. *If X and Y are discrete random variables that are independent with the supports satisfying $2 \leq |\mathcal{X}| < \infty, 2 \leq |\mathcal{Y}| < \infty$, then*

$$H(X + Y) > \max\{H(X), H(Y)\}. \tag{A11}$$

Proof. From Lemma A1 and due to the symmetry of X and Y, we only need to prove $H(X + Y) \neq H(X)$. The proof is by contradiction. Suppose $H(X + Y) = H(X)$, then from Equation (A7) it follows that $I(X + Y, Y) = 0$, thus $X + Y \perp Y$. However, $\mathbb{P}(Y = y_{max} | X + Y = x_{max} + y_{max}) = 1$, which is different from $\mathbb{P}(Y = y_{max}) < 1$ (due to $|Y| \geq 2$). This contradicts $X + Y \perp Y$. $\square$

Lemma A4. *If X and Y are continuous random variables that are independent and have a bounded support, then*

$$h(X + Y) > \max\{h(X), h(Y)\}. \tag{A12}$$

Proof. From Lemma A2 and due to the symmetry of X and Y, we only need to prove $h(X + Y) \neq h(X)$. The proof is by contradiction. Suppose $h(X + Y) = h(X)$, then from Equation (A10) it follows that $I(X + Y, Y) = 0$, thus $X + Y \perp Y$. For any $\delta > 0$, define an event $\mathcal{M} : x_{max} + y_{max} - \delta \leq X + Y \leq x_{max} + y_{max}$. If $\mathcal{M}$ occurs, then $Y \geq y_{max} - \delta$ and $X \geq x_{max} - \delta$. Thus, $\mathbb{P}_Y(Y \leq y_{max} - \delta | \mathcal{M}) = 0$. However, we can always choose a $\delta > 0$ that is small enough to make $\mathbb{P}_Y(Y \leq y_{max} - \delta) > 0$. This contradicts $X + Y \perp Y$. $\square$

Appendix B.2. Proof of Theorem 2

Proof. The proof is trivial. Since two sets $\cup_{e \in \mathcal{E}_{tr}} \mathcal{Z}^e_{spu}(Y^e = 1)$ and $\cup_{e \in \mathcal{E}_{tr}} \mathcal{Z}^e_{spu}(Y^e = 0)$ are linearly separable, there exists a linear classifier w that only relies on spurious features and can achieve zero classification error on each environment. Therefore, w is an invariant predictor across different training environments. In addition, $H(Z^e_{inv}) > H(Z^e_{spu})$ would make IB-IRM prefer to choose these spurious features. Therefore, w would be an optimal solution of IB-IRM, ERM, IRM, and IB-ERM. However, since w relies on spurious features which may change arbitrary in unseen environments, it thus fails to solve OOD generalization. $\square$

Appendix B.3. Proof of Theorem 3

Proof. Assume $\Phi^* \in \mathbb{R}^{c \times d}$ and w^* are the feature extractor and classifier learned by IB-ERM. Consider the feature variable extracted by Φ^* as

$$\Phi^* X^e = \Phi^* S(Z^e_{inv}, Z^e_{spu}) = \Phi_{inv} Z^e_{inv} + \Phi_{spu} Z^e_{spu}. \tag{A13}$$

We first show that $\Phi_{inv} = \mathbf{0}$ or $\Phi_{spu} = \mathbf{0}$. We prove this by contradiction. Assume $\Phi_{inv} \neq \mathbf{0}$ and $\Phi_{spu} \neq \mathbf{0}$. By observing that a solution of $\Phi_{inv} = \mathbf{1}, \Phi_{spu} = \mathbf{0}, w^* = w^*_{inv}$ could make the average training error to q; therefore any solution returned by IB-ERM should also achieve the error no larger than q (because $rth = q$ in the constraint of Equation (12)). Therefore $w^* \neq \mathbf{0}$.

1. In the case when each $e \in \mathcal{E}_{tr}$ follows Assumption 4 of $Z^e_{spu} \leftarrow A Z^e_{inv} + W^e$, we have

$$w^* \cdot (\Phi_{inv} Z^e_{inv} + \Phi_{spu} Z^e_{spu}) = w^* \cdot \Phi_{inv} Z^e_{inv} + w^* \cdot \Phi_{spu}(A Z^e_{inv} + W^e)$$
$$= w^* \cdot (\Phi_{inv} + \Phi_{spu} A) Z^e_{inv} + w^* \cdot \Phi_{spu} W^e.$$

Then, for any $z = (z^e_{inv}, z^e_{spu})$ of $\mathbf{1}(w^*_{inv} \cdot z^e_{inv}) = 1$, we must have $w^* \cdot (\Phi_{inv} + \Phi_{spu} A) z^e_{inv} + w^* \cdot \Phi_{spu} w^e \geq 0$ for any w^e to make error no larger than q. Since W^e is zero mean with at least two distinct points in each component, we can conclude that $w^* \cdot (\Phi_{inv} + \Phi_{spu} A) z^e_{inv} \geq 0$. Similarly, for any $z = (z^e_{inv}, z^e_{spu})$ of $\mathbf{1}(w^*_{inv} \cdot z^e_{inv}) = 0$, we have $w^* \cdot (\Phi_{inv} + \Phi_{spu} A) z^e_{inv} < 0$. From Lemma A3 or Lemma A4, we obtain $H((\Phi_{inv} + \Phi_{spu} A) Z^e_{inv} + \Phi_{spu} W^e) > H((\Phi_{inv} + \Phi_{spu} A) Z^e_{inv})$. Therefore, there exists a more optimal solution to IB-ERM with zero weight to Z^e_{spu}, which contradicts the assumption.

2. In the case when each $e \in \mathcal{E}_{tr}$ follows Assumption 5 of $Z^e_{inv} \leftarrow A Z^e_{spu} + W^e$, we have

$$w^* \cdot (\Phi_{inv} Z^e_{inv} + \Phi_{spu} Z^e_{spu}) = w^* \cdot \Phi_{inv}(A Z^e_{spu} + W^e) + w^* \cdot \Phi_{spu} Z^e_{spu}$$
$$= w^* \cdot (\Phi_{spu} + \Phi_{inv} A) Z^e_{spu} + w^* \cdot \Phi_{inv} W^e.$$

From Lemma A3 or Lemma A4, we obtain $H((\Phi_{spu} + \Phi_{inv} A) Z^e_{spu} + \Phi_{inv} W^e) > H((\Phi_{spu} + \Phi_{inv} A) Z^e_{spu})$. In addition, the spurious features are assumed to be linearly separable. Therefore, there exists a more optimal solution to IB-ERM with zero weight to Z^e_{inv}, which contradicts the assumption.

3. In the case when each $e \in \mathcal{E}_{tr}$ follows Assumption 6 of $Z_{spu}^e \leftarrow W_1^e Y^e + W_0^e(1 - Y^e)$, we have

$$
\begin{aligned}
w^* \cdot (\Phi_{inv} Z_{inv}^e + \Phi_{spu} Z_{spu}^e) &= w^* \cdot \Phi_{inv} Z_{inv}^e + w^* \cdot \Phi_{spu}(W_1^e Y^e + W_0^e(1 - Y^e)) \\
&= w^* \cdot \Phi_{inv} Z_{inv}^e + w^* \cdot \Phi_{spu} W_1^e Y^e + w^* \cdot \Phi_{spu} W_0^e(1 - Y^e).
\end{aligned}
$$

Then, for any $z = (z_{inv}^e, z_{spu}^e)$ of $\mathbf{1}(w_{inv}^* \cdot z_{inv}^e) = 1$, we must have $w^* \cdot \Phi_{inv} z_{inv}^e + w^* \cdot \Phi_{spu} w_1^e y^e + w^* \cdot \Phi_{spu} w_0^e(1 - y^e) \geq 0$ for any w_1^e and w_0^e to make error no larger than q. Since W_1^e and W_0^e are both zero mean variables with at least two distinct points in each component, we can conclude that $w^* \cdot \Phi_{inv} z_{inv}^e \geq 0$; Similarly, for any $z = (z_{inv}^e, z_{spu}^e)$ of $\mathbf{1}(w_{inv}^* \cdot z_{inv}^e) = 0$, we have $w^* \cdot \Phi_{inv} z_{inv}^e < 0$. From Lemma A3 or Lemma A4, we obtain $H(\Phi_{inv} Z_{inv}^e + \Phi_{spu} W_1^e Y^e + \Phi_{spu} W_0^e(1 - Y^e)) > H(\Phi_{inv} Z_{inv}^e)$. Therefore, there exists a more optimal solution to IB-ERM with zero weight to Z_{spu}^e, which contradicts the assumption.

So far, we have proved that the feature extractor Φ^* learned by IB-ERM would never extract both spurious features and invariant features together. Then, we perform singular value decomposition (SVD) to the Φ^* as

$$
\Phi^* = U \Lambda V^T = [U_1, U_2][\Lambda_1, \mathbf{0}; \mathbf{0}, \mathbf{0}][V_1^T; V_2^T] = U_1 \Lambda_1 V_1^T \tag{A14}
$$

Let $S \in \mathbb{R}^{d \times d}$ be the orthogonal matrix. Set r as the rank of the matrix Φ^*, i.e., $r = Rank(\Phi^*)$, and let $V_1^T S = [V_1', V_2']$ with $V_1' \in \mathbb{R}^{r \times m}$ and $V_2' \in \mathbb{R}^{r \times o}$, and $V_2^T S = [V_1'', V_2'']$ with $V_1'' \in \mathbb{R}^{(d-r) \times m}$ and $V_2'' \in \mathbb{R}^{(d-r) \times o}$, then

$$
\Phi^* X^e = U_1 \Lambda_1 V_1^T S[Z_{inv}^e; Z_{spu}^e] = U_1 \Lambda_1 (V_1' Z_{inv}^e + V_2' Z_{spu}^e). \tag{A15}
$$

Since $\Phi^* X^e$ contains the information either from spurious features or from invariant features, we must have $U_1 \Lambda_1 V_1' = \mathbf{0}$ or $U_1 \Lambda_1 V_2' = \mathbf{0}$, and thus, $V_1' = \mathbf{0}$ or $V_2' = \mathbf{0}$ due to $Rank(U_1 \Lambda_1) = r$. If $V_2' = \mathbf{0}$, then Φ^* extract invariant features only. Otherwise when $V_1' = \mathbf{0}$, we decompose the $V^T S$ by

$$
V^T S = [V_1^T; V_2^T] S = [V_1^T S; V_2^T S] = [V_1', V_2'; V_1'', V_2'']. \tag{A16}
$$

Since V^T and S are both the orthogonal matrix, $V^T S$ is also orthogonal; thus $V_1' = \mathbf{0} \Rightarrow V_2'^T V_2'' = \mathbf{0}$, and then $Rank(V_2'') = Rank([V_2'; V_2'']) - Rank(V_2') = o - r$ (note that $r \leq \min\{m, o\}$). Then,

$$
V_2^T X^e = V_2^T S[Z_{inv}^e; Z_{spu}^e] = [V_1'', V_2''][Z_{inv}^e; Z_{spu}^e] = V_1'' Z_{inv}^e + V_2'' Z_{spu}^e. \tag{A17}
$$

Therefore, by running the CSIB for one iteration, the rank of spurious features would be decreased by $r > 0$. This would result in zero weight to spurious features by finite runs of CSIB.

Then, we intend to show why the counterfactual supervision step could help to distinguish whether V_1' is $\mathbf{0}$ or not. For a specific instance $x = S[z_{inv}; z_{spu}]$, let two new features be z^1 and z^2, then $do(z_{1:r}^1) = [-M, ..., -M]$ and $do(z_{r+1:d}^1) = V_2^T x$; $do(z_{1:r}^2) = [M, ..., M]$ and $do(z_{r+1:d}^2) = V_2^T x$. Back the new features z^1 and z^2 to the input space as $x^1 = V z^1$ and $x^2 = V z^2$. If $V_1' = \mathbf{0}$, then

$$
\begin{aligned}
S^{-1} x^1 = S^{-1} V z^1 &= S^{-1} V[z_{1:r}^1; V_1'' z_{inv} + V_2'' z_{spu}] \\
&= (V^T S)^T [z_{1:r}^1; V_1'' z_{inv} + V_2'' z_{spu}] \\
&= [V_1'^T, V_1''^T; V_2'^T, V_2''^T][z_{1:r}^1; V_1'' z_{inv} + V_2'' z_{spu}] \\
&= [V_1'^T z_{1:r}^1 + V_1''^T (V_1'' z_{inv} + V_2'' z_{spu}); V_2'^T z_{1:r}^1 + V_2''^T (V_1'' z_{inv} + V_2'' z_{spu})] \\
&= [z_{inv}; V_2'^T z_{1:r}^1 + V_2''^T V_2'' z_{spu}],
\end{aligned}
$$

and similarly we have $S^{-1}x^2 = [z_{inv}; V_2'^T z_{1:r}^2 + V_2''^T V_2'' z_{spu}]$. Therefore, the ground truths of x^1 and x^2 are the same. On other hand, if $V_1' \neq \mathbf{0}$, then $V_2' = \mathbf{0}$, and

$$
\begin{aligned}
S^{-1}x^1 = S^{-1}Vz^1 &= S^{-1}V[z_{1:r}^1; V_1'' z_{inv} + V_2'' z_{spu}] \\
&= (V^T S)^T [z_{1:r}^1; V_1'' z_{inv} + V_2'' z_{spu}] \\
&= [V_1'^T, V_1''^T; V_2'^T, V_2''^T][z_{1:r}^1; V_1'' z_{inv} + V_2'' z_{spu}] \\
&= [V_1'^T z_{1:r}^1 + V_1''^T (V_1'' z_{inv} + V_2'' z_{spu}); V_2'^T z_{1:r}^1 + V_2''^T (V_1'' z_{inv} + V_2'' z_{spu})] \\
&= [V_1'^T z_{1:r}^1 + V_1''^T V_1'' z_{inv}; z_{spu}],
\end{aligned}
$$

and similarly we have $S^{-1}x^2 = [V_1'^T z_{1:r}^2 + V_1''^T V_1'' z_{inv}; z_{spu}]$. Since $z_{1:r}^1 = -z_{1:r}^2$ and their magnitudes are large enough to make $\mathbf{sgn}(w_{inv}^* \cdot (V_1'^T z_{1:r}^1 + V_1''^T V_1'' z_{inv})) \neq \mathbf{sgn}(w_{inv}^* \cdot (V_1'^T z_{1:r}^2 + V_1''^T V_1'' z_{inv}))$; thus the ground truths of x^1 and x^2 would be different. Therefore, the counterfactual supervision step could help to detect whether invariant features or spurious features are extracted by using a single sample only.

Finally, when only invariant features are extracted by Φ, the training error is minimized, i.e., $w^* \Phi_{inv} \in \arg\min_f \mathbb{E}_{\mathbb{P}}[l(f(Z_{inv}^{tr}), Y^{tr})]$. Then, based on our assumption to the OOD environments (Assumptions 8), i.e., $\forall e \in \mathcal{E}_{ood}, F_l(\mathbb{P}(Z_{inv}^{tr}, Y^{tr})) \subseteq F_l(\mathbb{P}(Z_{inv}^e, Y^e))$, therefore, for any $e \in \mathcal{E}_{ood}$, we have $\mathbb{E}_{\mathbb{P}}[l((X^e, Y^e), w^* \Phi)] = \mathbb{E}_{\mathbb{P}}[l((Z_{inv}^e, Y^e), w^* \Phi_{inv})] = \mathbb{E}_{\mathbb{P}}[l((Z_{inv}^{tr}, Y^{tr}), w^* \Phi_{inv})] = q.$ $\square$

It is worth noting that the proof of Theorem 3 does not rely on how many labels there would be, so it is easily extended to the multi-class classification case as long as the corresponding assumptions and conditions are satisfied.

References

1. Szegedy, C.; Zaremba, W.; Sutskever, I.; Bruna, J.; Erhan, D.; Goodfellow, I.; Fergus, R. Intriguing properties of neural networks. *arXiv* **2013**, arXiv:1312.6199.
2. Rosenfeld, A.; Zemel, R.; Tsotsos, J.K. The elephant in the room. *arXiv* **2018**, arXiv:1808.03305.
3. Geirhos, R.; Rubisch, P.; Michaelis, C.; Bethge, M.; Wichmann, F.A.; Brendel, W. ImageNet-trained CNNs are biased towards texture; increasing shape bias improves accuracy and robustness. In Proceedings of the International Conference on Learning Representations, New Orleans, LA, USA, 6–9 May 2019.
4. Nguyen, A.; Yosinski, J.; Clune, J. Deep neural networks are easily fooled: High confidence predictions for unrecognizable images. In Proceedings of the Computer Vision and Pattern Recognition Conference, Boston, MA, USA, 7–12 June 2015; pp. 427–436.
5. Gururangan, S.; Swayamdipta, S.; Levy, O.; Schwartz, R.; Bowman, S.R.; Smith, N.A. Annotation Artifacts in Natural Language Inference Data. In *Proceedings of the 2018 Conference of the North American Chapter of the Association for Computational Linguistics: Human Language Technologies, Volume 2 (Short Papers)*; Association for Computational Linguistics: New Orleans, LA, USA, 2018.
6. Geirhos, R.; Jacobsen, J.H.; Michaelis, C.; Zemel, R.; Brendel, W.; Bethge, M.; Wichmann, F.A. Shortcut learning in deep neural networks. *Nat. Mach. Intell.* **2020**, *2*, 665–673. [CrossRef]
7. Beery, S.; Van Horn, G.; Perona, P. Recognition in terra incognita. In Proceedings of the European Conference on Computer Vision, Munich, Germany, 8–14 September 2018; pp. 456–473.
8. Arjovsky, M.; Bottou, L.; Gulrajani, I.; Lopez-Paz, D. Invariant risk minimization. *arXiv* **2019**, arXiv:1907.02893.
9. Krueger, D.; Caballero, E.; Jacobsen, J.H.; Zhang, A.; Binas, J.; Zhang, D.; Le Priol, R.; Courville, A. Out-of-distribution generalization via risk extrapolation (rex). In Proceedings of the International Conference on Machine Learning, PMLR, Virtual, 18–24 July 2021; pp. 5815–5826.
10. Ahuja, K.; Shanmugam, K.; Varshney, K.; Dhurandhar, A. Invariant risk minimization games. In Proceedings of the International Conference on Machine Learning, PMLR, Virtual, 13–18 July 2020; pp. 145–155.
11. Pezeshki, M.; Kaba, O.; Bengio, Y.; Courville, A.C.; Precup, D.; Lajoie, G. Gradient starvation: A learning proclivity in neural networks. In Proceedings of the Neural Information Processing Systems, Virtual, 6–14 December 2021; Volume 34.
12. Ahuja, K.; Caballero, E.; Zhang, D.; Gagnon-Audet, J.C.; Bengio, Y.; Mitliagkas, I.; Rish, I. Invariance principle meets information bottleneck for out-of-distribution generalization. In Proceedings of the Neural Information Processing Systems, Virtual, 6–14 December 2021; Volume 34.
13. Pearl, J. *Causality*; Cambridge University Press: Cambridge, UK, 2009.
14. Peters, J.; Janzing, D.; Schölkopf, B. *Elements of Causal Inference: Foundations and Learning Algorithms*; The MIT Press: Cambridge, MA, USA, 2017.
15. Peters, J.; Bühlmann, P.; Meinshausen, N. Causal inference by using invariant prediction: Identification and confidence intervals. *J. R. Stat. Soc. Ser. B* **2016**, *78*, 947–1012. [CrossRef]

16. Tishby, N. The information bottleneck method. In Proceedings of the Annual Allerton Conference on Communications, Control and Computing, Monticello, IL, USA, 22–24 September 1999; pp. 368–377.

17. Aubin, B.; Słowik, A.; Arjovsky, M.; Bottou, L.; Lopez-Paz, D. Linear unit-tests for invariance discovery. *arXiv* **2021**, arXiv:2102.10867.

18. Soudry, D.; Hoffer, E.; Nacson, M.S.; Gunasekar, S.; Srebro, N. The implicit bias of gradient descent on separable data. *J. Mach. Learn. Res.* **2018**, *19*, 2822–2878.

19. Heinze-Deml, C.; Peters, J.; Meinshausen, N. Invariant causal prediction for nonlinear models. *arXiv* **2018**, arXiv:1706.08576.

20. Rojas-Carulla, M.; Schölkopf, B.; Turner, R.; Peters, J. Invariant models for causal transfer learning. *J. Mach. Learn. Res.* **2018**, *19*, 1309–1342.

21. Rosenfeld, E.; Ravikumar, P.K.; Risteski, A. The Risks of Invariant Risk Minimization. In Proceedings of the International Conference on Learning Representations, Virtual, 3–7 May 2021.

22. Kamath, P.; Tangella, A.; Sutherland, D.; Srebro, N. Does invariant risk minimization capture invariance? In Proceedings of the International Conference on Artificial Intelligence and Statistics, PMLR, San Diego, CA, USA, 13–15 April 2021; pp. 4069–4077.

23. Lu, C.; Wu, Y.; Hernández-Lobato, J.M.; Schölkopf, B. Invariant Causal Representation Learning for Out-of-Distribution Generalization. In Proceedings of the International Conference on Learning Representations, Virtual, 25–29 December 2022.

24. Liu, C.; Sun, X.; Wang, J.; Tang, H.; Li, T.; Qin, T.; Chen, W.; Liu, T.Y. Learning causal semantic representation for out-of-distribution prediction. In Proceedings of the Neural Information Processing Systems, Virtual, 6–14 December 2021; Volume 34.

25. Kingma, D.P.; Welling, M. Auto-encoding variational bayes. *arXiv* **2013**, arXiv:1312.6114.

26. Rezende, D.J.; Mohamed, S.; Wierstra, D. Stochastic backpropagation and approximate inference in deep generative models. In Proceedings of the International Conference on Machine Learning, PMLR, Beijing, China, 21–26 June 2014; pp. 1278–1286.

27. Lu, C.; Wu, Y.; Hernández-Lobato, J.M.; Schölkopf, B. Nonlinear invariant risk minimization: A causal approach. *arXiv* **2021**, arXiv:2102.12353.

28. Bengio, Y.; Courville, A.; Vincent, P. Representation learning: A review and new perspectives. *IEEE Trans. Pattern Anal. Mach. Intell.* **2013**, *35*, 1798–1828. [CrossRef] [PubMed]

29. Schölkopf, B.; Locatello, F.; Bauer, S.; Ke, N.R.; Kalchbrenner, N.; Goyal, A.; Bengio, Y. Toward causal representation learning. *Proc. IEEE* **2021**, *109*, 612–634. [CrossRef]

30. Namkoong, H.; Duchi, J.C. Stochastic gradient methods for distributionally robust optimization with f-divergences. In Proceedings of the Neural Information processing Systems, Barcelona, Spain, 5–10 December 2016; Volume 29.

31. Sinha, A.; Namkoong, H.; Volpi, R.; Duchi, J. Certifying some distributional robustness with principled adversarial training. *arXiv* **2017**, arXiv:1710.10571.

32. Lee, J.; Raginsky, M. Minimax statistical learning with wasserstein distances. In Proceedings of the Neural Information Processing Systems, Montreal, Canada, 3–8 December 2018; Volume 31.

33. Duchi, J.C.; Namkoong, H. Learning models with uniform performance via distributionally robust optimization. *Ann. Stat.* **2021**, *49*, 1378–1406. [CrossRef]

34. Bühlmann, P. Invariance, causality and robustness. *Stat. Sci.* **2020**, *35*, 404–426. [CrossRef]

35. Blanchard, G.; Lee, G.; Scott, C. Generalizing from several related classification tasks to a new unlabeled sample. In Proceedings of the Neural Information Processing Systems, Granada, Spain, 12–15 December 2011; Volume 24.

36. Muandet, K.; Balduzzi, D.; Schölkopf, B. Domain generalization via invariant feature representation. In Proceedings of the International Conference on Machine Learning, PMLR, Atlanta, GA, USA, 16–21 June 2013; pp. 10–18.

37. Deshmukh, A.A.; Lei, Y.; Sharma, S.; Dogan, U.; Cutler, J.W.; Scott, C. A generalization error bound for multi-class domain generalization. *arXiv* **2019**, arXiv:1905.10392.

38. Ye, H.; Xie, C.; Cai, T.; Li, R.; Li, Z.; Wang, L. Towards a Theoretical Framework of Out-of-Distribution Generalization. In Proceedings of the Neural Information Processing Systems, Virtual, 6–14 December 2021.

39. Xie, C.; Chen, F.; Liu, Y.; Li, Z. Risk variance penalization: From distributional robustness to causality. *arXiv* **2020**, arXiv:2006.07544.

40. Jin, W.; Barzilay, R.; Jaakkola, T. Domain extrapolation via regret minimization. *arXiv* **2020**, arXiv:2006.03908.

41. Mahajan, D.; Tople, S.; Sharma, A. Domain generalization using causal matching. In Proceedings of the International Conference on Machine Learning, PMLR, Virtual, 18–24 July 2021; pp. 7313–7324.

42. Bellot, A.; van der Schaar, M. Generalization and invariances in the presence of unobserved confounding. *arXiv* **2020**, arXiv:2007.10653.

43. Li, B.; Shen, Y.; Wang, Y.; Zhu, W.; Reed, C.J.; Zhang, J.; Li, D.; Keutzer, K.; Zhao, H. Invariant information bottleneck for domain generalization. In Proceedings of the Association for the Advancement of Artificial Intelligence, Virtual, 22 Februay–1 March 2022.

44. Alesiani, F.; Yu, S.; Yu, X. Gated information bottleneck for generalization in sequential environments. *Knowl. Informat. Syst.* **2022**, 1–23, *in press*. [CrossRef]

45. Wang, H.; Si, H.; Li, B.; Zhao, H. Provable Domain Generalization via Invariant-Feature Subspace Recovery. In Proceedings of the International Conference on Machine Learning, Baltimore, MD, USA, 17–23 July 2022.

46. Ganin, Y.; Lempitsky, V. Unsupervised domain adaptation by backpropagation. In Proceedings of the International conference on machine learning, PMLR, Lille, France, 6–11 July 2015; pp. 1180–1189.

47. Li, Y.; Tian, X.; Gong, M.; Liu, Y.; Liu, T.; Zhang, K.; Tao, D. Deep domain generalization via conditional invariant adversarial networks. In Proceedings of the European Conference on Computer Vision, Munich, Germany, 8–14 September 2018; pp. 624–639.

48. Zhao, S.; Gong, M.; Liu, T.; Fu, H.; Tao, D. Domain generalization via entropy regularization. In Proceedings of the Neural Information Processing Systems, Virtual, 6–12 December 2020; Volume 33, pp. 16096–16107.

49. Ben-David, S.; Blitzer, J.; Crammer, K.; Pereira, F. Analysis of representations for domain adaptation. In Proceedings of the Neural Information Processing Systems, Hong Kong, China, 3–6 October 2006; Volume 19.

50. Ben-David, S.; Blitzer, J.; Crammer, K.; Kulesza, A.; Pereira, F.; Vaughan, J.W. A theory of learning from different domains. *Mach. Learn.* **2010**, *79*, 151–175. [CrossRef]

51. Zhao, H.; Des Combes, R.T.; Zhang, K.; Gordon, G. On learning invariant representations for domain adaptation. In Proceedings of the International Conference on Machine Learning. PMLR, Long Beach, CA, USA, 10–15 June 2019; pp. 7523–7532.

52. Xu, Q.; Zhang, R.; Zhang, Y.; Wang, Y.; Tian, Q. A fourier-based framework for domain generalization. In Proceedings of the IEEE/CVF Conference on Computer Vision and Pattern Recognition, Nashville, TN, USA, 20–25 June 2021; pp. 14383–14392.

53. Zhou, K.; Yang, Y.; Qiao, Y.; Xiang, T. Domain Generalization with MixStyle. In Proceedings of the International Conference on Learning Representations, Vienna, Austria, 3–7 May 2021.

54. Zhang, X.; Cui, P.; Xu, R.; Zhou, L.; He, Y.; Shen, Z. Deep stable learning for out-of-distribution generalization. In Proceedings of the IEEE/CVF Conference on Computer Vision and Pattern Recognition, Nashville, TN, USA, 21–25 June 2021; pp. 5372–5382.

55. Wang, H.; Ge, S.; Lipton, Z.; Xing, E.P. Learning robust global representations by penalizing local predictive power. In Proceedings of the Neural Information Processing Systems, Vancouver, BC, Canada, 8–14 December 2019; Volume 32.

56. Gulrajani, I.; Lopez-Paz, D. In Search of Lost Domain Generalization. In Proceedings of the International Conference on Learning Representations, Virtual, 2–4 December 2020.

57. Wiles, O.; Gowal, S.; Stimberg, F.; Rebuffi, S.A.; Ktena, I.; Dvijotham, K.D.; Cemgil, A.T. A Fine-Grained Analysis on Distribution Shift. In Proceedings of the International Conference on Learning Representations, Virtual, 25–29 April 2022.

58. Thomas, M.; Joy, A.T. *Elements of Information Theory*; Wiley-Interscience: Hoboken, NJ, USA, 2006.

Article

On Neural Networks Fitting, Compression, and Generalization Behavior via Information-Bottleneck-like Approaches

Zhaoyan Lyu [1],* , Gholamali Aminian [2] and Miguel R. D. Rodrigues [1]

1 Department of Electronic and Electrical Engineering, University College London, Gower St., London WC1E 6BT, UK; m.rodrigues@ucl.ac.uk
2 The Alan Turing Institute, British Library, 96 Euston Rd., London NW1 2DB, UK; gaminian@turing.ac.uk
* Correspondence: z.lyu.17@ucl.ac.uk

Abstract: It is well-known that a neural network learning process—along with its connections to fitting, compression, and generalization—is not yet well understood. In this paper, we propose a novel approach to capturing such neural network dynamics using information-bottleneck-type techniques, involving the replacement of mutual information measures (which are notoriously difficult to estimate in high-dimensional spaces) by other more tractable ones, including (1) the minimum mean-squared error associated with the reconstruction of the network input data from some intermediate network representation and (2) the cross-entropy associated with a certain class label given some network representation. We then conducted an empirical study in order to ascertain how different network models, network learning algorithms, and datasets may affect the learning dynamics. Our experiments show that our proposed approach appears to be more reliable in comparison with classical information bottleneck ones in capturing network dynamics during both the training and testing phases. Our experiments also reveal that the fitting and compression phases exist regardless of the choice of activation function. Additionally, our findings suggest that model architectures, training algorithms, and datasets that lead to better generalization tend to exhibit more pronounced fitting and compression phases.

Keywords: deep learning; information theory; information bottleneck; generalization; fitting; compression

Citation: Lyu, Z.; Aminian, G.; Rodrigues, M.R.D. On Neural Networks Fitting, Compression, and Generalization Behavior via Information-Bottleneck-like Approaches. *Entropy* **2023**, *25*, 1063. https://doi.org/10.3390/e25071063

Academic Editors: Gerhard Bauch and Jan Lewandowsky

Received: 30 April 2023
Revised: 11 July 2023
Accepted: 12 July 2023
Published: 14 July 2023

1. Introduction

Deep learning models have gained enormous attention thanks to their impressive performance compared with traditional learning models in a variety of areas, such as computer vision, speech processing, natural language processing, and many more [1,2]. However, despite their stunning performance, we still do not fully understand how deep neural networks work [3].

A number of recent approaches have been proposed to study the generalization/optimization properties of over-parameterized models, such as deep neural networks [4,5]. However, these approaches do not fully capture certain neural network representation properties, including how these evolve during the neural network training procedure. Such an understanding of the role of different components of the model and their impact on the learning process can be essential for selecting or designing better neural network models and associated learning algorithms.

Another popular approach to studying the generalization/optimization dynamics of deep neural networks has been the information bottleneck (IB). This approach, which is based on the information bottleneck theory [6,7], employs the mutual information (MI) between the data and their neural network representation, as well as MI between labels and the neural network representation to capture neural network behavior. In particular, in classification problems, it is typical to model the relationship between the data label

Y, the data themselves X, and some neural network intermediate data representation Z via a Markov chain $Y \to X \to Z$, where Y, X, and Z represent random variables/vectors associated with these different objects. Then, the IB principle is described via two MIs: (1) $I(Z; X)$ to measure the amount of information contained in the data representation about the input data, and (2) $I(Z; Y)$ to measure the information in the data representation that could contribute to the prediction of ground-truth labels. One can capture how the value of $I(Z; X)$ and $I(Z; Y)$ evolve as a function of the number of training epochs for a neural network by plotting pairs of these mutual information values on a two-dimensional plane [8]. The plane defined by these MI terms is called the information plane (IP), and the trace of the MI value versus training epoch is called the information plane dynamic (IP-dynamic).

This approach has led to the identification of some trends associated with the optimization of neural networks. In particular, by observing the IP-dynamic of the networks trained on a synthetic dataset and the MNIST dataset, ref. [8] found that, in early epochs, both $I(Z; X)$ and $I(Z; Y)$ increase; and, in later epochs, $I(Z; Y)$ will keep increasing while $I(Z; X)$ decreases. This led to the conjecture that the training of a neural network contains two different phases: (1) a **fitting phase,** where the network representation Z fits the input data X as much as possible, and (2) a subsequent **compression phase** in which the network compresses the useless information in the representation Z about the labels Y.

However, the IB approach requires estimating $I(Z; X)$ and $I(Z; Y)$, which is notoriously difficult to accomplish because the inputs and representations typically lie in very high-dimensional spaces. For example, non-parametric mutual information estimators—such as [9,10]—suffer from either high bias or high variance, especially in high-dimensional settings [10]. This will directly affect any conclusions extracted from the IP-dynamics because high bias prevents recognizing the existence of fitting or compression phases, whereas high variance leads to inconsistent results across different numerical experiments. Indeed, with different mutual information estimators, researchers drew diverse or opposite conclusions about trends in IP-dynamics [8,11–25]. For instance, Saxe et al. [24] argued that the reported phenomena of fitting and compression in Shwartz et al.'s study [8] are highly dependent on the simple binning MI estimator setup adopted.

Therefore, the trends that one often extracts from an IB analysis may not always hold.

1.1. Paper Contributions

This paper attempts to resolve these issues by introducing a different approach to studying the dynamics of neural networks. Our main contributions are as follows:

1. First, we propose to use more tractable measures to capture the relationship between an intermediate network data representation and the original data or the intermediate network representation and the data label. In particular, we used the minimum mean-squared error between the intermediate data representation and the original data to try to capture fitting and compression phenomena occurring in a neural network; we also used the well-known cross-entropy between the intermediate data representation and the data label to capture performance.
2. Second, by building upon the variational representations of these quantities, we also propose to estimate such measures using neural networks. In particular, our experimental results demonstrate that such an approach leads to consistent estimates of the measures using different estimator neural network architectures and initializations.
3. Finally, using our proposed approach, we conducted an empirical study to reveal the influence of various factors on neural network learning processing, including compression, fitting, and generalization phenomena. Specifically, we considered the impact of (1) the machine learning model, (2) the learning algorithm (optimizer and regularization techniques), and (3) the data.

The main findings deriving from our empirical study—along with the literature that explored similar network architecture, training algorithm, or data setups—are summarized in Table 1. In particular, we highlight that our study suggests that (1) a neural network generalization performance improves with the magnitude of the network's fitting and compression phase; (2) a network tends to undergo a fitting phase followed by a compression phase, regardless of the activation function; and (3) the specific behavior of the fitting/compression phases depends on a number of factors, including the network architecture, the learning algorithm, and the nature of the data.

Table 1. Overview of our main results and related literature results. Fit., Com., and Gen. are abbreviations for fitting, compression, and generalization, respectively. Note that the related literature listed explored the information bottleneck under similar setups but may report different observations or focus on different phenomena in the dynamics.

Study	Model	Training Algorithm	Dataset	Section	Our Observation	Related Literature
Effects of model architectures	Tishby-nets with saturated or non-saturated activation functions	SGD	Tishby-dataset	Section 5.2.1	Fit./Com. phases exist regardless of the type of activation function.	[8,11,14, 24,26,27]
	MLPs with more or fewer neurons per layer		MNIST	Section 5.2.2	MLPs with more neurons per layer exhibit faster Fit., more Com., and better Gen.	-
	MLPs with more or fewer layers				MLPs with more layers have less Fit. but more Com. The MLP that exhibits more pronounced Fit. and Com. also tends to Gen. better.	[27]
	CNNs with more or fewer kernels	Adam	CIFAR-10	Section 5.2.3	CNNs with fewer kernels cannot Fit. and Com. effectively, and do not Gen. well. Increasing the number of kernels on a well-generalized CNN does not have a significant impact on Fit., Com., or Gen.	-
	CNNs with bigger or smaller kernels				Both very large and very small kernel sizes tend to result in less Fit. and Com., and can harm Gen.	
	ResCNN			Section 5.2.4	The representations at the outputs of residual blocks do not exhibit Fit./Com. phases, while the representations in the residual blocks exhibit Fit./Com. phases.	[13,18,28]
Effects of training algorithms	CNN	SGD, SGD-momentum, RMSprop, Adam		Section 5.3.1	Adaptive optimizers compress more on layers closer to the input.	[29]
	MLP	SGD with or without weight decay	MNIST	Section 5.3.2	Weight decay does not significantly affect the Fit. phase, but it can increase the Com. capability of the model and improve its Gen. performance.	[11,18,21]
	CNN	Adam with or without dropout	CIFAR-10		A low dropout rate does not significantly impact Fit., but it can enhance Com. and improve Gen. In contrast, a high dropout rate can lead to less Fit. and Com., resulting in worse Gen.	-
Effects of dataset size		Adam	CIFAR-10, CINIC	Section 5.4	CINIC dataset enhances Fit., Com., and Gen. CIFAR-10 subset has less Com. and worse Gen..	[8,14]

1.2. Scope of Study

Finally, we note that the information bottleneck technique has been used as a tool to cast insight into other machine learning paradigms, including semi-supervised learning [30] and unsupervised learning [31–33]. However, we focused exclusively on supervised learning settings—with an emphasis on neural networks—in order to contribute to a deeper understanding of deep learning techniques.

1.3. Paper Organization

This paper is organized as follows: Section 2 offers an overview of the literature that relates to our work. Section 3 proposes our approach to studying the compression, fitting, and generalization dynamics of neural networks, whereas Section 4 discusses practical implementation details associated with our proposed approach. Section 5 leverages our approach to conducting an empirical study of the impact of various factors on the compression, fitting, and generalization behavior of a neural network, including the underlying architecture, learning algorithm, and nature of the data. Finally, we summarize the paper, discuss its limitations, and propose future directions in Section 6.

1.4. Paper Notation

We adopt the following convention for random variables and their distributions throughout the paper. A random variable (or vector) is denoted by an upper-case letter (e.g., Z), and its space of possible values is denoted with the corresponding calligraphic letter (e.g., $\mathcal{Z}$). The probability distribution of the random variable Z is denoted by P_Z. The joint distribution of a pair of random variables (Z_1, Z_2) is denoted by P_{Z_1, Z_2}. $H(Z)$ represents the entropy (or differential entropy) of random variable Z, $H(Z_1|Z_2)$ represents the entropy (or differential entropy) of random variable Z_1 given random variable Z_2, and $I(Z_1; Z_2)$ represents the mutual information between random variables Z_1 and Z_2. We denote the set of integers from 1 to n by $[n] \triangleq \{1, \cdots, n\}$.

2. Related Work

There are various lines of research that connect to our work.

Information bottleneck (IB) and information plane (IP) dynamics: Many works have adopted the IB and the IP to study the optimization dynamics of neural networks. Refs. [8,18,19,26,28] concluded that there is a different fitting and compression phase during the training of a deep neural network, while [24,34] claim that neural networks with saturating activation functions exhibit a fitting phase but do not exhibit a compression phase. Ref. [11] conveyed that the network may occasionally compress only for some random initializations. On the other hand, ref. [11] found that weight decay regularization will increase the magnitude of the compression, while [14] did not observe compression unless weight decay is applied. Finally, overfitting was observed from the IP associated with hidden layers in [8,23,34].

While these works mentioned above explore various aspects of deep learning techniques, such as how network behaviors are affected by varying training dataset sizes and regularization techniques, their conclusions may not always be reliable due to the fact that MI estimation can be inaccurate and unstable in high-dimensional settings, as argued in [12].

IB and IP based on other information measures: Many works have also adopted IBs/IPs based on other information measures to study the dynamics of neural networks. Motivated by source coding, ref. [35] proposes to replace the $I(Z; X)$ with the entropy of the representation Z. The authors in [36] introduced a generalized IB based on f-divergence. The authors also proposed an estimation bottleneck based on χ^2-information, but this quantity is difficult to estimate in practice, preventing its applicability in various problems. The paper [37] proposed an information bottleneck approach based on MMSE and Fisher information to develop robust neural networks. However, the authors utilized MMSE to substitute mutual information between the representation and ground truth label, whereas we employed it to evaluate the association between representation and data. Inspired by [38], ref. [39] introduced a new IB—called the $\mathcal{V}$-information bottleneck—that articulates the amount of useful information a representation embodies about a target usable by a classifier drawn from a family of classifiers $\mathcal{V}$. Recently, refs. [40,41] have used sliced mutual information to study fitting in neural networks. However, their work mainly focused on the fitting phase and did not explore the role of compression and its relationship with generalization.

Mutual information estimation: Relying on mutual information to study the dynamics of neural networks leads to various challenges. The first challenge relates to the fact that the MI between two quantities that lie in continuous space and are linked by a functional relationship, such as the input and the output of a neural network, is theoretically infinite [42]. This limits its use since a neural network representation is typically a deterministic function of the neural network input [8,11,21,24]. Many works have circumvented this issue by adding additional noise to the random variables. For instance, kernel density estimation (KDE) [43,44] was used by [11,13,24,45], and the k-nearest-neighbor based Kraskov estimator [46] was used in [18,24,47]. Other works using variational mutual information estimators address the challenge by adding noise to the neural network representations [14,19]. However, adding noise to the representations of a neural network is not a widespread practice in most deep learning implementations. An alternative measure of dependence between two variables is sliced mutual information, which was proposed by [48]. This method involves random projections and the averaging of mutual information across pairs of projected scalar variables. Our approach differs from this method as we directly processed the random variables in high-dimensional space.

The second challenge relates to the fact that many mutual information estimators exhibit high bias and/or high variance in a high-dimensional setting. For example, simple binning methods [8,49] are known to lead to mutual information estimates that vary greatly depending on the bin size choice. Further, variational mutual information estimators, such as MINE [9], are also known to produce mutual information estimates that suffer from high bias or high variance [10,50].

Our work departs from existing work because we propose to study the evolution of two more stable measures during a neural network optimization process: (1) the minimum mean-squared error associated with the estimation of the original data given some intermediate network representation and (2) the cross-entropy associated with the original data label given an intermediate data representation. This offers a more reliable lens for studying compression, fitting, and generalization phenomena occurring in neural networks.

3. Proposed Framework

We now introduce our approach to studying the compression, fitting, and generalization dynamics of neural networks. We focused exclusively on classification problems characterized by a pair of random variables $\{(X, Y)|X \in \mathcal{X}, Y \in \mathcal{Y}\}$, where X is the input data and Y is the ground-truth label, that follow a distribution $P_{X,Y}$. We delivered an estimate of the ground-truth label $\hat{Y} \in \mathcal{Y}$ given the data $X \in \mathcal{X}$ using an L-layer neural network as follows:

$$\hat{Y} = f_\theta(X) = f_{\theta_L}^{(L)}\left(f_{\theta_{L-1}}^{(L-1)}\left(\cdots f_{\theta_1}^{(1)}(X)\right)\right) \tag{1}$$

where $f_{\theta_l}^{(l)}(\cdot)$ models the operation of the l-th ($l \in [L]$) network layer, where θ_l represents the parameters of this layer (the weights and biases). The network parameters were optimized using standard procedures given a (training) dataset containing various (training) samples.

The optimized network can then be used to make new output predictions $\hat{Y}$ given new input data X.

The network optimization procedure involves the application of iterative learning algorithms such as stochastic gradient descent. Therefore, at a certain epoch i associated with the learning algorithm, we can model the flow of information in the neural network via a Markov chain as follows:

$$Y \to X \to Z_1^{(i)} \to Z_2^{(i)} \to \cdots \to Z_L^{(i)} \to \hat{Y} \tag{2}$$

where the random variable $Z_l^{(i)} = f_{\theta_l}^{(l)}\left(Z_{l-1}^{(i)}\right) \in \mathcal{R}^{n_l}$ represents the network representation at layer l at epoch i in the n_l-dimension (with a convention that $Z_0^{(i)} = X$). Our goal was to examine how certain quantities—capturing the compression, fitting, and generalization behavior—associated with the network optimization process evolve as a function of the number of algorithm training epochs.

Z-X measure: Our first quantity describes the difficulty in recovering the original data X from some intermediate network representation $Z_l^{(i)}$ as follows:

$$m_{Z_l^{(i)};X} = \inf_{f_x \in \mathcal{C}(\mathcal{R}^{n_l} \to \mathcal{X})} \mathbb{E}\left[\ell_X\left(f_x\left(Z_l^{(i)}\right); X\right)\right] \tag{3}$$

where $f_x(\cdot) : \mathcal{R}^{n_l} \to \mathcal{X}$ is an estimator living in the function space $\mathcal{C}(\mathcal{R}^{n_l} \to \mathcal{X})$ and $\ell_X(\cdot; \cdot)$ is a loss function. We will take the loss function to correspond to the squared error so that the Z-X measure reduces to the well-known minimum mean-squared error given by:

$$m_{Z_l^{(i)};X} = \mathrm{mmse}\left(X|Z_l^{(i)}\right) = \mathbb{E}\left[\left(X - \mathbb{E}\left[X|Z_l^{(i)}\right]\right)^2\right] \tag{4}$$

where the function $f_x(\cdot)$ that minimizes the right-hand side of Equation (3) is the well-known conditional mean estimator. Our rationale for adopting this quantity to capture the relationship between the network representation and the data in *lieu* of mutual information—which is used in the conventional IB—is manifold:

- First, the minimum mean-squared error can act as a proxy to capture fitting—the lower the MMSE, the easier it is to recover the data from the representation—and compression—the higher the MMSE, the more difficult it is to estimate the data from the representation.
- Second, this quantity is also easier to estimate than mutual information, allowing us to capture the phenomena above reliably (see Section 5.1).
- Finally, the minimum mean-squared error is also connected to mutual information (see Section 3.1).

Z-Y measure: Our second quantity describes the difficulty in recovering the original label Y from some intermediate network representation $Z_l^{(i)}$ as follows:

$$m_{Z_l^{(i)};Y} = \inf_{f_y \in \mathcal{C}(\mathcal{R}^{n_l} \to \mathcal{Y})} \mathbb{E}\left[\ell_Y\left(f_y\left((Z_l^{(i)}); Y\right)\right)\right] \tag{5}$$

where $f_y(\cdot) : \mathcal{R}^{n_l} \to \mathcal{Y}$ is an estimator living in the function space $\mathcal{C}(\mathcal{R}^{n_l} \to \mathcal{Y})$ and $\ell_Y(\cdot; \cdot)$ is a loss function. We will take the loss function to correspond to the cross-entropy so that the Z-Y measure reduces to the well-known conditional entropy given by:

$$m_{Z_l^{(i)};Y} = H(Y|Z_l^{(i)}) \tag{6}$$

where the function $f_y(\cdot)$ that minimizes the right-hand side of Equation (5) should model the distribution of the label given the representation. We also adopted this measure because it connects directly to performance—hence the ability of the network to generalize—but also to mutual information (see Section 3.1).

Plane and Dynamics of the Z-X and Z-Y Measures: Equipped with the measures in Equations (4) and (6), one can immediately construct a two-dimensional plane plotting the Z-X measure $m_{Z_l^{(i)};X}$ against the Z-Y measure $m_{Z_l^{(i)};Y}$ as a function of the number of network training epochs $i = 1, 2, 3, \ldots$ in order to understand (empirically) how a particular neural network operates. Such a plane and the associated dynamics are the analogue of the IB plane and the IB dynamics introduced in [8].

3.1. Connecting our Approach to the Information Bottleneck

Our approach is also intimately connected to the conventional information bottleneck because—as alluded to earlier—our adopted measures are also connected to mutual information. First, in accordance to [51] (Theorem 10), we can bound the mutual information between the data X and the representation $Z_l^{(i)}$ as follows:

$$\frac{1}{2} I(X; Z_l^{(i)}) \geq \text{var}(X) - \text{mmse}(X | Z_l^{(i)}) \tag{7}$$

where $\text{var}(\cdot)$ represents the variance of the random variable.

Second, we can also trivially express the mutual information between the data Y and the representation $Z_l^{(i)}$ as follows:

$$I(Z_l^{(i)}; Y) = H(Y) - H(Y | Z_l^{(i)}) \tag{8}$$

However, the main advantage of our approach in relation to the traditional IB is that it is much easier to estimate the proposed Z-X and Z-Y measures than the corresponding mutual information in high-dimensional settings; see Section 5.1.

4. Implementation Aspects

4.1. Experimental Procedure

The crux of our approach involves tracking how the Z-X and Z-Y measures evolve during the network optimization process as a function of the learning algorithm epochs. However, we cannot estimate the measures in Equations (4) and (6) directly because we do not have access to the relevant probability distributions. Instead, we will leverage the variational representations of the Z-X measure in Equation (3) and the Z-Y measure in Equation (5) to approximate the measures in a data-driven manner, given access to a dataset $S = \{X(k), Y(k)\}_{k=1}^{n}$ consisting of various input–label pairs.

In particular, given this dataset $S = \{X(k), Y(k)\}_{k=1}^{n}$, we used learnable functions $f_\phi : \mathcal{R}^{n_l} \to \mathcal{X}$ and $f_\psi : \mathcal{R}^{n_l} \to \mathcal{Y}$—which are neural networks parameterized by ϕ and ψ, respectively—to approximate the measures in Equations (4) and (6) as follows:

$$m_{Z_l^{(i)}; X} = \inf_{f_x} \mathbb{E}\left[\ell_x\left(f_x(Z_l^{(i)}); X \right) \right] \leq \min_{\phi} \mathbb{E}\left[\ell_x\left(f_\phi(Z_l^{(i)}); X \right) \right] \approx \min_{\phi} \frac{1}{n} \sum_{k=1}^{n} \ell_x\left(f_\phi(Z_l^{(i)}(k)); X(k) \right)$$

$$= \hat{m}_{Z_l^{(i)}; X} \tag{9}$$

and

$$m_{Z_l^{(i)}; Y} = \inf_{f_y} \mathbb{E}\left[\ell_y\left(f_y(Z_l^{(i)}); Y \right) \right] \leq \min_{\psi} \mathbb{E}\left[\ell_y\left(f_\psi(Z_l^{(i)}); Y \right) \right] \approx \min_{\psi} \frac{1}{n} \sum_{k=1}^{n} \ell_y\left(f_\psi(Z_l^{(i)}(k)); Y(k) \right)$$

$$= \hat{m}_{Z_l^{(i)}; X} \tag{10}$$

respectively, where the learnable function parameters ϕ and ψ are drawn from Φ and Ψ, $Z_l^{(i)} = f_{\theta_l}^{(l)}\left(Z_{l-1}^{(i)} \right)$, and $Z_l^{(i)}(k) = f_{\theta_l}^{(l)}\left(Z_{l-1}^{(i)}(k) \right)$. Note that—in view of the fact that $\{f_\phi : \phi \in \Phi\} \subset \mathcal{C}(\mathcal{R}^{n_l} \to \mathcal{X})$ and $\{f_\psi : \psi \in \Psi\} \subset \mathcal{C}(\mathcal{R}^{n_l} \to \mathcal{Y})$—one is confronted with an immediate trade-off: the higher the number of parameters in the learnable functions, the closer the upper bounds to the measures in Equations (9) and (10) are to the actual measure but also the higher the number of samples that may be required to approximate the upper bound reliably. This will be further discussed in Section 5.1.

Our setup is summarized in Figure 1. We re-emphasize that there were three neural networks involved in our study: (1) $f_\theta(\cdot)$ is the network whose dynamics we wish to study (in the green box of Figure 1), (2) $f_\phi(\cdot)$ represents the neural network used to approximate the Z-X measure (in the blue box of Figure 1), and (3) $f_\psi(\cdot)$ represents the neural network used to estimate the Z-Y measure (in the yellow box of Figure 1)

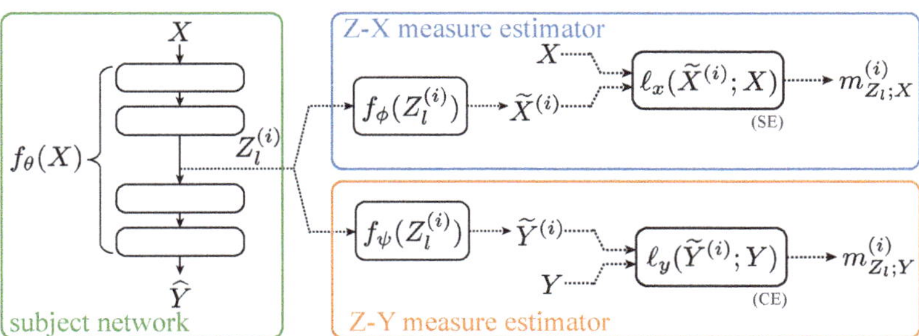

Figure 1. Proposed approach. We used two estimator neural networks $f_\phi(\cdot)$ and $f_\psi(\cdot)$ to study the behavior of the Z-X measure and the Z-Y measure associated with the different representations of the subject network $f_\theta(\cdot)$. The $\ell_x(\cdot)$ and $\ell_y(\cdot)$ are squared loss and cross-entropy, respectively.

We optimized these networks using the procedure outlined in Algorithms 1 and 2. The algorithm used to optimize the neural network f_θ can be used with different neural network models, different learning algorithms, or different datasets. Note that this algorithm saves the neural network learnable parameters as checkpoints every several epochs (as shown in Algorithm 1), where we used T to control the total number of checkpoints to limit computational overhead.

In turn, the algorithm used to train the estimator networks $f_\phi(\cdot)$ and $f_\psi(\cdot)$ uses the Adam optimizer with a learning rate of 0.01 for efficient and stable estimation. The estimator networks were initialized using the standard Xavier [52] initialization (unless otherwise specified), and the estimator networks were also optimized until convergence (which is identified by the increase in loss value on the validation set).

We note that we trained the subject network on a training set, but we trained the estimator networks on a different (independent) validation set in order to obtain estimates of the Z-X and Z-Y measures that can also capture generalization behavior. The Tishby-dataset is an exception since it does not have a separate validation set. Note, however, that, in the IB literature, few studies have reported differences in trends by estimating the relevant mutual information quantities on the training set or an independent validation set. Some studies (e.g., [8]) also do not specify the dataset used to compute the mutual information measures.

We will be referring for simplicity in the sequel to the network whose dynamics we wish to study (i.e., $f_\theta(\cdot)$) as the *subject network* and to the networks whose purpose is to estimate the relevant measures (i.e., $f_\phi(\cdot)$ and $f_\psi(\cdot)$) as the *estimator networks*.

Algorithm 1: Train the subject network

Input: number of epochs I, number of checkpoints T
Data: $\{(X(k); Y(k))\}_{k=1}^{n}$
Output: T checkpoint files containing network parameters associated with the
        different intermediate points
initialize f_θ with random $\theta^{(0)}$;
$i \leftarrow 1, t \leftarrow 1$;
while $i \leq I$ **do**
    if $i\%\lfloor I/T \rfloor = 0$ **then**
        save $\theta^{(i)}$ to a checkpoint file $\theta^{(t)}$;
        $t \leftarrow t+1$;
    end
    optimize the parameters $\theta^{(i)}$ of the subject neural network with a standard
    learning algorithm given dataset $\{(X(k); Y(k))\}_{k=1}^{n}$;
end

Algorithm 2: Estimate Z-X measure and Z-Y measure

Input: $f_{\theta^{(t)}}$ from Algorithm 1, t, random seed s
Data: $\{(X(k); Y(k))\}_{k=1}^{n}$
Output: $\{\hat{m}_{Z_1;X}, \ldots, \hat{m}_{Z_L;X}\}$, $\{\hat{m}_{Z_1;Y}, \ldots, \hat{m}_{Z_L;Y}\}$
Obtain subject network representations $\{\{Z_l(k)\}_{k=1}^{n}\}_{l=1}^{L} \leftarrow f_{\theta^{(t)}}(\{X(k)\})|_{k=1}^{n}$;
$l \leftarrow 1$;
while $l \leq L$;   `// This loop is parallelizable.`
do
    initialize f_{ϕ_l}, f_{ψ_l} with random seed s;
    while f_{ϕ_l} *not converge*;   `// Estimating Z-X measure.`
    do
        $\hat{m}_{Z_l;X}(\phi_l) \leftarrow \frac{1}{n}\sum_{k=1}^{n} \ell_x(f_{\phi_l}(Z_l(k)); X(k))$;
        update ϕ_l to minimize $\hat{m}_{Z_l;X}$ with a standard learning algorithm;
    end
    while f_{ψ_l} *not converge*;   `// Estimating Z-Y measure.`
    do
        $\hat{m}_{Z_l;Y}(\psi_l) \leftarrow \frac{1}{n}\sum_{k=1}^{n} \ell_y(f_{\psi_l}(Z_l(k)); Y(k))$;
        update ψ_l to minimize $\hat{m}_{Z_l;Y}$ with a standard learning algorithm;
    end
end

4.2. Experimental Setups

Our experiments studied the effect of the (subject) network model architecture, the (subject) network learning algorithm, and the dataset on key aspects, such as network fitting, compression, and generalization, via the Z-X and the Z-Y dynamics; see also Table 1. We therefore summarize next the main models, learning algorithms, and datasets used in the study reported in Section 5.

Subject Network Models: We adopted a series of neural network models, including: (1) the Tishby-net, proposed by [8], with the Tishby-dataset, consisting of 4096 samples with binary labels; (2) MLP models with varying number of layers and varying width per layer with an MNIST dataset [53], which has 60,000 grayscale handwritten digit images for training and 10,000 for validation; (3) a convolutional neural network (CNN) with VGG-like [54] architecture trained on the CIFAR-10 [55] and CINIC [56] dataset, where the CIFAR-10 dataset comprises 50,000 RGB images categorized into 10 classes for training and 10,000 for validation, and the CINIC dataset consists of 900,000 training samples labeled in the same way as the CIFAR-10 dataset; and (4) a ResCNN model on the CIFAR-10 dataset, which is a CNN architecture with residual connections modified from the original CNN. The various models and datasets will allow us to study the effect of model architectures and datasets on network dynamics. These models are illustrated in Figure 2. Note that, for MNIST and CIFAR-10 datasets, we separated the validation sets into two halves of the same size: one half was used for plotting the dynamics, and the other half served as the test set for evaluating generalization performance.

Subject Network Learning Algorithm: We also adopted a series of learning algorithms, including (1) training a CNN on the CIFAR-10 dataset using different optimizers, such as non-adaptive (SGD, SGD-momentum) and adaptive (RMSprop, Adam); (2) training an MLP on the MNIST dataset with or without weight decay regularization, where the regularization hyper-parameter was set to 0.001; and (3) training a CNN on the CIFAR-10 dataset with or without dropout regularization, where the dropout was only applied on the fully connected layer of the CNN as implemented in [54]. These setups allowed us to study the effect of different optimization algorithms and regularization methods on network dynamics.

Estimator Neural Network Model and Algorithms: We deployed a variety of estimator network architectures that depend on the architecture of the subject network (namely, the specific shape of the subject network representations in the different layers) as follows:

- For Tishby-net and MLP WxL models, the models for both the Z-X measure estimator and Z-Y measure estimator are fully connected neural networks. The input layer of the estimator networks matches the dimension of the representation (Z_l), while the output layer has a dimension equivalent to either the input vector (for Z-X measures) or label length (for Z-Y measures). If the estimator network has multiple layers, its hidden layers will be connected using ReLU non-linearity and have a number of neurons equal to the dimension of representation (Z_l).

- To estimate the Z-Y measure for CNN and ResCNN, we flattened the representation into a vector and employed the same network architecture as for the Z-Y measure estimator of Tishby-net and MLP WxL models. In turn, to estimate the Z-X measure, we used a convolution layer with a 3×3 kernel size to map the representation into the input space of $32 \times 32 \times 3$. However, if the representation is down-sampled by a pooling layer (e.g., Figure 2 CNN Z_2), we up-sampled it using a transposed convolutional layer with a 2×2 kernel size before feeding it into the convolutional layer. The number of transposed convolutional layers equals the number of pooling layers that the representation has gone through since each transposed convolutional layer can only up-sample the representation by a factor of 2. ReLU non-linearity exists between all hidden layers. For example, when the representation is generated by a layer with two pooling layers before it (e.g., Figure 2 CNN Z_3), the estimator for the Z-X measure would contain two transposed convolutional layers.

These estimators have been shown to be computationally efficient, offering stable results.

4.3. Other Practical Considerations

In view of the fact that we computed the relevant measures for different layers of the subject network at different learning epochs, we also adopted various other practical tricks to improve the computation efficiency as follows:

1. **Parallelize checkpoint enumeration** $t \in \{1, 2, \ldots, T\}$**:** To plot the Z-X / Z-Y measures dynamics, we need to calculate these quantities at different checkpoints saved from various epochs during the training of the subject network.
   We can easily deploy multiple Algorithm 2 instances on different checkpoints saved per Algorithm 1 in parallel;

2. **Parallelize layer iteration** $l \in \{1, 2, \ldots, L\}$**:** We can also break up the iteration of l layers in Algorithm 2 into parallel processes since the estimations of the measures on different layers are independent;

3. **Parallelize estimation of Z-X measure and Z-Y measure:** We can also deploy the Z-X measure estimator and the Z-Y measure estimator on different processes because they are also independent;

4. **Warm-start:** Moreover, we can accelerate the convergence of estimator networks by using warm-start. We randomly initialized and trained the estimators from scratch in the first checkpoint for Tishby-net and MLP WxL models. We then used the learned parameters as initialization for the estimators in subsequent checkpoints. However, we did not use warm-start in CNN and ResCNN estimator networks as it does not noticeably accelerate convergence in these cases.

We deployed our algorithms on a server equipped with one NVIDIA Tesla V100 GPU.

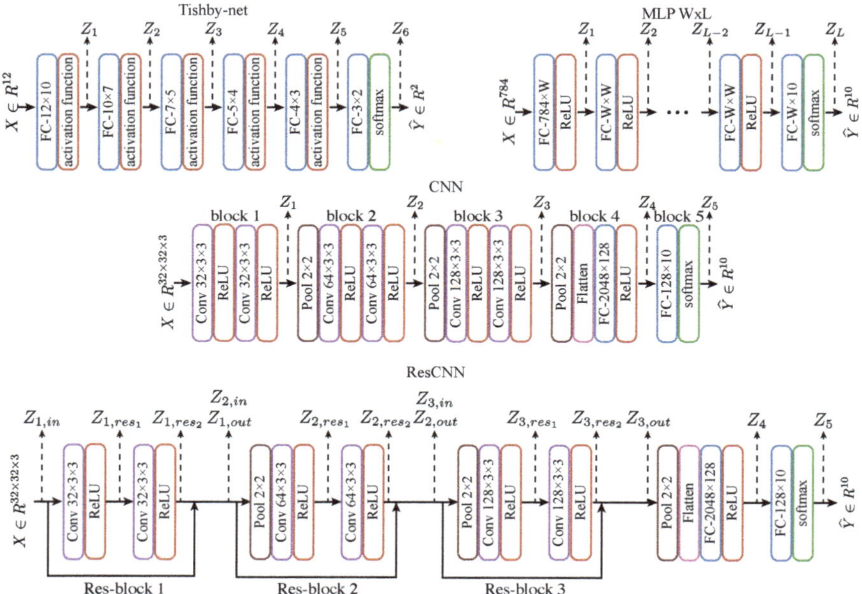

Figure 2. The architectures of subject neural networks involved in this paper. Tishby-net will be trained on the Tishby-dataset proposed in [8], MLP W×L will be trained by MNIST dataset [53], and CNN and ResCNN will be trained on CIFAR-10 dataset [55]. FC stands for fully connected layer, Conv represents the convolutional layer, and Pool refers to the max pooling layer. Note that we intentionally kept the architecture of the CNN as close to ResCNN as possible to enable a better-controlled comparison in later experiments.

5. Results

We now build upon the proposed framework to explore the dynamics of the Z-X and Z-Y measures and their relationship with fitting/compression (F/C) phases and generalization in a range of neural network models. In particular, the fitting phase refers to the initial phase of training where the Z-X measure decreases with the number of epochs, indicating that the network is attempting to fit the dataset. This phase commonly occurs during early training. On the other hand, the compression phases refer to the subsequent increase in the Z-X measure, indicating the compression of information in the network.

Firstly, we experimentally examined whether the estimation of the proposed measures is stable. Then, we examined the impact of (1) the model architecture; (2) the learning algorithm including optimizer and regularization techniques; and (3) the data on the dynamics of the measures.

The results will be presented using Z-X and/or Z-Y dynamics, and the tables show the losses, accuracy, and generalization error of each experiment. In the figures, the x-axes or y-axes will be shared unless specified otherwise by the presence of ticks.

5.1. Z-X and Z-Y Measures Estimation Stability

The reliability of the estimation of the proposed measures is critical for extracting robust conclusions about the behavior of the Z-X and Z-Y dynamics in a neural network. Such studies are, however, largely absent in the information bottleneck literature [12].

5.1.1. Criteria to Describe the Stability of Estimated Measures

We assessed the stability of the Z-X and Z-Y measures estimation using two criteria:

- **Stability with regard to the initialization of estimator networks:** First, we explored how different initializations of an estimator network affect the Z-X and Z-Y measures.
- **Stability with regard to the architecture of estimator networks:** Second, we also explored how (estimator) neural network architectures—with different depths—affect the estimation of the Z-X and Z-Y measures.

5.1.2. Subject Networks, Estimator Networks, and Datasets Involved

We examined the stability of Z-X and Z-Y measures estimates in both fully connected and convolutional subject networks. In particular, we used: (1) a Tishby-net (which has an MLP-like architecture) trained on the Tishby-dataset classification task with a standard stochastic gradient descent (SGD) optimizer, and (2) a CNN trained on the CIFAR-10 classification task trained with an Adam optimizer. However, we noticed that the Tishby-net may not always converge due to its simple architecture and small dataset size of 4096 samples. Therefore, we repeated the training process multiple times with different initializations and only retained converged subject networks to ensure meaningful results. We built estimator networks as elaborated in the previous sections, and their architectures are detailed in Appendix A.

To verify the first stability criterion, we tested different initializations by modifying the random seed of the Xavier initializer. For the second stability criterion, we experimented with estimators at different depths.

5.1.3. Are the Measures Stable in the MLP-like Subject Neural Networks?

Figure 3 depicts the Z-X and Z-Y measures estimates on the Tishby-net. Specifically, panels (a) and (b) display the behavior of such measures under different initializations of a one-layer and two-layer estimator network, respectively. Our results indicate that these measures are robust to changes in the initialization of the estimator network (for a given estimator network architecture).

In turn, panels (c) and (d) depict the behavior of the Z-X and Z-Y measure estimates for different estimator network architectures. It is clear that the capacity of the estimator (which depends on the number of estimator network layers) may affect the exact value of the Z-X and Z-Y measures estimate, indicating the presence of a bias; however, such estimators can still capture consistent trends (such as increases and decreases in the measures that are critical to identifying fitting or compression behavior; see panel (d)).

We however note—as we had elaborated previously—that the estimator networks need to be sufficiently complex to emulate a conditional mean estimator—to estimate the Z-X measure—or to emulate the conditional distribution of the label given the representation—to estimate the Z-Y measure. This may not always be possible depending on the complexity/capacity of the estimator network e.g., one-layer estimator networks are only capable of representing linear estimators whereas two-layer networks can represent more complex estimators (therefore, linear one-layer networks cannot reliably estimate the minimum mean-squared error unless the random variables are Gaussian). However, our results suggest that, with a two-layer network, we may already obtain a reliable estimate since—except for some representations—the difference in the measures estimated using a two-layer network does not differ much from those using a three-layer network. Naturally, with an increase in the capacity of the estimator networks, one may also need additional data in order to optimize the estimator network to deliver a reliable network, but our results also suggest that the variance of the estimates is relatively low for both two-layer and three-layer estimators. Further, the results in [57] suggest that the difference between the estimated value and the true value for our Z-X measure decays rapidly with the number of points in the (validation) dataset (note, however, that these results only apply for scalar random variables). Therefore, we will adopt a two-layer estimator network in our study of MLPs in the sequel.

We conducted a more robust analysis of the efficacy of different estimators using a Gaussian mixture data model in Appendix B, where we can also directly analytically compute the mean-squared error for comparison purposes.

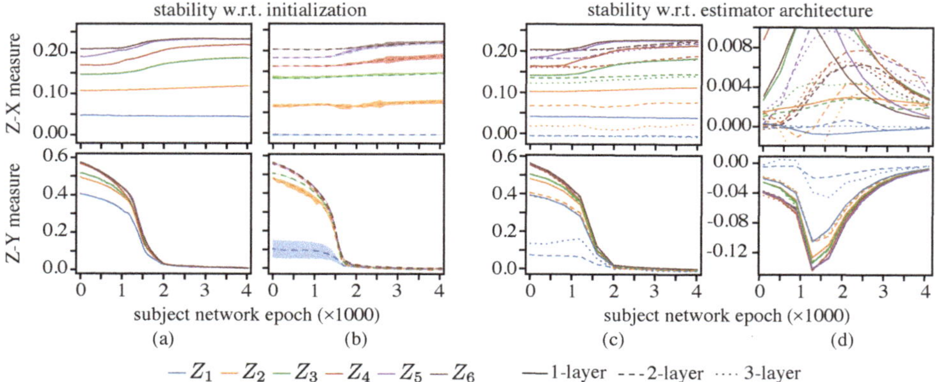

Figure 3. Z-X and Z-Y measures estimate on the Tishby-net: (**a**,**b**) stability with regard to the initialization of estimator networks, and (**c**,**d**) stability with regard to the architecture of estimator networks. The lines are averaged over five different initializations, and the shadow is *five times* the standard deviation. The representations (e.g., Z_1) are taken from the corresponding layer of the Tishby-net in Figure 2. The measures in (**a**) are estimated with 1-layer estimators with varying initializations, and measures in (**b**) are estimated with 2-layer estimators with different initializations. (**c**) compares the measures estimated by estimators with different depths, while the curves in (**d**) depict the measures increasing/decreasing trend, obtained by taking the derivative of (**c**).

5.1.4. Are the Measures Stable in the Convolutional Subject Neural Networks?

Figure 4 shows the Z-X and Z-Y measure estimates on the CNN. To test the stability criteria, we again used different estimator network initializations (varying the random seed of the Xavier initializer) and different estimator network architectures. We first plotted the Z-X dynamics and Z-Y dynamics based on the setup described in Section 4, and the results are shown in the left column of Figure 4. Then, for comparison, we added an extra convolutional layer to all Z-X estimators and a fully connected layer to all Z-Y estimators, and the results are displayed in the right column of Figure 4.

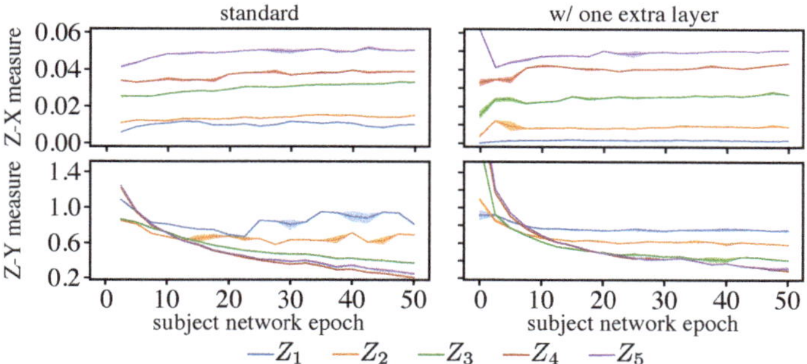

Figure 4. Z-X and Z-Y measures estimate on the CNN: the lines are averaged over five different initializations, and the shadow is *five times* the standard deviation. The representations (e.g., Z_1) are taken from the corresponding layer of the CNN in Figure 2. Note that the violation of the data processing inequality (DPI) observed in the Z-Y measure is attributed to the use of a pre-defined estimator model. This aspect is also acknowledged in the context of the $\mathcal{V}$-information framework, as discussed in [38].

The results show that both estimator networks lead to relatively consistent and stable measure estimates. This suggests that our proposed measures can be reliably inferred using such estimator networks—under different initializations—even in this high-dimensional setting that poses significant challenges to mutual information estimators. Comparing the dynamics estimated by the standard estimator architecture and the one with an extra layer, we observed that the trends of the dynamics are similar. Hence, we used the standard setup in the rest of the paper due to its higher computational efficiency, which is illustrated in Figure A3.

We next relied on this approach to estimate the Z-X and the Z-Y dynamics for different (subject) neural network models and algorithms in order to cast further insights into the compression, fitting, and generalization dynamics of deep learning.

5.2. The Impact of Model Architectures to the Network Dynamics

We started our study by investigating the effect of the neural network model on the Z-X and Z-Y dynamics of neural networks. We considered both MLPs with different activation functions, depths, and widths. We also considered CNN and res-net architectures. Our study will allow us to identify possible fitting, compression, and generalization behavior.

5.2.1. Does the Activation Function Affect the Existence of F/C Phases?

We began by examining whether the presence of the fitting and compression (F/C) phases is dependent on the activation function used in the network. This topic has been explored in previous studies using the IB approach [8,11,24,27], but different studies have led to different conclusions [27].

Setups: We deployed Tishby-net architecture with various activation functions, including both saturating (tanh and softsign [58]) and non-saturating (ReLU [59], ELU [60], GELU [58], swish [61,62], PELU [63], and leaky-ReLU [64]) options. The Tishby-net was trained on the Tishby-dataset using the same optimizer and hyper-parameter setups as described in the literature [8,24]. The Z-X and Z-Y measures were estimated using two-layer estimators, as argued in Section 5.1.

Results: Figure 5 reveals that the Z-X dynamics exhibit a consistent pattern among all Tishby-nets, characterized by an initial decrease in Z-X measures followed by an increase. Note that the initial decrease happens prior to the decrease in the subject network loss. There can be a longer period of epochs where the network struggles to converge and, during this phase, the changes in the Z-X measure may not be easily visible. The Z-X dynamics in some experimental setups, such as PELU, display fluctuation, which we attribute to the unstable convergence of the subject network, as evidenced by the fluctuations in the subject network loss. Moreover, the increases in Z-X measures coincide with epochs where the network experiences a decrease in loss. These observations suggest that the F/C phases are likely to occur in the network, regardless of the activation function employed. Our observation is in line with some of the previous studies that have used MI measures, such as [8,11].

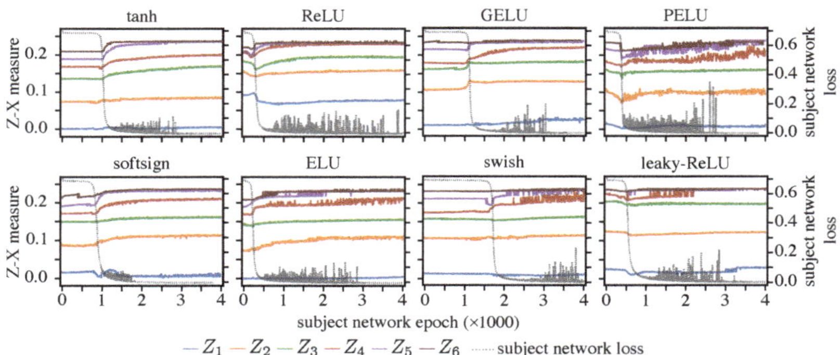

Figure 5. Z-X dynamics on Tishby-net with different activation functions. The left y-axes displays the Z-X measure estimate values, while the right y-axes represent the cross-entropy loss value of the subject network.

5.2.2. How Do the Width and Depth of an MLP Impact Network Dynamics?

We now examine the effect of the MLP width (number of neurons per layer) and depth on the Z-X and Z-Y dynamics.

Setups: For the MLP width analysis, we constructed four-layer MLPs with different numbers of neurons per layer: 16, 64, and 512. For the MLP depth experiment, we fixed the width of the subject network to 64 and varied its depth from two to six hidden layers. All models were trained on the full MNIST dataset using a standard SGD optimizer with a fixed learning rate of 0.001. We also used two-layer estimator networks to estimate the Z-X and Z-Y measures.

Figure 6 depicts the dynamics of the Z-X measure against the Z-Y measure for MLP networks with four layers and with different widths. As shown in Table 2, the best generalization performance is associated with the model MLP 512×4. We can observe that all MLP networks exhibit fitting and compression phases. However, wider networks (e.g., MLP 512×4) tend to begin compressing earlier, while the thinner ones (e.g., MLP 16×4) tend to have a longer fitting phase. This trend suggests that wider networks are able to fit data more quickly. We can also observe that the networks with more neurons per layer (MLP 512×4) exhibit more compression than network with fewer neurons per layer (MLP 16×4). Interestingly, the MLP 512×4 model also exhibits the best generalization performance, so one can potentially infer that significant compression may be necessary for good generalization [8,29].

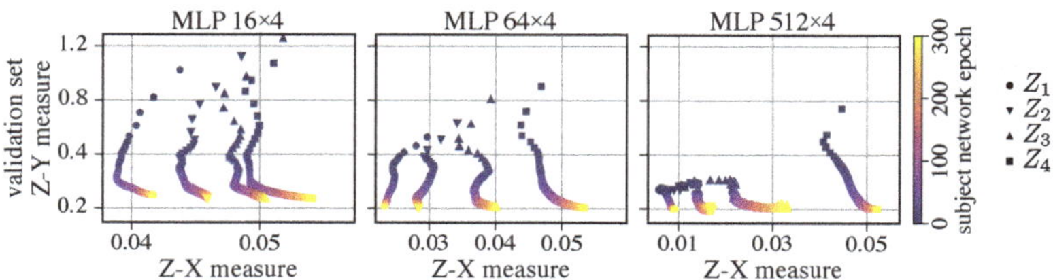

Figure 6. Z-X/Z-Y measures dynamics plane of MLP networks with different widths. The representations (e.g., Z_1) are taken from the corresponding layer of the MLP WxL network in Figure 2.

Table 2. The epoch that reached the minimum validation loss (ep.), the training losses, test losses (test loss), generalization error (GE), training accuracy (train acc.), and test accuracy (test acc.) of the MLPs with different widths and depths. The experiment with the best generalization error is highlighted using bold font.

Subject Network	ep.	Train Loss	Test Loss	GE	Train acc.	Test acc.
MLP 16 × 4	197	0.0890	0.1471	0.0581	0.9740	0.9572
MLP 64 × 4	168	0.0344	0.0967	0.0623	0.9919	0.9748
MLP 512 × 4	142	0.0191	0.0697	**0.0506**	0.9967	0.9800
MLP 64 × 2	299	0.0688	0.1247	0.0559	0.9815	0.9760
MLP 64 × 3	275	0.0338	0.0570	**0.0232**	0.9919	0.9762
MLP 64 × 4	142	0.0344	0.0967	0.0623	0.9919	0.9748
MLP 64 × 5	85	0.0659	0.1185	0.0526	0.9822	0.9672
MLP 64 × 6	68	0.0736	0.1320	0.0584	0.9798	0.9616

Figure 7 depicts the dynamics of the Z-X measure (associated with the first and last layers) of MLPs with a width of 64 and with different depths (we note that the best generalization performance is associated with the model MLP 64 × 3). In terms of fitting, we can observe that the different MLPs experience a fitting phase. However, deeper models such as MLP 64 × 5 and MLP 64 × 6 appear to experience a more pronounced fitting phase than shallower models, though deeper models still exhibit a higher Z-X measure than shallower ones toward the end of this fitting phase (see marker #1). In terms of compression, we find that deeper networks (e.g., MLP 64 × 5, MLP 64 × 6) compress data more aggressively than shallower ones. Indeed, the gap between the Z-X measure value between the last layer and the first layer of the network is much higher for a deeper model than for shallower ones (as indicated by marker #2).

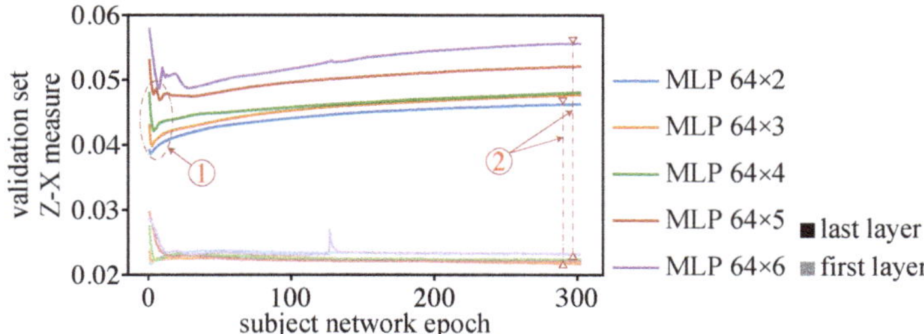

Figure 7. Z-X dynamics of the MLP 64 networks with different depths. The curves with higher saturation correspond to the last layer of the MLP model, while those with lower saturation belong to the first MLP layer.

We also highlight that the MLP 64 × 3 network, which demonstrated the best generalization performance (refer to Table 2), exhibited a significant fitting phase similar to MLP 64 × 2, as well as a notable compression phase close to MLP 64-4.

Overall, shallow networks may have difficulty compressing data effectively, while the layers close to the output in the deep networks may lose important information and cannot fit data well. We hypothesize that both of these phenomena—which are both present in the MLP 64 × 3 network—can have an impact on a network's ability to generalize effectively.

5.2.3. How Do the Number of Kernels and Kernel Size of a CNN Impact Network Dynamics?

We now examine the effect of the kernels, including their number and size, on the Z-X and Z-Y dynamics in a CNN.

Setups: To analyze the impact of the number of kernels on network F/C phases in CNNs, we adjusted the number of kernels by a factor derived from the baseline CNN architecture shown in Figure 2. To analyze the impact of the kernel size, we used 1×1, 3×3 (baseline), 5×5, and 7×7 kernel sizes for all convolutional layers The CNN models were trained on the CIFAR-10 dataset using the Adam optimizer with a learning rate of 0.001. We utilized minimal estimator networks, as described in the previous section.

Results: Figure 8 depicts the Z-X dynamics of our CNN network with different numbers of kernels. We observe that having a low number of kernels (e.g., /4, /8) seems to impair both the fitting and compression process, particularly in early layers (e.g., layers 1 and 2). In contrast, we observed that a high number of kernels do not significantly impact the F/C phases or the generalization performance. Indeed, as shown in Table 3, CNNs with more kernels (e.g., $\times 2$, $\times 4$) have a similar test loss performance to the baseline model (note that the best test loss performance corresponds to the $\times 4$ model, and that its generalization performance is also similar to that of the baseline model). This suggests that adding more kernels to a well-generalized CNN may not significantly impact the F/C phases and may not lead to an improved generalization.

Table 3. The epoch that reached the minimum validation loss (ep.), the training losses, test losses (test loss), generalization error (GE), training accuracy (train acc.), and test accuracy (test acc.) of the CNNs with a different number of kernels and kernel sizes. The experiment with the best generalization error is highlighted using bold font.

Subject Network	ep.	Train Loss	Test Loss	GE	Train acc.	Test acc.
CNN baseline	5	0.6747	0.8300	**0.1553**	0.7657	0.7190
CNN $\times 2$	7	0.3826	0.8303	0.4477	0.8637	0.7514
CNN $\times 4$	4	0.5667	0.7801	0.2135	0.8001	0.7332
CNN $/2$	11	0.6015	0.9055	0.3040	0.7871	0.7008
CNN $/4$	14	0.7704	1.0494	0.2790	0.7306	0.6492
CNN $/8$	26	0.9515	1.1353	0.1838	0.6589	0.6060
CNN 1×1	18	1.0307	1.1860	0.1553	0.6343	0.5978
CNN 3×3	5	0.6747	0.8065	**0.1318**	0.7657	0.7190
CNN 5×5	9	0.6001	0.9805	0.3804	0.7887	0.6958
CNN 7×7	6	0.8372	1.2011	0.3639	0.7031	0.6042

Figure 9 depicts the Z-X dynamics of our CNN network with different kernel sizes. It appears that networks with large kernels fail to fit and compress, but networks with small kernels also exhibit little fitting and compression. Indeed, the best test loss and generalization performance are associated with the CNN model with a 3×3 kernel size, which also exhibits a more pronounced fitting and compression phase (refer to Table 3).

Overall, we hypothesize that selecting an appropriate kernel size can improve a network's ability to both fit and compress data, leading to a better generalization performance, which is in line with the conclusion in [8,29].

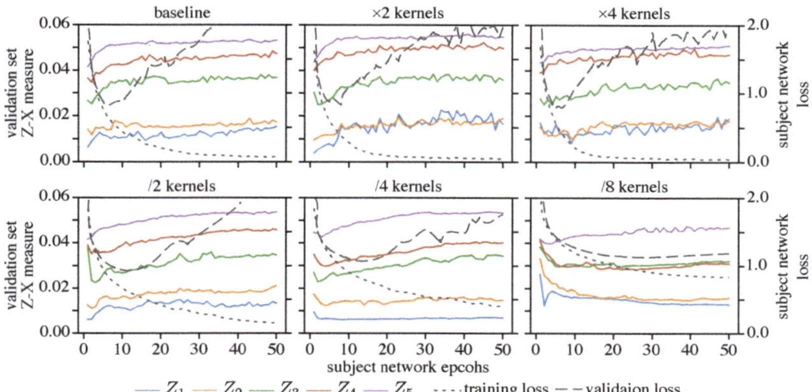

Figure 8. Z-X dynamics of the CNN network with different number of kernels on each layer. We make modifications based on the baseline CNN structure shown in Figure 2. For example, "×2" means doubling the number of kernels in each convolutional layer, while "/2" means halving the number of kernels in each convolutional layer. The representations (e.g., Z_1) are taken from the corresponding block of the CNN network in Figure 2.

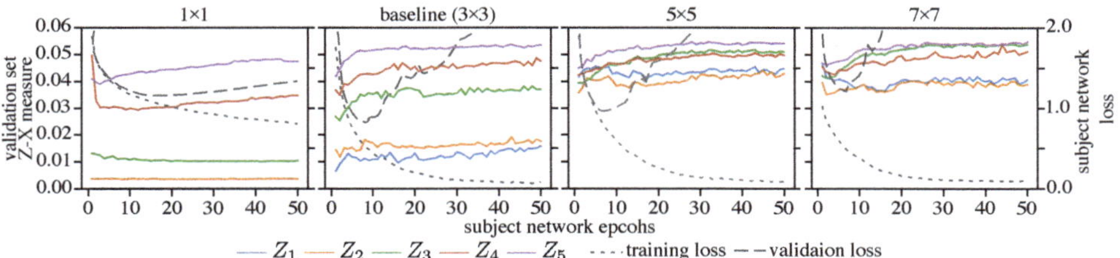

Figure 9. Z-X dynamics of the CNN network with different kernel sizes. The representations (e.g., Z_1) are taken from the corresponding block of the CNN network in Figure 2.

5.2.4. How Does Residual Connection Affect the Network Dynamics?

We finally assessed the impact of residual connections—introduced in [65]—on neural network learning dynamics, since these have been frequently used to address the gradient vanishing problem in very deep neural networks. We note that some works [13,18] have studied the behavior of ResNet or DenseNet (which also contain residual connections [66]). However, these studies did not delve into how residual connections may impact the information bottleneck of hidden layer representations and their relation to generalization.

Setup: We deployed a ResCNN, as elaborated in the previous section, that was trained using an Adam optimizer with a learning rate of 0.001 on the CIFAR-10 dataset. We also used the standard estimator network setups elaborated in Section 4.2 and shown in Appendix A Figure A3.

Results: We first analyzed the behavior of the Z-X dynamics at the output of the residual blocks (e.g., $Z_{1,out}$) and the fully connected layers, and compared it with the CNN with a similar architecture but without residual connections; see Figure 10.

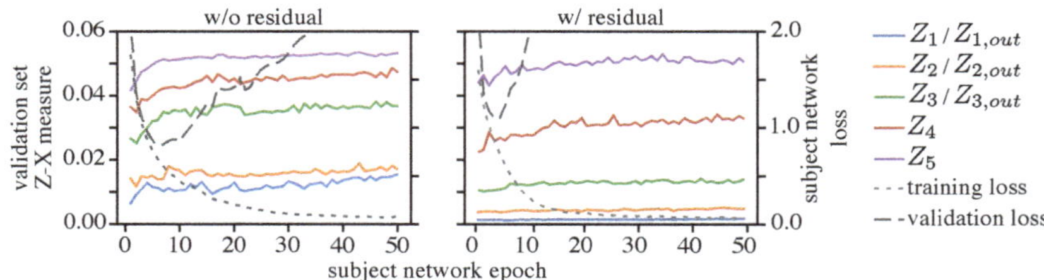

Figure 10. Z-X dynamics of CNNs with or without residual connections. The representations (e.g., Z_1, $Z_{1,out}$) are taken from the corresponding locations of the CNN or ResCNN network shown in Figure 2.

We notice that the ResCNN tends to have less pronounced compression in the (residual) convolutional blocks, e.g., the Z-X dynamic of Z_3 (without residual connection) shows a more pronounced increase than that of $Z_{3,out}$ (with residual connection). Additionally, we can see that the model with residual connection depends more on the fully connected layers to compress the Z-X measure, which is demonstrated by the significantly wider gap between representations Z_4 and Z_5, as well as between Z_4 and $Z_3/Z_{3,out}$ in the residual model.

We then inspected the behavior of the Z-X measure and the Z-Y measure within each residual block; see Figure 11 (note that the dynamics of the Z-X and Z-Y measures associated with $Z_{1,in}$ are flat because $Z_{1,in}$ corresponds to X).

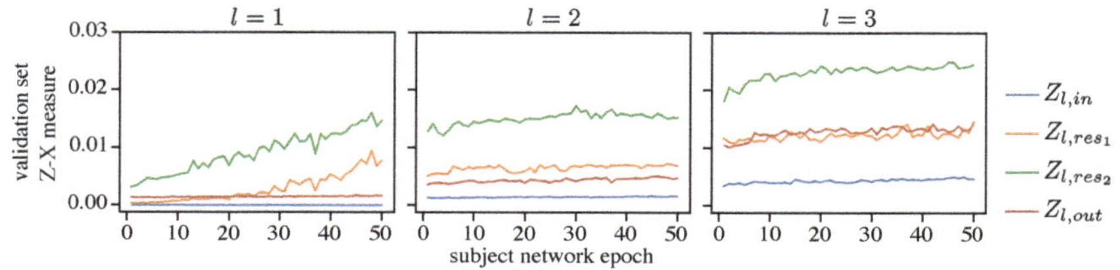

Figure 11. Z-X dynamics of the ResCNN in each residual block. l is the index of the residual block. The representations (e.g., $Z_{1,in}$) are taken from the corresponding block of the ResCNN network in Figure 2.

We can observe that, within each residual block (i.e., for a given index l), the Z-X measure of $Z_{l,out}$ is generally lower than that of Z_{l,res_1} and Z_{l,res_2}. This is because the representation $Z_{l,out}$ is the sum of Z_{l,res_2} and $Z_{l,in}$ and thus retains more information associated with the data.

We can also observe that, in every residual block, the Z-X dynamics of Z_{l,res_1} and Z_{l,res_2} have a pronounced increase over the epoch, while the Z-X dynamics of $Z_{l,in}$ and $Z_{l,out}$ are relatively stable. This suggests that each residual block may learn to form a mini-bottleneck. However, the overall network does not exhibit a visible compressing phase when observing the output of the residual blocks alone. Our experiments demonstrate the distinct behavior of networks with residual connections compared to those without.

5.3. The Impact of Training Algorithm to the Network Dynamics

A neural network generalization ability also tends to depend on the training procedure, including the learning algorithm and regularizers. Therefore, we now explore how different learning settings affect neural network Z-X and Z-Y measures dynamics.

5.3.1. How Does the Optimizer Impact the Network Dynamics?

It was suggested by [29] that the Adam optimizer leads to a better performance during the fitting phase, but it tends to perform worse during the compression phase.

We investigated, under the lens of our approach, the effect of Adam and various other optimizers on neural network learning dynamics.

Setup: Our experiments were conducted on CNNs (with the standard architecture illustrated in Figure 2) trained on the CIFAR-10 dataset using different optimizers. Specifically, we experimented with non-adaptive optimizers such as SGD and SGD-momentum [67], as well as adaptive optimizers such as RMSprop [68]. We also considered the Adam optimizer [69], which can be viewed as a combination of a momentum optimizer and RMSprop optimizer, representing a hybrid approach. We used standard hyper-parameters commonly used for CIFAR-10 classification tasks, setting the learning rate to 0.001 for all optimizers and a momentum parameter of 0.9 (if applicable). Our estimator networks are akin to those used in previous studies.

Results: Figure 12 shows the behavior of the normalized Z-X measure for CNNs trained with different optimizers. We normalized this measure using min-max normalization to allow for a better visualization of relative changes in performance. Specifically, each Z-X dynamic curve was normalized individually, and the minimum and maximum values were taken from the curve after the 50th epoch, as we observed that all Z-X dynamics enter the compression phase before this epoch.

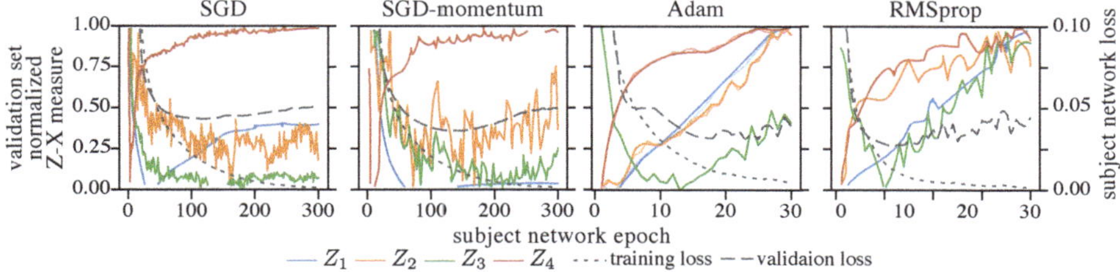

Figure 12. Z-X dynamics for a CNN trained with different optimizers. The representations (e.g., Z_1) are taken from the corresponding block of the CNN network in Figure 2.

We observe that SGD and SGD-momentum exhibit similar fitting phases, while Adam and RMSprop also display similar fitting phases. We can also note that, when trained on the Adam and RMSprop optimizer—which are adaptive optimizers—the representations associated with the various layers exhibit major compression; in contrast, when trained with the SGD optimizer, the representations $\{Z_2, Z_3\}$ do not show noticeable compression and, likewise, when trained with SGD-momentum optimizers, the representations $\{Z_2, Z_3\}$ also do not exhibit much compression. Note that, in our experiment with the CNN trained on the CIFAR classification task, we can see from Table 4 that the model trained with the RMSprop optimizer achieved the best generalization performance, followed closely by the model trained with Adam. Therefore, it appears that adaptive optimizers—which adjust the learning rate per parameter—may be critical for leading to network compression, and hence generalization [70].

Table 4. The epoch that reached the minimum validation loss (ep.), the training losses, test losses (test loss), generalization error (GE), training accuracy (train acc.), and test accuracy (test acc.) of the CNNs trained with different optimizers. The experiment with the best generalization error is highlighted using bold font.

Subject Network	ep.	Train Loss	Test Loss	GE	Train acc.	Test acc.
CNN SGD	106	0.0202	0.0429	0.0226	0.9945	0.9882
CNN SGD-momentum	131	0.0130	0.0356	0.0226	0.9972	0.9882
CNN Adam	24	0.0067	0.0275	0.0208	0.9979	0.9896
CNN RMSproop	11	0.0123	0.0263	**0.0139**	0.9965	0.9908

5.3.2. How Does Regularization Impact the Network Dynamics?

It has been suggested by [11,12] that weight decay regularization can significantly enhance the compression phase associated with a neural network learning dynamic. It has also been argued by others [18] that compression is only possible with regularization. Therefore, we also investigated, under the lens of our approach, the effect of regularization on the learning dynamics of MLPs and CNNs.

Setup: We deployed MLP 64×4 models trained on the MNIST dataset with or without weight decay (WD) regularization and CNN models trained with the CIFAR-10 dataset with or without dropout regularization. The weight decay was applied to all layers in the MLP 64×4 model with its hyper-parameter set to 0.001, while the dropout was only adopted in the first fully connected layer in the CNN with a 30%, 60%, or 90% dropout rate (which is a common approach in the literature [54]). The MLP with weight decay regularization requires more epochs to converge. Therefore, we trained the MLP 64×4 without weight decay for 300 epochs and the model with weight decay for 1200 epochs.

Results: We offer the dynamics of the Z-X and Z-Y measures associated with the MLP setting in Figure 13. We infer that weight decay regularization does not significantly impact the fitting phase; however, weight decay does seem to affect network compression, leading networks to compress more aggressively. Moreover, weight decay not only prevents the subject network from overfitting [2] but also prevents its representations from overfitting. Therefore, we conjecture that the weight decay regularization boosts the compression in MLPs (as also observed in [11]) and prevents the representation overfitting to improve the generalization performance (shown in Table 5), which is also in line with [11].

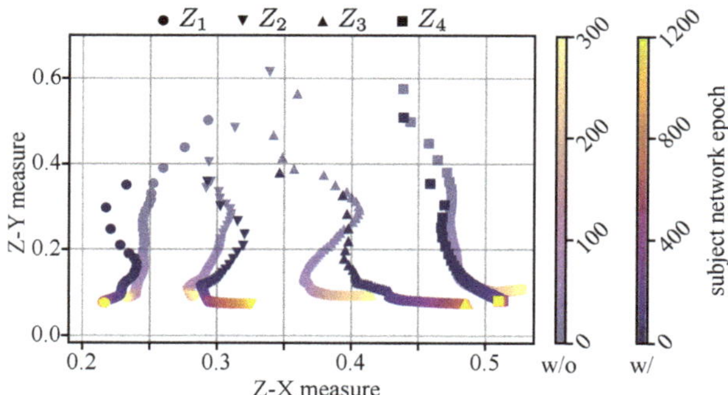

Figure 13. Z-X and Z-Y dynamics of MLP 64×4 trained on the MNIST dataset with or without weight decay regularization. The subject network regularized by weight decay gives relatively better test loss. The representations (e.g., Z_1) are taken from the corresponding block of the MLP 64×4 network in Figure 2.

We also offer the dynamics of the Z-X measure associated with the CNN setting in Figure 14 (Table 5 shows that the best generalization performance is obtained for a CNN with dropout regularization at a 60% dropout rate on the first fully connected layer). Our results suggest that tuning the dropout rate on the first fully connected layer affects not only the dynamics of its representation (Z_4) but also the dynamics of other layers. When a high dropout rate (e.g., 90%) is used, we observe less pronounced fitting and compression phases, which also lead to a worse generalization performance (refer to Table 5). Conversely, a low dropout rate (30%) showed similar fitting phases to the no-dropout group, but with more compression. These results support our conjecture that the F/C phases are linked to the generalization behavior of the model.

Table 5. The epoch that reached the minimum validation loss (ep.), the training losses, test losses (test loss), generalization error (GE), training accuracy (train acc.), and test accuracy (test acc.) of the MLPs and CNNs trained w/ or w/o regularization algorithms. The experiment with the best generalization error is highlighted using bold font.

Subject Network	ep.	Train Loss	Test Loss	GE	Train acc.	Test acc.
MLP w/o WD	168	0.0344	0.0967	0.0623	0.9919	0.9748
MLP w/ WD	626	0.0216	0.0722	**0.0505**	0.9976	0.9784
CNN 0% dropout	5	0.6747	0.8300	0.1553	0.7657	0.7190
CNN 30% dropout	12	0.4993	0.7985	0.2992	0.7608	0.7398
CNN 60% dropout	10	0.6888	0.7606	**0.0718**	0.8237	0.7510
CNN 90% dropout	19	1.0768	0.8765	0.2003	0.5959	0.7000

On the other hand, it can be observed that adopting dropout regularization diminishes the visibility of fitting phases across multiple layers. This suggests that the training algorithm effectively leverages the neurons and connections within the model, enabling rapid dataset fitting.

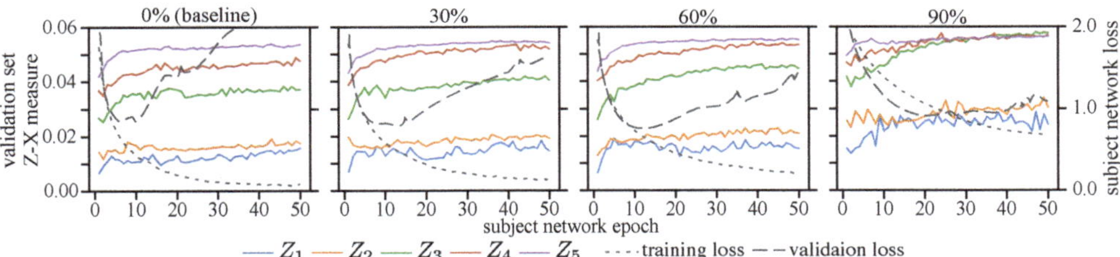

Figure 14. The Z-X dynamics for a CNN trained on the CIFAR-10 dataset with different amounts of dropout in its fully connected layers. The subject network regularized with a 60% dropout rate provides the best test loss and generalization error. The representations (e.g., Z_1) are taken from the corresponding block of the CNN network in Figure 2.

5.4. The Impact of Dataset to the Network Dynamics

It is well established that the size of the training set directly affects a machine learning model's generalization performance [71]. Our goal was to also understand how the dataset size affects neural network model learning dynamics, including its fitting and compression phases.

Setup: We compared the learning dynamics of CNN models trained on three different datasets: 1% of CIFAR-10 (0.5k samples), CINIC [56] (which has the same classes as CIFAR-10 but contains 180k samples), and the full CIFAR-10 dataset (50k samples).

We used the Adam optimizer with a learning rate of 0.001 to train the neural networks. We also estimated the Z-X and Z-Y measures using the network in Figure A3 using the CIFAR-10 validation and test sets.

Results: Figure 15 shows the Z-X dynamics of CNNs trained on datasets of different sizes. We can observe from Table 6 that the model trained on the CINIC dataset achieves the best generalization performance, while the model trained on the smallest dataset (1% CIFAR-10) performs the worst.

Table 6. The epoch that reached the minimum validation loss (ep.), the training losses, test losses (test loss), generalization error (GE), training accuracy (train acc.), and test accuracy (test acc.) of the CNNs trained on different datasets or dataset sizes, including 1% CIFAR-10 dataset (w/0.5k training samples), 100% CIFAR-10 dataset (w/50k training samples), and CINIC dataset (180k training samples). The experiment with the best generalization error is highlighted using bold font.

Subject Network	ep.	Train Loss	Test Loss	GE	Train acc.	Test acc.
CNN 1% CIFAR-10	38	1.2497	1.9902	0.7405	0.5380	0.3248
CNN 100% CIFAR-10	5	0.6747	0.8300	0.1553	0.7657	0.7190
CNN CINIC	12	0.5952	0.6395	**0.0443**	0.7744	0.7872

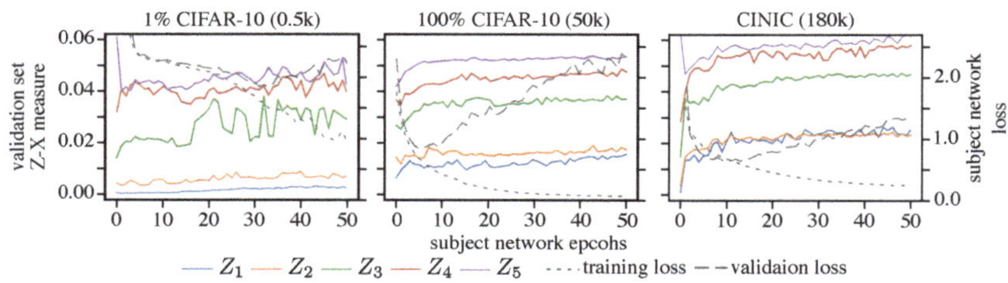

Figure 15. Z-X measures of representations in CNN trained on 1% CIFAR-10 dataset, full CIFAR-10 dataset, and CINIC dataset. The representations (e.g., Z_1) are taken from the corresponding block of the CNN network in Figure 2.

Our experiments show that the fitting behavior of the network trained on the small dataset is identical to that of the network trained on the standard CIFAR-10 dataset. However, the degree of compression exhibited by the network optimized on the 1% CIFAR-10 dataset was much less pronounced than that of the model trained on richer datasets. This suggests that compression may only be possible for sufficiently large datasets. Our experiments also show that the behavior of the Z-X measure associated with the network trained on the CINIC dataset rapidly increases during the optimization process. This indicates a significant F/C phase that may also justify the superior generalization performance.

Overall, these observations suggest that providing sufficient training data can amplify the magnitude of compression. This in turn helps the model learn to abstract key information for predicting labels more effectively, leading to a better generalization performance. Therefore, we conclude that compression may be a crucial factor for effective generalization in neural networks, and providing sufficient training data is essential for amplifying this phase [8].

6. Conclusions

In this paper, we proposed to replace the mutual information measures associated with information bottleneck studies with other measures capable of capturing fitting, compression, and generalization behavior. The proposed method includes: (1) the Z-X measure corresponding to the approximation of the minimum mean-squared error associated with the recovery of the network input (X) from some intermediate network representation (Z) and (2) the Z-Y measure associated with the cross-entropy of the data label/target (Y) given some intermediate data representation (Z). We also proposed to estimate such measures using neural-network-based estimators. The proposed approach can handle representations in high-dimension space, is computationally stable, and is also computationally affordable.

Our series of experiments explored—via the dynamics between the Z-X and Z-Y measure estimates—the interplay between network fitting, compression, and generalization on different neural networks, with varying architectures, learning algorithms, and datasets, that are as complex or more complex than those used in traditional IB studies [12]. Our main findings are as follows:

- **Impact of Neural Network Architecture:**
  - We have found that MLPs appear to compress regardless of the non-linear activation function.
  - We have observed that MLP generalization, fitting, and compression behavior depend on the number of neurons per layer and the number of layers. In general, the MLPS offering the best generalization performance exhibit more pronounced fitting and compression phases.
  - We have also observed that CNN generalization, fitting, and compression behavior also depend on the kernel's number/size. In general, CNNs exhibiting the best generalization performance also exhibit pronounced fitting and compression phases.
  - Finally, we have seen that the fitting/compression behavior exhibited by networks with residual connections is rather distinct from that shown in networks without such connections.

- **Impact of Neural Network Algorithms:** We have observed that adaptive optimizers seem to lead to more compression/better generalization in relation to non-adaptive ones. Likewise, we have also observed that regulation can help with compression/generalization.

- **Impact of Dataset:** Our main observation is that insufficient training data may prevent a model from compressing and hence generalizing; in turn, models trained with sufficient training data exhibit both a fitting phase followed by a compression phase, resulting in a higher generalization performance.

Overall, our findings are in line with an open conjecture that good neural network generalization is associated with the presence of a neural network fitting phase followed by a compression phase during the learning process [8,11,29].

There are some interesting directions for further research. First, it would be intriguing to explore the dynamics of state-of-the-art machine learning models, including transformers, which have demonstrated exceptional performance in various tasks. By analyzing the behavior of transformers under the lens of the information bottleneck theory, we may be able to gain additional insights into how these advanced models learn, compress information, and generalize.

Second, it would also be interesting to extend the study to other learning paradigms such as semi-supervised or unsupervised tasks. In semi-supervised learning, where a limited amount of labeled data are available along with a larger unlabeled dataset, using the proposed approach to study the learning process may help to uncover effective strategies for leveraging unlabeled data. Similarly, in unsupervised learning tasks, where the goal is to discover patterns and structure in unlabeled data, a similar approach could potentially uncover the interplay between compression and fitting and their implications in leading up to meaningful representations capturing essential information.

Finally, although our study has shed some light on the interplay between compression and generalization using the proposed method, conducting a specialized study and analysis to obtain a more comprehensive understanding of the relationship between these two factors would be interesting.

Author Contributions: Methodology, Z.L.; Software, Z.L.; Validation, Z.L. and G.A.; Formal analysis, Z.L., G.A. and M.R.D.R.; Writing—original draft, Z.L. and G.A.; Writing—review & editing, Z.L. and M.R.D.R.; Supervision, M.R.D.R. All authors have read and agreed to the published version of the manuscript.

Funding: The APC was funded by UCL Dean's Prize and China Scholarship Council.

Institutional Review Board Statement: Not applicable.

Data Availability Statement: Not applicable.

Conflicts of Interest: The authors declare no conflict of interest.

Appendix A. Estimator Network Architectures

This appendix lists the architectures of the Z-X and Z-Y estimator networks.

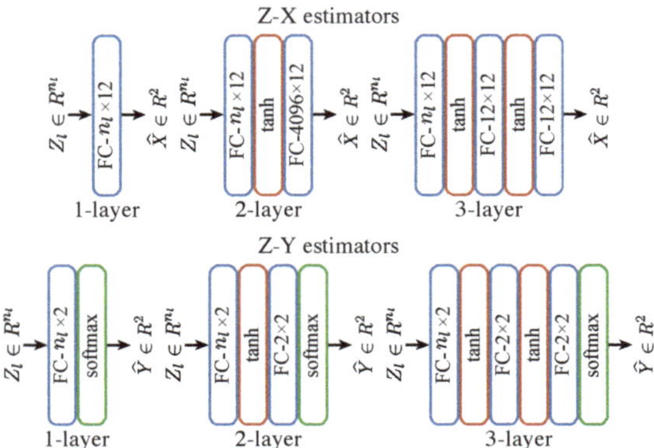

Figure A1. Architecture of the Z-X and Z-Y estimator networks for Tishby-net models. Note that estimators with different depths are used for the stability tests.

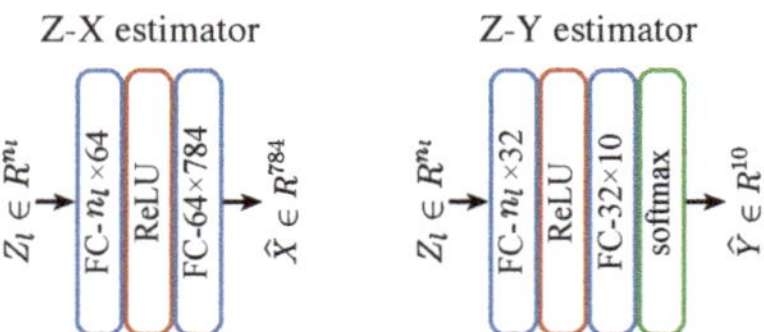

Figure A2. Architecture of the Z-X and Z-Y estimator networks for MLP WxL models.

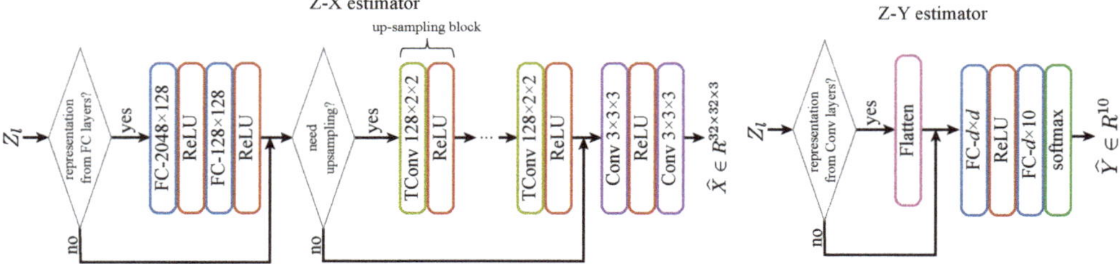

Figure A3. Architectures of the Z-X and Z-Y estimator networks for CNN and ResNet. The number of up-sampling blocks in the Z-X estimator is set to be equal to the number of down-sampling blocks in the network being analyzed, and the dimension of the fully connected layer d in the Z-Y measure is determined by the shape of the flattened input vector length. To test the stability of the estimators at different depths (as described in Section 5.1), we added an extra convolutional layer with ReLU non-linearity at the end of the Z-X estimator, and a fully connected layer with a width of d before the first fully connected layer in the Z-Y estimator.

Appendix B. Empirical Comparison of MMSE Estimator and MI Estimator for Multivariant Gaussian Random Variables

We experimentally compared the minimal mean-squared estimator and mutual information estimator for multivariant Gaussian random variables.

Consider a simple case where random vector $X \in \mathcal{R}^d$ (target) and $Y \in \mathcal{R}^d$ (observation) follow a multivariate normal distribution with correlation $\Sigma_{XY} = \rho\mathbf{I}$, i.e., $Y = \rho X + \sqrt{1-\rho^2}N$. Under this setup, the mutual information between X and Y is $I(X;Y) = -\frac{d}{2}\log(1-\rho^2)$, and the minimal mean-squared error is $\mathrm{mmse}(X|Y) = 1-\rho^2$.

Now, we estimate $I(X;Y)$ with MI estimators from the literature and estimate minimal mean-squared error with neural-network-based mean-squared error estimators. The results are shown in Figure A4. We show the case where $d = 20$ and change ρ from -0.99 to 0.99. Each test takes 4000 randomly generalized samples.

We can see from Figure A4a that the variational estimators tend to have high biases when the mutual information is high. The simple binning method failed to estimate the correct value of mutual information. Although the Kraskov estimator shows a relatively consistent trend, the time consumption of this algorithm grows exponentially as the dimension and number of samples increase [46]. On the other hand, the results of the estimated minimal mean-squared error are shown in Figure A4b. We can see that the estimated values are very close to the ground-truth values.

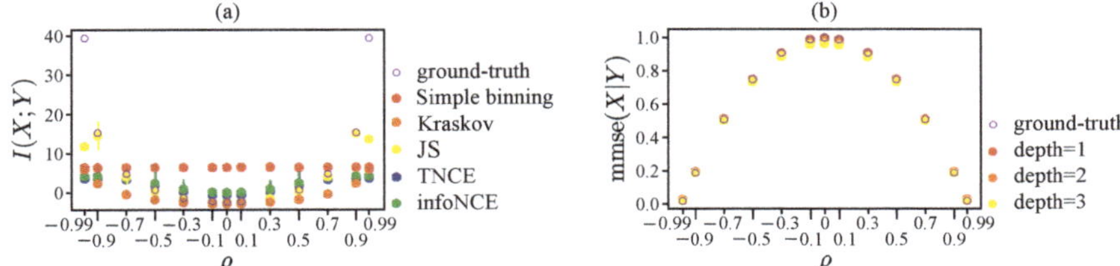

Figure A4. (**a**) The estimated mutual information, and (**b**) the neural network estimated MMSE for 20-dimension correlated Gaussian random variables. In both panels, the hollow purple circles are ground-truth values. In panel (**a**), the infoNCE [10], TNCE [72], and JS [73] are variational estimators, and the error bar is the variance. The simple binning method is adopted via the same implementation as in [24] and the Kraskov estimator is implemented based on the original paper [46]. For panel (**b**), we use a 1-layer (linear), 2-layer, and 3-layer network to estimate the minimal mean-squared error, respectively. The networks have 20 neurons per layer, and the activation function is the hyperbolic tangent function. The estimators are trained with Adam optimizer for 5000 epochs, and the learning rate is set to 0.001.

References

1. Shalev-Shwartz, S.; Ben-David, S. *Understanding Machine Learning: From Theory to Algorithms*; Cambridge University Press: Cambridge, UK, 2014.
2. Bengio, Y.; Goodfellow, I.; Courville, A. *Deep Learning*; MIT Press: Cambridge, MA, USA, 2017; Volume 1,
3. Raukur, T.; Ho, A.C.; Casper, S.; Hadfield-Menell, D. Toward Transparent AI: A Survey on Interpreting the Inner Structures of Deep Neural Networks. *arXiv* **2022**, arXiv:2207.13243.
4. Ma, S.; Bassily, R.; Belkin, M. The Power of Interpolation: Understanding the Effectiveness of SGD in Modern Over-parametrized Learning. In Proceedings of the International Conference on Machine Learning, Sydney, Australia, 6–11 August 2017.
5. Frei, S.; Chatterji, N.S.; Bartlett, P.L. Benign Overfitting without Linearity: Neural Network Classifiers Trained by Gradient Descent for Noisy Linear Data. *arXiv* **2022**, arXiv:2202.05928.
6. Tishby, N.; Pereira, F.C.; Bialek, W. The information bottleneck method. *arXiv* **2000**, arXiv:physics/0004057.
7. Tishby, N.; Zaslavsky, N. Deep learning and the information bottleneck principle. In Proceedings of the 2015 IEEE Information Theory Workshop (ITW), Jerusalem, Israel, 26 April–1 May 2015; pp. 1–5.
8. Shwartz-Ziv, R.; Tishby, N. Opening the black box of deep neural networks via information. *arXiv* **2017**, arXiv:1703.00810.
9. Belghazi, M.I.; Baratin, A.; Rajeshwar, S.; Ozair, S.; Bengio, Y.; Courville, A.; Hjelm, D. Mutual information neural estimation. In Proceedings of the International Conference on Machine Learning, PMLR, Vienna, Austria, 25–31 July 2018; pp. 531–540.
10. Poole, B.; Ozair, S.; Van Den Oord, A.; Alemi, A.; Tucker, G. On variational bounds of mutual information. In Proceedings of the International Conference on Machine Learning, PMLR, Long Beach, CA, USA, 9–15 June 2019; pp. 5171–5180.
11. Chelombiev, I.; Houghton, C.; O'Donnell, C. Adaptive estimators show information compression in deep neural networks. *arXiv* **2019**, arXiv:1902.09037.
12. Geiger, B.C. On Information Plane Analyses of Neural Network Classifiers–A Review. *IEEE Trans. Neural Netw. Learn. Syst.* **2021**, *33*, 7039–7051. [CrossRef]
13. Fang, H.; Wang, V.; Yamaguchi, M. Dissecting deep learning networks—Visualizing mutual information. *Entropy* **2018**, *20*, 823. [CrossRef]
14. Elad, A.; Haviv, D.; Blau, Y.; Michaeli, T. Direct validation of the information bottleneck principle for deep nets. In Proceedings of the IEEE/CVF International Conference on Computer Vision Workshops, Seoul, Republic of Korea, 27–28 October 2019.

15. Yu, S.; Wickstrøm, K.; Jenssen, R.; Principe, J.C. Understanding convolutional neural networks with information theory: An initial exploration. *IEEE Trans. Neural Netw. Learn. Syst.* **2020**, *32*, 435–442. [CrossRef]

16. Elidan, G.; Friedman, N.; Chickering, D.M. Learning Hidden Variable Networks: The Information Bottleneck Approach. *J. Mach. Learn. Res.* **2005**, *6*, 81–127.

17. Wickstrøm, K.; Løkse, S.; Kampffmeyer, M.; Yu, S.; Principe, J.; Jenssen, R. Information plane analysis of deep neural networks via matrix-based Renyi's entropy and tensor kernels. *arXiv* **2019**, arXiv:1909.11396.

18. Kirsch, A.; Lyle, C.; Gal, Y. Scalable training with information bottleneck objectives. In Proceedings of the International Conference on Machine Learning (ICML): Workshop on Uncertainty and Robustness in Deep Learning, Virtual, 17–18 July 2020.

19. Jónsson, H.; Cherubini, G.; Eleftheriou, E. Convergence behavior of DNNs with mutual-information-based regularization. *Entropy* **2020**, *22*, 727. [CrossRef]

20. Schiemer, M.; Ye, J. Revisiting the Information Plane. 2020. Available online: https://openreview.net/forum?id=Hyljn1SFwr (accessed on 5 May 2023).

21. Goldfeld, Z.; Berg, E.v.d.; Greenewald, K.; Melnyk, I.; Nguyen, N.; Kingsbury, B.; Polyanskiy, Y. Estimating information flow in deep neural networks. *arXiv* **2018**, arXiv:1810.05728.

22. Lorenzen, S.S.; Igel, C.; Nielsen, M. Information Bottleneck: Exact Analysis of (Quantized) Neural Networks. *arXiv* **2021**, arXiv:2106.12912.

23. Shwartz-Ziv, R.; Alemi, A.A. Information in infinite ensembles of infinitely-wide neural networks. In Proceedings of the Symposium on Advances in Approximate Bayesian Inference, PMLR, Vancouver, BC, Canada, 8 December 2020; pp. 1–17.

24. Saxe, A.M.; Bansal, Y.; Dapello, J.; Advani, M.; Kolchinsky, A.; Tracey, B.D.; Cox, D.D. On the information bottleneck theory of deep learning. *J. Stat. Mech. Theory Exp.* **2019**, *2019*, 124020. [CrossRef]

25. Zeitler, G.; Koetter, R.; Bauch, G.; Widmer, J. Design of network coding functions in multihop relay networks. In Proceedings of the 2008 5th International Symposium on Turbo Codes and Related Topics, Lausanne, Switzerland, 1–5 September 2008; pp. 249–254.

26. Noshad, M.; Zeng, Y.; Hero, A.O. Scalable mutual information estimation using dependence graphs. In Proceedings of the ICASSP 2019-2019 IEEE International Conference on Acoustics, Speech and Signal Processing (ICASSP), Brighton, UK, 12–17 May 2019; pp. 2962–2966.

27. Abrol, V.; Tanner, J. Information-bottleneck under mean field initialization. In Proceedings of the ICML 2020 Workshop on Uncertainty and Robustness in Deep Learning, Virtual, 17–18 July 2020.

28. Darlow, L.N.; Storkey, A. What Information Does a ResNet Compress? *arXiv* **2020**, arXiv:2003.06254.

29. Cheng, H.; Lian, D.; Gao, S.; Geng, Y. Utilizing Information Bottleneck to Evaluate the Capability of Deep Neural Networks for Image Classification †. *Entropy* **2019**, *21*, 456. [CrossRef]

30. Voloshynovskiy, S.; Taran, O.; Kondah, M.; Holotyak, T.; Rezende, D.J. Variational Information Bottleneck for Semi-Supervised Classification. *Entropy* **2020**, *22*, 943. [CrossRef]

31. Yu, S.; Príncipe, J.C. Understanding Autoencoders with Information Theoretic Concepts. *Neural Netw.* **2018**, *117*, 104–123. [CrossRef]

32. Tapia, N.I.; Est'evez, P.A. On the Information Plane of Autoencoders. In Proceedings of the 2020 International Joint Conference on Neural Networks (IJCNN), Glasgow, UK, 19–24 July 2020; pp. 1–8.

33. Lee, S.; Jo, J. Compression phase is not necessary for generalization in representation learning. *arXiv* **2021**, arXiv:2102.07402.

34. Raj, V.; Nayak, N.; Kalyani, S. Understanding learning dynamics of binary neural networks via information bottleneck. *arXiv* **2020**, arXiv:2006.07522.

35. Strouse, D.; Schwab, D.J. The deterministic information bottleneck. *Neural Comput.* **2017**, *29*, 1611–1630. [CrossRef]

36. Hsu, H.; Asoodeh, S.; Salamatian, S.; Calmon, F.P. Generalizing bottleneck problems. In Proceedings of the 2018 IEEE International Symposium on Information Theory (ISIT), Vail, CO, USA, 17–22 June 2018; pp. 531–535.

37. Pensia, A.; Jog, V.; Loh, P.L. Extracting robust and accurate features via a robust information bottleneck. *IEEE J. Sel. Areas Inf. Theory* **2020**, *1*, 131–144. [CrossRef]

38. Xu, Y.; Zhao, S.; Song, J.; Stewart, R.; Ermon, S. A Theory of Usable Information under Computational Constraints. In Proceedings of the International Conference on Learning Representations, New Orleans, LA, USA, 6–9 May 2019.

39. Dubois, Y.; Kiela, D.; Schwab, D.J.; Vedantam, R. Learning optimal representations with the decodable information bottleneck. *Adv. Neural Inf. Process. Syst.* **2020**, *33*, 18674–18690.

40. Wongso, S.; Ghosh, R.; Motani, M. Using Sliced Mutual Information to Study Memorization and Generalization in Deep Neural Networks. In Proceedings of the International Conference on Artificial Intelligence and Statistics, PMLR, Valencia, Spain, 25–27 April 2023; pp. 11608–11629.

41. Wongso, S.; Ghosh, R.; Motani, M. Understanding Deep Neural Networks Using Sliced Mutual Information. In Proceedings of the 2022 IEEE International Symposium on Information Theory (ISIT), Espoo, Finland, 26 June–1 July 2022; pp. 133–138.

42. Polyanskiy, Y.; Wu, Y. Lecture notes on information theory. *Lect. Notes ECE563 (UIUC)* **2014**, *6*, 7.

43. Kolchinsky, A.; Tracey, B.D. Estimating mixture entropy with pairwise distances. *Entropy* **2017**, *19*, 361. [CrossRef]

44. Moon, Y.I.; Rajagopalan, B.; Lall, U. Estimation of mutual information using kernel density estimators. *Phys. Rev. E* **1995**, *52*, 2318. [CrossRef]

45. Kolchinsky, A.; Tracey, B.D.; Wolpert, D.H. Nonlinear information bottleneck. *Entropy* **2019**, *21*, 1181. [CrossRef]

46. Kraskov, A.; Stögbauer, H.; Grassberger, P. Estimating mutual information. *Phys. Rev. E* **2004**, *69*, 066138. [CrossRef]

47. Kirsch, A.; Lyle, C.; Gal, Y. Learning CIFAR-10 with a simple entropy estimator using information bottleneck objectives. In Proceedings of the Workshop Uncertainty and Robustness in Deep Learning at International Conference on Machine Learning, ICML, Virtual, 17–18 July 2020.

48. Goldfeld, Z.; Greenewald, K. Sliced mutual information: A scalable measure of statistical dependence. *Adv. Neural Inf. Process. Syst.* **2021**, *34*, 17567–17578.

49. Li, J.; Liu, D. Information Bottleneck Theory on Convolutional Neural Networks. *arXiv* **2019**, arXiv:1911.03722.

50. Song, J.; Ermon, S. Understanding the Limitations of Variational Mutual Information Estimators. In Proceedings of the International Conference on Learning Representations, New Orleans, LA, USA, 6–9 May 2019.

51. Wu, Y.; Verdú, S. Functional properties of minimum mean-square error and mutual information. *IEEE Trans. Inf. Theory* **2011**, *58*, 1289–1301. [CrossRef]

52. Glorot, X.; Bengio, Y. Understanding the difficulty of training deep feedforward neural networks. In Proceedings of the International Conference on Artificial Intelligence and Statistics, Sardinia, Italy, 13–15 May 2010.

53. LeCun, Y.; Cortes, C.; Burges, C. MNIST Handwritten Digit Database. ATT Labs [Online]. 2010, Volume 2 Available online: http://yann.lecun.com/exdb/mnist (accessed on 1 May 2023).

54. Simonyan, K.; Zisserman, A. Very Deep Convolutional Networks for Large-Scale Image Recognition. *arXiv* **2014**, arXiv:1409.1556.

55. Krizhevsky, A.; Hinton, G.; *Learning Multiple Layers of Features from Tiny Images*; Technical Report; University of Toronto: Toronto, ON, Canada, 2009.

56. Darlow, L.N.; Crowley, E.J.; Antoniou, A.; Storkey, A.J. CINIC-10 is not ImageNet or CIFAR-10. *arXiv* **2018**, arXiv:1810.03505.

57. Díaz, M.; Kairouz, P.; Liao, J.; Sankar, L. Neural Network-based Estimation of the MMSE. In Proceedings of the 2021 IEEE International Symposium on Information Theory (ISIT), Melbourne, Australia, 12–20 July 2021; pp. 1023–1028.

58. Hendrycks, D.; Gimpel, K. Gaussian Error Linear Units (GELUs). *arXiv* **2020**, arXiv:1606.08415.

59. Glorot, X.; Bordes, A.; Bengio, Y. Deep sparse rectifier neural networks. In Proceedings of the Fourteenth International Conference on Artificial Intelligence and Statistics. JMLR Workshop and Conference Proceedings, Fort Lauderdale, FL, USA, 11–13 April 2011; pp. 315–323.

60. Clevert, D.A.; Unterthiner, T.; Hochreiter, S. Fast and accurate deep network learning by exponential linear units (elus). *arXiv* **2015**, arXiv:1511.07289.

61. Ramachandran, P.; Zoph, B.; Le, Q.V. Searching for activation functions. *arXiv* **2017**, arXiv:1710.05941.

62. Elfwing, S.; Uchibe, E.; Doya, K. Sigmoid-weighted linear units for neural network function approximation in reinforcement learning. *Neural Netw.* **2018**, *107*, 3–11. [CrossRef] [PubMed]

63. He, K.; Zhang, X.; Ren, S.; Sun, J. Delving deep into rectifiers: Surpassing human-level performance on imagenet classification. In Proceedings of the Proceedings of the IEEE International Conference on Computer Vision, Santiago, Chile, 7–13 December 2015; pp. 1026–1034.

64. Maas, A.L.; Hannun, A.Y.; Ng, A.Y. Rectifier nonlinearities improve neural network acoustic models. In Proceedings of the International Conference on Machine Learning, Atlanta, GA, USA, 16–21 June 2013; Volume 30, p. 3.

65. He, K.; Zhang, X.; Ren, S.; Sun, J. Deep Residual Learning for Image Recognition. In Proceedings of the 2016 IEEE Conference on Computer Vision and Pattern Recognition (CVPR), Las Vegas, NV, USA, 27–30 June 2016; pp. 770–778.

66. Huang, G.; Liu, Z.; Weinberger, K.Q. Densely Connected Convolutional Networks. In Proceedings of the 2017 IEEE Conference on Computer Vision and Pattern Recognition (CVPR), Honolulu, HI, USA, 21–26 July 2017; pp. 2261–2269.

67. Qian, N. On the momentum term in gradient descent learning algorithms. *Neural Netw.* **1999**, *12*, 145–151. [CrossRef]

68. Hinton, G. *Coursera Neural Networks for Machine Learning*; Lecture 6; University of Toronto: Toronto, ON, Canada, 2018.

69. Kingma, D.P.; Ba, J. Adam: A method for stochastic optimization. *arXiv* **2014**, arXiv:1412.6980.

70. Choi, D.; Shallue, C.J.; Nado, Z.; Lee, J.; Maddison, C.J.; Dahl, G.E. On Empirical Comparisons of Optimizers for Deep Learning. *arXiv* **2019**, arXiv:1910.05446.

71. Bianchini, B.; Halm, M.; Matni, N.; Posa, M. Generalization Bounded Implicit Learning of Nearly Discontinuous Functions. In Proceedings of the Conference on Learning for Dynamics & Control, Virtual Event, 7–8 June 2021.

72. Oord, A.v.d.; Li, Y.; Vinyals, O. Representation learning with contrastive predictive coding. *arXiv* **2018**, arXiv:1807.03748.

73. Hjelm, R.D.; Fedorov, A.; Lavoie-Marchildon, S.; Grewal, K.; Bachman, P.; Trischler, A.; Bengio, Y. Learning deep representations by mutual information estimation and maximization. *arXiv* **2018** arXiv:1808.06670.

Article

In-Network Learning: Distributed Training and Inference in Networks [†]

Matei Moldoveanu [1,2] and Abdellatif Zaidi [1,2,*]

[1] Laboratoire d'Informatique Gaspard-Monge, Université Paris-Est, 77454 Marne-la-Vallée, France
[2] Mathematical and Algorithmic Sciences Lab, Paris Research Center, Huawei Technologies, 92100 Boulogne-Billancourt, France
* Correspondence: abdellatif.zaidi@univ-eiffel.fr
[†] This paper is an extended version of our paper published in 2021 IEEE Globecom Workshops, Madrid, Spain, 7–11 December 2021.

Abstract: In this paper, we study distributed inference and learning over networks which can be modeled by a directed graph. A subset of the nodes observes different features, which are all relevant/required for the inference task that needs to be performed at some distant end (fusion) node. We develop a learning algorithm and an architecture that can combine the information from the observed distributed features, using the processing units available across the networks. In particular, we employ information-theoretic tools to analyze how inference propagates and fuses across a network. Based on the insights gained from this analysis, we derive a loss function that effectively balances the model's performance with the amount of information transmitted across the network. We study the design criterion of our proposed architecture and its bandwidth requirements. Furthermore, we discuss implementation aspects using neural networks in typical wireless radio access and provide experiments that illustrate benefits over state-of-the-art techniques.

Keywords: distributed learning; AI at the edge; inference over graphs

Citation: Moldoveanu, M.; Zaidi, A. In-Network Learning: Distributed Training and Inference in Networks. *Entropy* **2023**, *25*, 920. https://doi.org/10.3390/e25060920

Academic Editors: Gerhard Bauch and Jan Lewandowsky

Received: 27 April 2023
Revised: 2 June 2023
Accepted: 6 June 2023
Published: 10 June 2023

1. Introduction

The unprecedented success of modern machine learning (ML) techniques in areas such as computer vision [1], neuroscience [2], image processing [3], robotics [4] and natural language processing [5] has led to an increasing interest for their application to wireless communication systems in recent years.

Early efforts along this line of work fall into what is sometimes referred to as the "learning to communicate" paradigm, in which the goal is to automate one or more communication modules such as the modulator-demodulator, the channel coder-decoder, or others, by replacing them with suitable ML algorithms. Although important progress has been made for some particular communication systems, such as the molecular one [6], it is still not yet clear whether ML techniques can offer a reliable alternate solution to model-based approaches, especially as typical wireless environments suffer from time-varying noise and interference.

Wireless networks have other important intrinsic features which may pave the way for more cross-fertilization between ML and communication, as opposed to applying ML algorithms as black boxes in replacement of one or more communication modules. For example, while in areas such as computer vision, neuroscience, and others, relevant data is generally available at one point, it is typically highly distributed across several nodes in wireless networks.

Examples include self-driving cars where multiple sensors, both external and internal to the car can be used to help the car navigate its environment, medical applications to diagnose a patient based on data from different medical institutions or environmental monitoring to detect hazardous events or pollution, and others, see [7,8] for more information. We give more details of the usefulness of such setups in Examples 1 and 2. A prevalent approach for the implementation of ML solutions in such cases would consist of collecting all relevant data at one point (a cloud server) and then training a suitable ML model using all available data and processing power. Because the volumes of data needed for training are generally large, and with the scarcity of network resources (e.g., power and bandwidth), that approach might not be appropriate in many cases, however. In addition, some applications might have stringent latency requirements which are incompatible with sharing the data, such as in automatic vehicle driving. In other cases, it might be desired not to share the raw data for the sake of enhancing the privacy of the solution, in the sense that infringing the user's privacy is generally more easily accomplished from the raw data itself than from the output of a neural network (NN) that takes the raw data as input.

The above has called for a new paradigm in which intelligence moves from the heart of the network to its edge, which is sometimes referred to as "Edge Learning". In this new paradigm, communication plays a central role in the design of efficient ML algorithms and architectures because both data and computational resources, which are the main ingredients of an efficient ML solution, are highly distributed. A key aspect towards building suitable ML-based solutions is whether the setting assumes only the training phase involves distributed data, sometimes referred to as *distributed learning*, such as the Federated Learning (FL) of [9] or if the inference (or test) phase also involves distributed data.

The considered problem setup is strongly related to the problems of distributed estimation and detection (see, e.g., [10–13] and references therein). We differentiate ourselves from these problems as we assume no prior knowledge of distribution of the data. This is a common setup in many practical applications, such as image or speech processing, or text analysis, where the distribution between the observed data and the target variable is unknown or too complex to model.

In particular, of those most closely related to this paper, a growing line of works focus on developing distributed learning algorithms and architectures. The works of [14,15] address the problem of distributed learning using kernel methods when each node observes independent samples drawn from the same distribution. In our specific setup, however, the nodes observe correlated data, necessitating collaboration among all nodes during inference. On the other hand, works such as [16,17] are focused on the narrower problem of detection and impose certain restrictions on the scope of their investigation. However, perhaps most popular and related to our work is the FL of [9] which, as we already mentioned, is most suitable for scenarios in which the training phase has to be performed distributively, while the inference phase has to be performed centrally at one node. To this end, during the training phase, nodes (e.g., base stations) that possess data are all equipped with copies of a single NN model which they simultaneously train on their locally available data-sets. The learned weight parameters are then sent to a cloud or parameter server (PS) which aggregates them, e.g., by simply computing their average. The process is repeated, every time re-initializing using the obtained aggregated model, until convergence. The rationale is that, this way, the model is progressively adjusted to account for all variations in the data, not only those of the local data-set. For recent advances on FL and applications in wireless settings, the reader may refer to [18–20] and references therein. Another relevant work is the Split Learning (SL) of [21] in which, for a multiaccess type network topology, a two-part NN model, split into an encoder part and a decoder part, is learned sequentially. The decoder does not have its own data and in every round the NN encoder part is fed with a distinct data-set and its parameters are initialized using those learned from the previous round. The learned two-part model is then used as follows during the inference: one part of this model is used by an encoder, and the other one by a decoder. Another variation of SL, sometimes called "vertical SL", was proposed recently in [22]. The approach uses vertical partitioning of the data; in the special case of a multi-access topology, it is similar to the in-network learning solution that we propose in this paper.

Compared to both SL and FL, which consider only the training phase to be distributed, in this paper we focus on the problem in which the inference phase also takes place distributively. More specifically, in this paper, we study a network inference problem in which some of the nodes possess each, or can acquire, part of the data that is relevant for inference on a random variable Y. The node at which the inference needs to be performed is connected to the nodes that possess the relevant data through a number of intermediate other nodes. We assume that the network topology is fixed and known. This may model, e.g., a setting in which a macro BS needs to make inference on the position of a user on the basis of summary information obtained from correlated CSI measurements $X_1, \ldots, X_J$ that are acquired at some proximity edge BSs. Each of the edge nodes is connected with the central node either directly, via an error free link of given finite capacity, or via intermediary nodes. While in some cases it might be enough to process only a subset of the J nodes, we assume that processing only a (any) strict subset of the measurements cannot yield the desired inference accuracy and, as such, the J measurements $X_1, \ldots, X_J$ need to be processed during the inference or test phase.

Example 1. *(Autonomous Driving) One basic requirement of the problem of autonomous driving is the ability to cope with problematic roadway situations, such as those involving construction, road hazards, hand signals, and reckless drivers. Current approaches mainly depend on equipping the vehicle with more on-board sensors. Clearly, while this can only allow a better coverage of the navigation environment, it seems unlikely to successfully cope with the problem of blind spots due, e.g., to obstruction or hidden obstacles. In such contexts, external sensors such as other vehicles' sensors, cameras installed on the roofs of proximity buildings or wireless towers may help perform a more precise inference, by offering a complementary, possibly better, view of the navigation scene. An example scenario is shown in Figure 1. The application requires real-time inference which might be incompatible with current cellular radio standards, thus precluding the option of sharing the sensors' raw data and processing it locally, e.g., at some on-board server. When equipped with suitable intelligence capabilities, each sensor can successfully identify and extract those features of its measurement data that are not captured by other sensors' data. Then, it only needs to communicate those, not its entire data.*

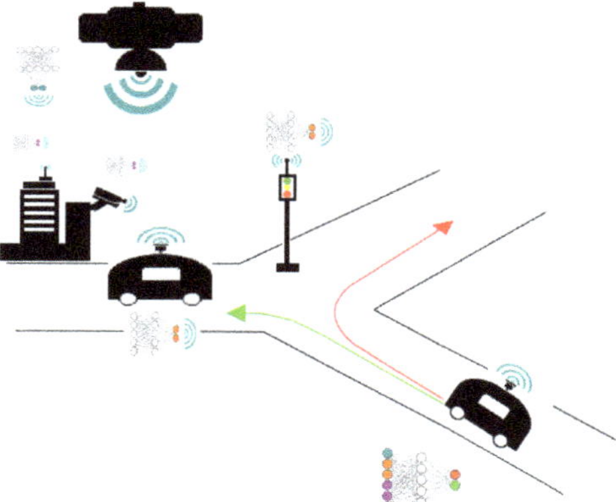

Figure 1. Fusion of inference from on-board and external sensors for automatic vehicle navigation.

Example 2. *(Public Health) One of the early applications of machine learning is in the area of medical imaging and public health. In this context, various institutions can hold different modalities of patient data in the form of electronic health records, pathology test results, radiology, and other sensitive imaging data such as genetic markers for disease. The correct diagnosis may be contingent on being able to using all relevant data from all institutions. However, these institutions may not be authorized to share their raw data. Thus, it is desired to distributively train machine learning models without sharing the patient's raw data in order to prevent illegal, unethical or unauthorized usage of it [23]. Local hospitals or tele-health screening centers seldom acquire enough diagnostic* *images on their own; collaborative distributed learning in this setting would enable each individual center to contribute data to an aggregate model without sharing any raw data.*

1.1. Contributions

In this paper, we study the aforementioned network inference problem in which the network is modeled as a weighted acyclic graph and inference about a random variable is performed on the basis of summary information obtained from possibly correlated variables at a subset of the nodes. Following an information-theoretic approach in which we measure discrepancies between true values and their estimated fits using average logarithmic loss, we first develop a bound on the best achievable accuracy given the network communication constraints. Then, considering a supervised setting in which nodes are equipped with NNs and their mappings need to be learned from distributively available training data-sets, we propose a distributed learning and inference architecture and we show that it can be optimized using a distributed version of the well-known stochastic gradient descent (SGD) algorithm that we develop here. The resulting distributed architecture and algorithm, which we herein name "in-network (INL) learning", generalize those introduced in [24] (see also [25,26]) for a specific case, multiaccess type, network topology. We investigate in more detail what the various nodes need to exchange during both the training and inference phases, as well as associated requirements in bandwidth. Finally, we provide a comparative study with (an adaptation of) the FL and the SL algorithms, and experiments that illustrate our results. Part of the results this paper have also been presented in [27,28]. However, in this paper, we go beyond those works by offering a more comprehensive and detailed review of the state-of-the-art. Additionally, we provide proofs for the theorem and lemmas presented in this paper, which were not included in the previous publications. Furthermore, we introduce additional insights and conclusions that further contribute to the overall understanding and significance of the research findings.

1.2. Outline and Notation

In Section 2 we describe the studied network inference problem formally. In Section 3 we present our in-network inference architecture, as well as a distributed algorithm for training it distributively. Section 4 contains a comparative study with FL and SL in terms of bandwidth requirements; as well as some experimental results. Finally, in Section 5 we summarize the insights and results presented in this paper.

Throughout the paper, the following notation will be used. Upper case letters denote random variables, e.g., X; lower case letters denote realizations of random variables, e.g., x, and calligraphic letters denote sets, e.g., $\mathcal{X}$. The cardinality of a set is denoted by $|\mathcal{X}|$. For a random variable X with probability mass function P_X, the shorthand $p(x) = P_X(x)$, $x \in X$ is used. Boldface letters denote matrices or vectors, e.g., $\mathbf{X}$ or $\mathbf{x}$. For random variables $(X_1, X_2, \ldots)$ and a set of integers $\mathcal{K} \subseteq \mathrm{N}$, the notation $X_\mathcal{K}$ designates the vector of random variables with indices in the set $\mathcal{K}$, i.e., $X_\mathcal{K} \triangleq \{X_k : k \in \mathcal{K}\}$. If $\mathcal{K} = \emptyset$ then $X_\mathcal{K} = \emptyset$. In addition, for zero-mean random vectors $\mathbf{x}$ and $\mathbf{y}$, the quantities $\Sigma_\mathbf{x}$, $\Sigma_{\mathbf{x},\mathbf{y}}$ and $\Sigma_{\mathbf{x}|\mathbf{y}}$ denote, respectively, the covariance matrix of the vector $\mathbf{x}$, the covariance matrix of vector $(\mathbf{x}, \mathbf{y})$ and the conditional covariance of $\mathbf{x}$ given $\mathbf{y}$. Finally, for two probability measures P_X and Q_X over the same alphabet $\mathcal{X}$, the relative entropy or Kullback-Leibler divergence is denoted as $D_{KL}(P_X\|Q_X)$. That is, if P_X is absolutely continuous with respect to Q_X, then $D_{KL}(P_X\|Q_X) = \mathbb{E}_{P_X}[\log(P_X(X)/Q_X(X))]$, otherwise $D_{KL}(P_X\|Q_X) = \infty$.

2. Network Inference: Problem Formulation

We consider the distributed supervised learning setup, in which multiple nodes observe different features relating to the same sample, sometimes refered to as distributed learning with vertically partitioned dataset, see [8,29]. We additionally assume the learning takes place over a communication constrained network. Specifically, consider an N node distributed network. Of these N nodes, $J \geq 1$ nodes possess or can acquire data that is relevant for inference on a random variable (r.v.) of interest Y, with alphabet $\mathcal{Y}$. Let

$\mathcal{J} = \{1, \ldots, J\}$ denote the set of such nodes, with node $j \in \mathcal{J}$ observing samples from the random variable X_j, with alphabet $\mathcal{X}_j$. The relationship between the r.v. of interest Y and the observed ones, $X_1, \ldots, X_J$, is given by the joint probability mass function $P_{X_{\mathcal{J}},Y} := P_{X_1,\ldots,X_J,Y}(x_1, \ldots x_J, y)$, with $(x_1, \ldots, x_j) \in \mathcal{X}_1 \times \cdots \times \mathcal{X}_J$ and $y \in \mathcal{Y}$. For simplicity, we assume that random variables are discreet, however our technique can be applied to continuous variables as well. Inference on Y needs to be performed at some node N which is connected to the nodes that possess the relevant data through a number of intermediate other nodes. It has to be performed without any sharing of raw data. The network is modeled as a weighted directed acyclic graph and may represent, for example, a wired network or a wireless mesh network operated in time or frequency division, where the nodes may be servers, handsets, sensors, base stations or routers. We assume that the network graph is fixed and known. The edges in the graph represent point-to-point communication links that use channel coding to achieve close to error-free communication at rates below their respective capacities. For a given loss function $\ell(\cdot, \cdot)$ that measures discrepancies between true values of Y and their estimated fits, what is the best precision for the estimation of Y? Clearly, discarding any of the relevant data X_j can only lead to a reduced precision. Thus, intuitively features that collectively maximize information about Y need to be extracted distributively by the nodes from the set $\mathcal{J}$, without explicit coordination between them and they then need to propagate and combine appropriately at the node N. How should that be performed optimally without sharing raw data? In particular, how should each node process information from the incoming edges (if any) and what should it transmit on every one of its outgoing edges? Furthermore, how should the information be fused optimally at Node N?

More formally, we model an N-node network by a directed acyclic graph $\mathcal{G} = (\mathcal{N}, \mathcal{E}, C)$, where $\mathcal{N} = [1 : N]$ is the set of nodes, $\mathcal{E} \subset \mathcal{N} \times \mathcal{N}$ is the set of edges and $C = \{C_{jk} : (j,k) \in \mathcal{E}\}$ is the set of edge weights. Each node represents a device and each edge represents a noiseless communication link with capacity C_{jk}. See Figure 2. The processing at the nodes of the set $\mathcal{J}$ is such that each of them assigns an index $m_{jl} \in [1, M_{jl}]$ to each $x_j \in \mathcal{X}_j$ and each received index tuple $(m_{ij} : (i,j) \in \mathcal{E})$, for each edge $(j,l) \in \mathcal{E}$. Specifically, let for $j \in \mathcal{J}$ and l such that $(j,l) \in \mathcal{E}$, the set $\mathcal{M}_{jl} = [1 : M_{jl}]$. The encoding function at node j is

$$\omega_j : \mathcal{X}_j \times \left\{ \Pi_{i : (i,j) \in \mathcal{E}} \mathcal{M}_{ij} \right\} \longrightarrow \Pi_{l : (j,l) \in \mathcal{E}} \mathcal{M}_{jl}, \tag{1}$$

where Π designates the Cartesian product of sets. Similarly, for $k \in [1 : N - 1]/\mathcal{J}$, node k assigns an index $m_{kl} \in [1, M_{kl}]$ to each index tuple $(m_{ik} : (i,k) \in \mathcal{E})$ for each edge $(k,l) \in \mathcal{E}$. That is,

$$\omega_k : \Pi_{i : (i,k) \in \mathcal{E}} \mathcal{M}_{ik} \longrightarrow \Pi_{l : (k,l) \in \mathcal{E}} \mathcal{M}_{kl}. \tag{2}$$

The range of the encoding functions $\{\omega_i\}$ are restricted in size, as

$$\log |\mathcal{M}_{ij}| \leq C_{ij} \quad \forall i \in [1, N-1] \quad \text{and} \quad \forall j : (i,j) \in \mathcal{E}. \tag{3}$$

Node N needs to infer on the random variable $Y \in \mathcal{Y}$ using all incoming messages, i.e.,

$$\psi : \Pi_{i : (i,N) \in \mathcal{E}} \mathcal{M}_{iN} \longrightarrow \hat{\mathcal{Y}}. \tag{4}$$

In this paper, we choose the reconstruction set $\hat{\mathcal{Y}}$ to be the set of distributions on $\mathcal{Y}$, i.e., $\hat{\mathcal{Y}} = \mathcal{P}(\mathcal{Y})$ and we measure discrepancies between true values of $Y \in \mathcal{Y}$ and their estimated fits in terms of average logarithmic loss, i.e., for $(y, \hat{P}) \in \mathcal{Y} \times \mathcal{P}(\mathcal{Y})$

$$d(y, \hat{P}) = \log \frac{1}{\hat{P}(y)}. \tag{5}$$

As such, the performance of a distributed inference scheme $\left((\omega_j)_{j\in\mathcal{J}}, (\omega_k)_{k\in[1,N-1]/\mathcal{J}}, \psi\right)$ for which (3) is fulfilled is given by its achievable *relevance* given by

$$\Delta = H(Y) - \mathbb{E}\big[d(Y,\hat{Y})\big], \tag{6}$$

which, for a discrete set $\mathcal{Y}$, is directly related to the error of misclassifying the variable $Y \in \mathcal{Y}$. It is imporant to note that $H(Y)$ is problem specific constant and as such the relavance given by (6) is simply a another form of the logarithmic loss.

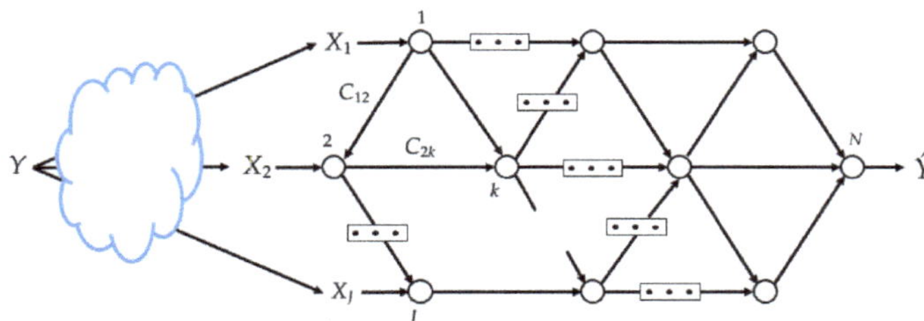

Figure 2. Studied network inference model.

In practice, in a supervised setting, the mappings given by (1), (2) and (4) need to be learned from a set of training data samples $\{(x_{1,i},\ldots,x_{J,i},y_i)\}_{i=1}^n$. The data is distributed such that the samples $\mathbf{x}_j := (x_{j,1},\ldots,x_{j,n})$ are available at node j for $j \in \mathcal{J}$ and the desired predictions $\mathbf{y} := (y_1,\ldots,y_n)$ are available at the end decision node N. We parametrize the possibly stochastic mappings (1), (2) and (4) using NNs. This is depicted in Figure 3. We denote the parameters of the NNs that parameterize the encoding function at each node $i \in [1 : (N-1)]$ with $\boldsymbol{\theta}_i$ and the parameters of the NN that parameterizes the decoding function at node N with $\boldsymbol{\phi}$. Let $\boldsymbol{\theta} = [\boldsymbol{\theta}_1,\ldots,\boldsymbol{\theta}_{N-1}]$, we aim to find the parameters $\boldsymbol{\theta}, \boldsymbol{\phi}$ that maximize the relevance of the network, given the network constraints of (3). Given that the actual distribution is unknown and we only have access to a dataset, the loss function needs to strike a balance between its performance on the dataset, given by empirical estimate of the relevance, and the network's ability to perform well on samples outside the dataset.

The NNs at the various nodes are arbitrary and can be chosen independently—for instance, they need *not* be identical as in FL. It is only required that the following mild condition which, as will become clearer from what follows, facilitates the back-propagation be met. Specifically, for every $j \in \mathcal{J}$ and $x_j \in \mathcal{X}_j$, under the assumtion that all elements of $\mathcal{X}_j$ have the same dimension, it holds that

$$\text{Size of first layer of NN } (j) =$$
$$\text{Dimension } (x_j) + \sum_{i\,:\,(i,j)\,\in\,\mathcal{E}} (\text{Size of last layer of NN } (i)). \tag{7}$$

Similarly, for $k \in [1 : N]/\mathcal{J}$ we have

$$\text{Size of first layer of NN } (k) =$$
$$\sum_{i\,:\,(i,k)\,\in\,\mathcal{E}} (\text{Size of last layer of NN } (i)). \tag{8}$$

Remark 1. *Conditions (7) and (8) were imposed only for the sake of ease of implementation of the training algorithm; the techniques present in this paper, including optimal trade-offs between relevance and complexity for the given topology, the associated loss function, the variational lower bound, how to parameterize it using NNs and so on, do not require (7) and (8) to hold. Alternative aggregation techniques, such as element-wise multiplication or element-wise averaging, can be* *employed to combine the information received by each node, in replacement to concatenation. The impact of these aggregation techniques has been analyzed in [22].*

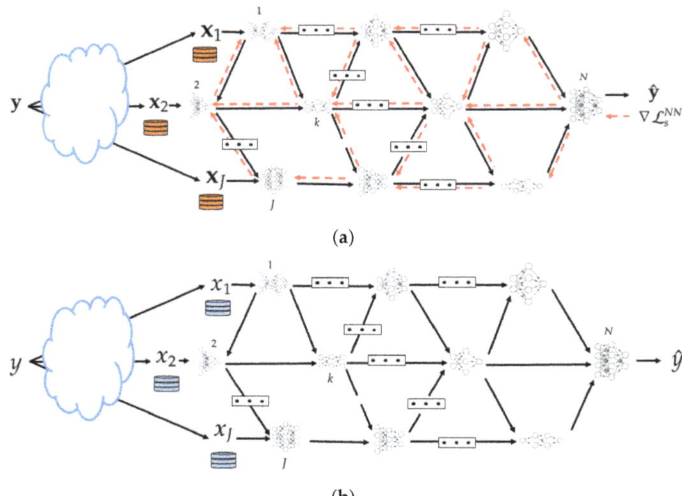

(a)

(b)

Figure 3. In-network learning and inference using neural networks. (**a**) Training phase. (**b**) Inference phase.

3. Proposed Solution: In-Network Learning and Inference

For convenience, we first consider a specific setting of the model of network inference problem of Figure 3 in which $J = N - 1$ and all the nodes that observe data are only connected to the end decision node, but not among them.

3.1. A Specific Model: Fusing of Inference

In this case, a possible suitable loss function was shown by [25] to be:

$$\mathcal{L}_s^{\text{NN}}(n) = \frac{1}{n} \sum_{i=1}^{n} \log Q_{\phi_{\mathcal{J}}}(y_i | u_{1,i}, \ldots, u_{J,i})$$

$$+ \frac{s}{n} \sum_{i=1}^{n} \sum_{j=1}^{J} \left(\log Q_{\phi_j}(y_i | u_{j,i}) - \log \left(\frac{P_{\theta_j}(u_{j,i} | x_{j,i})}{Q_{\varphi_j}(u_{j,i})} \right) \right), \tag{9}$$

where s is a Lagrange parameter and for $j \in \mathcal{J}$ the distributions $P_{\theta_j}(u_j | x_j)$, $Q_{\phi_j}(y | u_j)$, $Q_{\phi_{\mathcal{J}}}(y | u_{\mathcal{J}})$ are variational ones whose parameters are determined by the chosen NNs using the re-parametrization trick of [30] and $Q_{\varphi_j}(u_j)$ are priors known to the encoders. For example, denoting by f_{θ_j} the NN used at node $j \in \mathcal{J}$ whose (weight and bias) parameters are given by θ_j, for regression problems the conditional distribution $P_{\theta_j}(u_j | x_j)$ can be chosen to be multivariate Gaussian, i.e., $P_{\theta_j}(u_j | x_j) = \mathcal{N}(u_j; \mu_j^\theta, \Sigma_j^\theta)$, where $\mu_j^\theta, \Sigma_j^\theta$ are outputs of $f_{\theta_j}(x_j)$. For discrete data, concrete variables (i.e., Gumbel-Softmax) can be used instead.

The rationale behind the choice of loss function (9) is that in the regime of large n, if the encoders and decoder are not restricted to use NNs under some conditions. The optimality is proved therein under the assumption that for every subset $\mathcal{S} \subseteq \mathcal{J}$, it holds that $X_{\mathcal{S}} \multimap Y \multimap X_{\mathcal{S}^c}$. The RHS of (10) is achievable for arbitrary distributions, however, regardless of such an assumption; the optimal stochastic mappings $P_{U_j | X_j}$, P_U, $P_{Y | U_j}$ and

$P_{Y|U_{\mathcal{J}}}$ are found by marginalizing the joint distribution that maximizes the following Lagrange cost function [25] (Proposition 2)

$$\mathcal{L}_s^{\text{optimal}} = -H(Y|U_{\mathcal{J}}) - s \sum_{j=1}^{J} \left[H(Y|U_j) + I(U_j; X_j) \right]. \tag{10}$$

where the maximization is over all joint distributions of the form $P_Y \prod_{j=1}^{J} P_{X_j|Y} \prod_{j=1}^{J} P_{U_j|X_j}$.

3.1.1. Inference Phase

During this phase node j observes a new sample x_j. It uses its NN to output an encoded value u_j which it sends to the decoder. After collecting $(u_1, \ldots, u_J)$ from all input NNs, node $(J+1)$ uses its NN to output an estimate of Y in the form of soft output $Q_{\phi_{\mathcal{J}}}(Y|u_1, \ldots, u_J)$. The procedure is depicted in Figure 4b.

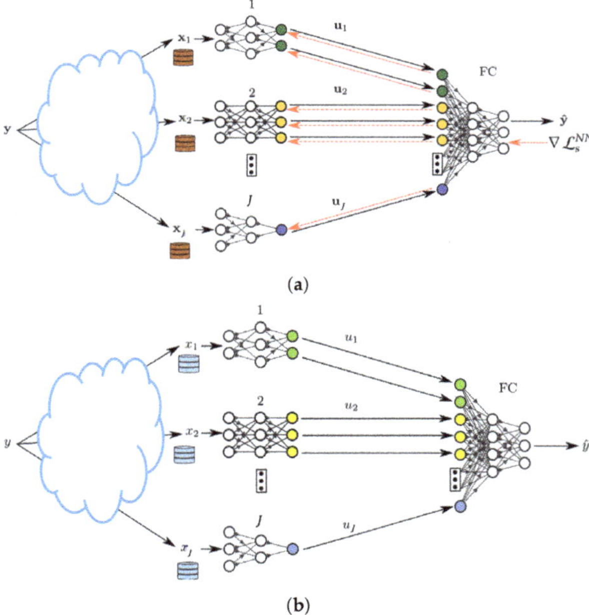

Figure 4. In-network learning for the network model for the case without hops. (**a**) Training phase. (**b**) Inference phase.

Remark 2. *One can combine our proposed technique with an appropriate transmission scheme and channel coding. One possible suitable practical implementation in wireless settings can be obtained using Orthogonal Frequency-Division Multiple Access (OFDMA). That is, the J input nodes are allocated non-overlapping bandwidth segments and the output layers of the corresponding NNs are chosen accordingly. The encoding of the activation values can be performed, e.g., using entropy type coding [31].*

3.1.2. Training Phase

During the forward pass, every node $j \in \mathcal{J}$ processes mini-batches of size, say, b_j of its training data-set x_j. Node $j \in \mathcal{J}$ then sends a vector, $\mathbf{u}_j$, whose elements are the activation values of the last layer of (NN j), see Figure 4a. Due to (8) the activation vectors are concatenated vertically at the input layer of NN $(J+1)$. The forward pass continues on the NN $(J+1)$ until the last layer of the latter. The parameters of NN $(J+1)$ are updated using standard backpropagation. Specifically, let L_{J+1} denote the index of the last layer of NN $(J+1)$. Additionally, let $\mathbf{w}_{J+1}^{[l]}$, $\mathbf{b}_{J+1}^{[l]}$ and $\mathbf{a}_{J+1}^{[l]}$ denote the weights, biases and activation values at layer $l \in [2 : L_{J+1}]$ for the NN $(J+1)$ and σ is the activation function, respectively. Node $(J+1)$ computes the error vectors

$$\delta_{J+1}^{[L_{J+1}]} = \nabla_{\mathbf{a}_{J+1}^{[L_{J+1}]}} \mathcal{L}_s^{NN}(b) \odot \sigma'(\mathbf{w}_{J+1}^{[L_{J+1}]} \mathbf{a}_{J+1}^{[L_{(J+1)}-1]} + \mathbf{b}_{J+1}^{[L_{J+1}]}) \tag{11a}$$

$$\delta_{J+1}^{[l]} = [(\mathbf{w}_{J+1}^{[l+1]})^T \delta_{J+1}^{[l+1]}] \odot \sigma'(\mathbf{w}_{J+1}^{[l]} \mathbf{a}_{J+1}^{[l-1]} + \mathbf{b}_{J+1}^{[l]}) \ \forall \ l \in [2, L_{J+1} - 1], \tag{11b}$$

$$\delta_{J+1}^{[1]} = [(\mathbf{w}_{J+1}^{[2]})^T \delta_{J+1}^{[2]}] \tag{11c}$$

and then updates its weight- and bias parameters as

$$\mathbf{w}_{J+1}^{[l]} \to \mathbf{w}_{J+1}^{[l]} - \eta \delta_{J+1}^{[l]} (\mathbf{a}_{J+1}^{[l-1]})^T, \tag{12a}$$

$$\mathbf{b}_{J+1}^{[l]} \to \mathbf{b}_{J+1}^{[l]} - \eta \delta_{J+1}^{[l]}, \tag{12b}$$

where η designates the learning parameter; for simplicity, η and σ are assumed here to be identical for all NNs.

Remark 3. *It is important to note that for the computation of the RHS of (11a) node $(J+1)$, which knows $Q_{\phi_J}(y_i|u_{1,i}, \ldots, u_{J,i})$ and $Q_{\phi_j}(y_i|u_{j,i})$ for all $i \in [1 : n]$ and all $j \in \mathcal{J}$, only the derivative of $\mathcal{L}_s^{NN}(n)$ w.r.t. the activation vector $\mathbf{a}_{J+1}^{L_{J+1}}$ is required. For instance, node $(J+1)$ does not need to know any of the conditional variationals $P_{\theta_j}(u_j|x_j)$ or the priors $Q_{\varphi_j}(u_j)$.*

The backward propagation of the error vector from node $(J+1)$ to the nodes j, $j \in \{1, \ldots, J\}$, is as follows. Node $(J+1)$ horizontally splits the error vector of its input layer into J sub-vectors with sub-error vector j having the same size as the dimension of the last layer of NN j [recall (8) and that the activation vectors are concatenated vertically during the forward pass]. See Figure 4a. The backward propagation then continues on each of the J input NNs simultaneously, each of them essentially applying operations similar to (11) and (12).

Remark 4. *Let $\delta_{J+1}^{[1]}(j)$ denote the sub-error vector sent back from node $(J+1)$ to node $j \in \mathcal{J}$. It is easy to see that, for every $j \in \mathcal{J}$,*

$$\nabla_{\mathbf{a}_j^{L_j}} \mathcal{L}_s^{NN}(b_j) = \delta_{J+1}^{[1]}(j) - s \nabla_{\mathbf{a}_j^{L_j}} \left(\sum_{i=1}^{b} \log \left(\frac{P_{\theta_j}(u_{j,i}|x_{j,i})}{Q_{\varphi_j}(u_{j,i})} \right) \right); \tag{13}$$

and this explains why node $j \in \mathcal{J}$ needs only the part $\delta_{J+1}^{[1]}(j)$, not the entire error vector at node $(J+1)$.

3.2. General Model: Fusion and Propagation of Inference

Consider now the general network inference model of Figure 2. Part of the difficulty of this problem is in finding a suitable loss function which can be optimized distributively via NNs that only have access to local data-sets each. The next theorem provides a bound on the achievable relevance (under some assumptions) for an arbitrary network topology $(\mathcal{E}, \mathcal{N})$. The result of Theorem 1 is asymptotic in the size of the training data-sets, while the inference problem is a one-shot problem. One-shot results for this problem can be obtained, e.g., along the approach of [32]. For convenience, we define for $\mathcal{S} \subseteq [1, \ldots, N-1]$ and non-negative $(C_{ij} : (i,j) \in \mathcal{E})$ the quantity

$$C(\mathcal{S}) = \sum_{(i,j) \,:\, i \in \mathcal{S}, j \in \mathcal{S}^c} C_{ij}. \tag{14}$$

Theorem 1. *For the network inference model of Figure 2, in the regime of large data-sets the following relevance is achievable,*

$$\Delta = \max I(U_1, \ldots, U_J; Y) \tag{15}$$

where the maximization is over joint measures of the form

$$P_Q P_{X_1, \ldots, X_J, Y} \prod_{j=1}^{J} P_{U_j | X_j, Q} \tag{16}$$

for which there exist non-negative $R_1, \ldots, R_J$ that satisfy

$$\sum_{j \in \mathcal{S}} R_j \geq I(U_{\mathcal{S}}; X_{\mathcal{S}} | U_{\mathcal{S}^c}, Q), \quad \text{for all} \quad \mathcal{S} \subseteq \mathcal{J}$$

$$\sum_{j \in \mathcal{S} \cap \mathcal{J}} R_j \leq C(\mathcal{S}) \quad \text{for all} \quad \mathcal{S} \subseteq [1 : N-1] \quad \text{with} \quad \mathcal{S} \cap \mathcal{J} \neq \emptyset.$$

Proof. The proof of Theorem 1 appears in Appendix A. An outline is as follows. The result is achieved using a separate compression-transmission-estimation scheme in which the observations $(\mathbf{x}_1, \ldots, \mathbf{x}_J)$ are first compressed distributively using Berger-Tung coding [33] into representations $(\mathbf{u}_1, \ldots, \mathbf{u}_J)$ and then the bin indices are transmitted as independent messages over the network $\mathcal{G}$ using linear-network coding [34] (Section 15.5). The decision node N first recovers the representation codewords $(\mathbf{u}_1, \ldots, \mathbf{u}_J)$ and then produces an estimate of the label $\mathbf{y}$. The scheme is illustrated in Figure 5. $\quad\square$

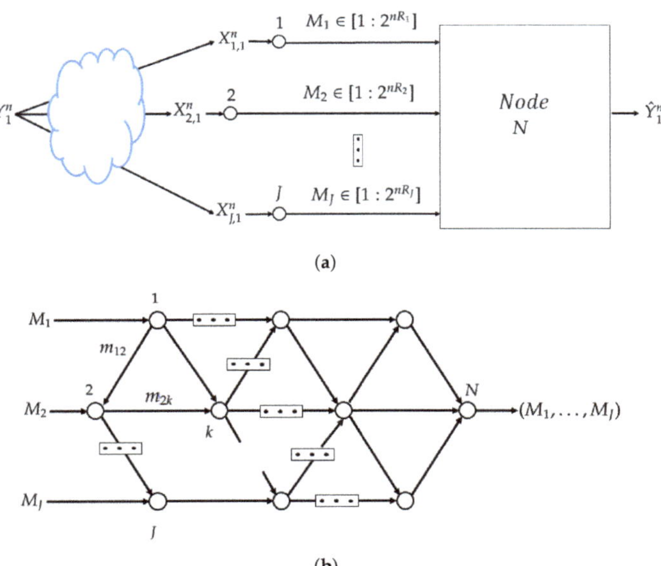

Figure 5. Block diagram of the separate compression-transmission-estimation scheme of Theorem 1. (**a**) Compression using Berger-Tung coding. (**b**) Transmission of the bin indices using linear coding.

Part of the utility of the loss function of Theorem 1 is in that it accounts explicitly for the network topology for inference fusion and propagation. In addition, although as seen from its proof the setting of Theorem 1 assumes knowledge of the joint distribution of the tuple $(X_1, \ldots, X_J, Y)$, the result can be used to train, distributively, NNs from a set of available date-sets. To do so, we first derive a Lagrangian function, from Theorem 1, which can be used as an objective function to find the desired set of encoders and decoder. Afterwards, we use a variational approximation to avoid the computation of marginal distributions, which can be costly in practice. Finally, we parameterize the distributions suing NNs. For a given network topology in essence, the approach generalizes that of Section 3.1 to more general networks that involve hops. For simplicity, in what follows, this is illustrated for the example architecture of Figure 6. While the example is simple, it showcases the important aspect of any such topology, the fusion of the data at an intermediary nodes, i.e., a hop. Firstly, we leverage Theorem 1 to establish a feasible trade-off between the performance of the network illustrated in Figure 6, quantified by its *relevance*, and the quantity of information that must be communicated between the nodes. Subsequently, employing the aforementioned approach, we derive a loss function tailored for the scenarios where the nodes are equipped with neural networks, as depicted in Figure 7.

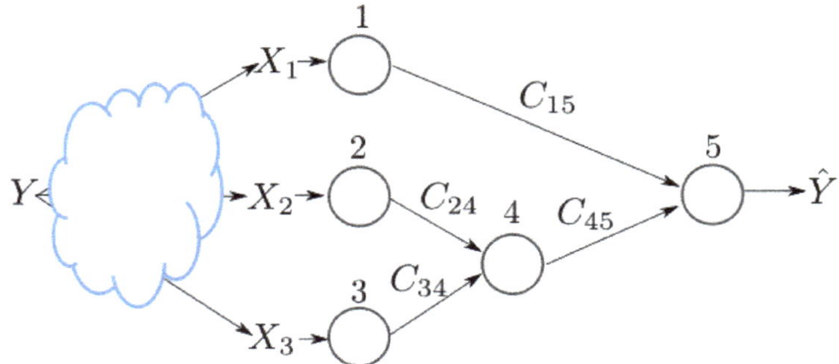

Figure 6. An example in-network learning with inference fusion and propogation.

Setting $\mathcal{N} = \{1, 2, 3, 4, 5\}$ and $\mathcal{E} = \{(3, 4), (2, 4), (4, 5), (1, 5)\}$ in Theorem 1, we obtain that

$$\Delta = \max I(U_1, U_2, U_3; Y) \tag{17}$$

where the maximization is over joint measures of the form

$$P_Q P_{X_1, X_2, X_3, Y} P_{U_1 | X_1, Q} P_{U_2 | X_2, Q} P_{U_3 | X_3, Q} \tag{18}$$

for which the following holds for some $R_1 \geq 0$, $R_2 \geq 2$ and $R_3 \geq 0$:

$$C_{15} \geq R_1, \ C_{24} \geq R_2, \ C_{34} \geq R_3, \ C_{45} \geq R_2 + R_3 \tag{19a}$$
$$R_1 \geq I(U_1; X_1 | U_2, U_3, Q), \tag{19b}$$
$$R_2 \geq I(U_2; X_2 | U_1, U_3, Q), \tag{19c}$$
$$R_3 \geq I(U_3; X_3 | U_1, U_2, Q) \tag{19d}$$
$$R_3 + R_2 \geq I(X_2, X_3; U_2, U_3 | U_1, Q), \tag{19e}$$
$$R_3 + R_1 \geq I(X_1, X_3; U_1, U_3 | U_2, Q) \tag{19f}$$
$$R_2 + R_1 \geq I(X_1, X_2; U_1, U_2 | U_3, Q), \tag{19g}$$
$$R_2 + R_1 + R_3 \geq I(X_1, X_2, X_3; U_1, U_2, U_3 | Q). \tag{19h}$$

Let $C_{\text{sum}} = C_{15} + C_{24} + C_{34} + C_{45}$; consider the region of all pairs $(\Delta, C_{\text{sum}}) \in \mathbb{R}_+^2$ for which the relevance level Δ as given by the RHS of (17) is achievable for some $C_{15} \geq 0$, $C_{24} \geq 0$, $C_{34} \geq 0$ and $C_{45} \geq 0$ such that $C_{\text{sum}} = C_{15} + C_{24} + C_{34} + C_{45}$. Hereafter, we denote such region as $\mathcal{RI}_{\text{sum}}$. Applying Fourier-Motzkin elimination on the region defined by (17) and (19), we obtain that the region $\mathcal{RI}_{\text{sum}}$ is given by the union of pairs $(\Delta, C_{\text{sum}}) \in \mathbb{R}^2_+$ for which (the time sharing random variable is set to a constant for simplicity)

$$\Delta \leq I(Y; U_1, U_2, U_3) \tag{20a}$$

$$C_{\text{sum}} \geq I(X_1, X_2, X_3; U_1, U_2, U_3) + I(X_2, X_3; U_2, U_3|U_1) \tag{20b}$$

for some measure of the form

$$P_Y P_{X_1, X_2, X_3|Y} P_{U_1|X_1} P_{U_2|X_2} P_{U_3|X_3}. \tag{21}$$

The next proposition gives a useful parameterization of the region $\mathcal{RI}_{\text{sum}}$ as described by (20) and (21).

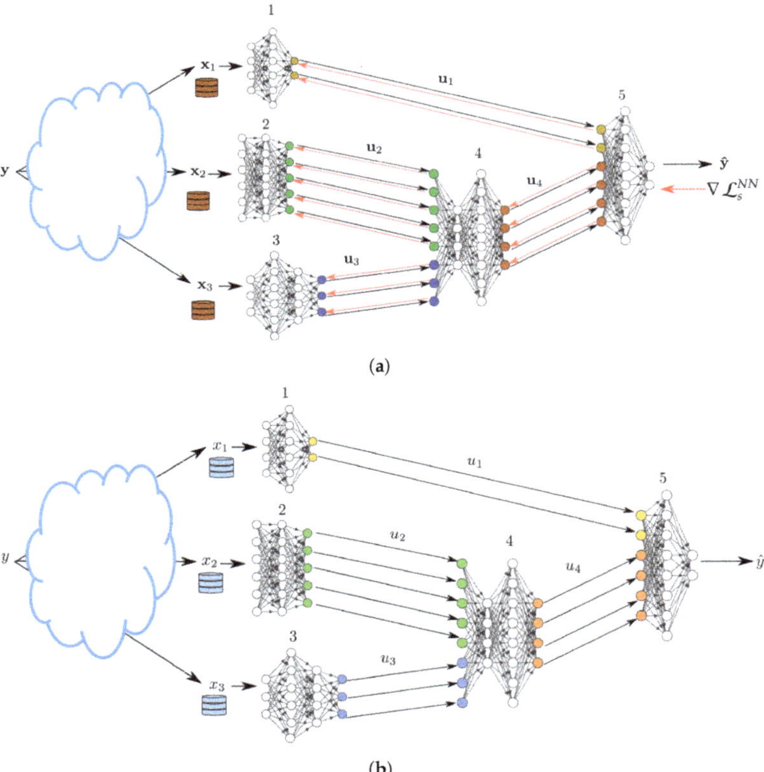

Figure 7. Forward and backward passes for the inference problem of Figure 6. (**a**) Training phase. (**b**) Inference phase.

Proposition 1. *For every pair (Δ, C_{sum}) that lies on the boundary of the region described by (20) and (21) there exists $s \geq 0$ such that $(\Delta, C_{sum}) = (\Delta_s, C_s)$, with*

$$\Delta_s = H(Y) + \max_{\mathbf{P}} \mathcal{L}_s(\mathbf{P}) + sC_s \tag{22a}$$

$$C_s = I(X_1, X_2, X_3; U_1^*, U_2^*, U_3^*) + I(X_2, X_3; U_2^*, U_3^*|U_1^*), \tag{22b}$$

and $\mathbf{P}^*$ is the set of pmfs $\mathbf{P} := \{P_{U_1|X_1}, P_{U_2|X_2}, P_{U_3|X_3}\}$ that maximize the cost function

$$\mathcal{L}_s(\mathbf{P}) := -H(Y|U_1, U_2, U_3) - sI(X_1, X_2, X_3; U_1, U_2, U_3)$$

$$-sI(X_2, X_3; U_2, U_3|U_1).\tag{23}$$

Proof. See Appendix B. □

In accordance with the studied example network inference problem shown in Figure 6, let a random variable U_4 be such that $U_4 \multimap (U_2, U_3) \multimap (X_1, X_2, X_3, Y, U_1)$. That is, the joint distribution factorizes as

$$P_{X_1,X_2,X_3,Y,U_1,U_2,U_3,U_4} = P_{X_1,X_2,X_3,Y}P_{U_1|X_1}P_{U_2|X_2}P_{U_3|X_3}P_{U_4|U_2,U_3}.\tag{24}$$

Let for given $s \geq 0$ and conditional $P_{U_4|U_2,U_3}$ the Lagrange term

$$\mathcal{L}_s^{\text{low}}(\mathbf{P}, P_{U_4|U_2,U_3}) = -H(Y|U_1, U_4) - sI(X_1; U_1) - 2sI(X_2; U_2)$$
$$- 2s\Big[I(X_3; U_3) - I(U_2; U_1) - I(U_3; U_1, U_2)\Big].\tag{25}$$

The following lemma shows that $\mathcal{L}_s^{\text{low}}(\mathbf{P}, P_{U_4|U_2,U_3})$ lower bounds $\mathcal{L}_s(\mathbf{P})$ as given by (23).

Lemma 1. *For every $s \geq 0$ and joint measure that factorizes as (24), we have*

$$\mathcal{L}_s(\mathbf{P}) \geq \mathcal{L}_s^{low}(\mathbf{P}, P_{U_4|U_2,U_3}),\tag{26}$$

Proof. See Appendix C. □

For convenience let $\mathbf{P}_+ := \{P_{U_1|X_1}, P_{U_2|X_2}, P_{U_3|X_3}, P_{U_4|U_2,U_3}\}$. The optimization of (25) generally requires the computation of marginal distributions, which can be costly in practice. Hereafter, we derive a variational lower bound on $\mathcal{L}_s^{\text{low}}$ with respect to some arbitrary (variational) distributions. Specifically, let

$$\mathbf{Q} := \{Q_{Y|U_1,U_4}, Q_{U_3}, Q_{U_2}, Q_{U_1}\},\tag{27}$$

where $Q_{Y|U_1,U_4}$ represents variational (possibly stochastic) decoders and Q_{U_3}, Q_{U_2} and Q_{U_1} represent priors. Additionally, let

$$\mathcal{L}_s^{\text{v-low}}(\mathbf{P}_+, \mathbf{Q}) := \mathbb{E}[\log Q_{Y|U_1,U_4}(Y|U_1, U_4)] - sD_{\text{KL}}(P_{U_1|X_1}\|Q_{U_1})$$
$$- 2sD_{\text{KL}}(P_{U_2|X_2}\|Q_{U_2}) - 2sD_{\text{KL}}(P_{U_3|X_3}\|Q_{U_3}).\tag{28}$$

The following lemma, the proof of which is essentially similar to that of [25] (Lemma 1), shows that for every $s \geq 0$, the cost function $\mathcal{L}_s^{\text{low}}(\mathbf{P}, P_{U_4|U_2,U_3})$ is lower-bounded by $\mathcal{L}_s^{\text{v-low}}(\mathbf{P}_+, \mathbf{Q})$ as given by (28).

Lemma 2. *For fixed $\mathbf{P}_+$, we have*

$$\mathcal{L}_s^{low}(\mathbf{P}_+) \geq \mathcal{L}_s^{v-low}(\mathbf{P}_+, \mathbf{Q})\tag{29}$$

for all pmfs $\mathbf{Q}$, with equality when:

$$Q_{Y|U_1,U_4} = P_{Y|U_1,U_4},\tag{30}$$
$$Q_{U_3} = P_{U_3|U_2,U_1},\tag{31}$$
$$Q_{U_2} = P_{U_2|U_1},\tag{32}$$
$$Q_{U_1} = P_{U_1},\tag{33}$$

where $P_{Y|U_1,U_4}$, $P_{U_3|U_2,U_1}$, $P_{U_2|U_1}$, P_{U_1} are calculated using (24).

Proof. See Appendix D. □

From the above, we get that

$$\max_{\mathbf{P}_+} \mathcal{L}_s^{\text{low}}(\mathbf{P}_+) = \max_{\mathbf{P}_+} \max_{\mathbf{Q}} \mathcal{L}_s^{\text{v-low}}(\mathbf{P}_+, \mathbf{Q}). \tag{34}$$

Since, as described in Section 2, the distribution of the data is not known, but only a set of samples is available $\{(x_{1,i}, \ldots, x_{J,i}, y_i)\}_{i=1}^{n}$, we restrict the optimization of (28) to the family of distributions that can be parameterized by NNs. Thus, we obtain the following loss function which can be optimized empirically, in a distributed manner, using gradient based techniques,

$$\mathcal{L}_s^{\text{NN}}(n) := \frac{1}{n} \sum_{i=1}^{n} \left[\log Q_{\phi_5}(y_i | u_{1,i}, u_{4,i}) - s \log\left(\frac{P_{\theta_1}(u_{1,i}|x_{1,i})}{Q_{\varphi_1}(u_{1,i})} \right) \right]$$
$$- \frac{2s}{n} \sum_{i=1}^{n} \left[\log\left(\frac{P_{\theta_2}(u_{2,i}|x_{2,i})}{Q_{\varphi_2}(u_{2,i})} \right) + \log\left(\frac{P_{\theta_3}(u_{3,i}|x_{3,i})}{Q_{\varphi_3}(u_{3,i})} \right) \right], \tag{35}$$

with s stands for a Lagrange multiplier and the distributions $Q_{\phi_5}, P_{\theta_4}, P_{\theta_3}, P_{\theta_2}, P_{\theta_1}$ are variational ones whose parameters are determined by the chosen NNs using the re-parametrization trick of [30] and $\{Q_{\varphi_i} : i \in \{1,2,3\}\}$ are priors known to the encoders. The parameterization of the distributions with NNs is performed similarly to that for the setting of Section 3.1.

3.2.1. Inference Phase

During this phase, nodes 1, 2 and 3 each observe (or measure) a new sample. Let x_1 be the sample observed by node 1 and x_2 and x_3 those observed by node 2 and node 3, respectively. Node 1 processes x_1 using its NN and sends an encoded value u_1 to node 5 and so do nodes 2 and 3 towards node 4. Upon receiving u_2 and u_3 from nodes 2 and 3, node 4 concatenates them vertically and processes the obtained vector using its NN. The output u_4 is then sent to node 5. The latter performs similar operations on the activation values u_1 and u_4 and outputs an estimate of the label y in the form of a soft output $Q_{\phi_5}(y|u_1, u_4)$.

3.2.2. Training Phase

During the forward pass, every node $j \in \{1, 2, 3\}$ processes mini-batches of size, b_j of its training data set $\mathbf{x}_j$. Nodes 2 and 3 send their vector formed of the activation values of the last layer of their NNs to node 4. Because the sizes of the last layers of the NNs of nodes 2 and 3 are chosen according to (8) the sent activation vectors are concatenated vertically at the input layer of NN 4. The forward pass continues on the NN at node 4 until its last layer. Next, nodes 1 and 4 send the activation values of their last layers to node 5. Again, as the sizes of the last layers of the NNs of nodes 1 and 4 satisfy (8) the sent activation vectors are concatenated vertically at the input layer of NN 5 and the forward pass continues until the last layer of NN 5.

During the backward pass, each of the NNs updates its parameters according to (11) and (12). Node 5 is the first to apply the back propagation procedure in order update the parameters of its NN. It applies (11) and (12) sequentially, starting from its last layer.

Remark 5. *It is important to note that, similar to the setting of Section III-A, for the computation of the RHS of (11a) for node 5, only the derivative of $\mathcal{L}_s^{\text{NN}}(n)$ w.r.t. the activation vector $\mathbf{a}_5^{L_5}$ is required, which depends only on $Q_{\phi_5}(y_i|u_{1,i}, u_{4,i})$. The distributions are known to node 5 given only $u_{1,i}$ and $u_{4,i}$.*

The error propagates back until it reaches the first layer of the NN of node 5. Node 5 then splits horizontally the error vector of its input layer into 2 sub-vectors with the top sub-error vector having as size that of the last layer of the NN of node 1 and the bottom sub-error vector having as size that of the last layer of the NN of node 4—see Figure 7a. Similarly, the two nodes 1 and 4 continue the backward propagation at their turns simultaneously. Node 4 then splits horizontally the error vector of its input layer into 2 sub-vectors with the top sub-error vector having as size that of the last layer of the NN of node 2 and the bottom sub-error vector having as size that of the last layer of the NN of node 3. Finally, the backward propagation continues on the NNs of nodes 2 and 3. The entire process continues until convergence.

Remark 6. *Let $\delta_J^{[1]}(j)$ denote the sub-error vector sent back from node J to node j. It is easy to see that, for every $j \in \mathcal{J}$,*

$$\nabla_{\mathbf{a}_4^{[L]}} \mathcal{L}_s^{NN}(b) = \delta_5^{[1]}(4),$$

$$\nabla_{\mathbf{a}_3^{[L]}} \mathcal{L}_s^{NN}(b) = \delta_4^{[1]}(3) - 2s\nabla_{\mathbf{a}_3^{[L]}}\left[\frac{1}{b}\sum_{i=1}^{b}\left[\log\left(\frac{P_{\theta_3}(u_{3,i}|x_{3,i})}{Q_{\varphi_3}(u_{3,i})}\right)\right]\right],$$

$$\nabla_{\mathbf{a}_2^{[L]}} \mathcal{L}_s^{NN}(b) = \delta_4^{[1]}(2) - 2s\nabla_{\mathbf{a}_2^{[L]}}\left[\frac{1}{b}\sum_{i=1}^{b}\left[\log\left(\frac{P_{\theta_2}(u_{2,i}|x_{2,i})}{Q_{\varphi_2}(u_{2,i})}\right)\right]\right],$$

$$\nabla_{\mathbf{a}_1^{[L]}} \mathcal{L}_s^{NN}(b) = \delta_5^{[1]}(1) - s\nabla_{\mathbf{a}_1^{[L]}}\left[\frac{1}{b}\sum_{i=1}^{b}\left[\log\left(\frac{P_{\theta_1}(u_{1,i}|x_{1,i})}{Q_{\varphi_1}(u_{1,i})}\right)\right]\right].$$

and this explains why, for back propagation, nodes $1, 2, 3, 4$ need only part of the error vector at the node they are connected to.

3.3. Bandwidth Requirements

In this section, we study the bandwidth requirements of our in-network learning. Let q denote the size of the entire data set (each input node has a local dataset of size $\frac{q}{J}$), $p = L_{J+1}$ the size of the input layer of NN $(J+1)$ and s the size in bits of a parameter. Since as per (8), the output of the last layers of the input NNs are concatenated at the input of NN $(J+1)$ whose size is p, and each activation value is s bits, one then needs $\frac{2sp}{J}$ bits for each data point—the factor 2 accounts for both the forward and backward passes and so, for an epoch, our in-network learning requires $\frac{2pqs}{J}$ bits.

Note that the bandwidth requirement of in-network learning does not depend on the sizes of the NNs used at the various nodes, but does depend on the size of the dataset. For comparison, notice that with FL one would require $2NJs$, where N designates the number of (weight- and bias) parameters of a NN at one node. For the SL of [21], assuming for simplicity that the NNs $j = 1, \ldots, J$ all have the same size ηN, where $\eta \in [0, 1]$, SL requires $(2pq + \eta NJ)s$ bits for an entire epoch.

The bandwidth requirements of the three schemes are summarized and compared in Table 1 for two popular NNs architectures, VGG16 ($N = 138{,}344{,}128$ parameters) and ResNet50 ($N = 25{,}636{,}712$ parameters) and two example datsets, $q = 50{,}000$ data points and $q = 500{,}000$ data points. The numerical values are set as $J = 500$, $p = 25{,}088$ and $\eta = 0.88$ for ResNet50 and 0.11 for VGG16.

Compared to FL and SL, INL has an advantage in that all nodes work jointly also during inference to make a prediction,not just during the training phase. As a consequence nodes only need to exchange latent representations, not model parameters, during training.

Table 1. Comparison of bandwidth requirements.

	Federated Learning	Split Learning	In-Network Learning
Bandwidth requirement	$2NJs$	$(2pq + \eta NJ)s$	$\frac{2pqs}{J}$
VGG 16 50,000 data points	4427 Gbits	324 Gbits	0.16 Gbits
ResNet 50 50,000 data points	820 Gbits	441 Gbits	0.16 Gbits
VGG 16 500,000 data points	4427 Gbits	1046 Gbits	1.6 Gbits
ResNet 50 500,000 data points	820 Gbits	1164 Gbits	1.6 Gbits

4. Experimental Results

We perform two series of experiments for which we compare the performance of our INL with those of FL and SL. The dataset used is the CIFAR-10 and there are five client nodes. In the first experiment, the three techniques are implemented in such a way such that during the inference phase the same NN is used to make the predictions. In the second experiment, the aim is to implement each of the techniques such that the data is spread in the same manner across the five client nodes for each of the techniques.

4.1. Experiment 1

In this setup, we create five sets of noisy versions of the images of CIFAR-10. To this end, the CIFAR images are first normalized, and then corrupted by additive Gaussian noise with standard deviation set respectively to $0.4, 1, 2, 3, 4$. For our INL each of the five input NNs is trained on a different noisy version of the same image. Each NN uses a variation of the VGG network of [35], with the categorical cross-entropy as the loss function, L2 regularization, and Dropout and BatchNormalization layers. Node $(J + 1)$ uses two dense layers. The architecture is shown in Figure 8. In the experiments, all five (noisy) versions of every CIFAR-10 image are processed simultaneously, each by a different NN at a distinct node, through a series of convolutional layers. The outputs are then concatenated and then passed through a series of dense layers at node $(J + 1)$.

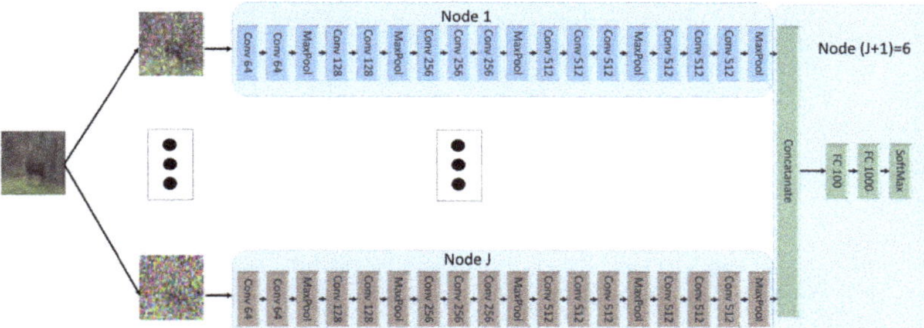

Figure 8. Network architecture. *Conv* stands for a convolutional layer, *Fc* stand for a fully connected layer.

For FL, each of the five client nodes is equipped with the *entire* network of Figure 8. The dataset is split into five sets of equal sizes and the split is now performed such that all five noisy versions of a same CIFAR-10 image are presented to the same client NN (distinct clients observe different images, however). For SL of [21], each input node is equipped with an NN formed by *all* fives branches with convolutional networks (i.e., all the network of

Figure 8, except the part at Node $(J + 1)$) and node $(J + 1)$ is equipped with fully connected layers at Node $(J + 1)$ in Figure 8. Here, the processing during training is such that each input NN concatenates vertically the outputs of all convolutional layers and then passes that to node $(J + 1)$, which then propagates back the error vector. After one epoch at one NN, the learned weights are passed to the next client, which performs the same operations on its part of the dataset.

The model depicted in Figure 8, which utilizes convolutional layers with a filter size of 3×3, comprises of approximately seventy-four million parameters, with 99.5% of these parameters constituting the encoding parts of the neural network. Table 2 presents the bandwidth requirements per epoch for the three techniques, considering the variation of the CIFAR-10 dataset used in the experiment, as well as the scenario where a dataset with ten times the amount of data is employed. It is observed that increasing the data size results in higher bandwidth requirements for both SL and INL, whereas the bandwidth requirements for FL remain unaffected.

Table 2. Experiment 1 bandwidth requirements of INL, FL and SL.

	Federated Learning	Split Learning	In-Network Learning
Bandwidth requirement	$2NJs$	$(2pq + \eta NJ)s$	$\dfrac{2pqs}{J}$
250,000 data points	2.96 GB	2.5 GB	0.2 GB
2,500,000 data points	2.96 GB	11.71 GB	2.05 GB

Figure 9a depicts the evolution of the classification accuracy on CIFAR-10 as a function of the number of training epochs, for the three schemes. As visible from the figure, the convergence of FL is relatively slower comparatively. The final result is also less accurate. Figure 9b shows the amount of data needed to be exchanged among the nodes (i.e., bandwidth resources) in order to get a prescribed value of classification accuracy. Observe that both our INL and SL require significantly less data exchange than FL and our INL is better than SL especially for small values of bandwidth. This experiment showcases that the INL framework can save bandwidth, compared to SL and FL, when training large models by exchanging latent representations as opposed to model parameters. This is particularly relevant as some works argue to overparametrizing models can result in better model performance [36].

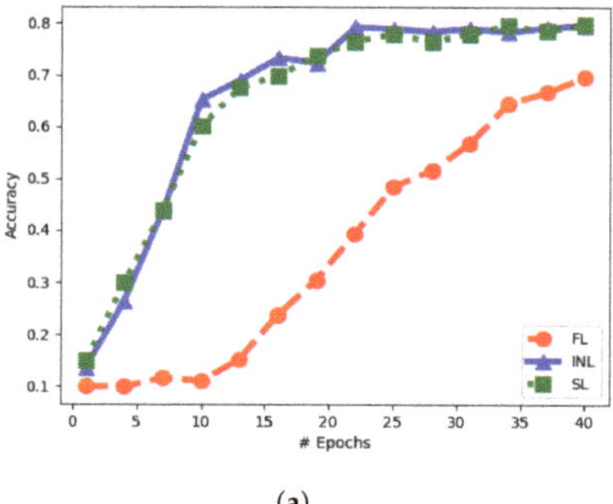

(a)

Figure 9. *Cont.*

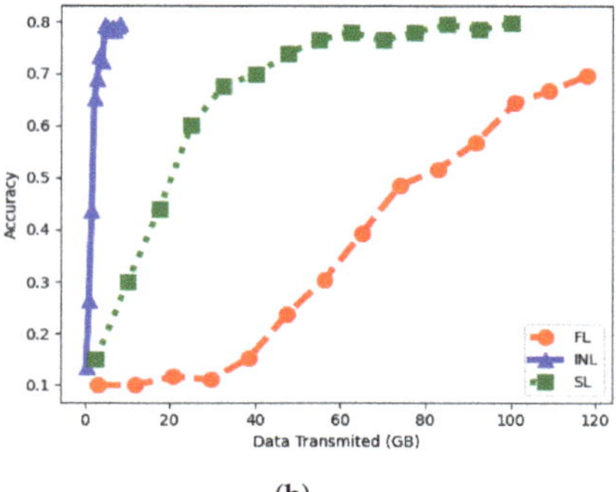

(b)

Figure 9. Comparison of INL, FL and SL—Experiment 1. (**a**) Accuracy vs. # of epochs. (**b**) Accuracy vs. bandwidth cost.

4.2. Experiment 2

In Experiment 1, the entire training dataset was partitioned differently for INL, FL and SL (in order to account for the particularities of the three). In this second experiment, they are all trained on the same data. Specifically, each client NN sees all CIFAR-10 images during training and its local dataset differs from those seen by other NNs only by the amount of added Gaussian noise (standard deviation chosen as $0.4, 1, 2, 3, 4$, respectively). Additionally, for the sake of a fair comparison between INL, FL and SL the nodes are set to utilize fairly the same NNs for the three of them (see, Figure 10).

Figure 10. Used NN architecture for FL in Experiment 2.

The model shown in Figure 10, for convolutional layers with filter of size 3×3, has approximately fifteen million parameters, with 97.6% of the parameters forming the decoding part of the network. Table 3 shows the bandwidth requierments for the three techniques per epoch for the variation of the CIFAR-10 dataset used in the experiment as well as for the case in which another dataset would be used that had ten times the amount of data. It is observed that increasing the data size results in higher bandwidth requirements for both SL and INL, whereas the bandwidth requirements for FL remain unaffected.

Table 3. Experiment 2 bandwidth requirements of INL, FL and SL.

	Federated Learning	Split Learning	In-Network Learning
Bandwidth requirement	$2NJs$	$(2pq + \eta NJ)s$	$\dfrac{2pqs}{J}$
250,000 data points	0.6 GB	1.32 GB	0.2 GB
2,500,000 data points	0.6 GB	10.53 GB	2.05 GB

Figure 11b shows the performance of the three schemes during the inference phase in this case (for FL the inference is performed on an image which has average quality of the five noisy input images for INL and SL). Again, observe the benefits of INL over FL and SL in terms of both achieved accuracy and bandwidth requirements. This experiment showacases INL's ability to make use of the correlations between the data observed by the different nodes, thus resulting in better network performance.

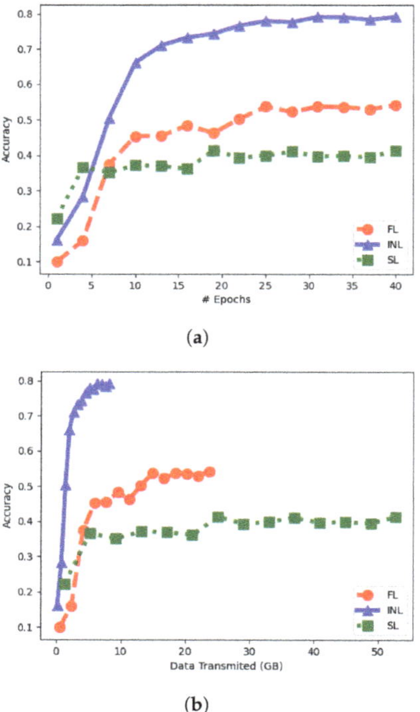

(**a**)

(**b**)

Figure 11. Comparison of INL, FL and SL—Experiment 2. (**a**) Accuracy vs. # of epochs. (**b**) Accuracy vs. bandwidth cost.

5. Conclusions

In this paper, our focus is on addressing the problem of distributed training and inference. We introduce INL, a novel framework which enables multiple nodes to collaboratively train a model that can be utilized in a distributed manner during the inference phase. Unlike existing works on distributed estimation and detection, our framework does not require prior knowledge of the data distribution; instead, it only necessitates access to a set of training samples. Furthermore, while other approaches to distributed training, such as FL and SL, assume local decision-making during the inference phase, we consider a scenario where the nodes observe data associated with the same event, thus enabling a joint decision that can lead to improved accuracy. The proposed INL algorithm offers a loss function derived through theoretical analysis, aiming to achieve the best trade-off between prediction accuracy, measured by logarithmic loss, and the amount of information exchanged among the nodes in the communication network.

Author Contributions: Conceptualization, A.Z.; methodology, A.Z.; software, M.M.; validation, M.M.; formal analysis, A.Z. and M.M.; investigation, M.M.; data curation, M.M.; writing—original draft preparation, A.Z. and M.M.; writing—review and editing, A.Z. and M.M.; visualization, M.M.; supervision, A.Z. All authors have read and agreed to the published version of the manuscript.

Funding: This research received no external funding.

Institutional Review Board Statement: Not applicable.

Informed Consent Statement: Not applicable.

Data Availability Statement: Data is contained within the article.

Conflicts of Interest: The authors declare no conflict of interest.

Appendix A. Proof of Theorem 1

The proof of Theorem 1 is based on a scheme in which the observations $\{x_j\}_{j \in \mathcal{J}}$ are compressed distributively using Berger-Tung coding [33], then, the compression bin indices are transmitted as independent messages over the network $\mathcal{G}$ using linear-network coding [34] (Section 15.4). The decision node N first decompresses the compression codewords and then uses them to produce an estimate $\hat{Y}$ of Y. In what follows, for simplicity we set the time-sharing random variable to be a constant, i.e., $Q = \emptyset$. Let $0 < \epsilon'' < \epsilon' < \epsilon$.

Appendix A.1. Codebook Generation

Fix a joint distribution $P_{X_1,\ldots,X_J,Y,U_1,\ldots,U_J}$ that factorizes as given by (16). Additionally, let $D = H(Y|U_1,\ldots,U_J)$, for $(u_1,\ldots,u_J) \in \mathcal{U}_1 \times \ldots \times \mathcal{U}_J$, the reconstruction function $\hat{y}(\cdot|u_1,\ldots,u_J) \in \mathcal{P}(\mathcal{Y})$ such that $\mathbb{E}\big[d(Y,\hat{Y})\big] \leq \dfrac{D}{1+\epsilon}$, where $d : \mathcal{Y} \times \mathcal{P}(\mathcal{Y}) \longrightarrow \mathbb{R}_+$ is the distortion measure given by (5). For every $j \in \mathcal{J}$, let $\tilde{R}_j \geq R_j$. In addition, randomly and independently generate $2^{n\tilde{R}_j}$ sequences $u_j^n(l_j), l_j \in [1 : 2^{n\tilde{R}_j}]$, each according to $\prod_{i=1}^{n} p_{U_j}(u_{ji})$. Partition the set of indices $l_j \in 2^{n\tilde{R}_j}$ into equal size bins $B_j(m_j) = \big[(m_j - 1)2^{n\tilde{R}_j - R_j} : m_j 2^{n\tilde{R}_j - R_j}\big]$, $m_j \in [1 : 2^{nR_j}]$. The codebook is revealed to all source nodes $j \in \mathcal{J}$ as well as to the decision node N, but not to the intermediary nodes.

Appendix A.2. Compression of the Observations

Node $j \in \mathcal{J}$ observes x_j^n and finds an index $l_j \in [1 : 2^{n\tilde{R}_j}]$ such that $(x_j^n, u_j^n(l_j)) \in \mathcal{T}_{\epsilon''}^{(n)}$. If there is more than one index the node selects one at random. If there is no such index, it selects one at random from $[1 : 2^{n\tilde{R}_j}]$. Let m_j be the index of the bin that contains the selected l_j, i.e., $l_j \in \mathcal{B}_j(m_j)$.

Appendix A.3. Transmission of the Compression Indices over the Graph Network

In order to transmit the bins indices $(M_1,\ldots,M_J) \in [1 : 2^{nR_1}] \times \ldots \times [1 : 2^{nR_J}]$ to the decision node N over the graph network $\mathcal{G} = (\mathcal{E}, \mathcal{N}, C)$, they are encoded as if they were independent-messages using the linear network coding scheme of [34] (Theorem 15.5) and then transmitted over the network. The transmission of the multimessage $(M_1,\ldots,M_J) \in [1 : 2^{nR_1}] \times \ldots \times [1 : 2^{nR_J}]$ to the decision node N is without error as long as for all $\mathcal{S} \subseteq [1 : N-1]$ we have

$$\sum_{j \in \mathcal{S} \cap \mathcal{J}} R_j \leq C(\mathcal{S}) \tag{A1}$$

where $C(\mathcal{S})$ is defined by (14).

Appendix A.4. Decompression and Estimation

The decision node N first looks for the unique tuple $(\hat{l}_1,\ldots,\hat{l}_J) \in \mathcal{B}_1(m_1) \times \ldots \times \mathcal{B}_J(m_J)$ such that $(u_1^n(\hat{l}_1),\ldots,u_J^n(\hat{l}_J)) \in \mathcal{T}_{\epsilon}^{(n)}$. With high probability, Node N finds such a unique tuple as long as n is large and for all $\mathcal{S} \subseteq \mathcal{J}$ it holds that [33] (see also [34] (Theorem 12.1))

$$\sum_{j \in \mathcal{S}} R_j \geq I(U_{\mathcal{S}}; X_{\mathcal{S}}|U_{\mathcal{S}^c}). \tag{A2}$$

The decision node N then produces an estimate $\hat{y}^n$ of y^n as $\hat{y}(u_1^n(\hat{I}_1),\ldots,u_J^n(\hat{I}_J))$. It can be shown easily that the per-sample relevance level achieved using the described scheme is $\Delta = I(U_1,\ldots,U_J;Y)$ and this completes the proof of Theorem 1.

Appendix B. Proof of Proposition 1

For $C_{sum} \geq 0$ fix $s \geq 0$ such that $C_s = C_{sum}$ and let $\mathbf{P}^* = \{P_{U_1^*|X_1}, P_{U_2^*|X_2}, P_{U_3^*|X_3}\}$ be the solution to (23) for the given s. By making the substitution in (22):

$$\Delta_s = I(Y; U_1^*, U_2^*, U_3^*) \tag{A3}$$
$$\leq \Delta \tag{A4}$$

where (A4) holds since Δ is the maximum $I(Y; U_1, U_2, U_3)$ over all distribution for which (20b) holds, which includes $\mathbf{P}^*$.

Conversely, let $\mathbf{P}^*$ be such that (Δ, C_{sum}) is on the bound of the $\mathcal{RI}_{sum}$ then:

$$\begin{aligned}
\Delta &= H(Y) - H(Y|U_1^*, U_2^*, U_3^*) \\
&\leq H(Y) - H(Y|U_1^*, U_2^*, U_3^*) + sC_{sum} \\
&\quad - s\Big[I(X_2, X_3; U_2^*, U_3^*|U_1^*) + I(X_1, X_2, X_3; U_1^*, U_2^*, U_3^*)\Big] \tag{A5} \\
&\leq H(Y) + \max_{\mathbf{P}} \mathcal{L}_s(\mathbf{P}) + sC_{sum} \tag{A6} \\
&= \Delta_s - sC_s + sC_{sum} \\
&= \Delta_s + s(C_{sum} - C_s). \tag{A7}
\end{aligned}$$

where (A5) follows from (20b). Inequality (A6) holds due to the fact that $\max_{\mathbf{P}} \mathcal{L}(\mathbf{P})$ takes place over all $\mathbf{P}$, including $\mathbf{P}^*$. Since (A7) is true for any $s \geq 0$ we take s such that $C_{sum} = C_s$, which implies $\Delta \leq \Delta_s$. Together with (A4) this completes the proof.

Appendix C. Proof of Lemma 1

We have

$$\begin{aligned}
\mathcal{L}_s(\mathbf{P}) &= -H(Y|U_1, U_2, U_3) - sI(X_1, X_2, X_3; U_1, U_2, U_3) \\
&\quad - sI(X_2, X_3; U_2, U_3|U_1) \tag{A8} \\
&= -H(Y|U_1, U_2, U_3) \\
&\quad - s\Big[I(X_1; U_1) + 2I(X_2, X_3; U_2, U_3|U_1)\Big] \tag{A9} \\
&= -H(Y|U_1, U_2, U_3) - sI(X_1; U_1) - 2sI(X_2; U_2) \\
&\quad - 2s\Big[I(X_3; U_3) - I(U_3; U_1, U_2) - I(U_2; U_1)\Big] \tag{A10} \\
&= -H(Y|U_1, U_2, U_3) - sI(X_1; U_1) - 2sI(X_2; U_2) \\
&\quad + 2s\Big[I(U_2; U_1) + I(U_3; U_1, U_2) - I(X_3; U_3)\Big] \tag{A11} \\
&\geq -H(Y|U_1, U_4) - sI(X_1; U_1) - 2s\Big[I(X_2; U_2) + I(X_3; U_3)\Big] \\
&\quad + 2s\Big[I(U_2; U_1) + I(U_3; U_1, U_2)\Big] \tag{A12}
\end{aligned}$$

where (A9) holds since $U_1 \,\text{--}\!\circ\!\text{--}\, X_1 \,\text{--}\!\circ\!\text{--}\, (X_2, X_3, U_2, U_3)$ and $(U_2, U_3) \,\text{--}\!\circ\!\text{--}\, (X_2, X_3) \,\text{--}\!\circ\!\text{--}\, (U_1, X_1)$ (A10) holds since $U_2 \,\text{--}\!\circ\!\text{--}\, X_2 \,\text{--}\!\circ\!\text{--}\, (U_1, X_3)$ and $U_3 \,\text{--}\!\circ\!\text{--}\, X_3 \,\text{--}\!\circ\!\text{--}\, (U_1, U_2, X_2)$; (A12) hold since $U_4 \,\text{--}\!\circ\!\text{--}\, (U_2, U_3) \,\text{--}\!\circ\!\text{--}\, (Y, U_1)$.

Appendix D. Proof of Lemma 2

From [25] (eq. (55)) it can be shown that for any pmf $Q_{Y|Z}(y|z)$, $y \in \mathcal{Y}$ and $z \in \mathcal{Z}$ the conditional entropy $H(Y|Z)$ is:

$$H(Y|Z) = \mathbb{E}[-\log Q_{Y|Z}(Y|Z)] - D_{\mathrm{KL}}(P_{Y|Z}\|Q_{Y|Z}). \tag{A13}$$

From [25] (eq. (81)):

$$\begin{aligned}
I(X;Z) &= H(Z) - H(Z|X) \\
&= D_{\mathrm{KL}}(P_{Z|X}\|Q_Z) - D_{\mathrm{KL}}(P_Z\|Q_Z). \tag{A14}
\end{aligned}$$

Now substituting Equations (A13) and (A14) in (28) the following result is obtained:

$$\begin{aligned}
\mathcal{L}_s^{\mathrm{low}}(\mathbf{P}_+) =& - H(Y|U_1, U_4) - sI(X_1; U_1) - 2sI(X_2; U_2) \\
& - 2sI(X_3; U_3) + 2s\Big[I(U_2; U_1) + I(U_3; U_1, U_2)\Big] \\
=& \mathbb{E}[\log Q_{Y|u_1,u_4}] + D_{\mathrm{KL}}(P_{Y|u_1,u_4}\|Q_{Y|u_1,u_4}) \\
& - sD_{\mathrm{KL}}(P_{U_1|X_1}\|Q_{U_1}) + sD_{\mathrm{KL}}(P_{U_1}\|Q_{U_1}) \\
& - 2sD_{\mathrm{KL}}(P_{U_2|X_2}\|Q_{U_2}) + 2sD_{\mathrm{KL}}(P_{U_2}\|Q_{U_2}) \\
& - 2sD_{\mathrm{KL}}(P_{U_3|X_3}\|Q_{U_3}) + 2sD_{\mathrm{KL}}(P_{U_3}\|Q_{U_3}) \\
& + 2sD_{\mathrm{KL}}(P_{U_2|U_1}\|Q_{U_2}) - 2sD_{\mathrm{KL}}(P_{U_2}\|Q_{U_2}) \\
& + 2sD_{\mathrm{KL}}(P_{U_3|U_1,U_2}\|Q_{U_3}) - 2sD_{\mathrm{KL}}(P_{U_3}\|Q_{U_3}) \\
=& \mathcal{L}_s^{\mathrm{v\text{-}low}} + sD_{\mathrm{KL}}(P_{U_1}\|Q_{U_1}) + 2sD_{\mathrm{KL}}(P_{U_2|U_1}\|Q_{U_2}) \\
& + 2sD_{\mathrm{KL}}(P_{U_3|U_1,U_2}\|Q_{U_3}) + D_{\mathrm{KL}}(P_{Y|u_1,u_4}\|Q_{Y|u_1,u_4}) \\
\geq& \mathcal{L}_s^{\mathrm{v\text{-}low}} \tag{A15}
\end{aligned}$$

The last inequality (A15) holds due to the fact that KL divergence is always positive and $s \geq 0$, thus proving the lemma.

References

1. Zou, Z.; Shi, Z.; Guo, Y.; Ye, J. Object Detection in 20 Years: A Survey. *arXiv* **2019**, arXiv:1905.05055.
2. Glaser, J.I.; Benjamin, A.S.; Farhoodi, R.; Kording, K.P. The roles of supervised machine learning in systems neuroscience. *Prog. Neurobiol.* **2019**, *175*, 126–137. [CrossRef]
3. Pluim, J.P.W.; Maintz, J.B.A.; Viergever, M.A. Mutual-information-based registration of medical images: A survey. *IEEE Trans. Med. Imaging* **2003**, *22*, 986–1004. [CrossRef]
4. Kober, J.; Bagnell, J.; Peters, J. Reinforcement Learning in Robotics: A Survey. *Int. J. Robot. Res.* **2013**, *32*, 1238–1274. [CrossRef]
5. Vinyals, O.; Le, Q.V. A Neural Conversational Model. *arXiv* **2015**, arXiv:1506.05869.
6. Farsad, N.; Yilmaz, H.B.; Eckford, A.; Chae, C.; Guo, W. A Comprehensive Survey of Recent Advancements in Molecular Communication. *IEEE Commun. Surv. Tutor.* **2016**, *18*, 1887–1919. [CrossRef]
7. Peter Hong, Y.W.; Wang, C.C. In-Network Learning via Over-the-Air Computation in Internet-of-Things. In Proceedings of the 2021 IEEE 22nd International Workshop on Signal Processing Advances in Wireless Communications (SPAWC), Lucca, Italy, 27–30 September 2021; pp. 141–145. [CrossRef]
8. Du, R.; Magnusson, S.; Fischione, C. The Internet of Things as a deep neural network. *IEEE Commun. Mag.* **2020**, *58*, 20–25. [CrossRef]
9. McMahan, B.; Moore, E.; Ramage, D.; Hampson, S.; y Arcas, B.A. Communication-Efficient Learning of Deep Networks from Decentralized Data. In Proceedings of the 20th International Conference on Artificial Intelligence and Statistics, AISTATS 2017, Fort Lauderdale, FL, USA, 20–22 April 2017; Volume 54, pp. 1273–1282.
10. Xiao, J.J.; Ribeiro, A.; Luo, Z.Q.; Giannakis, G. Distributed compression-estimation using wireless sensor networks. *IEEE Signal Process. Mag.* **2006**, *23*, 27–41.
11. Kreidl, O.P.; Tsitsiklis, J.N.; Zoumpoulis, S.I. Decentralized detection in sensor network architectures with feedback. In Proceedings of the 48th Annual Allerton Conference on Communication, Control, and Computing (Allerton), Monticello, IL, USA, 29 September–1 October 2010; pp. 1605–1609. [CrossRef]
12. Chamberland, J.f.; Veeravalli, V.V. Wireless Sensors in Distributed Detection Applications. *IEEE Signal Process. Mag.* **2007**, *24*, 16–25.

13. Tsitsiklis, J.N. Decentralized detection. In *Advances in Statistical Signal Processing*; JAI Press: Stamford, CT, USA, 1993; pp. 297–344.

14. Simic, S. A learning-theory approach to sensor networks. *IEEE Pervasive Comput.* **2003**, *2*, 44–49.

15. Predd, J.; Kulkarni, S.; Poor, H. Distributed learning in wireless sensor networks. *IEEE Signal Process. Mag.* **2006**, *23*, 56–69.

16. Nguyen, X.; Wainwright, M.; Jordan, M. Nonparametric decentralized detection using kernel methods. *IEEE Trans. Signal Process.* **2005**, *53*, 4053–4066. [CrossRef]

17. Jagyasi, B.; Raval, J. Data aggregation in multihop wireless mesh sensor Neural Networks. In Proceedings of the 2015 9th International Conference on Sensing Technology (ICST), Auckland, New Zealand, 8–10 December 2015; pp. 65–70. [CrossRef]

18. Tran, N.H.; Bao, W.; Zomaya, A.; Nguyen, M.N.H.; Hong, C.S. Federated Learning over Wireless Networks: Optimization Model Design and Analysis. In Proceedings of the IEEE INFOCOM 2019—IEEE Conference on Computer Communications, Paris, France, 29 April–2 May 2019; pp. 1387–1395. [CrossRef]

19. Amiri, M.M.; Gündüz, D. Federated learning over wireless fading channels. *IEEE Trans. Wirel. Commun.* **2020**, *19*, 3546–3557. [CrossRef]

20. Yang, H.H.; Liu, Z.; Quek, T.Q.S.; Poor, H.V. Scheduling Policies for Federated Learning in Wireless Networks. *IEEE Trans. Commun.* **2020**, *68*, 317–333. [CrossRef]

21. Gupta, O.; Raskar, R. Distributed learning of deep neural network over multiple agents. *J. Netw. Comput. Appl.* **2018**, *116*, 1–8. [CrossRef]

22. Ceballos, I.; Sharma, V.; Mugica, E.; Singh, A.; Roman, A.; Vepakomma, P.; Raskar, R. SplitNN-driven Vertical Partitioning. *arXiv* **2020**, arXiv:2008.04137.

23. National Institutes of Health. *NIH Data Sharing Policy and Implementation Guidance*; National Institutes of Health: Bethesda, MD, USA, 2003; Volume 18, p. 2009.

24. Aguerri, I.E.; Zaidi, A. Distributed Information Bottleneck Method for Discrete and Gaussian Sources. In Proceedings of the IEEE International Zurich Seminar on Information and Communications, Zurich, Switzerland, 21–23 February 2018.

25. Aguerri, I.E.; Zaidi, A. Distributed Variational Representation Learning. *IEEE Trans. Pattern Anal. Mach. Intell.* **2021**, *43*, 120–138. [CrossRef]

26. Zaidi, A.; Aguerri, I.E.; Shamai (Shitz), S. On the Information Bottleneck Problems: Models, Connections, Applications and Information Theoretic Views. *Entropy* **2020**, *22*, 151. [CrossRef]

27. Moldoveanu, M.; Zaidi, A. On in-network learning. A comparative study with federated and split learning. In Proceedings of the 2021 IEEE 22nd International Workshop on Signal Processing Advances in Wireless Communications (SPAWC), Lucca, Italy, 27–30 September 2021; pp. 221–225.

28. Moldoveanu, M.; Zaidi, A. In-network Learning for Distributed Training and Inference in Networks. In Proceedings of the IEEE Globecom 2021 Workshops, Madrid, Spain, 7–11 December 2021; pp. 1–6. [CrossRef]

29. Yang, Q.; Liu, Y.; Chen, T.; Tong, Y. Federated machine learning: Concept and applications. *ACM Trans. Intell. Syst. Technol. (TIST)* **2019**, *10*, 1–19. [CrossRef]

30. Kingma, D.P.; Welling, M. Auto-encoding variational bayes. *arXiv* **2013**, arXiv:1312.6114.

31. Flamich, G.; Havasi, M.; Hernández-Lobato, J.M. Compressing images by encoding their latent representations with relative entropy coding. *Adv. Neural Inf. Process. Syst.* **2020**, *33*, 16131–16141.

32. Li, C.T.; Gamal, A.E. Strong Functional Representation Lemma and Applications to Coding Theorems. *IEEE Trans. Inf. Theory* **2018**, *64*, 6967–6978. [CrossRef]

33. Berger, T.; Yeung, R. Multiterminal source encoding with one distortion criterion. *IEEE Trans. Inf. Theory* **1989**, *35*, 228–236. [CrossRef]

34. El Gamal, A.; Kim, Y.H. *Network Information Theory*; Cambridge University Press: Cambridge, UK, 2011. [CrossRef]

35. Liu, S.; Deng, W. Very deep convolutional neural network based image classification using small training sample size. In Proceedings of the 2015 3rd IAPR Asian Conference on Pattern Recognition (ACPR), Kuala Lumpur, Malaysia, 3–6 November 2015; pp. 730–734. [CrossRef]

36. Liu, H.; Chen, M.; Er, S.; Liao, W.; Zhang, T.; Zhao, T. Benefits of overparameterized convolutional residual networks: Function approximation under smoothness constraint. In Proceedings of the International Conference on Machine Learning. PMLR, Baltimore, MD, USA, 17–23 July 2022; pp. 13669–13703.

Article

Distributed Quantization for Partially Cooperating Sensors Using the Information Bottleneck Method

Steffen Steiner [1,*], **Abdulrahman Dayo Aminu** [2] **and Volker Kuehn** [1]

[1] Institute of Communications Engineering, University of Rostock, 18119 Rostock, Germany; volker.kuehn@uni-rostock.de

[2] Pydro GmbH, 18119 Rostock, Germany; aminuabdulrahmandayo@gmail.com

[*] Correspondence: steffen.steiner@uni-rostock.de

Abstract: This paper addresses the optimization of distributed compression in a sensor network with partial cooperation among sensors. The widely known Chief Executive Officer (CEO) problem, where each sensor has to compress its measurements locally in order to forward them over capacity limited links to a common receiver is extended by allowing sensors to mutually communicate. This extension comes along with modified statistical dependencies among involved random variables compared to the original CEO problem, such that well-known outer and inner bounds do not hold anymore. Three different inter-sensor communication protocols are investigated. The successive broadcast approach allows each sensor to exploit instantaneous side-information of all previously transmitting sensors. As this leads to dimensionality problems for larger networks, a sequential point-to-point communication scheme is considered forwarding instantaneous side-information to only one successor. Thirdly, a two-phase transmission protocol separates the information exchange between sensors and the communication with the common receiver. Inspired by algorithmic solutions for the original CEO problem, the sensors are optimized in a greedy manner. It turns out that partial communication among sensors improves the performance significantly. In particular, the two-phase transmission can reach the performance of a fully cooperative CEO scenario, where each sensor has access to all measurements and the knowledge about all channel conditions. Moreover, exchanging instantaneous side-information increases the robustness against bad Wyner–Ziv coding strategies, which can lead to significant performance losses in the original CEO problem.

Keywords: distributed compression; chief executive officer problem; cooperating sensors; distributed source coding; information bottleneck

Citation: Steiner, S.; Aminu, A.D.; Kuehn, V. Distributed Quantization for Partially Cooperating Sensors Using the Information Bottleneck Method. *Entropy* **2022**, *24*, 438. https://doi.org/10.3390/e24040438

Academic Editors: Gerhard Bauch and Jan Lewandowsky

Received: 24 February 2022
Accepted: 18 March 2022
Published: 22 March 2022

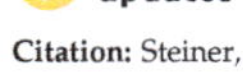

1. Introduction

This contribution considers a special case of the distributed source coding problem where each sensor observes the same source signal. In order to forward their measurements over capacity limited links to a common receiver, the sensors have to compress their measurements. In the case where a direct communication among operating sensors is not possible, this problem is termed as the CEO problem. Here, the compression at each sensor is optimized according to the Wyner–Ziv coding principle exploiting only statistical side-information. Within this paper, the term CEO problem always stands for this non-cooperative CEO problem, which means that sensors cannot communicate with each other during runtime.

We extend this scenario and allow sensors to cooperate with each other by exchanging instantaneous side-information. The fully cooperative Chief Executive Officer (fcCEO) problem is obtained if sensors can forward their uncompressed observations over inter-sensor links to all other sensors. The partially cooperative Chief Executive Officer (pcCEO) problem represents a scenario where instantaneous side-information is compressed before it is forwarded to other sensors.

1.1. The CEO Problem

The CEO problem has been investigated for various system assumptions which mainly differ in the distribution of the variable of interest and the applied distortion measure. The quadratic Gaussian CEO problem considers jointly Gaussian signals and the mean squared error (MSE) distortion measure [1–6]. Using an infinite number of encoders, Oohama analytically derived an asymptotic version of the sum-rate distortion in [1]. In [2], the authors investigated the influence of cooperating and non-cooperating encoders on the distortion measure. It turns out that the distortion decreases asymptotically with the reciprocal sum-rate R for non-cooperating encoder. In scenarios with cooperating sensors, the MSE distortion decays exponentially with 2^{-2R}. The non-asymptotic case was first investigated in [3] where the authors derived an upper bound on the sum-rate distortion function. Moreover, they showed that this bound is tight when each encoder has the same measurement SNR. Prabhakaran et al. [4], Oohama [5] and Wagner et al. [6] characterized the complete rate-region for this quadratic Gaussian CEO problem. Results also exist for multivariate Gaussian relevant processes [7,8]. The CEO problem for arbitrary discrete source distributions and the logarithmic loss distortion measure has been analyzed in [9–11]. For this scenario, Courtade and Weissman completely characterized the CEO rate-region in [9]. In [10,11], asymptotic analyses for an infinite number of sensors have been performed. Using the Hamming distance as a distortion measure, in [10] an inevitable loss in error rate performance due to non-cooperating sensors has been discovered. For arbitrary distortion measures, a scaling law on the sum-rate distortion function has been derived in [11]. In [12,13], the authors developed a variational bound on the optimal trade-off between relevance and complexity of the CEO setup and used neural networks for encoder and decoder to compute this bound.

There exist several algorithmic approaches to solve the CEO problem [8,14–17]. In our previous work [14,15], a Greedy Distributed Information Bottleneck (GDIB) algorithm based on the inner bound of the CEO problem with log-loss distortion measure defined in [9] has been introduced to determine extreme points of the contra-polymatroid solution space. Equivalently to Wyner–Ziv coding, the algorithm optimizes the quantizer of a specific sensor using the mappings of previously designed quantizers as statistical side-information. We showed that the GDIB algorithm outperforms an individual scalar information bottleneck optimization of sensors without Wyner–Ziv coding, especially for larger networks. For asymmetric scenarios, we demonstrated the dependency of the performance on the optimization order, i.e., the Wyner–Ziv coding strategy. Since the memory complexity of this optimization algorithm depends exponentially on the network size, we introduced a way to compress the Wyner–Ziv side-information by means of the information bottleneck principle. However, there still remains a large performance gap between non-cooperative and fully cooperative distributed compression.

1.2. Partially Cooperating Sensors

In order to close the gap between non-cooperative and fully cooperative distributed compression, we consider an extension of the CEO system model allowing partial cooperation among sensors. While a rich literature can be found on the classical CEO problem, less results are known for partially cooperating sensors. Most work has been done for jointly Gaussian signals because they allow an analytical treatment at least in parts. In [18], it is shown that cooperation among sensors can reduce the compression sum-rate except for the quadratic Gaussian CEO problem. In [19], the authors consider estimation problems under communication constraints and propose coding strategies for tree-structured sensor networks. Exploiting the Wyner–Ziv coding principle, the authors developed solutions for general trees and provide particular results for serial and parallel networks. A two sensor system with a Gaussian source was investigated in [20] for two transmission scenarios: orthogonal but rate-limited links between sensors and the common receiver as well as the Gaussian multiple access channel. It was shown that cooperation between the sensors over rate-limited inter-sensor links leads in both cases to substantial gains in terms of the compression sum-rate. Finally, a simple three node network consisting of encoder, helper and decoder was analyzed in [21]. The authors showed that side-information provided by the helper to the encoder need not exceed that given to the decoder.

1.3. Structure and Notation

This paper is structured as follows: Section 2 gives a short introduction to the information bottleneck principle. Sections 3 and 4 introduce the non-cooperative distributed sensing scenario with the Greedy Distributed Information Bottleneck as an algorithmic solution defined in [14,15] and the fully cooperative distributed sensing scenario being equivalent to a centralized quantization approach, respectively. The main contribution of this paper can be found in Section 5, which introduces the partially cooperative distributed sensing scenario containing three different inter-sensor communication protocols, i.e., successive broadcasting, successive point to point transmission and two-phase transmission. Section 6 concludes this paper.

Throughout this paper, the following notation is used. Calligraphic letters $\mathcal{X}, \mathcal{Y}, \mathcal{Z}$ denote random variables with realizations x, y, z, which are elements of the sets $\mathbb{X}, \mathbb{Y}, \mathbb{Z}$ with cardinalities $|\mathbb{X}|, |\mathbb{Y}|$ and $|\mathbb{Z}|$, respectively. Bold letters $\mathbf{y} = [y_1 \ \ldots \ y_M]^{\mathrm{T}}$ denote vectors while boldface calligraphic letters $\boldsymbol{\mathcal{Y}}, \boldsymbol{\mathcal{Z}}$ denote multivariate random variables. Note that $\boldsymbol{\mathcal{Z}}_{<m}$ covers only the processes $\mathcal{Z}_1$ to $\mathcal{Z}_{m-1}$. $I(\mathcal{X}; \mathcal{Y})$ represents the mutual information between the random variables $\mathcal{X}$ and $\mathcal{Y}$. Conditional and joint probability mass functions (pmfs) are termed $p(y|x)$ and $p(x, y)$, respectively. The Kullback-Leibler (KL) divergence is given as $D_{\mathrm{KL}}[\cdot \| \cdot]$. Finally, the expectation of a function $f(x)$ with respect to the random variable $\mathcal{X}$ is denoted as $\mathbb{E}_{\mathcal{X}}[f(\mathcal{X})]$.

2. The Information Bottleneck Principle

The information bottleneck (IB) principle was first introduced by Tishby et al. in [22,23] and defines a clustering framework based on information theoretic measures. An overview about algorithmic solutions for this basic optimization problem is given in [23,24]. The IB principle finds application in various fields in communications [25–30].

The general IB setup is depicted in Figure 1. It contains the relevant process $\mathcal{X}$, a noisy observation $\mathcal{Y}$ of $\mathcal{X}$ and a compressed version $\mathcal{Z}$ of $\mathcal{Y}$. The IB approach aims to optimize the mapping $p(z|y)$ in order to preserve as much information about the relevant process $\mathcal{X}$ in $\mathcal{Z}$ as possible. More precisely, it tries to maximize the relevant mutual information $I(\mathcal{X}; \mathcal{Z})$ while fulfilling a rate constraint $I(\mathcal{Y}; \mathcal{Z}) \leq C$. This general goal is summarized in Figure 2a. The optimization can be formulated as a maximization of the Lagrangian function

$$L_{\mathrm{IB}} = I(\mathcal{X}; \mathcal{Z}) - \beta I(\mathcal{Y}; \mathcal{Z}) \,. \tag{1}$$

It turns out to be a non-convex optimization problem, since $I(\mathcal{X}; \mathcal{Z})$ and $I(\mathcal{Y}; \mathcal{Z})$ are both convex functions of the mapping $p(z|y)$. The parameter β is a trade-off parameter steering the focus between the preservation of relevant information and the compression of the observation. In the case of $\beta = 0$, the focus only lies on preservation of relevant information. By increasing β the compression becomes more and more important up to the case of $\beta \to \infty$. Here, the functional in (1) becomes maximal if $I(\mathcal{Y}; \mathcal{Z}) = 0$, which means that all information is compressed to a single cluster. Therefore, the parameter β can be used to adjust the compression rate $I(\mathcal{Y}; \mathcal{Z})$ in order to fulfill a desired rate constraint $I(\mathcal{Y}; \mathcal{Z}) \leq C$. Since the compression-rate curve is a monotonic increasing function in $\frac{1}{\beta}$, a simple bisection search can be applied. The optimization problem in (1) can be solved by taking the derivative with respect to the mapping $p(z|y)$ and equating it to zero. It results in the implicit update equation

$$p(z|y) = \frac{e^{-d_\beta(y,z)}}{\sum_z e^{-d_\beta(y,z)}} \tag{2}$$

with

$$d_\beta(y,z) = \frac{1}{\beta} D_{\mathrm{KL}}[p(x|y)\|p(x|z)] - \log p(z) \tag{3}$$

$$= \frac{1}{\beta} \mathbb{E}_{\mathcal{X}|y}\left[\log \frac{p(x|y)}{p(x|z)}\right] - \log p(z).$$

In (3), $D_{\mathrm{KL}}[p(x|y)\|p(x|z)]$ denotes the Kullback–Leibler divergence. This implicit solution can be solved by an iterative Blahut–Arimoto-like algorithm.

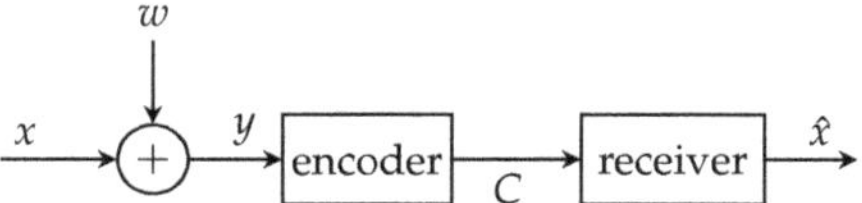

Figure 1. Illustration of a remote sensing problem for a single sensor.

(a)

(b)

Figure 2. (a) Illustration of the IB setup, (b) Exemplary IB graph.

In the case of focusing solely on preservation of relevant information with $\beta = 0$, the optimization algorithm yields a deterministic clustering $p(z|y) \in \{0,1\}$. For $\beta > 0$, the clustering $p(z|y) \in [0,1]$ is generally stochastic. The IB method can easily be extended to multiple input values. A graphical tool for visualization are IB graphs [31]. Figure 2b illustrates an example where the observations y_1, y_2 are compressed into the cluster index z. The trapezoid represents the IB compression with respect to the relevant variable written inside the trapezoid.

3. Non-Cooperative Distributed Sensing System

Figure 3 illustrates the CEO system model without communication among operating sensors. Here, M sensors observe noisy versions y_m of the same relevant signal x. The measurement processes can be modeled as statistically independent memoryless channels (MCs).

Exemplarily, a measurement $y_m = x + w_m$ represents the relevant signal x corrupted by zero mean white Gaussian measurement noise w_m with measurement signal-to-noise-ratio (SNR) $\gamma_m = \frac{\sigma_x^2}{\sigma_{w_m}^2}$, where σ_x^2, $\sigma_{w_m}^2$ denote signal and noise variances, respectively. In order to be able to forward the measurements over capacity limited links with capacities $C_1, \ldots, C_M$, each sensor has to compress its observations using a specific encoding process. More precisely, each sensor compresses its observations y_m to a cluster index z_m using the mapping $p(z_m|y_m)$ leading to the Markov property:

$$p(\mathbf{z}, \mathbf{y}, x) = \prod_{m=1}^{M} p(z_m|y_m) p(y_m|x) p(x). \tag{4}$$

The encoding process contains a second lossy compression step if the mapping $p(z_m|y_m)$ is stochastic and lossless entropy coding if the mapping $p(z_m|y_m)$ is deterministic. Therefore, a compressed version of the index z_m is transmitted without any further loss to the common receiver. The optimization of $p(z_m|y_m)$ for each sensor is done offline.

The mathematical analysis of the CEO problem and the structure of its rate-region for discrete input alphabets and the log-loss distortion measure was presented in [9] and exploits (4). It was proved that the extreme points of the contra-polymatroid solution space can be determined by greedy algorithms as the one described next. Since the communication among sensors during run-time is not possible in this approach, the solution represents a lower bound on the performance of cooperative distributed compression in this paper.

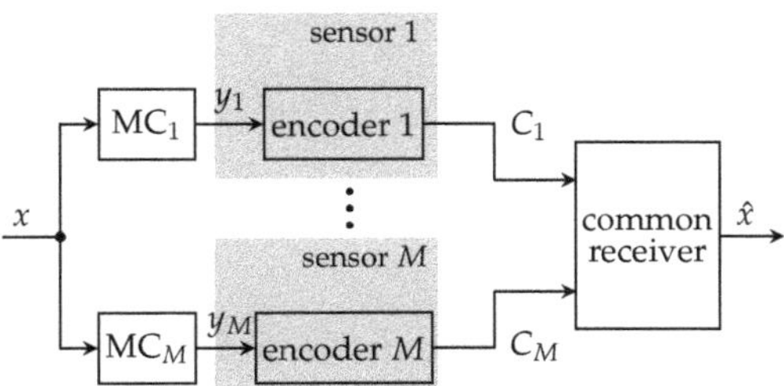

Figure 3. Non-cooperative distributed sensing system with M sensors, a common receiver and individual link capacities C_m.

An algorithmic solution to solve the CEO problem has previously been proposed in [14,15] as the so called Greedy Distributed Information Bottleneck (GDIB) algorithm. It is based on the inner bound of the CEO rate-region for the logarithmic loss distortion measure [9] and optimizes the quantization at the sensors successively. Replacing the logarithmic loss function $H(\mathcal{X}|\mathcal{Z})$ by the relevant mutual information $I(\mathcal{X};\mathcal{Z})$ delivers the optimization problem

$$\max_{\mathbf{P}} I(\mathcal{X};\mathcal{Z}) \text{ s.t. } I(\mathcal{Y}_\mathbb{S};\mathcal{Z}_\mathbb{S}|\mathcal{Z}_{\overline{\mathbb{S}}}) \leq \sum_{m \in \mathbb{S}} C_m$$
$$\forall \quad \mathbb{S} \subseteq \{1,2,\ldots,M\}. \tag{5}$$

The set $\mathbf{P} = \begin{bmatrix} p(z_1|y_1) & \cdots & p(z_M|y_M) \end{bmatrix}$ defines the set of all mappings. According to [9], the compression rates $I(\mathcal{Y}_\mathbb{S};\mathcal{Z}_\mathbb{S}|\mathcal{Z}_{\overline{\mathbb{S}}})$ are supermodular set functions with respect to the sets $\mathbb{S}$ [32], while the relevant information $I(\mathcal{X};\mathcal{Z})$ does not depend on $\mathbb{S}$. Therefore, the greedy optimization structure of the GDIB algorithm is optimal and finds the extreme points of the solution space. It has to be emphasized that since the GDIB algorithm is based on the inner bound of the rate-region, it does not find the complete rate-region of the CEO problem. Following this approach, M IB related Lagrangian optimization problems are obtained, one for each sensor.

$$L_{\text{GDIB}}^{(1)} = I(\mathcal{X};\mathcal{Z}_1) - \beta_1 I(\mathcal{Y}_1;\mathcal{Z}_1) \tag{6}$$

$$\vdots$$

$$L_{\text{GDIB}}^{(M)} = I(\mathcal{X};\mathcal{Z}_M|\mathcal{Z}_{<M}) - \beta_M I(\mathcal{Y}_M;\mathcal{Z}_M|\mathcal{Z}_{<M}) \tag{7}$$

Obviously, the optimization problem of the first sensor resembles the optimization problem for the scalar IB problem given in (1) since there is no predecessor. Subsequent sensors exploit the mappings of previously designed quantizers as statistical side-information leading to the well-known Wyner–Ziv coding strategy. Naturally, each Lagrange multiplier β_m has to be chosen such that the corresponding compression rate fulfills the individual rate constraint $I(\mathcal{Y}_m; \mathcal{Z}_m | \mathbf{Z}_{<m}) \leq C_m$. The objectives in (6) and (7) can be solved by equating the derivative with respect to the mapping $p(z_m | y_m)$ to zero delivering the update rule

$$p(z_m | y_m) = \frac{e^{-d_{\beta_m}(y_m, z_m)}}{\sum_{z_m} e^{-d_{\beta_m}(y_m, z_m)}} \tag{8}$$

with the exponent

$$d_{\beta_m}(y_m, z_m) := \mathbb{E}_{\mathbf{Z}_{<m}|y_m}\left[\frac{1}{\beta_m} D_{\mathrm{KL}}[p(x|y_m, \mathbf{z}_{<m}) \| p(x|\mathbf{z}_{\leq m})] - \log p(z_m | \mathbf{z}_{<m})\right]. \tag{9}$$

Similar to the scalar IB optimization, the implicit expression in (8) can be solved using a Blahut–Arimoto like algorithm, providing local optimal solutions. It has to be mentioned that for asymmetric scenarios, this optimization has to be performed for all $M!$ possible permutations of the optimization order to find the best solution.

A detailed derivation and performance analysis of this algorithm can be found in [14,15]. If the capacity is equally distributed over all sensors in the network, e.g., sensors share the same channel in an orthogonal way and a round robin fashion, numerical results demonstrate that the GDIB algorithm outperforms an individual scalar IB optimization at each sensor. However, there is still a large gap to the performance of a fcCEO scenario, which is defined in Section 4. Moreover, in asymmetric scenarios, the performance highly depends on the optimization order. Although no clear conclusion about the optimal Wyner–Ziv coding strategy can be drawn, a good solution can be expected when starting the optimization with the best forward channel conditions, i.e., the lowest compression (highest compression rate).

4. Fully Cooperative Distributed Sensing—A Centralized Quantization Approach

This section introduces the fcCEO scenario, which considers distributed sensors being able to forward their uncompressed observations to all other sensors in the network over ideal noiseless inter-sensor links. In this case, sensors can perfectly exchange their measurements y_m before they jointly compress the received signals taking into account the rate constraints of all individual forward channels. Naturally, the exchange has to be done by a two-phase transmission protocol, consisting of a cooperation phase and a transmission phase. During the cooperation phase sensors exchange information until every sensor knows measurements $\mathbf{y} = [y_1 \ldots y_M]^{\mathrm{T}}$ of all M sensors. The actual forwarding of the compressed observations to the common receiver is performed during the transmission phase. This full cooperation is equivalent to a single central quantizer having access to all measurements $\mathbf{y}$ as depicted in Figure 4. Applying the IB principle, this central quantizer can be designed in order to compress the vector $\mathbf{y}$ onto a cluster index z using the mapping $p(z|\mathbf{y})$, which motivates the name centralized IB (CIB) for the algorithmic solution in a fcCEO scenario. The optimization problem can be formulated as the maximization of

$$L_{\mathrm{CIB}} = I(\mathcal{X}; \mathcal{Z}) - \beta I(\mathcal{Y}; \mathcal{Z}) \tag{10}$$

and is solved using update Equation (2) with (3) substituting the scalar y by vector $\mathbf{y}$. The number of output clusters $|\mathbb{Z}|$ has to be chosen to $|\mathbb{Z}| = \prod_{m=1}^{M} |\mathbb{Z}_m|$ while the single link from the imaginary central quantizer to the receiver in Figure 4 has a channel capacity of $C_{\mathrm{sum}} = \sum_{m=1}^{M} C_m$. The actual transmission over the M links has to be coordinated such that each sensor m transmits a specific part of the bits corresponding to its link capacity C_m.

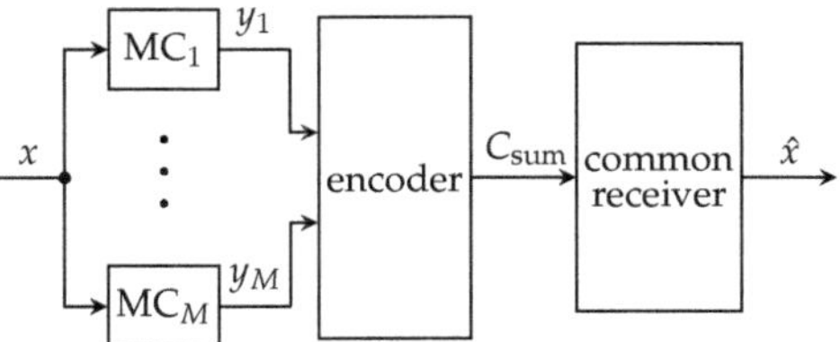

Figure 4. Model of a centralized compression approach representing the fully cooperative Chief Executive Officer scenario.

In the special case of the measurement process being modeled as additive noise, the algorithm can be simplified to a scalar optimization problem where maximum ratio combining of all inputs y_m delivers a scalar sufficient statistics

$$\bar{y} = \sum_{m=1}^{M} \gamma_m \cdot y_m$$

of the desired relevant signal x with an overall SNR $\gamma = \sum_m \gamma_m$. The solution of the fcCEO scenario serves as an upper bound in this paper.

5. Partially Cooperative Distributed Sensing

In order to investigate how the gap between non-cooperative and fully-cooperative distributed compression can be reduced, partially cooperating sensors shall now be considered. Partial cooperation means a limited exchange of instantaneous side-information among the sensors during runtime due to a rate-limitation of inter-sensor links. Non-cooperative CEO and fully-cooperative CEO problems represent the extreme cases for zero rate and unlimited rate inter-sensor links, respectively. The rate limitation requires the compression of instantaneous side-information before forwarding it to other sensors.

In this paper, only deterministic mappings are considered for this compression, while indexes z_m are still obtained by stochastic mappings. This is motivated by the fact that deterministic mappings do not require further lossy compression and the resulting side-information indices s_m can be exploited at other sensors by choosing a particular mapping $p(z_m|y_m)$ from a list of possible mappings designed offline in advance. As a consequence, the compression rates for instantaneous side-information can only be adjusted by changing the cardinalities $|\mathbb{S}_m|$. For all results presented below, inter-sensor links are modeled as bit pipes being able to deliver s_m reliably.

The GDIB algorithm to solve the non-cooperative CEO problem is based on the inner bound (5) of the CEO rate-region. Moreover, a greedy optimization approach is optimal due to the supermodularity of the compression rates in (5). Both require the Markovian structure in (4). However, cooperation among sensors changes the Markovian structure and implies different statistical dependencies among involved random variables. As (4) does not hold anymore in pcCEO scenarios, the inner bound on the rate-region in (5) cannot be utilized to find solutions of the pcCEO scenario. To the knowledge of the authors, tight bounds on the rate-region are not available for the cooperative case. Therefore, a heuristic approach based on the greedy optimization structure of the GDIB algorithm will be applied to solve the pcCEO scenario, which is not proven to be optimal. Nevertheless, the numerical evaluation of the found solutions demonstrate their usefulness. However, the computation of required pmfs becomes more challenging and results in recursive calculations given in Appendices A.1 and A.2 because the Markovian structure of (4) does not hold anymore in pcCEO scenarios.

This paper introduces three different inter-sensor communication protocols for exchanging this instantaneous side-information: successive broadcasting, a successive point-to-point transmission and a two-phase transmission. The first two protocols perform the exchange of instantaneous side-information s_m with other sensors and the forwarding of compressed versions of z_m to the common receiver in the same time slot. Contrarily, the two-phase transmission protocol separates the exchange of instantaneous side-information among sensors and the communication with the common receiver into two distinct phases. The latter starts after the exchange among sensors has been completed such that all sensors have (approximately) the same amount of side-information.

5.1. Successive Broadcasting Protocol

The system model for the successive broadcasting protocol is illustrated in Figure 5. In the same time slot, sensor $m - 1$ not only forwards a compressed version of the quantization index z_{m-1} to the common receiver, but also broadcasts instantaneous side-information s_{m-1} to all other sensors. However, due to the greedy optimization structure, only subsequent sensors can exploit this instantaneous side-information. Thus, sensor m can exploit indices $\mathbf{s}_{<m}$ of all previously transmitting sensors in order to select its quantization index z_m as well as a new instantaneous side-information index s_m. This scenario leads to the Markov model

$$p(x, \mathbf{y}, \mathbf{z}, \mathbf{s}) = \prod_{m=1}^{M} p(z_m | y_m, \mathbf{s}_{<m}) p(s_m | y_m, \mathbf{s}_{<m}) p(y_m | x) p(x). \tag{11}$$

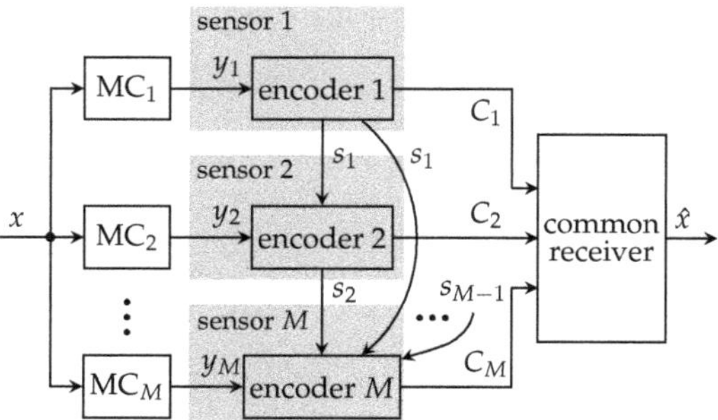

Figure 5. Partially cooperative CEO scenario using broadcast exchange of side-information among sensors.

The indices s_m and z_m are obtained by deterministic mappings $p(s_m | y_m, \mathbf{s}_{<m}) \in \{0, 1\}$ and stochastic mappings $p(z_m | y_m, \mathbf{s}_{<m}) \in [0, 1]$, respectively. The design of these mappings can be performed offline leveraging the IB principle as illustrated in Figure 6. It combines the observation y_m and the instantaneous side-information $\mathbf{s}_{<m}$ to indexes s_m and z_m while maintaining as much information as possible about the relevant signal x.

Figure 6. Graphical illustration of IB fusion of involved inputs to determine instantaneous side-information s_m (**a**) and the quantizer sensor output z_m (**b**) for a broadcast exchange of instantaneous side-information.

5.1.1. Generation of Broadcast Side-Information

The design of $p(s_m|y_m, \mathbf{s}_{<m})$ is inspired by the general GDIB algorithm, i.e., the optimization is done in a greedy manner. Again, there emerges one optimization problem for each sensor:

$$L^{(1)}_{\text{BC-SIDE}} = I(\mathcal{X}; \mathcal{S}_1) - \beta I(\mathcal{Y}_1; \mathcal{S}_1) \tag{12}$$

$$\vdots$$

$$L^{(M-1)}_{\text{BC-SIDE}} = I(\mathcal{X}; \mathcal{S}_{M-1}|\boldsymbol{S}_{<M-1}) - \beta I(\mathcal{Y}_{M-1}; \mathcal{S}_{M-1}|\boldsymbol{S}_{<M-1}). \tag{13}$$

The optimization problem of the first sensor equals the individual scalar optimization without any side-information at all, as described in Section 2. Subsequent sensors combine the instantaneous side-information of all previously transmitting sensors $\mathbf{s}_{<m}$ with its observation y_m. The relevant mutual information and the compression rate of sensor m are conditioned on $\boldsymbol{S}_{<m}$ since broadcasting instantaneous side-information ensures all successive sensors to have access to $\mathbf{s}_{<m}$ allowing Wyner–Ziv coding for generating s_m. Each optimization problem given in (12) and (13) can be solved by taking the derivative with respect to the mapping $p(s_m|y_m, \mathbf{s}_{<m})$ and equating it to zero. This results in the implicit update equation

$$p(s_m|y_m, \mathbf{s}_{<m}) = \frac{e^{-d_{\beta m}(y_m, s_m, \mathbf{s}_{<m})}}{\sum_{s_m} e^{-d_{\beta m}(y_m, s_m, \mathbf{s}_{<m})}} \tag{14}$$

with

$$d_{\beta m}(y_m, s_m, \mathbf{s}_{<m}) := \frac{1}{\beta_m} D_{\text{KL}}[p(x|y_m, \mathbf{s}_{<m}) \| p(x|\mathbf{s}_{\leq m})] - \log p(s_m|\mathbf{s}_{<m}). \tag{15}$$

As in the general GDIB algorithm, the implicit update equation in (14) can be solved using a Blahut–Arimoto like algorithm resulting in local optimal solutions.

Algorithmic pcCEO Solution for the Successive Broadcasting Protocol

After designing the mapping for the instantaneous side-information, the mapping $p(z_m|y_m, \mathbf{s}_{<m})$ can be optimized, again by means of the IB principle. Therefore, the original GDIB algorithm is modified to exploit the broadcasted instantaneous side-information, defining the GDIB-BC algorithm. The optimization problem for each sensor is given as

$$L^{(1)}_{\text{GDIB-BC}} = I(\mathcal{X}; \mathcal{Z}_1) - \beta_1 I(\mathcal{Y}_1; \mathcal{Z}_1) \tag{16}$$

$$\vdots$$

$$L^{(M)}_{\text{GDIB-BC}} = I(\mathcal{X}; \mathcal{Z}_M|\boldsymbol{Z}_{<M}) - \beta_M I(\mathcal{Y}_M, \boldsymbol{S}_{<M}; \mathcal{Z}_M|\boldsymbol{Z}_{<M}). \tag{17}$$

The main difference to the original GDIB optimization problem in (6) and (7) lies in the definition of the compression rate $I(\mathcal{Y}_m, \boldsymbol{S}_{<m}; \mathcal{Z}_m|\boldsymbol{Z}_{<m})$ which emerges from the combination of the observation y_m and the instantaneous side-information $\mathbf{s}_{<m}$. Taking the derivative of the optimization problem for sensor m with respect to the mapping $p(z_m|y_m, \mathbf{s}_{<m})$ and equating it to zero delivers

$$p(z_m|y_m, \mathbf{s}_{<m}) = \frac{e^{-d_{\beta m}(y_m, z_m, \mathbf{s}_{<m})}}{\sum_{z_m} e^{-d_{\beta m}(y_m, z_m, \mathbf{s}_{<m})}} \tag{18}$$

with

$$d_{\beta_m}(y_m, z_m, \mathbf{s}_{<m}) := \mathbb{E}_{\mathbf{z}_{<m}|y_m,\mathbf{s}_{<m}}\left[\frac{1}{\beta_m}\cdot\right.$$

$$\left.D_{\mathrm{KL}}[p(x|y_m,\mathbf{s}_{<m},\mathbf{z}_{<m})\|p(x|\mathbf{z}_{\leq m})] - \log p(z_m|\mathbf{z}_{<m})\right]. \tag{19}$$

Again, the implicit update equation in (18) can be solved using a Blahut–Arimoto like algorithm. The extended Blahut–Arimoto like algorithm to design the mapping $p(z_m|y_m, \mathbf{s}_{<m})$ of sensor m for a specific Lagrange parameter β_m and instantaneous side-information $\mathbf{s}_{<m}$ is given in Algorithm 1. The input pmf $p(y_{m-1}, \mathbf{s}_{<m-1}, \mathbf{z}_{<m-1}, x)$ can be computed during the optimization of previous sensors. Lines 3 to 5 determine the required pmfs for the calculation of the KL-divergence of (19) in lines 6 to 9. The statistical distance of (19) is determined in lines 10 to 14. It is used to update the quantizer mapping $p(z_m|y_m, \mathbf{s}_{<m})$ of sensor m. This procedure is repeated until no significant changes of the desired mappings occur anymore. The algorithm returns the updated mapping $p(z_m|y_m, \mathbf{s}_{<m})$ as well as the pmf $p(y_m, \mathbf{s}_{<m}, \mathbf{z}_{<m}, x)$, which is used as an input for the successive sensor.

Algorithm 1: Extended Blahut–Arimoto algorithm for broadcast cooperating sensors.

---

input : $m, p^{\mathrm{init}}(z_m|y_m, \mathbf{s}_{<m}), p(y_i, x), \beta_m, \epsilon$
$\quad\quad\quad p(s_i|y_i, \mathbf{s}_{<i}) \; \forall i \leq m$
$\quad\quad\quad$ recursively calculated input from previous
$\quad\quad\quad$ sensor optimizations:
$\quad\quad\quad p(y_{m-1}, \mathbf{s}_{<m-1}, \mathbf{z}_{<m-1}, x)$

output : $p(z_m|y_m, \mathbf{s}_{<m}) \in [0,1], p(y_m, \mathbf{s}_{<m}, \mathbf{z}_{<m}, x)$

1 **begin**
$\quad$ **initialization:**
$$p(z_m|y_m, \mathbf{s}_{<m})^{(0)} \leftarrow p^{\mathrm{init}}(z_m|y_m, \mathbf{s}_{<m}),$$
$$l \leftarrow 1$$

2 $\quad$ **do**

3 $\quad\quad$ // calculate required pmfs (see Appendix A.1)

4 $\quad\quad p(y_m, \mathbf{s}_{<m}, \mathbf{z}_{<m}, x) = \sum_{y_{m-1}} p(z_{m-1}|y_{m-1}, \mathbf{s}_{<m-1}) p(y_m|x)$
$\quad\quad\quad\quad \cdot p(s_{m-1}|y_{m-1}, \mathbf{s}_{<m-1}) p(y_{m-1}, \mathbf{s}_{<m-1}, \mathbf{z}_{<m-1}, x)$

5 $\quad\quad p(\mathbf{z}_{\leq m}, x) = \sum_{\mathbf{s}_{<m}} \sum_{y_m} p(z_m|y_m, \mathbf{s}_{<m}) p(y_m, \mathbf{s}_{<m}, \mathbf{z}_{<m}, x)$

6 $\quad\quad$ // KL-Divergence D_{KL} of (19)

7 $\quad\quad p(x|y_m, \mathbf{s}_{<m}, \mathbf{z}_{<m}) = \frac{p(y_m, \mathbf{s}_{<m}, \mathbf{z}_{<m}, x)}{\sum_x p(y_m, \mathbf{s}_{<m}, \mathbf{z}_{<m}, x)}$

8 $\quad\quad p(x|\mathbf{z}_{\leq m}) = \frac{p(\mathbf{z}_{\leq m}, x)}{\sum_x p(\mathbf{z}_{\leq m}, x)}$

9 $\quad\quad D_{\mathrm{KL}} = \sum_x p(x|y_m, \mathbf{s}_{<m}, \mathbf{z}_{<m}) \cdot \log \frac{p(x|y_m, \mathbf{s}_{<m}, \mathbf{z}_{<m})}{p(x|\mathbf{z}_{\leq m})}$

10 $\quad\quad$ // distance $d_{\beta_m}(y_m, z_m, \mathbf{s}_{<m})$ (19)

11 $\quad\quad p(\mathbf{z}_{<m}|y_m, \mathbf{s}_{<m}) = \frac{\sum_x p(y_m, \mathbf{s}_{<m}, \mathbf{z}_{<m}, x)}{\sum_x \sum_{\mathbf{z}_{<m}} p(y_m, \mathbf{s}_{<m}, \mathbf{z}_{<m}, x)}$

12 $\quad\quad p(\mathbf{z}_{\leq m}) = \sum_x p(\mathbf{z}_{\leq m}, x)$

13 $\quad\quad p(z_m|\mathbf{z}_{<m}) = \frac{p(\mathbf{z}_{\leq m})}{\sum_{z_m} p(\mathbf{z}_{\leq m})}$

14 $\quad\quad d_{\beta_m}(z_m, y_m, \mathbf{s}_{<m}) = \sum_{\mathbf{z}_{<m}} p(\mathbf{z}_{<m}|y_m, \mathbf{s}_{<m}) \cdot \left[\frac{1}{\beta_m} D_{\mathrm{KL}} - \log p(z_m|\mathbf{z}_{<m})\right]$

15 $\quad\quad$ // update quantizer $p(z_m|y_m, \mathbf{s}_{<m})$

16 $\quad\quad p(z_m|y_m, \mathbf{s}_{<m})^{(l)} = \frac{e^{-d_{\beta_m}(y_m, z_m, \mathbf{s}_{<m})}}{\sum_z e^{-d_{\beta_m}(y_m, z_m, \mathbf{s}_{<m})}}$

17 $\quad\quad l \leftarrow l + 1$

18 $\quad$ **while** $D_{\mathrm{JS}}[p^{(l)}(z_m|y_m, \mathbf{s}_{<m}) \| p^{(l-1)}(z_m|y_m, \mathbf{s}_{<m})] < \epsilon$

The parameter β_m, which determines the compression rate at sensor m, has to be adjusted such that $I(\mathcal{Y}_m, \boldsymbol{S}_{<m}; \mathcal{Z}_m | \boldsymbol{Z}_{<m}) \leq C_m$ is fulfilled. Similar to the original GDIB algorithm, the GDIB-BC algorithm has to be performed for each sensor and all possible optimization orders.

5.1.2. Evolution of Instantaneous Side-Information

Figure 7 illustrates the amount of instantaneous side-information available at the different sensors in a network of size $M = 6$ considering the broadcast of side-information. It depicts the relevant mutual information $I(\mathcal{X}; \boldsymbol{S}_{\leq m})$ versus the sensor number m for different cardinalities $|\mathbb{S}_m|$ and SNRs γ_m. The relevant signal is chosen to be a uniformly distributed 4-ASK signal leading to $|\mathbb{X}| = 4$. As expected, the amount of available instantaneous side-information increases with each additional sensor for all $|\mathbb{S}_m|$ and γ_m. To be more specific, the resolution and the quality of instantaneous side-information available at sensor m increases with growing m. In the considered symmetric scenario, the amount of information $I(\mathcal{X}; \mathcal{S}_m | \boldsymbol{S}_{<m})$ a sensor can contribute to $I(\mathcal{X}; \boldsymbol{S}_{\leq m})$ gets smaller for each additional sensor and the slopes of the curves decrease. Since one bit is not enough to represent the information of $|\mathbb{X}| = 4$, the largest gain can be observed between $|\mathbb{S}_m| = 2$ and $|\mathbb{S}_m| = 4$. Increasing the cardinality further to $|\mathbb{S}_m| = 8$ results only in a small additional improvement. Certainly, this observation depends on the relevant signal $\mathcal{X}$ and can not be generalized.

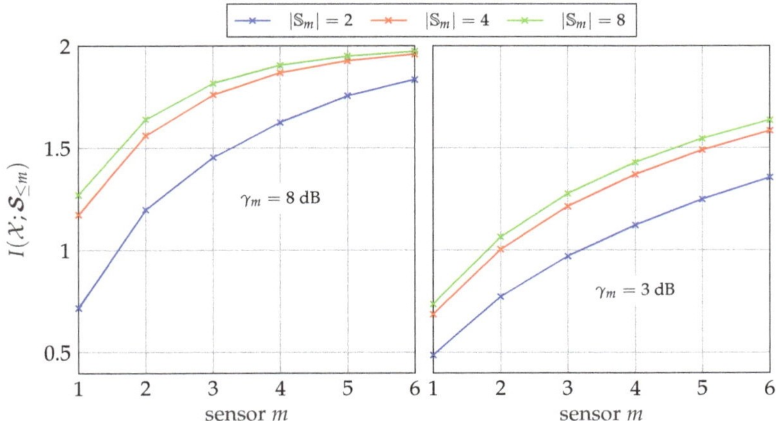

Figure 7. Available mutual information $I(\mathcal{X}; \boldsymbol{S}_{\leq m})$ for sensor m in a network with $M = 6$ sensors and different cardinalities $|\mathbb{S}_m|$ using the successive broadcasting protocol; $|\mathbb{X}| = 4$, $|\mathbb{Y}_m| = 64$.

5.1.3. Performance for Different Network Sizes

Figures 8 and 9 illustrate the overall performance of the GDIB-BC approach when broadcasting instantaneous side-information for different network sizes. The relevant mutual information $I(\mathcal{X}; \boldsymbol{Z})$ is depicted versus the number of sensors M in the network. The gray colored area represents the non-achievable region, since $I(\mathcal{X}; \boldsymbol{Z})$ cannot exceed $I(\mathcal{X}; \boldsymbol{\mathcal{Y}})$ due to the data-processing inequality. Both figures consider a scenario where all sensors in the network share the same channel to the common receiver with a fixed sum-rate C_{sum} in an orthogonal way and a round robin fashion. Consequently, larger network sizes correspond to smaller individual capacities $C_m = \frac{C_{\text{sum}}}{M}$ for each forward link. The performance of partially cooperating sensors broadcasting instantaneous side-information (pcCEO-BC) is compared to the non-cooperative case (CEO) of Section 3 and the fully cooperative case (fcCEO) of Section 4. As already mentioned, these two scenarios provide upper and lower bounds. In general, it can be observed that increasing the number of sensors in the network also increases the overall relevant mutual information $I(\mathcal{X}; \boldsymbol{Z})$.

This holds even for the case without cooperation, since each sensor applies Wyner–Ziv coding and exploits the mapping of previously designed quantizers as statistical side-information [14]. Independent of the cardinality $|\mathbb{S}_m|$, the performance of the pcCEO-BC scenario is superior to the case without cooperation among sensors. This difference grows for larger network sizes because the amount of information $\mathbf{s}_{<m}$ has about the relevant variable x increases. As expected from Figure 7, increasing the cardinality $|\mathbb{S}_m|$ not only improves the relevant information $I(\mathcal{X}; \boldsymbol{\mathcal{S}}_{\leq m})$, but also the overall performance measured by $I(\mathcal{X}; \boldsymbol{\mathcal{Z}})$. However, it can be observed that even for large $|\mathbb{S}_m|$ there remains a gap to the fcCEO upper bound, especially for smaller network sizes or lower SNRs. This gap can be explained by the successive transmission protocol resulting in a gradually increasing amount of instantaneous side-information at the sensors. For instance, the first sensors does not profit at all from the partial cooperation in contrast to the fcCEO scenario where all sensors exploit almost the same amount of side-information.

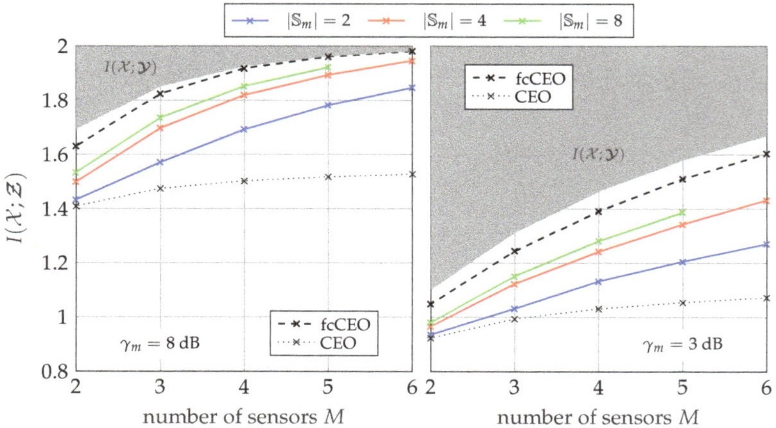

Figure 8. Relevant mutual information $I(\mathcal{X}; \boldsymbol{\mathcal{Z}})$ versus the network size for a fixed sum-rate of $C_{\text{sum}} = 2.5$ bit/s/Hz and $C_m = \frac{C_{\text{sum}}}{M}$ using the successive broadcasting protocol with different cardinalities $|\mathbb{S}_m|$; $|\mathbb{X}| = 4$, $|\mathbb{Y}_m| = 64$, $|\mathbb{Z}_m| = 4$.

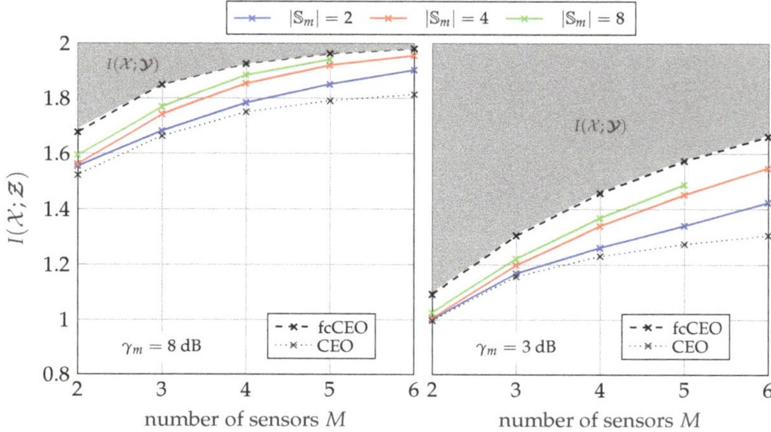

Figure 9. Relevant mutual information $I(\mathcal{X}; \boldsymbol{\mathcal{Z}})$ versus the network size for a fixed sum-rate of $C_{\text{sum}} = 4$ bit/s/Hz and $C_m = \frac{C_{\text{sum}}}{M}$ using the successive broadcasting protocol with different cardinalities $|\mathbb{S}_m|$; $|\mathbb{X}| = 4$, $|\mathbb{Y}_m| = 64$, $|\mathbb{Z}_m| = 4$.

Considering the pmfs in Algorithm 1, it becomes obvious that larger networks might suffer from the curse of dimensionality. More precisely, pmfs like $p(y_m, \mathbf{s}_{<m}, \mathbf{z}_{<m}, x)$ can become very large during the optimization for larger network sizes. Moreover, the mapping $p(z_m | y_m, \mathbf{s}_{<m})$ also depends on the network size, i.e., this problem does not only occur during the optimization, but also when storing the already optimized mapping. This numerical issue is the reason why there is no result for $|\mathbb{S}_m| = 8$ and a network size of $M = 6$ in Figures 8 and 9. In this case, it requires 2024 GiB (1 GiB = 1024 MiB, 1 MiB = 1024 KiB, 1 KiB = 1024 byte) just for storing a single instance of the pmf $p(y_m, \mathbf{s}_{<m}, \mathbf{z}_{<m}, x)$.

5.2. Successive Point-to-Point Protocol

For larger network sizes, broadcasting side-information might not be feasible anymore, since the dimensions of the mappings $p(z_m | y_m, \mathbf{s}_{<m})$ and $p(s_m | y_m, \mathbf{s}_{<m})$ as well as intermediate pmfs used within the optimization become huge. In order to relax this curse of dimensionality, the successive way of cooperation is exploited and the instantaneous side-information of sensor m shall only be forwarded to the direct successor $m + 1$ as depicted in Figure 10. Hence, a sequential chain is established from the first to the last sensor leading to the Markov Model:

$$p(x, \mathbf{y}, \mathbf{z}, \mathbf{s}) = \prod_{m=1}^{M} p(z_m | y_m, s_{m-1}) p(s_m | y_m, s_{m-1}) p(y_m | x) p(x). \tag{20}$$

Again, the instantaneous side-information is obtained by a deterministic mapping optimized by means of the information bottleneck principle, illustrated in Figure 11. With each step in the sequential chain, the information s_m has about the relevant signal x increases.

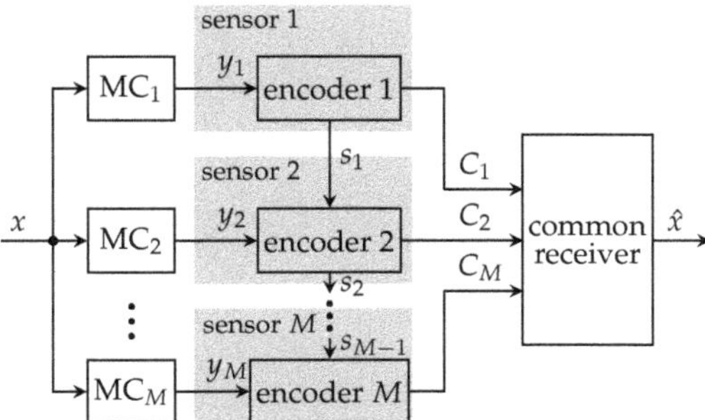

Figure 10. Partially cooperative CEO scenario using successive point-to-point transmission of side-information.

Figure 11. Graphical illustration of IB fusion of two inputs to determine instantaneous side-information s_m (**a**) and the quantizer sensor output z_m (**b**) for a successive point-to-point transmission of instantaneous side-information.

5.2.1. Generation of Point-to-Point Side-Information

Similar to the broadcast case, the design of $p(s_m|y_m, s_{m-1})$ is inspired by the original GDIB algorithm. The optimization problem can be formulated in a greedy manner as

$$L_{\text{PTP-SIDE}}^{(1)} = I(\mathcal{X}; \mathcal{S}_1) - \beta I(\mathcal{Y}_1; \mathcal{S}_1) \tag{21}$$

$$\vdots$$

$$L_{\text{PTP-SIDE}}^{(M-1)} = I(\mathcal{X}; \mathcal{S}_{M-1}) - \beta I(\mathcal{Y}_{M-1}, \mathcal{S}_{M-2}; \mathcal{S}_{M-1}), \tag{22}$$

where Equation (21) equals the individual scalar optimization without any side-information. Subsequent sensors combine the instantaneous side-information s_{m-1} sent by the previous sensor with its observation y_m. In contrast to the broadcast case, the relevant mutual information is not conditioned on $\mathcal{S}_{<m}$ as in (12) and (13) because sensor m will only have access to s_{m-1} and not to indices of any other sensor. Therefore, Wyner–Ziv coding cannot be applied for exchanging instantaneous side-information with the successive point-to-point protocol. The optimization problems can be solved by taking the derivative with respect to the mapping $p(s_m|y_m, s_{m-1})$ and equating it to zero, resulting in the implicit update equation

$$p(s_m|y_m, s_{m-1}) = \frac{e^{-d_{\beta_m}(y_m, s_m, s_{m-1})}}{\sum_{s_m} e^{-d_{\beta_m}(y_m, s_m, s_{m-1})}} \tag{23}$$

with

$$d_{\beta_m}(y_m, s_m, s_{m-1}) := \frac{1}{\beta_m} \cdot D_{\text{KL}}[p(x|y_m, s_{m-1}) \| p(x|s_m)] - \log p(s_m) . \tag{24}$$

As in the broadcast case, using a Blahut–Arimoto like algorithm to solve the update Equation (23) results in local optimal solutions.

5.2.2. Algorithmic pcCEO Solution Applying the Successive Point-to-Point Protocol

After the optimization of the mapping for instantaneous side-information, the mapping $p(z_m|y_m, s_{m-1})$ can be designed by means of the information bottleneck principle. Inspired by the original GDIB algorithm, the optimization problem can be formulated as

$$L_{\text{GDIB-PTP}}^{(1)} = I(\mathcal{X}; \mathcal{Z}_1) - \beta_1 I(\mathcal{Y}_1; \mathcal{Z}_1) \tag{25}$$

$$\vdots$$

$$L_{\text{GDIB-PTP}}^{(M)} = I(\mathcal{X}; \mathcal{Z}_M|\mathcal{Z}_{<M}) - \beta_M I(\mathcal{Y}_M, \mathcal{S}_{M-1}; \mathcal{Z}_M|\mathcal{Z}_{<M}). \tag{26}$$

The main difference to the original GDIB optimization problem in (6) and (7) now lies in the compression rate $I(\mathcal{Y}_m, \mathcal{S}_{m-1}; \mathcal{Z}_m|\mathcal{Z}_{<m})$, which emerges from the combination of the instantaneous side-information s_{m-1} of the previous sensor $m-1$ and the observation y_m of sensor m. Taking the derivative with respect to the mapping $p(z_m|y_m, s_{m-1})$ and equating it to zero, the optimization problem for sensor m can be solved, leading to the implicit update equation

$$p(z_m|y_m, s_{m-1}) = \frac{e^{-d_{\beta_m}(y_m, z_m, s_{m-1})}}{\sum_{z_m} e^{-d_{\beta_m}(y_m, z_m, s_{m-1})}} \tag{27}$$

with

$$d_{\beta_m}(y_m, z_m, s_{m-1}) := \mathbb{E}_{\mathbf{z}_{<m}|y_m, s_{m-1}}\left[\frac{1}{\beta_m}\cdot\right.$$

$$\left. D_{\mathrm{KL}}[p(x|y_m, s_{m-1}, \mathbf{z}_{<m})\|p(x|\mathbf{z}_{\leq m})] - \log p(z_m|\mathbf{z}_{<m})\right]. \tag{28}$$

Thus, the mapping $p(z_m|y_m, s_{m-1})$ can be optimized using a Blahut–Arimoto-like algorithm. The specific algorithm for a given sensor m and a Lagrange parameter β_m is given in Algorithm 2. The input pmfs $p(z_i|y_i, s_{i-1})$ $\forall i < m$ and $p(s_i|\mathbf{z}_{\leq i}, x)$ $\forall i < m$ as well as $p(\mathbf{z}_{<m-1}, x)$ are calculated in advance by previous sensor optimizations. Lines 3 to 7 calculates required pmfs as given the Appendix A.2. The KL-divergence is calculated in lines 8 to 11. Using this, the statistical distance $d_{\beta_m}(z_m, y_m, s_{m-1})$ of (28) can be calculated in lines 12 to 16, which is then used to update the quantizer mapping $p(z_m|y_m, s_{m-1})$. The algorithm stops if this mapping does not change significantly anymore during subsequent iterations. Finally, the output pmfs $p(s_m|\mathbf{z}_{\leq m}, x)$ and $p(\mathbf{z}_{<m}, x)$ need to be calculated in lines 21 to 25 for their usage in the optimization of the next sensor.

Similar to the original GDIB algorithm, the optimization needs to be done for all possible optimization orders. A simple bisection search can be applied to find the rate-fulfilling parameter β_m, such that $I(\mathcal{Y}_m, \mathcal{S}_{m-1}; \mathcal{Z}_m|\mathcal{Z}_{<m}) \leq C_m$ holds.

5.2.3. Evolution of Instantaneous Side-Information

Figure 12 illustrates the amount of instantaneous side-information $I(\mathcal{X}; \mathcal{S}_m)$ at a specific sensor m in a network of size $M = 6$ using the successive point-to-point transmission protocol for different cardinalities $|\mathbb{S}_m|$. Obviously, $I(\mathcal{X}; \mathcal{S}_m)$ increases with each further sensor. The main difference to the broadcast case is that the instantaneous side-information provided to sensor m is represented by a single highly compressed index s_{m-1} with cardinality $|\mathbb{S}_{m-1}|$. While the resolution $|\mathbb{S}_{<m}|$ of the available instantaneous side-information $\mathbf{s}_{<m}$ increases with m in the broadcast case, it remains the same for the successive point-to-point protocol. Therefore, a higher cardinality $|\mathbb{S}_m|$ is required compared to the broadcast case to avoid additional compression losses.

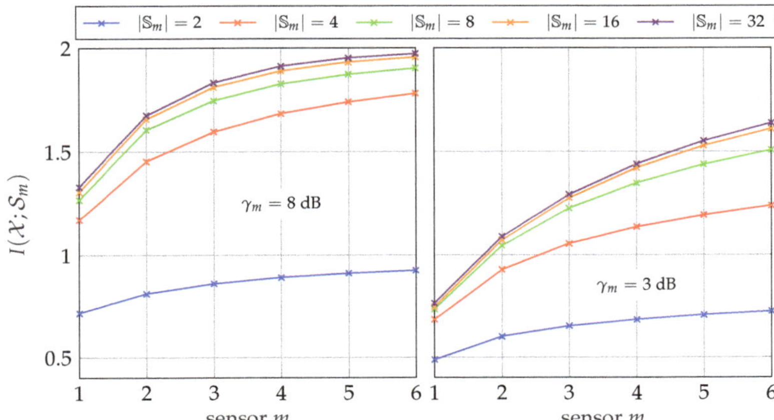

Figure 12. Evolution of $I(\mathcal{X}; \mathcal{S}_m)$ for sensor m in a network with $M = 6$ sensors and different cardinalities $|\mathbb{S}_m|$ for the successive point-to-point transmission protocol; $|\mathbb{X}| = 4$, $|\mathbb{Y}_m| = 64$.

Algorithm 2: Extended Blahut–Arimoto algorithm for the successive point-to-point protocol.

---

    input   : $m, p^{\text{init}}(z_m|y_m, s_{m-1}), p(y_i, x),$
                             $p(s_i|y_i, s_{i-1}) \; \forall i \leq m, \beta_m, \epsilon$
                             recursively calculated inputs from previous
                             sensor optimizations:
                             $p(z_i|y_i, s_{i-1}), p(s_i|\mathbf{z}_{\leq i}, x) \; \forall i < m,$
                             $p(\mathbf{z}_{<m-1}, x)$

    output   : $p(z_m|y_m, s_{m-1}) \in [0,1], p(\mathbf{z}_{<m}, x), p(s_m|\mathbf{z}_{\leq m}, x)$

1 begin

    initialization:
$$p(z_m|y_m, s_{m-1})^{(0)} \leftarrow p^{\text{init}}(\cdot|\cdot, \cdot),$$
$$l \leftarrow 1$$

2   **do**

3   `// calc. pmfs (see Appendix A.2)`

4 $p(z_{m-1}, s_{m-1}|x, \mathbf{z}_{<m-1}) = \sum_{s_{m-2}} \sum_{y_{m-1}} p(z_{m-1}|y_{m-1}, s_{m-2}) p(s_{m-2}|\mathbf{z}_{\leq m-2}, x)$
              $\cdot \, p(s_{m-1}|y_{m-1}, s_{m-2}) p(y_{m-1}|x)$

5 $p(\mathbf{z}_{<m-1}, y_m, x) = p(\mathbf{z}_{<m-1}, x) p(y_m|x)$

6 $p(y_m, s_{m-1}, \mathbf{z}_{<m}, x) = p(z_{m-1}, s_{m-1}|x, \mathbf{z}_{<m-1}) p(\mathbf{z}_{<m-1}, y_m, x)$

7 $p(\mathbf{z}_{\leq m}, x) = \sum_{s_{m-1}} \sum_{y_m} p(z_m|y_m, s_{m-1}) p(z_{m-1}, s_{m-1}|x, \mathbf{z}_{<m-1})$
              $\cdot \, p(\mathbf{z}_{<m-1}, x) p(y_m|x)$

8   `// KL-Divergence` D_{KL} `of (28)`

9 $p(x|y_m, s_{m-1}, \mathbf{z}_{<m}) = \frac{p(y_m, s_{m-1}, \mathbf{z}_{<m}, x)}{\sum_x p(y_m, s_{m-1}, \mathbf{z}_{<m}, x)}$

10 $p(x|\mathbf{z}_{\leq m}) = \frac{p(\mathbf{z}_{\leq m}, x)}{\sum_x p(\mathbf{z}_{\leq m}, x)}$

11 $D_{\text{KL}} = \sum_x p(x|y_m, s_{m-1}, \mathbf{z}_{<m}) \log \frac{p(x|y_m, s_{m-1}, \mathbf{z}_{<m})}{p(x|\mathbf{z}_{\leq m})}$

12   `// distance` $d_{\beta_m}(y_m, z_m, s_{m-1})$ `(28)`

13 $p(\mathbf{z}_{<m}|y_m, s_{m-1}) = \frac{\sum_x p(y_m, s_{m-1}, \mathbf{z}_{<m}, x)}{\sum_x \sum_{\mathbf{z}_{<m}} p(y_m, s_{m-1}, \mathbf{z}_{<m}, x)}$

14 $p(\mathbf{z}_{\leq m}) = \sum_x p(\mathbf{z}_{\leq m}, x)$

15 $p(z_m|\mathbf{z}_{<m}) = \frac{p(\mathbf{z}_{\leq m})}{\sum_{z_m} p(\mathbf{z}_{\leq m})}$

16 $d_{\beta_m}(z_m, y_m, s_{m-1}) = \sum_{\mathbf{z}_{<m}} p(\mathbf{z}_{<m}|y_m, s_{m-1}) \cdot \left[\frac{1}{\beta_m} D_{\text{KL}} - \log p(z_m|\mathbf{z}_{<m}) \right]$

17   `// update quantizer` $p(z_m|y_m, s_{m-1})$

18 $p(z_m|y_m, s_{m-1})^{(l)} = \frac{e^{-d_{\beta_m}(y_m, z_m, s_{m-1})}}{\sum_z e^{-d_{\beta_m}(y_m, z_m, s_{m-1})}}$

19 $l \leftarrow l + 1$

20   **while** $D_{\text{JS}}\big[p^{(l)}(z_m|y_m, s_{m-1}) \, || \, p^{(l-1)}(z_m|y_m, s_{m-1}) \big] < \epsilon$

21   `// cal.` $p(s_m|\mathbf{z}_{\leq m}, x)$ `and` $p(\mathbf{z}_{<m}, x)$ `for successive sensor`

22 $p(\mathbf{z}_{<m}, x) = \sum_{z_m} p(\mathbf{z}_{\leq m}, x)$

23 $p(y_m, x, z_m) = \sum_{s_{m-1}} p(y_m, x) p(z_m|y_m, s_{m-1}) p(s_{m-1}|x)$

24 $p(y_m|x, z_m) = \frac{p(y_m, x, z_m)}{\sum_{y_m} p(y_m, x, z_m)}$

25 $p(s_m|\mathbf{z}_{\leq m}, x) = \sum_{s_{m-1}} \sum_{y_m} p(s_{m-1}|\mathbf{z}_{\leq m-1}, x) p(y_m|x, z_m) p(s_m|y_m, s_{m-1})$

5.2.4. Performance for Different Network Sizes

Figures 13 and 14 illustrate the overall performance of the pcCEO system with point-to-point exchanged instantaneous side-information where all sensors share the same channel to the common receiver with a fixed sum-rate $C_{\text{sum}} = \sum_{m=1}^{M} C_m$ in an orthogonal way and a round robin fashion. Again, the black curves represent the non-cooperative CEO scenario and the fcCEO scenario. Hence, they serve as lower and upper bound, respectively. In general, the curves are very similar to those for broadcasting instantaneous side-information in Figures 8 and 9. Independent of the SNR or the sum-rate C_{sum}, the relevant mutual information $I(\mathcal{X};\mathcal{Z})$ increases for larger networks and even a single bit as instantaneous side-information $|\mathbb{S}_m| = 2$ leads to slight improvements compared to the non-cooperative case. However, there still remains a gap to the fcCEO scenario even for large $|\mathbb{S}_m|$, which results from the successive communication strategy, since sensors at the beginning of the optimization chain can exploit no or little instantaneous side-information.

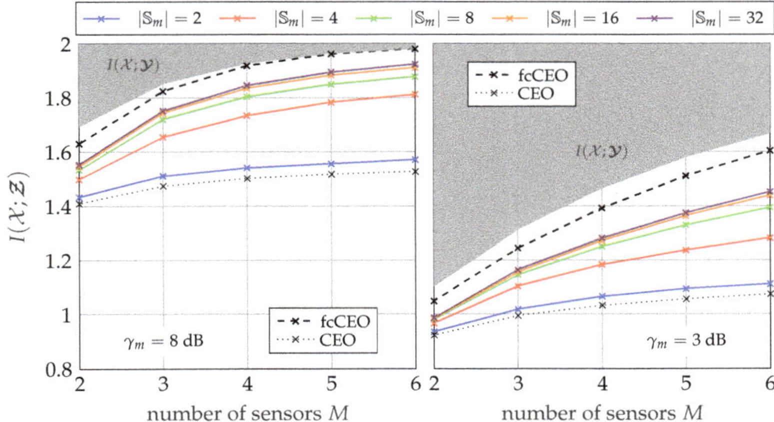

Figure 13. Relevant mutual information $I(\mathcal{X};\mathcal{Z})$ versus the network size for a fixed sum-rate of $C_{\text{sum}} = 2.5$ bit/s/Hz and $C_m = \frac{C_{\text{sum}}}{M}$ using the successive point-to-point transmission protocol with different cardinalities $|\mathbb{S}_m|$; $|\mathbb{X}| = 4$, $|\mathbb{Y}_m| = 64$, $|\mathbb{Z}_m| = 4$.

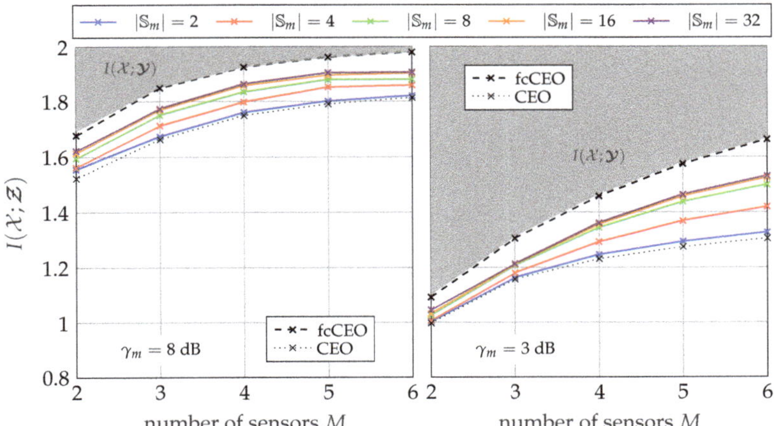

Figure 14. Relevant mutual information $I(\mathcal{X};\mathcal{Z})$ versus the network size for a fixed sum-rate of $C_{\text{sum}} = 4$ bit/s/Hz and $C_m = \frac{C_{\text{sum}}}{M}$ using the successive point-to-point transmission protocol with different cardinalities $|\mathbb{S}_m|$; $|\mathbb{X}| = 4$, $|\mathbb{Y}_m| = 64$, $|\mathbb{Z}_m| = 4$.

5.2.5. Performance for Different Sum-Rates

Figure 15 illustrates the influence of the sum-rate $C_{\text{sum}} = \sum_{m=1}^{M} C_m$ for a scenario with $M = 5$ sensors. Naturally, larger sum-rates correlate with higher individual link capacities. Again, CEO and fcCEO scenarios provide lower and upper bounds, respectively.

For a cardinality of $|\mathbb{S}_m| = 2$, only a small gain compared to the non-cooperative CEO scenario can be observed. However, the gain gets more and more significant with increasing $|\mathbb{S}_m|$. Comparing the results to the upper fcCEO bound illuminates the loss due to limited available side-information at early transmitting sensors. The largest difference can be observed for sum-rates between $2 \leq C_{\text{sum}} \leq 4\,\text{bit/s/Hz}$.

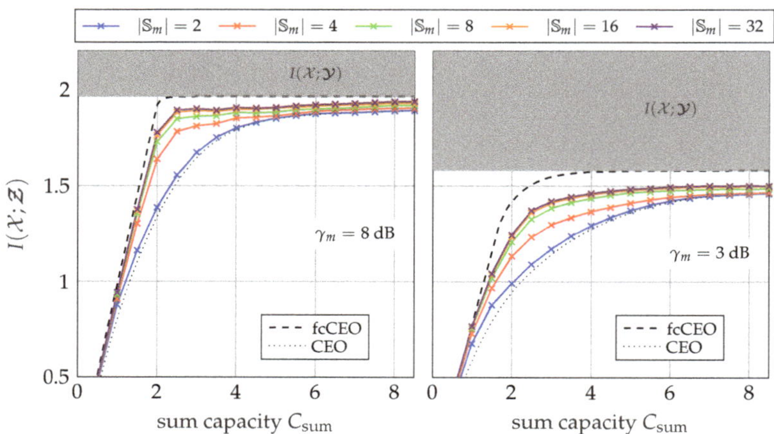

Figure 15. Relevant mutual information $I(\mathcal{X}; \mathcal{Z})$ versus sum-rate C_{sum} with $C_m = \frac{C_{\text{sum}}}{M}$ using the successive point-to-point transmission protocol with different cardinalities $|\mathbb{S}_m|$; $M = 5$, $|\mathbb{X}| = 4$, $|\mathbb{Y}_m| = 64$, $|\mathbb{Z}_m| = 4$.

5.2.6. Asymmetric Scenarios

A very important part is the investigation of asymmetric scenarios. As the achievable relevant information $I(\mathcal{X}; \mathcal{Z})$ of a non-cooperative CEO scenario is very sensitive to the optimization order, i.e., the Wyner–Ziv coding strategy, in asymmetric scenarios [14] the question arises if the exchange of instantaneous side-information can improve the robustness against bad optimization orders. Therefore, the same two asymmetric setups as in [14] are analyzed. Scenario 1 considers the case where sensors with low SNRs γ_m have low link capacities C_m while sensors with high SNRs γ_m have high link capacities C_m. Scenario 2 considers the opposite case, where sensors with low SNRs have high link capacities and vice versa.

Figure 16 illustrates the relevant mutual information $I(\mathcal{X}; \mathcal{Z})$ for all $M! = 24$ sensor permutations for a network of $M = 4$ sensors. The dots represent the results from [14] for a non-cooperative CEO scenario. Blue dots show Scenario 1 while the red dots represent Scenario 2. The results for the pcCEO scenario with successive point-to-point side-information exchange is depicted as bars.

Comparing the non-cooperative case with the successive point-to-point exchange of side-information for Scenario 1, we observe a slight increase of the overall relevant mutual information $I(\mathcal{X}; \mathcal{Z})$ for partial cooperation and this particular scenario. Moreover, the influence of the Wyner–Ziv coding strategy (optimization order) becomes smaller due to cooperation. The performance for Scenario 2 is worse than the performance for Scenario 1, again for both the cooperative and the non-cooperative case. In this scenario, accurate measurements have to be strongly compressed in order to forward them to the common receiver while unreliable measurements cannot contribute much to the overall performance although they can be forwarded to the common receiver at high rates. However, the loss due to bad optimization orders is much lower for partial cooperation. A sensor with a bad forward channel and a high SNR can still forward its information to the next sensor, which might have a better forward channel. Therefore, exchanging instantaneous side-information can improve the robustness against bad optimization orders.

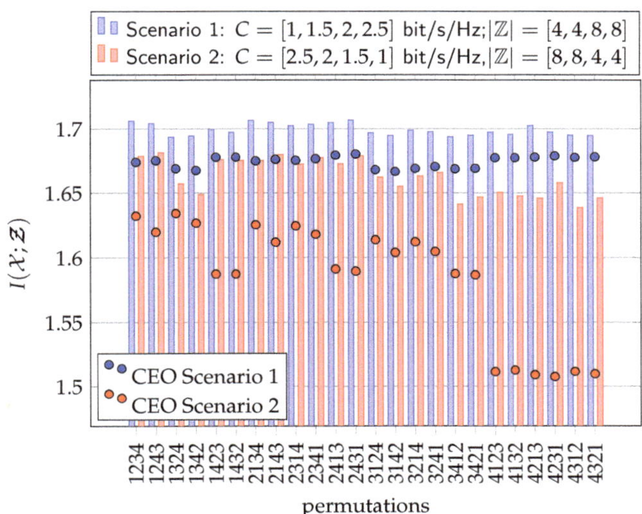

Figure 16. Relevant mutual information $I(\mathcal{X}; \mathcal{Z})$ for non-symmetric scenario with $M = 4$ sensors, SNRs $\gamma_m = [2,4,6,8]$ dB and $|\mathbb{X}| = 4$, $|\mathbb{Y}_m| = 64$, $|\mathbb{Z}_m| = 4$ using the successive point-to-point transmission protocol with $|\mathbb{S}_m| = 8$.

5.3. Two-Phase Transmission Protocol with Artificial Side-Information

Previous subsections revealed that partial cooperation by exchanging instantaneous side-information improves the overall performance. However, a gap to the fcCEO scenario still remains, and we claimed that the successive exchange of instantaneous side-information is the reason for this difference. Due to the sequential forwarding protocols considered so far, early sensors have no or little instantaneous side-information. They hardly profit from the cooperation as opposed to the full cooperation case where all sensors have access to the complete information. In order to substantiate this statement, a third transmission protocol consisting of two phases is considered. Inspired by the fcCEO scenario, the first cooperation phase is used to exchange instantaneous side-information between all sensors, while the transmission phase is used to forward the information to the common receiver in the usual way. The difference to the fcCEO scenario is that only compressed versions of the observations can be exchanged during the cooperation phase.

For simplicity, we assume that each sensor obtains the same instantaneous side-information represented by s^*, independent of its position in the optimization chain, see Figure 17. Moreover, we pursue the EXIT chart philosophy [33], where extrinsic information is artificially created to analyze the information exchange between decoders in concatenated coding schemes. In the pcCEO context, the artificial side-information can be interpreted as extrinsic information about the relevant signal x being generated by adding AWGN to x. The noise variance is adapted to obtain a specific SNR γ_{extr} or equivalently a desired mutual side-information $I(\mathcal{X}; \mathcal{S}^*)$. It has to be emphasized that γ_{extr} can be chosen independently from the measurement SNRs at the sensors in order to obtain general conclusions. Since the instantaneous side-information is created artificially, s^* is assumed to be independent of the indexes y_m given the relevant signal x, i.e., $p(y_m, s^*|x) = p(y_m|x)p(s^*|x)$ holds. This simplifies the Markovian structure of the optimization problem which equals the one of the original CEO problem. With the same argumentation as in the original CEO problem, we claim that the supermodularity holds and the greedy optimization structure is optimal. This model leads to the modified optimization problem

$$L^{(1)}_{\text{GDIB-TP}} = I(\mathcal{X}; \mathcal{Z}_1) - \beta_1 I(\mathcal{Y}_1, \mathcal{S}^*; \mathcal{Z}_1) \tag{29}$$

$$\vdots$$

$$L^{(M)}_{\text{GDIB-TP}} = I(\mathcal{X}; \mathcal{Z}_M | \boldsymbol{Z}_{<M}) - \beta_M I(\mathcal{Y}_M, \mathcal{S}^*; \mathcal{Z}_M | \boldsymbol{Z}_{<M}). \tag{30}$$

Theoptimization problem can be solved using the same strategy as described in previous Sections 5.1 and 5.2. This leads to the implicit update equation

$$p(z_m | y_m, s^*) = \frac{e^{-d_{\beta_m}(y_m, z_m, s^*)}}{\sum_{z_m} e^{-d_{\beta_m}(y_m, z_m, s^*)}} \tag{31}$$

with

$$d_{\beta_m}(y_m, z_m, s^*) := \mathbb{E}_{\boldsymbol{Z}_{<m} | y_m, s^*} \left[\frac{1}{\beta_m} \cdot \right.$$

$$\left. D_{\text{KL}}\left[p(x | y_m, s^*, \boldsymbol{z}_{<m}) \| p(x | \boldsymbol{z}_{\leq m}) \right] - \log p(z_m | \boldsymbol{z}_{<m}) \right]. \tag{32}$$

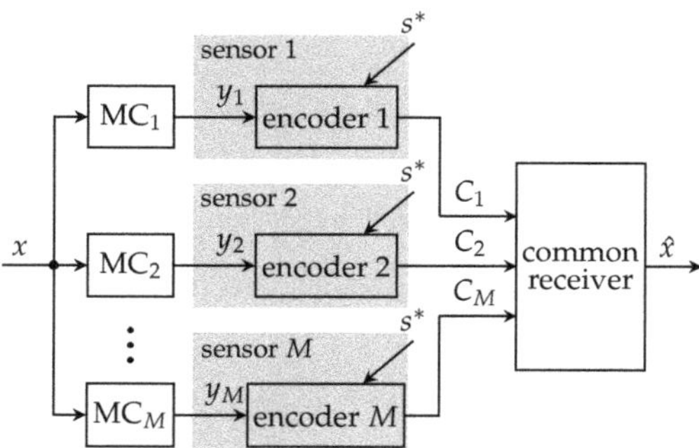

Figure 17. Partially cooperative CEO scenario using the two-phase transmission protocol.

5.3.1. Performance of Two-Phase Transmission

Figure 18 illustrates the same experiment as in Figure 8 or Figure 13, but for the two-phase transmission protocol. The extrinsic information is chosen independent of the measurement SNR and has its own SNR γ_{ext} represented by different colors in Figure 18. The cardinality of the extrinsic information is chosen as $|\mathbb{S}^*| = 512$ to not introduce any compression losses. As before, the black dashed line represents the fcCEO scenario. The curve for $\gamma_{\text{extr}} = \gamma_m = 8$ dB represents the case where each sensor forwards instantaneous side-information whose quality corresponds to its measurement SNR. We observe the same performance as for the fcCEO scenario. This demonstrates that the remaining performance gap to the fcCEO scenario disappears completely for appropriate cooperation among sensors. Naturally, decreasing the SNR of the extrinsic information γ_{ext} or equivalently $I(\mathcal{X}; \mathcal{S}^*)$ leads to a lower overall performance $I(\mathcal{X}; \boldsymbol{Z})$.

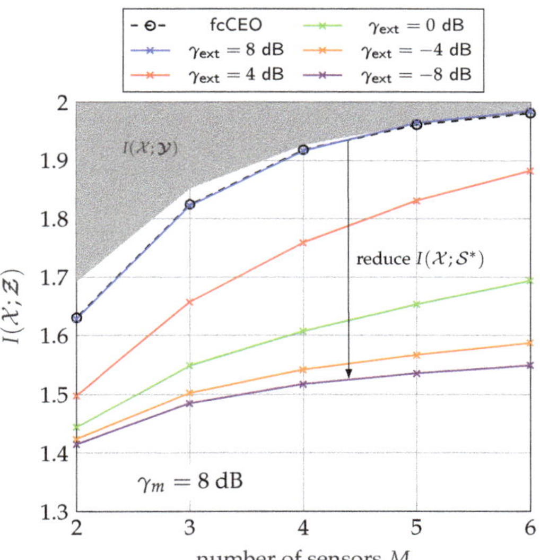

Figure 18. Relevant mutual information $I(\mathcal{X};\mathcal{Z})$ versus the network size for a fixed sum-rate of $C_{\text{sum}} = 2.5$ bit/s/Hz and $C_m = \frac{C_{\text{sum}}}{M}$ using a two-phase transmission protocol for artificially decoupled extrinsic information with different γ_{ext}; $\gamma_m = 8$ dB, $|\mathbb{X}| = 4$, $|\mathbb{Y}_m| = 64$, $|\mathbb{Z}_m| = 4$, $|\mathbb{S}^*| = 512$.

5.3.2. Influence of Extrinsic Information

The influence of extrinsic information is depicted in Figure 19 for $\gamma_m = 8$ dB and $\gamma_m = 3$ dB. Therefore, the overall relevant mutual information $I(\mathcal{X};\mathcal{Z})$ is depicted versus the mutual information of the extrinsic information $I(\mathcal{X};\mathcal{S}^*)$ for different network sizes. As before, all sensors share the same forward channel with $C_{\text{sum}} = 2.5$ bit/s/Hz and $C_m = \frac{C_{\text{sum}}}{M}$. Providing no extrinsic information, i.e., $I(\mathcal{X};\mathcal{S}^*) = 0$ delivers the same result as the non-cooperative CEO scenario of Section 3. Naturally, enhancing the quality of the extrinsic information increases the overall relevant mutual information $I(\mathcal{X};\mathcal{Z})$ up to the maximum of 2 bit/s/Hz.

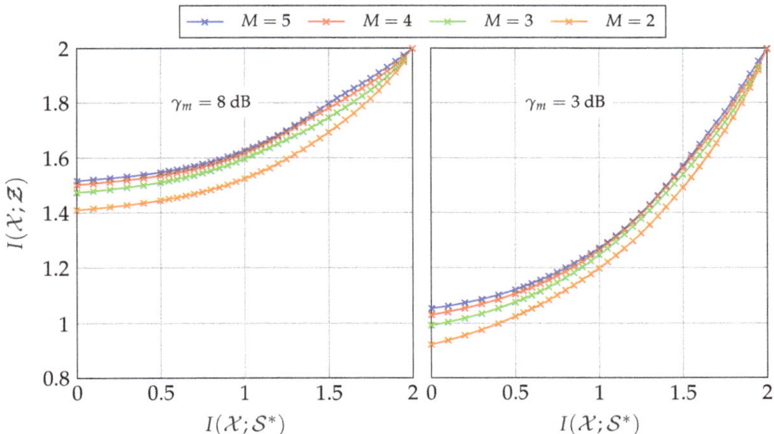

Figure 19. Relevant mutual information $I(\mathcal{X};\mathcal{Z})$ versus extrinsic mutual information $I(\mathcal{X};\mathcal{S}^*)$ for different network sizes and a fixed sum-rate of $C_{\text{sum}} = 2.5$ bit/s/Hz and $C_m = \frac{C_{\text{sum}}}{M}$ and $|\mathbb{X}| = 4$, $|\mathbb{Y}_m| = 64$, $|\mathbb{Z}_m| = 4$.

6. Conclusions

This paper extends the non-cooperative CEO scenario allowing partial cooperation among sensors in the network. Therefore, it extends the algorithmic solution introduced in [14] for three different inter-sensor communication protocols: successive broadcasting, successive point-to-point communication and a two-phase transmission protocol. The first two protocols perform the exchange of instantaneous side-information and forwarding information to the common receiver at the same time step. Therefore, successive broadcasting exploits the instantaneous side-information of all previous sensors within the optimization chain. Since this may cause dimensionality problems during the optimization, the successive point-to-point transmission protocol forwards the instantaneous side-information only to the next sensor. It turns out that allowing this partial communication outperforms the non-cooperative compression where no communication among sensors is possible. Moreover, cooperative compression shows a larger robustness to suboptimal Wyner–Ziv coding strategies in asymmetric scenarios. However, a small performance gap to the fcCEO scenario still remains for the proposed successive broadcasting and successive point-to-point transmission protocols. This gap can be closed by a third protocol separating the cooperation from the forwarding phase and allowing each sensor to access the maximal available side-information. Although no formal conclusion about the optimality of the pcCEO can be drawn, the closeness to the fcCEO scenario in the investigated simulations reveals that solutions found by the proposed greedy algorithms are at least close to optimal.

Author Contributions: Conceptualization, S.S. and V.K.; formal analysis, S.S. and V.K.; investigation, S.S., A.D.A. and V.K.; methodology, S.S.; software, S.S. and A.D.A.; supervision, V.K.; validation, S.S.; visualization, S.S.; writing—original draft, S.S.; writing—review and editing, S.S. and V.K. All authors have read and agreed to the published version of the manuscript.

Funding: This work was supported in part by the University of Rostock, and in part by the German Research Foundation (DFG) in the funding programme Open Access Publishing under Grant 325496636. Parts of the computation were done by using a compute cluster funded by DFG (grant: 440623123).

Institutional Review Board Statement: Not applicable.

Informed Consent Statement: Not applicable.

Data Availability Statement: Not applicable.

Conflicts of Interest: The authors declare no conflict of interest.

Appendix A

Appendix A.1. Optimization for Broadcasting Side-Information

According to Section 5.1, partial cooperation by broadcasting side-information requires the maximization of

$$L^{(m)}_{\text{GDIB-BC}} = I(\mathcal{X}; \mathcal{Z}_m | \boldsymbol{\mathcal{Z}}_{<m}) - \beta_m I(\mathcal{Y}_m, \boldsymbol{S}_{<m}; \mathcal{Z}_m | \boldsymbol{\mathcal{Z}}_{<m}). \tag{A1}$$

The variational problem can be solved by taking the derivative with respect to the mapping $p(z_m | y_m, s_{<m})$ and equating it to zero. In the following, the derivatives of both mutual information are given.

Appendix A.1.1. Derivative of $I(\mathcal{X}; \mathcal{Z}_m | \boldsymbol{\mathcal{Z}}_{<m})$

The relevant mutual information in (A1) can be rewritten such that the desired mapping occurs explicitly.

$$I(\mathcal{X}; \mathcal{Z}_m | \mathbf{\mathcal{Z}}_{<m}) = \mathbb{E}_{\mathcal{X}, \mathcal{Z}_m, \mathbf{\mathcal{Z}}_{<m}} \left[\log \frac{p(z_m | x, \mathbf{z}_{<m})}{p(z_m | \mathbf{z}_{<m})} \right]$$

$$= \sum_{z_m} \sum_{\mathbf{z}_{<m}} \sum_{x} \sum_{y_m} \sum_{\mathbf{s}_{<m}} p(z_m | y_m, \mathbf{s}_{<m}) p(y_m, \mathbf{s}_{<m}, \mathbf{z}_{<m}, x)$$

$$\cdot \log \sum_{a \in \mathbb{Y}_m} \sum_{\mathbf{b} \in \mathbb{S}_{<m}} p(z_m | a, \mathbf{b}) p(a, \mathbf{b} | \mathbf{z}_{<m}, x)$$

$$- \sum_{z_m} \sum_{\mathbf{z}_{<m}} \sum_{x} \sum_{y_m} \sum_{\mathbf{s}_{<m}} p(z_m | y_m, \mathbf{s}_{<m}) p(y_m, \mathbf{s}_{<m}, \mathbf{z}_{<m}, x)$$

$$\cdot \log \sum_{a \in \mathbb{Y}_m} \sum_{\mathbf{b} \in \mathbb{S}_{<m}} p(z_m | a, \mathbf{b}) p(a, \mathbf{b} | \mathbf{z}_{<m}) \tag{A2}$$

The derivative of (A2) delivers

$$\frac{\partial I(\mathcal{X}; \mathcal{Z}_m | \mathbf{\mathcal{Z}}_{<m})}{\partial p(z_m | y_m, \mathbf{s}_{<m})}$$

$$= \sum_{x} \sum_{\mathbf{z}_{<m}} p(y_m, \mathbf{s}_{<m}, \mathbf{z}_{<m}, x) \cdot \log p(z_m | x, \mathbf{z}_{<m})$$

$$+ \sum_{x} \sum_{\mathbf{z}_{<m}} \underbrace{\left[\sum_{\mathbf{s}_{<m}} \sum_{y_m} p(z_m | y_m, \mathbf{s}_{<m}) p(y_m, \mathbf{s}_{<m}, \mathbf{z}_{<m}, x) \right]}_{= p(z_m, \mathbf{z}_{<m}, x)} \cdot \frac{p(y_m, \mathbf{s}_{<m} | x, \mathbf{z}_{<m})}{p(z_m | x, \mathbf{z}_{<m})}$$

$$- \sum_{x} \sum_{\mathbf{z}_{<m}} p(y_m, \mathbf{s}_{<m}, \mathbf{z}_{<m}, x) \cdot \log p(z_m | \mathbf{z}_{<m})$$

$$- \sum_{x} \sum_{\mathbf{z}_{<m}} \underbrace{\left[\sum_{\mathbf{s}_{<m}} \sum_{y_m} p(z_m | y_m, \mathbf{s}_{<m}) p(y_m, \mathbf{s}_{<m}, \mathbf{z}_{<m}, x) \right]}_{= p(z_m, \mathbf{z}_{<m}, x)} \cdot \frac{p(y_m, \mathbf{s}_{<m} | \mathbf{z}_{<m})}{p(z_m | \mathbf{z}_{<m})}$$

$$= \sum_{\mathbf{z}_{<m}} p(y_m, \mathbf{s}_{<m}, \mathbf{z}_{<m}) \sum_{x} p(x | y_m, \mathbf{s}_{<m}, \mathbf{z}_{<m}) \log \frac{p(z_m | x, \mathbf{z}_{<m})}{p(z_m | \mathbf{z}_{<m})} . \tag{A3}$$

Exploiting Bayes' theorem, the argument of the logarithmic function can be rewritten

$$\frac{p(z_m | x, \mathbf{z}_{<m})}{p(z_m | \mathbf{z}_{<m})} = \frac{p(x | \mathbf{z}_{\leq m})}{p(x | \mathbf{z}_{<m})}$$

$$= \frac{p(x | \mathbf{z}_{\leq m})}{p(x | y_m, \mathbf{s}_{<m}, \mathbf{z}_{<m})} \cdot \frac{p(x | y_m, \mathbf{s}_{<m}, \mathbf{z}_{<m})}{p(x | \mathbf{z}_{<m})} . \tag{A4}$$

The last ratio in (A4) can be dropped because it does not depend on $p(z_m | y_m, \mathbf{s}_{<m})$ and its contribution can be incorporated into the Lagrange multiplier β_m. The insertion of the first ratio into (A3) yields the contribution of the derivative of the relevant mutual information

$$\frac{\partial I(\mathcal{X}; \mathcal{Z}_m | \mathbf{\mathcal{Z}}_{<m})}{\partial p(z_m | y_m, \mathbf{s}_{<m})} \rightarrow - \sum_{\mathbf{z}_{<m}} p(y_m, \mathbf{s}_{<m}, \mathbf{z}_{<m}) \sum_{x} p(x | y_m, \mathbf{s}_{<m}, \mathbf{z}_{<m}) \log \frac{p(x | y_m, \mathbf{s}_{<m}, \mathbf{z}_{<m})}{p(x | \mathbf{z}_{\leq m})}$$

$$= - \sum_{\mathbf{z}_{<m}} p(y_m, \mathbf{s}_{<m}, \mathbf{z}_{<m}) D_{\mathrm{KL}}[p(x | y_m, \mathbf{s}_{<m}, \mathbf{z}_{<m}) \| p(x | \mathbf{z}_{\leq m})] . \tag{A5}$$

Appendix A.1.2. Derivative of $I(\mathcal{Y}_m, \boldsymbol{S}_{<m}; \mathcal{Z}_m | \boldsymbol{\mathcal{Z}}_{<m})$

With the definition of the conditional compression rate

$$
\begin{aligned}
I(\mathcal{Y}_m, \boldsymbol{S}_{<m}; \mathcal{Z}_m | \boldsymbol{\mathcal{Z}}_{<m}) &= \mathbb{E}_{y_m, z_m, \boldsymbol{z}_{<m}, \boldsymbol{S}_{<m}} \left[\log \frac{p(z_m | y_m, \mathbf{s}_{<m}, \mathbf{z}_{<m})}{p(z_m | \mathbf{z}_{<m})} \right] \\
&= \sum_{z_m} \sum_{\mathbf{z}_{<m}} \sum_{\mathbf{s}_{<m}} \sum_{y_m} p(z_m | y_m, \mathbf{s}_{<m}) p(y_m, \mathbf{s}_{<m}, \mathbf{z}_{<m}) \log p(z_m | y_m, \mathbf{s}_{<m}) \\
&\quad - \sum_{z_m} \sum_{\mathbf{z}_{<m}} \sum_{\mathbf{s}_{<m}} \sum_{y_m} p(z_m | y_m, \mathbf{s}_{<m}) p(y_m, \mathbf{s}_{<m}, \mathbf{z}_{<m}) \log \sum_{a \in \mathbb{Y}_m} \sum_{b \in \mathbb{S}_{<m}} p(z_m | a, \mathbf{b}) p(a, \mathbf{b} | \mathbf{z}_{<m}) \quad \text{(A6)}
\end{aligned}
$$

its derivative becomes

$$
\begin{aligned}
\frac{\partial I(\mathcal{Y}_m, \boldsymbol{S}_{<m}; \mathcal{Z}_m | \boldsymbol{\mathcal{Z}}_{<m})}{\partial p(z_m | y_m, \mathbf{s}_{<m})} \\
= \sum_{\mathbf{z}_{<m}} p(y_m, \mathbf{s}_{<m}, \mathbf{z}_{<m}) \log p(z_m | y_m, \mathbf{s}_{<m}) \\
+ \sum_{\mathbf{z}_{<m}} p(y_m, \mathbf{s}_{<m}, \mathbf{z}_{<m}) \frac{p(z_m | y_m, \mathbf{s}_{<m})}{p(z_m | y_m, \mathbf{s}_{<m})} \\
- \sum_{\mathbf{z}_{<m}} p(y_m, \mathbf{s}_{<m}, \mathbf{z}_{<m}) \log p(z_m | \mathbf{z}_{<m}) \\
- \sum_{\mathbf{z}_{<m}} \underbrace{\left[\sum_{y_m} \sum_{\mathbf{s}_{<m}} p(z_m | y_m, \mathbf{s}_{<m}) p(y_m, \mathbf{s}_{<m}, \mathbf{z}_{<m}) \right]}_{= p(z_m, \mathbf{z}_{<m})} \cdot \frac{p(y_m, \mathbf{s}_{<m} | \mathbf{z}_{<m})}{p(z_m | \mathbf{z}_{<m})} \\
= \sum_{\mathbf{z}_{<m}} p(y_m, \mathbf{s}_{<m}, \mathbf{z}_{<m}) \log \frac{p(z_m | y_m, \mathbf{s}_{<m})}{p(z_m | \mathbf{z}_{<m})} \, . \quad \text{(A7)}
\end{aligned}
$$

Appendix A.1.3. Fusion of Derived Parts

Combining the result in (A5) and (A7) delivers the complete derivative

$$
\begin{aligned}
&- \sum_{\mathbf{z}_{<m}} p(y_m, \mathbf{s}_{<m}, \mathbf{z}_{<m}) D_{\text{KL}}[p(x | y_m, \mathbf{s}_{<m}, \mathbf{z}_{<m}) \| p(x | \mathbf{z}_{\leq m})] \\
&- \beta_m \sum_{\mathbf{z}_{<m}} p(y_m, \mathbf{s}_{<m}, \mathbf{z}_{<m}) \log p(z_m | y_m, \mathbf{s}_{<m}) \\
&+ \beta_m \sum_{\mathbf{z}_{<m}} p(y_m, \mathbf{s}_{<m}, \mathbf{z}_{<m}) \cdot \log p(z_m | \mathbf{z}_{<m}) = 0 \, . \quad \text{(A8)}
\end{aligned}
$$

Following the idea of Blahut and Arimoto [34], $p(x | \mathbf{z}_{\leq m})$ and $p(z_m | \mathbf{z}_{<m})$ are assumed to be independent of $p(z_m | y_m, \mathbf{s}_{<m})$. With this trick, (A8) can be resolved with respect to the desired mapping of sensor m leading to the implicit solution

$$
p(z_m | y_m, \mathbf{s}_{<m}) = \frac{e^{-d_{\beta_m}(y_m, z_m, \mathbf{s}_{<m})}}{\sum_{z_m} e^{-d_{\beta_m}(y_m, z_m, \mathbf{s}_{<m})}} \quad \text{(A9)}
$$

with

$$
\begin{aligned}
d_{\beta_m}(y_m, z_m, \mathbf{s}_{<m}) \\
:= \sum_{\mathbf{z}_{<m}} p(\mathbf{z}_{<m} | y_m, \mathbf{s}_{<m}) \left[\frac{1}{\beta_m} D_{\text{KL}}[p(x | y_m, \mathbf{s}_{<m}, \mathbf{z}_{<m}) \| p(x | \mathbf{z}_{\leq m})] - \log p(z_m | \mathbf{z}_{<m}) \right] \\
= \mathbb{E}_{\boldsymbol{z}_{<m} | y_m, \mathbf{s}_{<m}} \left[\frac{1}{\beta_m} D_{\text{KL}}[p(x | y_m, \mathbf{s}_{<m}, \mathbf{z}_{<m}) \| p(x | \mathbf{z}_{\leq m})] - \log p(z_m | \mathbf{z}_{<m}) \right] \, . \quad \text{(A10)}
\end{aligned}
$$

Appendix A.1.4. Calculating Required pmfs

This section covers the calculation of the required pmfs for the previously described algorithm. The first term in the KL divergence in (A10) is defined by

$$p(x|y_m, \mathbf{s}_{<m}, \mathbf{z}_{<m}) = \frac{p(x, y_m, \mathbf{s}_{<m}, \mathbf{z}_{<m})}{\sum_x p(x, y_m, \mathbf{s}_{<m}, \mathbf{z}_{<m})} \tag{A11}$$

where $p(x, y_m, \mathbf{s}_{<m}, \mathbf{z}_{<m})$ is determined recursively as

$$p(x, y_m, \mathbf{s}_{<m}, \mathbf{z}_{<m}) = \sum_{y_{m-1}} p(z_{m-1}|y_{m-1}, \mathbf{s}_{<m-1}) \cdot$$
$$p(s_{m-1}|y_{m-1}, \mathbf{s}_{<m-1}) p(y_m|x) p(x, y_{m-1}, \mathbf{s}_{<m-1}, \mathbf{z}_{<m-1}) . \tag{A12}$$

In (A12), the pmf $p(z_{m-1}|y_{m-1}, \mathbf{s}_{<m-1})$ represents the quantizer mapping of the previous sensor $m - 1$ while $p(s_{m-1}|y_{m-1}, \mathbf{s}_{<m-1})$ denotes its the mapping for the instantaneous side-information. Hence, both have already been determined when optimizing sensor $m - 1$ leading to a recursive computation. The second term in the KL divergence in (A10) is calculated by

$$p(x|\mathbf{z}_{\leq m}) = \frac{p(\mathbf{z}_{\leq m}, x)}{\sum_x p(\mathbf{z}_{\leq m}, x)} \tag{A13}$$

where $p(\mathbf{z}_{\leq m}, x)$ can be calculated by

$$p(\mathbf{z}_{\leq m}, x) = \sum_{\mathbf{s}_{<m}} \sum_{y_m} p(z_m|y_m, \mathbf{s}_{<m}) p(x, y_m, \mathbf{s}_{<m}, \mathbf{z}_{<m}) \tag{A14}$$

with $p(z_m|y_m, \mathbf{s}_{<m})$ being the quantizer mapping of the current sensor. The joint pmf $p(x, y_m, \mathbf{s}_{<m}, \mathbf{z}_{<m})$ has already been calculated in (A12). The argument of the logarithm in (A10) can be derived as

$$p(z_m|\mathbf{z}_{<m}) = \frac{\sum_x p(\mathbf{z}_{\leq m}, x)}{\sum_{z_m} \sum_x p(\mathbf{z}_{\leq m}, x)} . \tag{A15}$$

Finally, the pmf to calculate the conditional expectation in (A10) is determined as

$$p(\mathbf{z}_{<m}|y_m, \mathbf{s}_{<m}) = \frac{\sum_x p(x, y_m, \mathbf{s}_{<m}, \mathbf{z}_{<m})}{\sum_{\mathbf{z}_{<m}} \sum_x p(x, y_m, \mathbf{s}_{<m}, \mathbf{z}_{<m})} . \tag{A16}$$

Note that all above equations simplify to the scalar IB equations given in Section 2 when optimizing the first sensor.

Appendix A.2. Optimization for Point-to-Point Exchange of Side-Information

According to Section 5.2, partial cooperation using a point-to-point communication protocol requires the maximization of

$$L_{\text{GDIB-PTP}}^{(m)} = I(\mathcal{X}; \mathcal{Z}_m|\mathbf{\mathcal{Z}}_{<m}) - \beta_m I(\mathcal{Y}_m, \mathcal{S}_{m-1}; \mathcal{Z}_m|\mathbf{\mathcal{Z}}_{<m}). \tag{A17}$$

The variational problem can be solved by taking the derivative with respect to the mapping $p(z_m|y_m, s_{m-1})$ and equating it to zero. In the following, the derivatives of both mutual information are given.

Appendix A.2.1. Derivative of $I(\mathcal{X}; \mathcal{Z}_m | \boldsymbol{\mathcal{Z}}_{<m})$

The relevant mutual information in (A17) can be rewritten such that the desired mapping occurs explicitly.

$$
\begin{aligned}
I(\mathcal{X}; \mathcal{Z}_m | \boldsymbol{\mathcal{Z}}_{<m}) &= \mathbb{E}_{\mathcal{X}, \mathcal{Z}_m, \boldsymbol{\mathcal{Z}}_{<m}} \left[\log \frac{p(z_m | x, \mathbf{z}_{<m})}{p(z_m | \mathbf{z}_{<m})} \right] \\
&= \sum_{\mathbf{z}_{\le m}} \sum_{x} \sum_{y_m} \sum_{s_{m-1}} p(z_m | y_m, s_{m-1}) p(y_m, s_{m-1}, \mathbf{z}_{<m}, x) \\
&\quad \cdot \log \sum_{a \in \mathbb{Y}_m} \sum_{b \in \mathbb{S}_{m-1}} p(z_m | a, b) p(a, b | \mathbf{z}_{<m}, x) \\
&\quad - \sum_{\mathbf{z}_{\le m}} \sum_{x} \sum_{y_m} \sum_{s_{m-1}} p(z_m | y_m, s_{m-1}) p(y_m, s_{m-1}, \mathbf{z}_{<m}, x) \\
&\quad \cdot \log \sum_{a \in \mathbb{Y}_m} \sum_{b \in \mathbb{S}_{m-1}} p(z_m | a, b) p(a, b | \mathbf{z}_{<m})
\end{aligned}
\tag{A18}
$$

The derivative of (A18) delivers

$$
\begin{aligned}
&\frac{\partial I(\mathcal{X}; \mathcal{Z}_m | \boldsymbol{\mathcal{Z}}_{<m})}{\partial p(z_m | y_m, s_{m-1})} \\
&= \sum_{x} \sum_{\mathbf{z}_{<m}} p(y_m, s_{m-1}, \mathbf{z}_{<m}, x) \cdot \log p(z_m | x, \mathbf{z}_{<m}) \\
&\quad + \sum_{x} \sum_{\mathbf{z}_{<m}} \underbrace{\left[\sum_{s_{m-1}} \sum_{y_m} p(z_m | y_m, s_{m-1}) p(y_m, s_{m-1}, \mathbf{z}_{<m}, x) \right]}_{= p(z_m, \mathbf{z}_{<m}, x)} \cdot \frac{p(y_m, s_{m-1} | x, \mathbf{z}_{<m})}{p(z_m | x, \mathbf{z}_{<m})} \\
&\quad - \sum_{x} \sum_{\mathbf{z}_{<m}} p(y_m, s_{m-1}, \mathbf{z}_{<m}, x) \cdot \log p(z_m | \mathbf{z}_{<m}) \\
&\quad - \sum_{x} \sum_{\mathbf{z}_{<m}} \underbrace{\left[\sum_{s_{m-1}} \sum_{y_m} p(z_m | y_m, s_{m-1}) p(y_m, s_{m-1}, \mathbf{z}_{<m}, x) \right]}_{= p(z_m, \mathbf{z}_{<m}, x)} \cdot \frac{p(y_m, s_{m-1} | \mathbf{z}_{<m})}{p(z_m | \mathbf{z}_{<m})} \\
&= \sum_{x} \sum_{\mathbf{z}_{<m}} p(y_m, s_{m-1}, \mathbf{z}_{<m}, x) \cdot \log \frac{p(z_m | x, \mathbf{z}_{<m})}{p(z_m | \mathbf{z}_{<m})} \\
&= \sum_{\mathbf{z}_{<m}} p(y_m, s_{m-1}, \mathbf{z}_{<m}) \sum_{x} p(x | y_m, s_{m-1}, \mathbf{z}_{<m}) \log \frac{p(z_m | x, \mathbf{z}_{<m})}{p(z_m | \mathbf{z}_{<m})} .
\end{aligned}
\tag{A19}
$$

Exploiting Bayes' theorem the argument of the logarithmic function can be rewritten

$$
\begin{aligned}
\frac{p(z_m | x, \mathbf{z}_{<m})}{p(z_m | \mathbf{z}_{<m})} &= \frac{p(x | \mathbf{z}_{\le m})}{p(x | \mathbf{z}_{<m})} \\
&= \frac{p(x | \mathbf{z}_{\le m})}{p(x | y_m, s_{m-1}, \mathbf{z}_{<m})} \frac{p(x | y_m, s_{m-1}, \mathbf{z}_{<m})}{p(x | \mathbf{z}_{<m})} .
\end{aligned}
\tag{A20}
$$

The last ratio in (A20) can be dropped because it does not depend on $p(z_m | y_m, s_{m-1})$ and its contribution can be incorporated into the Lagrange multiplier β_m. The insertion of the first ratio into (A19) yields the contribution of the derivative of the relevant mutual information

$$
\begin{aligned}
\frac{\partial I(\mathcal{X}; \mathcal{Z}_m | \boldsymbol{\mathcal{Z}}_{<m})}{\partial p(z_m | y_m, s_{m-1})} &\rightarrow - \sum_{\mathbf{z}_{<m}} p(y_m, s_{m-1}, \mathbf{z}_{<m}) \sum_{x} p(x | y_m, s_{m-1}, \mathbf{z}_{<m}) \log \frac{p(x | y_m, s_{m-1}, \mathbf{z}_{<m})}{p(x | \mathbf{z}_{\le m})} \\
&= - \sum_{\mathbf{z}_{<m}} p(y_m, s_{m-1}, \mathbf{z}_{<m}) D_{\mathrm{KL}}[p(x | y_m, s_{m-1}, \mathbf{z}_{<m}) \| p(x | \mathbf{z}_{\le m})] .
\end{aligned}
\tag{A21}
$$

Appendix A.2.2. Derivative of $I(\mathcal{Y}_m, \mathcal{S}_{m-1}; \mathcal{Z}_m | \mathcal{Z}_{<m})$

With the definition of the conditional compression rate

$$
\begin{aligned}
I(\mathcal{Y}_m, \mathcal{S}_{m-1}; \mathcal{Z}_m | \mathcal{Z}_{<m}) &= \mathbb{E}_{y_m, z_m, \mathbf{z}_{<m}, s_{m-1}} \left[\log \frac{p(z_m | y_m, s_{m-1}, \mathbf{z}_{<m})}{p(z_m | \mathbf{z}_{<m})} \right] \\
&= \sum_{z_m} \sum_{\mathbf{z}_{<m}} \sum_{s_{m-1}} \sum_{y_m} p(z_m | y_m, s_{m-1}) p(y_m, s_{m-1}, \mathbf{z}_{<m}) \log p(z_m | y_m, s_{m-1}) \\
&- \sum_{z_m} \sum_{\mathbf{z}_{<m}} \sum_{s_{m-1}} \sum_{y_m} p(z_m | y_m, s_{m-1}) p(y_m, s_{m-1}, \mathbf{z}_{<m}) \log \sum_{a \in \mathbb{Y}_m} \sum_{b \in \mathbb{S}_{m-1}} p(z_m | a, b) p(a, b | \mathbf{z}_{<m})
\end{aligned}
\tag{A22}
$$

its derivative becomes

$$
\begin{aligned}
\frac{\partial I(\mathcal{Y}_m, \mathcal{S}_{m-1}; \mathcal{Z}_m | \mathcal{Z}_{<m})}{\partial p(z_m | y_m, s_{m-1})} \\
= \sum_{\mathbf{z}_{<m}} p(y_m, s_{m-1}, \mathbf{z}_{<m}) \log p(z_m | y_m, s_{m-1}) \\
+ \sum_{\mathbf{z}_{<m}} p(y_m, s_{m-1}, \mathbf{z}_{<m}) \frac{p(z_m | y_m, s_{m-1})}{p(z_m | y_m, s_{m-1})} \\
- \sum_{\mathbf{z}_{<m}} p(y_m, s_{m-1}, \mathbf{z}_{<m}) \log p(z_m | \mathbf{z}_{<m}) \\
- \sum_{\mathbf{z}_{<m}} \underbrace{\left[\sum_{y_m} \sum_{s_{m-1}} p(z_m | y_m, s_{m-1}) p(y_m, s_{m-1}, \mathbf{z}_{<m}) \right]}_{= p(z_m, \mathbf{z}_{<m})} \cdot \frac{p(y_m, s_{m-1} | \mathbf{z}_{<m})}{p(z_m | \mathbf{z}_{<m})} \\
= \sum_{\mathbf{z}_{<m}} p(y_m, s_{m-1}, \mathbf{z}_{<m}) \log \frac{p(z_m | y_m, s_{m-1})}{p(z_m | \mathbf{z}_{<m})} \; .
\end{aligned}
\tag{A23}
$$

Appendix A.2.3. Fusion of Derived Parts

Combining the result in (A21) and (A23) delivers the complete derivative

$$
\begin{aligned}
&- \sum_{\mathbf{z}_{<m}} p(y_m, s_{m-1}, \mathbf{z}_{<m}) D_{\mathrm{KL}}[p(x | y_m, s_{m-1}, \mathbf{z}_{<m}) \| p(x | \mathbf{z}_{\leq m})] \\
&- \beta_m \sum_{\mathbf{z}_{<m}} p(y_m, s_{m-1}, \mathbf{z}_{<m}) \log p(z_m | y_m, s_{m-1}) \\
&+ \beta_m \sum_{\mathbf{z}_{<m}} p(y_m, s_{m-1}, \mathbf{z}_{<m}) \cdot \log p(z_m | \mathbf{z}_{<m}) = 0 \; .
\end{aligned}
\tag{A24}
$$

Following the idea of Blahut and Arimoto [34], $p(x | \mathbf{z}_{\leq m})$ and $p(z_m | \mathbf{z}_{<m})$ are assumed to be independent of $p(z_m | y_m, s_{m-1})$. With this trick, (A24) can be resolved with respect to the desired mapping of sensor m leading to the implicit solution

$$
p(z_m | y_m, s_{m-1}) = \frac{e^{-d_{\beta_m}(y_m, z_m, s_{m-1})}}{\sum_{z_m} e^{-d_{\beta_m}(y_m, z_m, s_{m-1})}}
\tag{A25}
$$

with

$$
\begin{aligned}
d_{\beta_m}&(y_m, z_m, s_{m-1}) \\
&:= \sum_{\mathbf{z}_{<m}} p(\mathbf{z}_{<m} | y_m, s_{m-1}) \left[\frac{1}{\beta_m} D_{\mathrm{KL}}[p(x | y_m, s_{m-1}, \mathbf{z}_{<m}) \| p(x | \mathbf{z}_{\leq m})] - \log p(z_m | \mathbf{z}_{<m}) \right] \\
&= \mathbb{E}_{\mathbf{z}_{<m} | y_m, s_{m-1}} \left[\frac{1}{\beta_m} D_{\mathrm{KL}}[p(x | y_m, s_{m-1}, \mathbf{z}_{<m}) \| p(x | \mathbf{z}_{\leq m})] - \log p(z_m | \mathbf{z}_{<m}) \right] \; .
\end{aligned}
\tag{A26}
$$

Appendix A.2.4. Calculating Required pmfs

This section covers the calculation of the required pmfs for the previously described algorithm. The first term in the KL divergence in (A26) is defined by

$$p(x|y_m, s_{m-1}, \mathbf{z}_{<m}) = \frac{p(x, y_m, s_{m-1}, \mathbf{z}_{<m})}{\sum_x p(x, y_m, s_{m-1}, \mathbf{z}_{<m})} \tag{A27}$$

with

$$p(x, y_m, s_{m-1}, \mathbf{z}_{<m}) = p(z_{m-1}, s_{m-1}|x, \mathbf{z}_{<m-1})p(x, y_m, \mathbf{z}_{<m-1}) . \tag{A28}$$

The first term on the right hand side in (A28) can be calculated by

$$p(z_{m-1}, s_{m-1}|x, \mathbf{z}_{<m-1}) = \sum_{y_{m-1}} \sum_{s_{m-2}} p(z_{m-1}|y_{m-1}, s_{m-2})$$
$$p(s_{m-1}|y_{m-1}, s_{m-2})p(y_{m-1}|x)p(s_{m-2}|x, \mathbf{z}_{<m-1}) \tag{A29}$$

where $p(z_{m-1}|y_{m-1}, s_{m-2})$ represents the mapping of previously designed quantizer $m - 1$, $p(s_{m-1}|y_{m-1}, s_{m-2})$ denotes its mapping of instantaneous side-information and $p(y_{m-1}|x)$ statistically describes the corresponding known measurement channel. The last term $p(s_{m-2}|x, \mathbf{z}_{<m-1})$ has already been calculated when optimizing previous sensors (in this case the pre-predecessor) in a recursive way

$$p(s_m|x, \mathbf{z}_{<m}) = \sum_{y_m} \sum_{s_{m-1}} p(s_m|y_m, s_{m-1})p(y_m|x, z_m)p(s_{m-1}|x, \mathbf{z}_{<m-1}) \tag{A30}$$

with

$$p(y_m|x, z_m) = \frac{p(y_m, x, z_m)}{\sum_{y_m} p(y_m, x, z_m)} \tag{A31}$$

and

$$p(y_m, x, z_m) = \sum_{s_{m-1}} p(z_m|y_m, s_{m-1})p(y_m|x)p(s_{m-1}|x)p(x) . \tag{A32}$$

The second term on the right hand side in (A28) can be determined by

$$p(x, y_m, \mathbf{z}_{<m-1}) = p(y_m|x)p(\mathbf{z}_{<m-1}, x) . \tag{A33}$$

Again, the last term in (A33) is obtained as side product when optimizing the previous sensor by

$$p(\mathbf{z}_{\leq m}, x) = \sum_{y_m} \sum_{s_{m-1}} p(z_m|y_m, s_{m-1})p(z_{m-1}, s_{m-1}|x, \mathbf{z}_{<m-1})p(x, y_m, \mathbf{z}_{<m-1}) \tag{A34}$$

where $p(z_{m-1}, s_{m-1}|x, \mathbf{z}_{<m-1})$ and $p(x, y_m, \mathbf{z}_{<m-1})$ are already defined in (A29) and (A33), respectively. The second term in the KL divergence in (A26) is calculated by

$$p(x|\mathbf{z}_{\leq m}) = \frac{p(\mathbf{z}_{\leq m}, x)}{\sum_x p(\mathbf{z}_{\leq m}, x)} \tag{A35}$$

where $p(\mathbf{z}_{\leq m}, x)$ is already defined in (A34). The term in the logarithm in (A26) can be expressed as

$$p(z_m|\mathbf{z}_{<m}) = \frac{\sum_x p(\mathbf{z}_{\leq m}, x)}{\sum_x \sum_{z_m} p(\mathbf{z}_{\leq m}, x)} . \tag{A36}$$

Finally, the required pmf to calculate the conditional expectation in (A26) can be determined by

$$p(\mathbf{z}_{<m}|y_m, s_{m-1}) = \frac{\sum_x p(x, y_m, s_{m-1}, \mathbf{z}_{<m})}{\sum_x \sum_{\mathbf{z}_{<m}} p(x, y_m, s_{m-1}, \mathbf{z}_{<m})} \tag{A37}$$

with $p(x, y_m, s_{m-1}, \mathbf{z}_{<m})$ being already defined in (A28). Note that all above equations simplify to the scalar IB equations given in Section 2 when optimizing the first sensor. Moreover, when optimizing the second sensor, there is no pre-predecessor $m - 2$ and its impact on the above equations can be omitted.

References

1. Oohama, Y. The rate-distortion function for the quadratic Gaussian CEO problem. *IEEE Trans. Inf. Theory* **1998**, *44*, 1057–1070. [CrossRef]
2. Viswanathan, H.; Berger, T. The Quadratic Gaussian CEO Problem. *IEEE Trans. Inf. Theory* **1997**, *43*, 1549–1559. [CrossRef]
3. Chen, J.; Zhang, X.; Berger, T.; Wicker, S.B. An upper bound on the sum-rate distortion function and its corresponding rate allocation schemes for the CEO problem. *IEEE J. Sel. Areas Commun.* **2004**, *22*, 977–987. [CrossRef]
4. Prabhakaran, V.; Tse, D.; Ramachandran, K. Rate region of the quadratic Gaussian CEO problem. In Proceedings of the International Symposium on Information Theory (ISIT 2004), Chicago, IL, USA, 27 June–2 July 2004; p. 119. [CrossRef]
5. Oohama, Y. Rate-distortion theory for Gaussian multiterminal source coding systems with several side informations at the decoder. *IEEE Trans. Inf. Theory* **2005**, *51*, 2577–2593. [CrossRef]
6. Wagner, A.; Tavildar, S.; Viswanath, P. Rate Region of the Quadratic Gaussian Two-Encoder Source-Coding Problem. *IEEE Trans. Inf. Theory* **2008**, *54*, 1938–1961. [CrossRef]
7. Ugur, Y.; Aguerri, I.E.; Zaidi, A. Vector Gaussian CEO problem under logarithmic loss. In Proceedings of the 2018 IEEE Information Theory Workshop (ITW), Guangzhou, China, 25–29 November 2018; pp. 1–5.
8. Uğur, Y.; Aguerri, I.E.; Zaidi, A. Vector Gaussian CEO Problem Under Logarithmic Loss and Applications. *IEEE Trans. Inf. Theory* **2020**, *66*, 4183–4202. [CrossRef]
9. Courtade, T.A.; Weissman, T. Multiterminal Source Coding Under Logarithmic Loss. *IEEE Trans. Inf. Theory* **2014**, *60*, 740–761. [CrossRef]
10. Berger, T.; Zhang, Z.; Viswanathan, H. The CEO Problem [Multiterminal Source Coding]. *IEEE Trans. Inf. Theory* **1996**, *42*, 887–902. [CrossRef]
11. Eswaran, K.; Gastpar, M. Remote Source Coding under Gaussian Noise: Dueling Roles of Power and Entropy Power. *arXiv* **2018**, arXiv:1805.06515v2.
12. Zaidi, A.; Aguerri, I.E. Distributed Deep Variational Information Bottleneck. In Proceedings of the 2020 IEEE 21st International Workshop on Signal Processing Advances in Wireless Communications (SPAWC), Atlanta, GA, USA, 26–29 May 2020; pp. 1–5. [CrossRef]
13. Aguerri, I.E.; Zaidi, A. Distributed Variational Representation Learning. *IEEE Trans. Pattern Anal. Mach. Intell.* **2021**, *43*, 120–138. [CrossRef] [PubMed]
14. Steiner, S.; Kuehn, V.; Stark, M.; Bauch, G. Reduced-Complexity Optimization of Distributed Quantization Using the Information Bottleneck Principle. *IEEE Open J. Commun. Soc.* **2021**, *2*, 1267–1278. [CrossRef]
15. Steiner, S.; Kuehn, V. Distributed Compression using the Information Bottleneck Principle. In Proceedings of the ICC 2021—IEEE International Conference on Communications, Montreal, QC, Canada, 14–23 June 2021; pp. 1–6. [CrossRef]
16. Estella Aguerri, I.; Zaidi, A. Distributed information bottleneck method for discrete and Gaussian sources. In Proceedings of the International Zurich Seminar on Information and Communication (IZS 2018) Proceedings, Zurich, Switzerland, 21–23 February 2018; ETH Zurich: Zurich, Switzerland, 2018; pp. 35–39.
17. Uğur, Y.; Aguerri, I.E.; Zaidi, A. A generalization of blahut-arimoto algorithm to compute rate-distortion regions of multiterminal source coding under logarithmic loss. In Proceedings of the 2017 IEEE Information Theory Workshop (ITW), Kaohsiung, Taiwan, 6–10 November 2017; pp. 349–353.
18. Prabhakaran, V.; Ramchandran, K.; Tse, D. On the Role of Interaction Between Sensors in the CEO Problem. In Proceedings of the 42nd Annual Allerton Conference on Communication, Control, and Computing, Monticello, IL, USA, 29 September–1 October 2004.
19. Draper, S.; Wornell, G. Side Information Aware Coding Strategies for Sensor Networks. *IEEE J. Sel. Areas Commun.* **2004**, *22*, 966–976. [CrossRef]
20. Simeone, O. Source and Channel Coding for Homogeneous Sensor Networks with Partial Cooperation. *IEEE Trans. Wirel. Commun.* **2009**, *8*, 1113–1117. [CrossRef]
21. Permuter, H.; Steinberg, Y.; Weissman, T. Problems we can solve with a helper. In Proceedings of the 2009 IEEE Information Theory Workshop on Networking and Information Theory, Volos, Greece, 10–12 June 2009; pp. 266–270. [CrossRef]

22. Tishby, N.; Pereira, F.C.; Bialek, W. The Information Bottleneck Method. In Proceedings of the 37th Annual Allerton Conference on Communication, Control, and Computing, Monticello, IL, USA, 22–24 September 1999; pp. 368–377.
23. Slonim, N. The Information Bottleneck Theory and Applications. Ph.D. Thesis, Hebrew University of Jerusalem, Jerusalem, Israel, 2002.
24. Hassanpour, S.; Wuebben, D.; Dekorsy, A. Overview and Investigation of Algorithms for the Information Bottleneck Method. In Proceedings of the SCC 2017—11th International ITG Conference on Systems, Communications and Coding, Hamburg, Germany, 6–9 February 2017.
25. Lewandowsky, J.; Bauch, G. Information-Optimum LDPC Decoders Based on the Information Bottleneck Method. *IEEE Access* **2018**, *6*, 4054–4071. [CrossRef]
26. Zeitler, G. Low-precision analog-to-digital conversion and mutual information in channels with memory. In Proceedings of the 48th Annual Allerton Conference on Communication, Control and Computing, Monticello, IL, USA, 29 September–1 October 2010; pp. 745–752.
27. Meidlinger, M.; Matz, G. On Irregular LDPC Codes with Quantized Message Passing Decoding. In Proceedings of the 2017 IEEE 18th International Workshop on Signal Processing Advances in Wireless Communications (SPAWC), Sapporo, Japan, 3–6 July 2017; IEEE: Piscataway, NJ, USA, 2017; pp. 1–5. [CrossRef]
28. Romero, F.; Kurkoski, B. LDPC Decoding Mappings That Maximize Mutual Information. *IEEE J. Sel. Areas Commun.* **2016**, *34*, 2391–2401. [CrossRef]
29. Zeitler, G. Low-Precision Quantizer Design for Communication Problems. Ph.D. Thesis, Technische Universitaet Muenchen, Muenchen, Germany, 2012.
30. Chen, D.; Kuehn, V. Alternating information bottleneck optimization for the compression in the uplink of C-RAN. In Proceedings of the 2016 IEEE International Conference on Communications (ICC), Kuala Lumpur, Malaysia, 23–27 May 2016; pp. 1–7. [CrossRef]
31. Lewandowsky, J.; Stark, M.; Bauch, G. Information Bottleneck Graphs for Receiver Design. In Proceedings of the 2016 IEEE International Symposium on Information Theory (ISIT), Barcelona, Spain, 10–15 July 2016; pp. 2888–2892. [CrossRef]
32. Fujishige, S. *Submodular Functions and Optimization*; Elsevier: Amsterdam, The Netherlands, 2005.
33. ten Brink, S. Convergence Behavior of Iteratively Decoded Parallel Concatenated Codes. *IEEE Trans. Commun.* **2001**, *49*, 1727–1737. [CrossRef]
34. Cover, T.; Thomas, J. *Elements of Information Theory*, 2nd ed.; Wiley & Sons: New York, NY, USA, 2006.

entropy

Article

Revisiting Sequential Information Bottleneck: New Implementation and Evaluation

Assaf Toledo *, **Elad Venezian** and **Noam Slonim** *

IBM Research AI, Haifa University Campus, Mount Carmel Haifa, Haifa 3498825, Israel
* Correspondence: assaf.toledo@ibm.com (A.T.); noams@il.ibm.com (N.S.)

Abstract: We introduce a modern, optimized, and publicly available implementation of the sequential Information Bottleneck clustering algorithm, which strikes a highly competitive balance between clustering quality and speed. We describe a set of optimizations that make the algorithm computation more efficient, particularly for the common case of sparse data representation. The results are substantiated by an extensive evaluation that compares the algorithm to commonly used alternatives, focusing on the practically important use case of text clustering. The evaluation covers a range of publicly available benchmark datasets and a set of clustering setups employing modern word and sentence embeddings obtained by state-of-the-art neural models. The results show that in spite of using the more basic Term-Frequency representation, the proposed implementation provides a highly attractive trade-off between quality and speed that outperforms the alternatives considered. This new release facilitates the use of the algorithm in real-world applications of text clustering.

Keywords: clustering; information bottleneck; sequential algorithm

Citation: Toledo, A.; Venezian, E.; Slonim, N. Revisiting Sequential Information Bottleneck: New Implementation and Evaluation. *Entropy* **2022**, *24*, 1132. https://doi.org/10.3390/e24081132

Academic Editors: Gerhard Bauch and Jan Lewandowsky

Received: 11 July 2022
Accepted: 4 August 2022
Published: 16 August 2022

Publisher's Note: MDPI stays neutral with regard to jurisdictional claims in published maps and institutional affiliations.

1. Introduction

Unsupervised clustering of texts is a central problem in the domain of Natural Language Processing (NLP) [1–3], which has various applications in contemporary data analysis. For example, in the field of customer aid, clustering support tickets is helpful for identifying classes of user complaints and estimating their volume [4,5]. In research on public opinion, clustering data from social media such as Twitter or Reddit is useful for discovering active topics and learning about user engagement [6–8].

The Lloyd K-Means algorithm [9] is perhaps the most common choice for text clustering. It is also readily available in a modern, fast, and free implementation as part of the popular Scikit-Learn package [10]. The algorithm can be executed on top of a range of vector representations for the texts at hand, offering different trade-offs between clustering quality and speed.

The Term Frequency–Inverse Document Frequency (TF/IDF) [11] is a traditional method for textual data representation that was developed in the field of Information Retrieval. In this method, a text is first represented by a vector of term frequency (count) over a fixed vocabulary, and then, the value of each term is weighted by the inverse of its frequency in the overall corpus. In this weighting scheme, words that are common in the full corpus will be given a lesser weight compared to rare words when representing the text instance. This method is very fast to compute and process but suffers from the curse of dimensionality [12]. In addition, as a Bag-of-Words (BoW) method, it ignores word order.

More modern approaches that tackle the high dimensionality issue rely on Word Embeddings, such as Word2Vec [13,14] and GloVe [15]. In these cases, every word is assigned a fixed-size dense representation (embedding) trained by a neural network based on contextual information such as word–word co-occurrence statistics. With these methods, a common practice for obtaining the representation of a text instance is by averaging the embeddings of its words. These representations, however, still ignore word order, and while being relatively fast to generate and to work with, they disregard the sentential context in which words are meant to be interpreted.

With the surge of Deep Neural Networks [16] and advanced network architectures [17] and Language Models such as BERT [18], new records have been set in benchmarks of Natural Language Understanding [19,20]. Sentence-BERT (S-BERT) [21] is one of the latest advancements in creating models for general-purpose text vector representation that serve tasks such as semantic similarity, semantic search, and clustering. A common technique for creating these models is by adding a pooling layer to a BERT neural network. The pooling layer averages the contextualized word embeddings that BERT outputs and returns a single fixed-size vector for the whole text.

However, using an advanced language model typically comes at the expense of requiring more processing power and in many cases necessitates a GPU. The main concern in relying on vector representations from this kind of neural models is that in the age of Big Data, when the volume of texts is high and grows ever so quickly, it is critical for a clustering setup to be not only of high quality but also very efficient.

The sequential Information Bottleneck (sIB) [22] algorithm has shown strong results that outperform K-Means by a large margin on benchmark datasets such as [23]. However, sIB has never been available in a fast implementation that makes it comparable to K-Means for practical applications. Thus, despite being superior in lab testing, sIB was not commonly used in practice.

This paper introduces an efficient implementation of sIB that leverages a set of optimizations for obtaining a substantial improvement in speed compared to the original Matlab implementation. This is achieved while maintaining the same quality of clustering analysis provided by the original implementation.

The optimizations are focused on the case of sparse data representation, as sIB is usually running on top of sparse TF vectors, and it includes a mathematical derivation that computes the Jensen–Shannon divergence (JS) more efficiently on this type of representation. Our work builds on the analysis proposed in [24] and extends it to the case of a weighted JS, as this is the divergence used in sIB. In addition, we show an optimization that reduces the computational load by means of caching.

We present an evaluation scheme for assessing the quality and speed of the new sIB implementation and for comparing it against a set of competing clustering setups and over a range of benchmark datasets. Empirical results indicate that sIB is as fast as the fastest setups on most datasets, and in some cases, it is even the fastest. Quality-wise, sIB outperforms most competing setups by a large margin, and it maintains a small edge even over the best-performing setup of K-Means—when this algorithm leverages the advanced S-Bert representation.

In this manner, the new implementation strikes an attractive trade-off between quality and speed of text clustering, and it facilitates the use of sIB in real-world applications. The implementation is released as an open-source Python package under a permissive license (https://github.com/IBM/sib, available since 14 July 2020).

2. Algorithm Overview

In this section, we review the main properties of sIB, highlight key-differences between sIB and Lloyd K-Means, and put the focus on the main part of the algorithm that the new implementation optimizes.

2.1. Theoretical Foundation

sIB builds on the work of Tishby, Pereira, and Bialek [25] and views the clustering task as an Information Bottleneck (IB) problem. The algorithm first appeared in [22] and was explained in detail in [26]. Before formulating it, we present some assumptions and definitions.

sIB adopts the *Bag-of-Words* (BoW) approach with TF (count) vector representation for the texts to cluster. We use the following notation:

- X—the list of vectors to cluster;
- Y—the vocabulary used for representing the texts;
- $p(X, Y)$—the estimated joint distribution between X and Y;
- K—the number of clusters to produce;

- T—a partition of X into K clusters.

According to the IB method, given the joint distribution $p(X, Y)$, we look for the partition T—in our case, a compressed representation of X into K clusters—that preserves as much information about Y as possible. Quoting from the original paper of sIB [22]:

"Intuitively, in this procedure the information contained in X about Y is 'squeezed' through a compact 'bottleneck' of clusters T, that is forced to represent the 'relevant' part in X with respect to Y" (p. 2)

Formally, the IB method is stated as:

$$\min_{p(t|x)} I(X; T) - \beta I(T; Y) \, ,$$

where $I(X; T)$ and $I(T; Y)$ are the mutual information (MI) between X and T and between T and Y, respectively. MI is defined as:

$$I(A; B) = \sum_{a \in A, b \in B} p(a) \cdot p(b|a) \cdot log \frac{p(b|a)}{p(b)} \, .$$

The optimization is over the conditional probability $p(t|x)$, and the non-negative parameter β is a Lagrange multiplier. The IB method [25] provides an exact optimal formal solution to this problem without any assumption about the origin of the joint distribution $p(X, Y)$. In this analysis, the compactness of the representation T is determined by $I(X; T)$, while the quality of T is measured by the fraction of the information that T and X capture about Y: $I(T; Y)/I(X; Y)$. For a more detailed and technical discussion, see [22,25,26].

2.2. Divergence Function

As shown in [25], minimizing the IB functional defined above is obtained by using the Kullback–Leibler (KL) divergence [27] as the clustering divergence function between the conditional distributions $p(y|x)$ and $p(y|t)$:

$$KL(p(y|x)\|p(y|t)) = \sum_{y} p(y|x) \frac{log\, p(y|x)}{p(y|t)} \, .$$

However, as shown in [26,28], in the hard-clustering setup—which is the focus of this work—to optimize the IB functional while merging x to t; one should use a weighted Jensen–Shannon divergence (JS) between $p(y|x)$ and $p(y|t)$. This divergence is defined based on the KL divergence with the additional features of being symmetric and always returning a finite value:

$$JS(p(y|x)\|p(y|t)) = \pi_1 \cdot KL(p(y|x)\|M) + \pi_2 \cdot KL(p(y|t)\|M) \, ,$$

where $M = \pi_1 \cdot p(y|x) + \pi_2 \cdot p(y|t)$, and $\pi_1, \pi_2 \in [0, 1]$ are two weights such that $\pi_1 + \pi_2 = 1$. For conciseness, we skip the definitions of π_1 and π_2 at this point and provide it in Section 3 where they are relevant for the formal analysis. The experimental setup in [22] confirms that the quality of the clustering analysis obtained when using JS as the clustering divergence function is superior to the clustering obtained when using KL. On the other hand, in terms of computational workload, JS is more demanding than KL as every computation of JS involves two computations of KL.

Let us compare this to K-Means. The traditional Lloyd K-Means algorithm is used with a geometrical distance function such as the Euclidean or cosine distance. This results in spatial clustering of the representations in a vector space. From a computational standpoint, the geometrical distances are lightweight and in some cases reduce to the computation of the dot product, which is fairly fast in comparison to the intensive log calculations as part of JS in sIB. Theoretically, this gives the traditional K-Means setup a substantial speed advantage over sIB.

The KL-Means algorithm [29–31] is a variant of the traditional algorithm with the distance function set to be the KL divergence. This effectively creates a version of K-

Means that performs distributional clustering. It has been shown by [32] that in this setup, K-Means is algorithmically equivalent to the IB method where $\beta \to \infty$. In this sense, KL-Means is more similar to sIB than the traditional K-Means algorithm. sIB is still unique, however, in using JS divergence rather than KL divergence.

In the next section we provide the pseudo code for the algorithm and then move to another distinctive feature of sIB—namely, its sequential nature.

2.3. Pseudo-Code

The pseudo-code of the algorithm's main loop is given in Algorithm 1, which is quoted from [22] with slight adjustments. The only modifications from the original pseudo-code are in the inner for-loop, where we explicitly mention the $shuffle$ function and use x instead of x_j. The pseudo-code outlines the sequential workflow in which sIB works. In this code, recall that K is the number of clusters to generate, n is the number of (random) initializations, $maxL$ is the maximal number of iterations per initialization, and ϵ is a lower bound threshold on the cluster updates for continuing to another iteration. In addition, $shuffle$ is a function that randomizes the order of elements, t is used as a cluster identifier, x as a sample identifier, c is a counter of cluster changes during an iteration over X and C is a counter of iterations per initialization. Using several random initializations is a common practice with many clustering algorithms, as each initialization converges to a local maximum/minimum.

Algorithm 1 Algorithm pseudo-code.

Input:

$|X|$ objects to be clustered
Parameters: $K,\ n,\ maxL,\ \epsilon$

Output:

A partition T of X into K clusters

Main Loop:

For $i = 1, \ldots, n$
    $T_i \leftarrow$ random partition of X.
    $c \leftarrow 0,\ C \leftarrow 0,\ done = FALSE$
    While not $done$
        For x in $shuffle(X)$
            draw x out of $t(x)$
            $t^{new}(x) = \arg\min_{t'} d_F(x, t')$
            If $t^{new}(x) \neq t(x)$ then $c \leftarrow c + 1$
            Merge x into $t^{new}(x)$
        $C \leftarrow C + 1$
        if $C \geq maxL$ or $c \leq \epsilon \cdot |X|$ then
            $done \leftarrow TRUE$
    $T \leftarrow \arg\max_{T_i} f(T_i)$

2.4. Sequential Clustering Algorithm

As shown in the pseudo-code, sIB is a sequential algorithm. This means that: (a) before selecting the new cluster for a sample, sIB withdraws that sample from its current cluster to prevent it from biasing the distance calculation toward keeping the sample in the same cluster, and (b) sIB updates the centroids while iterating over the samples and not only at the end of a full iteration over all samples.

Overall, while iterating over X, every sample is withdrawn from its cluster, the centroid of that cluster is updated, a new cluster is selected for that sample using the weighted JS divergence distance function, and then, the sample is added to the new cluster and the centroid of the new cluster is updated. In total, sIB performs $2 \cdot |X|$ centroid updates during a full iteration.

As discussed in detail in [33], this is a more powerful partition optimization method than the one employed by Lloyd K-Means, where there are no centroid updates while iterating over the samples during the assignment step. Lloyd K-Means performs only K centroid updates, which are all happening at the end of an iteration. Since $K << 2|X|$ under normal circumstances, this gives Lloyd K-Means another substantial advantage in terms of computational workload.

2.5. Vector Representation

We distinguish between two vector representations: (a) for the texts to cluster and (b) for the centroids of clusters. Typically, the number of unique terms found in a specific text is much smaller than the vocabulary size. Therefore, it is more efficient to represent texts using sparse vector representations, both in terms of memory usage and processing time. In the sparse representation, it is sufficient to hold the list of IDs of vocabulary items found in the text and their frequency rather than an array of the size of the full vocabulary in which most of the values are zero. With regard to centroid vectors, as a centroid-based clustering algorithm, sIB constructs a centroid vector from the vectors of the samples that are associated with that cluster. Therefore, centroid vectors refer to a large part of the vocabulary and are encoded as regular non-sparse vectors.

2.6. Focus of This Work

This work focuses on the inner *for*-loop of the pseudo-code, which is the partition optimization part, and more specifically the computation of $t^{new}(x)$. We investigate it in the next section.

3. Methods

In this section, we present mathematical derivations and code optimizations that are at the center of the new implementation of sIB.

3.1. Computation of $t^{new}(x)$ and Associated Intuition

Recall that finding the new cluster assignment for x relies on computing

$$t^{new}(x) = \arg\min_{t} d_F(x,t) \ , \tag{1}$$

where d_F is given by

$$d_F(x,t) = (p(x) + p(t)) \cdot JS(p(y|x), p(y|t)) \ , \tag{2}$$

and JS is a weighted *Jensen–Shannon divergence* defined with weights π_1 and π_2:

$$\pi_1 = \frac{p(x)}{p(x) + p(t)}, \quad \pi_2 = \frac{p(t)}{p(x) + p(t)} \ . \tag{3}$$

Intuitively, when selecting the new cluster assignment for x, we examine the distribution over the vocabulary induced by x ($p(y|x)$) and compare it to the distribution over the vocabulary induced by each cluster's centroid ($p(y|t)$) using the weighted JS divergence multiplied by $p(x) + p(t)$. The cluster t^{new} is selected as the cluster for which this multiplication is minimized.

In what follows, we use the following notation:

- $\hat{x} = p(y|x)$—the TF vector representing the sample x normalized by the L^1-norm. Let $u \in \mathbb{R}^n$, the L^1-norm $|u|_1$ of u is defined by: $|u|_1 = \sum_{i=1}^{n} |u_i|$;
- $\hat{t} = p(y|t)$—the vector representing the centroid of cluster t, normalized by L^1;
- $m = \pi_1 \cdot \hat{x} + \pi_2 \cdot \hat{t}$—the average of $\hat{x}$ and $\hat{t}$ weighted by π_1 and π_2, respectively.

Using these notations:

$$JS(\hat{x}, \hat{t}) = \pi_1 \cdot KL(\hat{x}\|m) + \pi_2 \cdot KL(\hat{t}\|m) \ , \tag{4}$$

where KL is the *Kullback–Leibler divergence* [27] defined as:

$$KL(u\|v) = \sum_i u[i] \cdot log(\frac{u[i]}{v[i]}) \ .$$ (5)

Following the analysis in [24], simple algebra gives the form in (6):

$$JS(\hat{x}, \hat{t}) = H(m) - \pi_1 \cdot H(\hat{x}) - \pi_2 \cdot H(\hat{t}) \ ,$$ (6)

where H is Shannon's entropy function: $H(u) = -\sum_i u[i] \cdot log(u[i])$.
Since $(p(x) + p(t)) \cdot \pi_1 = p(x)$ and $(p(x) + p(t)) \cdot \pi_2 = p(t)$:

$$d_F(x,t) = (p(x) + p(t)) \cdot H(m) - p(x) \cdot H(\hat{x}) - p(t) \cdot H(\hat{t}) \ .$$ (7)

Because $p(x) \cdot H(\hat{x})$ is a constant with respect to t, we get:

$$\arg\min_t (d_F(x,t)) = \arg\min_t \left((p(x) + p(t)) \cdot H(m) - p(t) \cdot H(\hat{t}) \right) \ .$$ (8)

To obtain some insight into how sIB selects the cluster t for a sample x, we examine two pairs of components in Equation (8)—(a) $H(m)$ and $H(\hat{t})$; and (b) $(p(x) + p(t))$ and $p(t)$. Starting with (a), since m is a weighted average of $\hat{x}$ and $\hat{t}$, a better fit of x to t implies lower discrepancy between m and $\hat{t}$, which in turn results in a smaller difference between $H(m)$ and $H(\hat{t})$. Thus, the preference is for selecting a cluster t that represents a good fit for x. Moving to (b), as the cluster t increases, the relative difference between $p(x) + p(t)$ and $p(t)$ decreases. Therefore, the components in (b) can be seen as balancing factors for the selection of t by taking into account the size of the cluster and giving preference to larger clusters. Typically, these two parts compete, since as t becomes larger, it often also becomes less distinctive; hence, it is harder for it to provide a good fit for x.

3.2. Optimization for Sparse Vector Representation

In this section, we show a computation of $t^{new}(x)$ that is optimized for sparse vector representation. Let x_{ind} be the indices of non-zero values in x. As explained in Section 2.5, a sparse representation is the natural choice for TF vectors since typically $|x_{ind}| << |Y|$.
We evaluate Equation (6) as:

$$JS(\hat{x}, \hat{t}) = \sum_{i \in x_{ind}} R_{i,\hat{x},\hat{t}} + \sum_{i \notin x_{ind}} R_{i,\hat{x},\hat{t}} \ ,$$ (9)

where $R_{i,\hat{x},\hat{t}}$ is defined as:

$$R_{i,\hat{x},\hat{t}} := \pi_1 \cdot \hat{x}_i \cdot log(\hat{x}_i) + \pi_2 \cdot \hat{t}_i \cdot log(\hat{t}_i) - (\pi_1 \cdot \hat{x}_i + \pi_2 \cdot \hat{t}_i) \cdot log(\pi_1 \cdot \hat{x}_i + \pi_2 \cdot \hat{t}_i) \ .$$ (10)

Since the computation of $R_{i,\hat{x},\hat{t}}$ involves a constant number of operations, the first component in (9) has a computational complexity of $O(|x_{ind}|)$. Let us now evaluate the second component and show that it has the same complexity.
By definition, $\forall i \notin x_{ind} \ \hat{x}_i = 0$. Consequently,

$$\sum_{i \notin x_{ind}} R_{i,\hat{x},\hat{t}} = \sum_{i \notin x_{ind}} \left[\pi_2 \cdot \hat{t}_i \cdot log(\hat{t}_i) - \pi_2 \cdot \hat{t}_i \cdot log(\pi_2 \cdot \hat{t}_i) \right] \ .$$ (11)

Since $log(\pi_2 \cdot \hat{t}_i) = log(\pi_2) + log(\hat{t}_i)$,

$$\sum_{i \notin x_{ind}} R_{i,\hat{x},\hat{t}} = \sum_{i \notin x_{ind}} \left[\pi_2 \cdot \hat{t}_i \cdot log(\hat{t}_i) - \pi_2 \cdot \hat{t}_i \cdot (log(\pi_2) + log(\hat{t}_i)) \right] \ .$$ (12)

With simple algebra, we obtain:

$$\sum_{i \notin x_{ind}} R_{i,\hat{x},\hat{t}} = -\pi_2 \cdot log(\pi_2) \cdot \sum_{i \notin x_{ind}} \hat{t}_i \ . \tag{13}$$

Since $\hat{t}$ is normalized by L^1-norm, $\sum_i \hat{t}_i = 1$. Therefore:

$$\sum_{i \notin x_{ind}} R_{i,\hat{x},\hat{t}} = -\pi_2 \cdot log(\pi_2) \cdot (1 - \sum_{i \in x_{ind}} \hat{t}_i) \ . \tag{14}$$

This means that the second component of (9) also has a computational complexity of $O(|x_{ind}|)$, which shows that (6) and consequently (2) have the same computational complexity of $O(|x_{ind}|)$. In the non-sparse case, the computational complexity is $O(|Y|)$, which is significantly higher.

With respect to (1), since there are K centroids to select from, the overall complexity of computing $t^{new}(x)$ is $O(K \cdot |x_{ind}|)$ in the sparse case and $O(K \cdot |Y|)$ in the non-sparse case.

3.3. Caching Log Computations

The most time-consuming operation in the computation of a new cluster for a sample is the *log* function. In this section, we show a way to reduce the number of *log* computations via caching. Let us recall Equation (8), which is repeated below:

$$\arg\min_t (d_F(x,t)) = \arg\min_t \left((p(x) + p(t)) \cdot H(m) - p(t) \cdot H(\hat{t}) \right) \ .$$

We observe that $H(\hat{t})$ is independent of the sample x for which we calculate the new cluster. Thus, we can cache this computation and reuse it when we iterate over the samples. When a sample is drawn out of a cluster or merged into a cluster, we update only the entries in the cache that refer to the clusters that have been updated. The gain can be summarized as follows. Given a sample x for which we compute a new cluster, instead of computing the entropy over all centroids, we compute it only for two centroids—the centroid of the cluster from where x is drawn out and the one to which x is merged into.

3.4. Implementation

The new implementation of sIB is based on the optimizations presented above. See Appendix A for information about the source-code availability and its Python packaging.

4. Experimental Setup

We evaluate the quality and speed of the new implementation of sIB against the robust open source implementation of Lloyd K-Means [9] from Scikit-Learn [10]. We use a set of five datasets that are common in text classification and text clustering benchmarks and measure the clustering quality by standard clustering metrics. We use multiple setups for K-Meams, each on top of a different vector representation type.

4.1. Materials

In order to cover various use cases, we employ datasets of different source, size, text length and number of classes. The datasets are described below, and statistical information is summarized in Table 1. All datasets are publicly available online at the locations specified in the Data Availability Statement.

- *BBC News* [34] consists of 2225 articles from the BBC news website. The articles are from 2004–2005 and cover stories in five topical areas: *business, entertainment, politics, sport,* and *tech*.
- *20 News Groups* [23] consists of 18,846 emails sent through 20 news groups. The topics are diverse and cover *tech, religion,* and *politics,* among others.
- *AG News* [35] consists of 127,600 pairs of titles and snippets of news articles from the AG corpus, covering four topical areas: *World, Sports, Business,* and *Sci/Tech*. The title and snippet of each article are concatenated when the data is clustered.

- *DBPedia* [35] consists of 630,000 pairs of titles and abstracts of documents from 14 non-overlapping ontology classes such as *Artist*, *Film*, and *Company*. The title and abstract are concatenated when the data are clustered.
- *Yahoo! Answers* [35] consists of 1,460,000 triplets of question title, question content, and best answer from the Yahoo! Answers Comprehensive Questions and Answers version 1.0 dataset. The data covers the 10 largest topical categories, such as *Society & Culture*, *Computers & Internet*, and *Health*. The question title, content, and best answer are concatenated when the data are clustered.

Table 1. Benchmark datasets for evaluation. The column *#Texts* indicates the number of texts in the dataset. The column *#Words* shows the average text length in terms of word count in the dataset, and *#Classes* shows the number of classes in the dataset.

Dataset	#Texts	#Words	#Classes
BBC News	2225	390	5
20 News Groups	18,846	284	20
AG NEWS	127,600	38	4
DBPedia	630,000	46	14
Yahoo! Answers	1,460,000	92	10

4.2. Clustering Metrics

We use five metrics to evaluate the clustering quality: (a) Adjusted Mutual-Information (AMI): the mutual-information corrected for chance [36,37], (b) Adjusted Rand-Index (ARI): the rand index corrected for chance [36], (c) V-Measure: the harmonic mean between homogeneity and completeness [38], (d) Micro-F1: the micro average of F1 scores over all classes in the dataset, and (e) Macro-F1: the macro average of F1 scores over all classes in the dataset. All metrics are calculated against the ground-truth labels of each dataset.

4.3. Clustering Setups

sIB runs on top of sparse TF representations. The encoding is done using a vocabulary of the 10,000 most common words in each dataset after stop-words filtering. We use the Scikit-Learn [10] TF encoder. The algorithm runs with 10 random partitions of equally sized clusters in parallel. Each initialization is optimized by up to 15 iterations or until the number of samples changing cluster is less than 2% (all are default values). This generates 10 partitions of the data, and the algorithm returns the partition that maximizes $I(T;Y)/I(X;Y)$ as explained in Section 2.1. The sIB version is 0.1.8, and the Scikit-Learn version is 1.1.1.

K-Means runs on top of several representations: TF, TF/IDF [11], GloVe [15] mean vectors and Sentence-Bert (S-Bert) [21]. The TF is the same as used for sIB (described above). The TF/IDF representation is generated using Scikit-Learn [10] TF/IDF encoder with the same settings as TF and is also sparse. For GloVe, each text is represented by averaging the embeddings of its words after punctuation and stop-words filtering. We use the *glove-840b-300d* pre-trained model, which was trained on 840 billion tokens and produces 300-dimensional dense vectors. For S-Bert, we employ the pre-trained model *all-MiniLM-L6-v2* which aims to provide a fine balance between quality and processing time. This model was trained on 1 billion sentence pairs and produces 384-dimensional dense vectors.

We use the Scikit-Learn [10] Lloyd K-Means implementation in its default settings. The algorithm runs with 10 random centroid initializations obtained by K-Means++ [39] in parallel, yielding 10 partitions of the data. Each initialization is optimized by up to 300 iterations or until the centroids movement between iterations, as measured by Frobenius norm, is less than 10^{-4}. The algorithm returns the partition that minimizes the sum of distances between each sample and the centroid of its cluster. All are default values. The distance function used in this version of K-Means is the squared Euclidean distance [40]. Minimizing the squared distance is equivalent to minimizing the Euclidean distance since squaring is a monotonic function of non-negative values. Oftentimes, the squared distance is preferred because it is faster to compute.

4.4. Robustness Considerations

Since both sIB and K-Means rely on random initialization, every run of these algorithms converges to a different local minimum and yields a different clustering result. For robustness, we run every setup described above 10 times and apply the metrics described in Section 4.2 to every such run. We obtain 10 scores for each metric for a given setup and report only the average score per metric. We also use the distribution of the metric scores for calculating confidence intervals.

4.5. Hardware

The hardware used is a MacBook Pro 2019 with an 8-Core Intel Core i9 running at 2.3 Ghz. Hyper-threads: 16. Memory is 64 GB 2667 MHz DDR4. This simulates a local run by a data scientist.

4.6. Code

The evaluation code is available on the sIB open source repository and can be extended and tweaked to cover more algorithms, representations, and settings.

5. Results

The results are detailed in Table 2. In terms of clustering quality, the metrics indicate that sIB has the edge over the setup of K-Mean on top of S-Bert on the 20 News Groups and AG News datasets. On the BBC News dataset they are even, and then the trend reverses and K-Means on top of S-Bert takes the lead by a relatively small margin on the DBPedia and Yahoo! Answers datasets. Overall, these two setups are roughly on par with a slight edge to sIB. The other K-Means setups are trailing behind by a large margin, with the GloVe setup being better than the TF/IDF setup, and the TF setup being the weakest. Figure 1 illustrates the results on the AMI and ARI metrics. We include charts also for the Micro-F1, Macro-F1 and V-Measure in Appendix B. As explained in Section 4.4, the reported result of every metric is the average of 10 runs of each setup. Error bars in the figures indicate the 95% confidence interval obtained by bias-corrected and accelerated (BCa) bootstrapping of the 10 results per metric.

As for run-time measurements, we can see in Table 2 that sIB is as fast as the quickest K-Means setups (TF and TF/IDF) on the datasets of 20 News Groups, AG News, BBC News and DBPedia, and it is the fastest setup on the Yahoo! Answers dataset. sIB is also faster than the setup of K-Means on top of GloVe by a noticeable margin.

The setup of K-Means on top of S-Bert, which is the only setup that is competitive with sIB quality-wise, is substantially slower due to the neural vectorization on CPU. On average, this setup is 200 times slower than sIB. More generally, the S-Bert model is more power demanding than any other representation type evaluated here, and for practical use cases, especially on large datasets such as DBPedia and Yahoo! Answers, it is likely to necessitate a GPU or even more than one. A chart of the total run-time measurements is included in Appendix B.

Discussion

The results emphasize the premise of the sIB implementation proposed in this work: delivering a clustering analysis that is as good as can be obtained by a state-of-the-art neural model while being far less demanding in terms of run-time. In this way, sIB offers a more attractive trade-off between quality and speed than the rest of the setups evaluated here.

Looking at the run-time measures in absolute terms, sIB is able to cluster the 630,000 texts of DBPedia in about 1 minute and the 1,460,000 texts of Yahoo! Answers in about 3.5 min using standard CPU hardware. Both are very practical and workable run-times for real-world applications.

Table 2. Assessment of clustering quality using the metrics: AMI, ARI, V-Measure (*VM*), Micro-F1 (*Mic-F1*) and Macro-F1 (*Mac-F1*), and of clustering speed based on measurements of the vectorization time (*Vector*), clustering time (*Cluster*), and their sum (*Total*).

Dataset	Algorithm	Embed	AMI	ARI	VM	Mic-F1	Mac-F1	Vector	Cluster	Total
20 News Groups	K-Means	TF	0.01	0.00	0.01	0.06	0.01	**00:03**	**00:07**	**00:10**
	K-Means	TF/IDF	0.36	0.14	0.36	0.35	0.32	**00:03**	00:10	00:14
	K-Means	GloVe	0.36	0.17	0.36	0.34	0.31	00:19	00:09	00:28
	K-Means	S-Bert	0.59	0.44	0.59	0.61	0.58	22:27	00:08	22:35
	sIB	TF	**0.65**	**0.53**	**0.65**	**0.66**	**0.61**	**00:03**	00:11	00:14
AG NEWS	K-Means	TF	0.03	0.00	0.03	0.29	0.20	**00:03**	**00:02**	**00:05**
	K-Means	TF/IDF	0.04	0.01	0.04	0.31	0.24	**00:03**	00:03	00:06
	K-Means	GloVe	0.53	0.55	0.53	0.80	0.80	00:19	00:07	00:26
	K-Means	S-Bert	0.60	0.63	0.60	0.84	0.84	38:18	00:11	38:29
	sIB	TF	**0.66**	**0.70**	**0.66**	**0.87**	**0.87**	**00:03**	00:03	00:06
BBC News	K-Means	TF	0.24	0.11	0.24	0.41	0.32	00:01	**00:00**	**00:01**
	K-Means	TF/IDF	0.70	0.62	0.70	0.83	0.83	**00:00**	**00:00**	**00:01**
	K-Means	GloVe	0.75	0.76	0.75	0.90	0.90	00:05	**00:00**	00:06
	K-Means	S-Bert	0.87	**0.90**	0.87	**0.96**	**0.96**	02:55	**00:00**	02:56
	sIB	TF	**0.88**	**0.90**	**0.88**	**0.96**	**0.96**	00:01	00:01	**00:01**
DBPedia	K-Means	TF	0.56	0.21	0.56	0.50	0.47	**00:20**	**00:43**	**01:04**
	K-Means	TF/IDF	0.61	0.24	0.61	0.56	0.55	**00:20**	00:45	01:06
	K-Means	GloVe	0.73	0.63	0.73	0.76	0.72	01:28	02:08	03:37
	K-Means	S-Bert	**0.79**	**0.71**	**0.79**	**0.82**	**0.79**	03:38:31	02:00	03:40:31
	sIB	TF	**0.79**	0.68	**0.79**	0.78	0.74	**00:20**	00:44	01:05
Yahoo! Answers	K-Means	TF	0.03	0.01	0.03	0.15	0.08	**01:15**	04:22	05:37
	K-Means	TF/IDF	0.16	0.05	0.16	0.29	0.25	01:16	03:44	05:01
	K-Means	GloVe	0.32	0.23	0.32	0.49	0.44	06:21	06:42	13:03
	K-Means	S-Bert	**0.41**	**0.33**	**0.41**	**0.59**	**0.56**	16:20:10	06:18	16:26:28
	sIB	TF	0.39	0.32	0.39	0.57	0.54	**01:15**	**02:20**	**03:35**

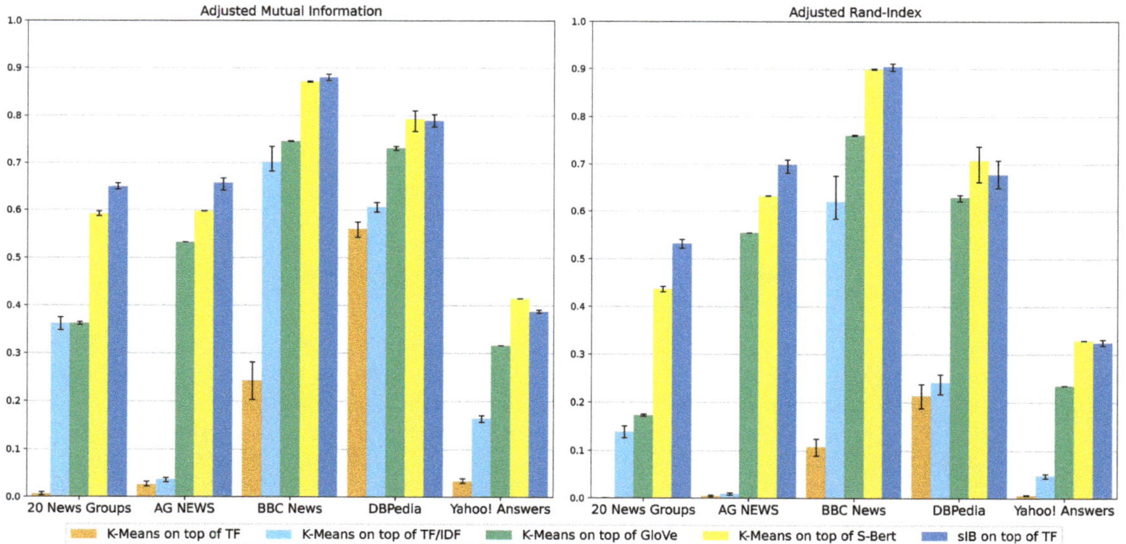

Figure 1. Adjusted Mutual-Information (**left**) and Adjusted Rand-Index (**right**) scores for the clustering setups over all benchmark datasets. The scores are means of 10 samples per metric. Error bars indicate the 95% confidence interval obtained by bias-corrected and accelerated bootstrapping.

A question can be raised as to how sIB can match or even improve on the K-Means run-time. Given that sIB is a more demanding algorithm in terms of computational workload, it would have been expected to show inferior run-time measurements compared to the more lightweight K-Means. We look into this in Appendix C and provide a hypothesis to this phenomenon. Note that we ignore here the discrepancy between sIB and K-Means with respect to the default maximal number of partition optimization iterations (15 for sIB, 300 for K-Means). This is because internal testing with fewer iterations for K-Means proved ineffective for reducing the algorithm run-time. We assume that this is because the algorithm declared convergence (to a local minimum) long before the iteration limit is reached.

6. Conclusions

The sIB algorithm was introduces more than 15 years before the rise of the language models revolution in NLP. Although sIB uses simple TF representations, it utilizes a powerful probabilistic framework and a robust optimization method. This work is the first to offer a highly efficient implementation of the algorithm and also to evaluate it on contemporary benchmark datasets against competing, more popular, clustering setups.

Empirical results indicate that sIB creates a high-quality clustering analysis, which is comparable to the level of analysis obtained when using representations from a state-of-the-art language model. Speed-wise, the results show that the new implementation enables users to easily run sIB on a standard CPU hardware, and that it is far less demanding than a neural solution. In this manner, sIB offers an attractive trade-off between quality and speed, outperforming the rest of the setups considered in this work.

In the future, we plan to look into new ways to reduce sIB's run-time further by creating "lossy" modes of the algorithm. In such modes, rather then iterating over all samples per iteration, the algorithm can allow certain samples to be skipped based on information from previous iterations. For example, if a sample remains in the same cluster for several consecutive iterations, or if it fits much better in one cluster compared to the others, it can be considered as locked-in in its current cluster. In this manner, one can further reduce the algorithm run-time and offer more control in tuning the desired trade-off between quality and speed, allowing sIB to fit an even broader set of use-cases and reach a wider audience.

The new implementation of sIB is released as open-source under a permissive license, and it can be integrated as part of a more complex pipeline of natural language processing in research projects as well as in real-world applications. We hope that practitioners of text clustering and researchers interested in the IB line of study will find this work and the released code valuable.

Author Contributions: Conceptualization, A.T. and E.V.; methodology, A.T. and N.S.; software, A.T.; validation, A.T.; formal analysis, A.T. and E.V.; investigation, A.T.; writing—original draft preparation, A.T.; writing—review and editing, N.S. and E.V.; visualization, A.T.; supervision, N.S.; All authors have read and agreed to the published version of the manuscript.

Funding: This research received no external funding.

Institutional Review Board Statement: Not applicable.

Data Availability Statement: Publicly available datasets were analyzed in this study. These datasets were accessed on 30 June 2022 at: 20 News Groups: https://scikit-learn.org/stable/datasets/real_world.html#the-20-newsgroups-text-dataset; AG News: https://huggingface.co/datasets/ag_news; BBC News: https://huggingface.co/datasets/SetFit/bbc-news; DBPedia: https://huggingface.co/datasets/dbpedia_14; Yahoo! Answers: https://huggingface.co/datasets/yahoo_answers_topics.

Conflicts of Interest: The authors declare no conflict of interest.

Appendix A. Package Distribution

The source code of the new implementation is publicly available under a permissive license (https://github.com/ibm/sib, available since 14 July 2020). The core computational part of partition optimization is written in C++ to obtain fast processing and direct memory access. This code is wrapped by Python and released as a Python package. The algorithm

API entry point, parallelism, iteration loop with stopping condition, and selection of best partition are all written in Python. We also include a Python implementation of the partition optimization part, but the C++ code is the default as it is faster.

The Python package is available on the Python Index for easy access and integration in research projects and applications (https://pypi.org/project/sib-clustering, available since 6 October 2020). Since the C++ code necessitates compilation to binary code, we release pre-compiled versions for popular operating systems: Windows, MacOS and Linux.

Appendix B. Illustrations of Micro-F1, Macro-F1, V-Measure and Total Run-Time

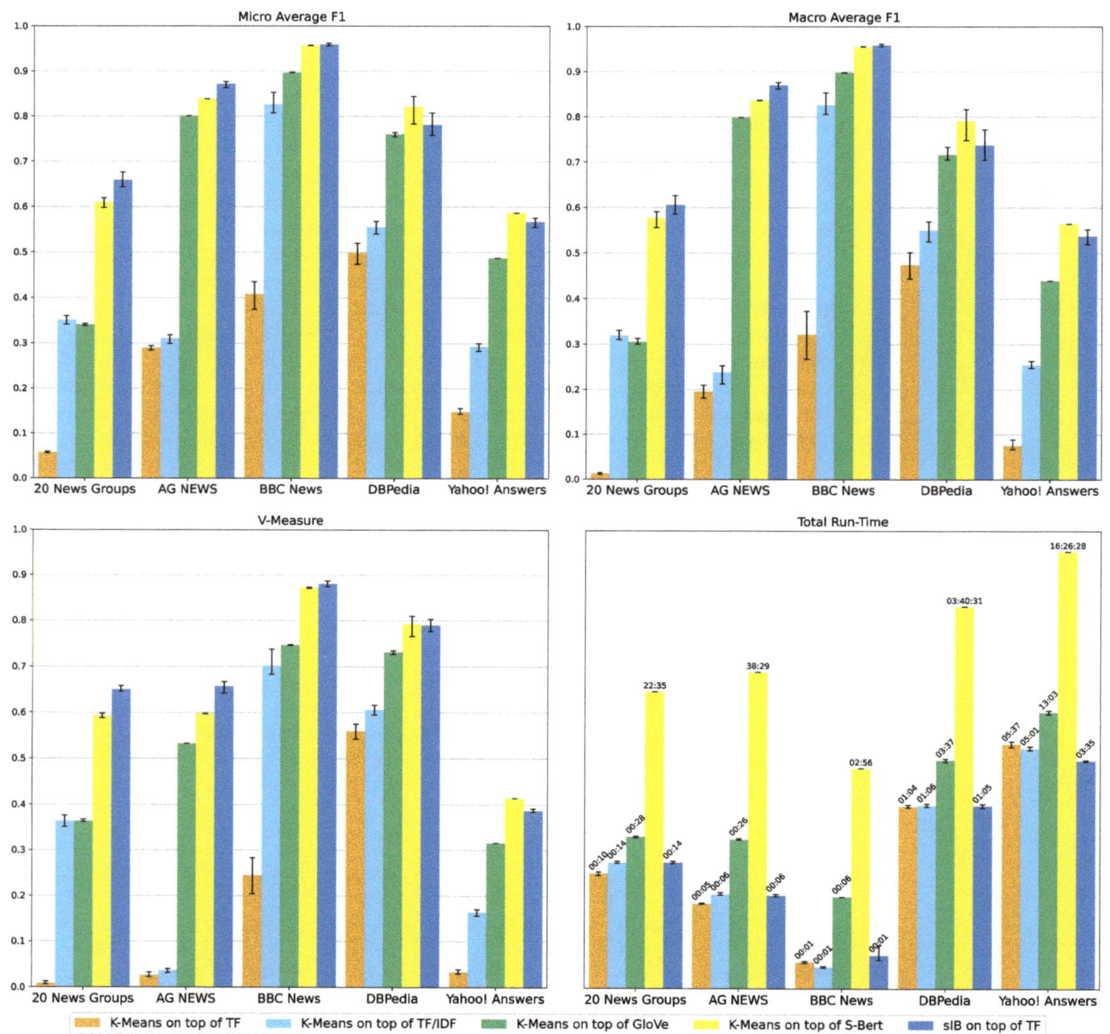

Figure A1. Micro-F1 (**top-left**), Macro-F1 (**top-right**) V-Measure (**bottom-left**) scores, and Total Run-Time (**bottom-right**) measurements for the clustering setups over all benchmark datasets. The Total Run-Time chart is presented on a log_2 scale for the y-axis. The scores and measurements are means of 10 samples per metric. Error bars indicate the 95% confidence interval obtained by bias-corrected and accelerated bootstrapping.

Appendix C. Comparing sIB and K-Means Parallelism Models

Given that sIB is inherently more demanding than K-Means in terms of computational load, as explained in Sections 2.1 and 2.4, it could be somewhat puzzling how sIB can even match K-Means run-time when working on top of the same representation type. In what follows, we aim to provide a hypothesis that explains this phenomenon. We focus on the distinctions between the models of parallelism that these algorithms employ.

As explained in Section 4.3, in every run, sIB starts off by generating 10 random partitions of the data and then optimizes them in parallel by several iterations over the samples. The number of random partitions is configurable. We demonstrate the logic with 10 since this is the default value. Since every partition is optimized independently of the rest, sIB allocates a single, dedicated, CPU core per partition and avoids any task switching. This means that a CPU core which is allocated to optimize partition i will never be interrupted and switched to partition j while i is still being processed. On the machine used for the clustering evaluation (see Section 4.5), which can be considered as having 16 cores (8 physical $\times$ 2 virtual), each of the 10 partitions will be allocated a core and perform the optimization work. The remaining six cores are left idle.

This approach has pros and cons. The pros are that the cores (hyper-threads) that perform partition optimization work without any interruption and no time is wasted on task switching. The cons are that idle cores cannot take some of the workload off the working cores. This also limits the number of cores that can be used simultaneously to the number of partitions to optimize.

The Scikit-Learn K-Means implementation that we use here works very differently. While it also starts off with 10 random partitions of the data, in contrast to sIB, it iterates over these 10 partitions sequentially and uses parallel computing to optimize each partition using all available cores.

The advantage of this approach is that it utilizes all available machine power. Therefore, even when running on a server with a high number of cores (e.g., 100), this K-Means implementation will be able to utilize all of them.

The downside here is that during the optimization of a partition, the assignment stage of K-Means is a stage consisting of a high number of relatively lightweight and short computations of Euclidean distance between a sample and a list of centroids. Thus, every core is utilized for a very short task and then immediately switched to another with no continuity.

Our hypothesis is that the overhead of intensive task switching in such a fine-grained parallelism model is what holds K-Means back. Had sIB and K-Means been implemented with the same parallelism, it would be expected from K-Means to be much faster. However, establishing this hypothesis is beyond the scope of this paper.

As a final remark, we note that even though a different, faster, K-Means implementation can be offered, the main reason for the slow performance of sIB's main competitor—the setup of K-Means on top of S-Bert—is its power-hungry neural vectorization stage and not the clustering itself.

References

1. Aggarwal, C.C.; Zhai, C. A Survey of Text Clustering Algorithms. In *Mining Text Data*; Aggarwal, C.C., Zhai, C., Eds.; Springer US: Boston, MA, USA, 2012; pp. 77–128. [CrossRef]
2. Huang, A. Similarity measures for text document clustering. In Proceedings of the Sixth New Zealand Computer Science Research Student Conference (NZCSRSC2008), Christchurch, New Zealand, 6–9 April 2008; Volume 4, pp. 9–56.
3. Abualigah, L.M.Q. *Feature Selection and Enhanced Krill Herd Algorithm for Text Document Clustering*, 1st ed.; Springer Publishing Company, Incorporated: Berlin/Heidelberg, Germany, 2018.
4. Roy, S.; Muni, D.P.; Tack Yan, J.J.Y.; Budhiraja, N.; Ceiler, F. Clustering and Labeling IT Maintenance Tickets. In Proceedings of the Service-Oriented Computing, Banff, AB, Canada, 10–13 October 2016; Sheng, Q.Z., Stroulia, E., Tata, S., Bhiri, S., Eds.; pp. 829–845.
5. Compton, J.E.; Adams, M.C. *Clustering Support Tickets with Natural Language Processing: K-Means Applied to Transformer Embeddings*; Sandia National Lab. (SNL-NM): Albuquerque, NM, USA, 2020.

6. Poomagal, S.; Visalakshi, P.; Hamsapriya, T. A novel method for clustering tweets in Twitter. *Int. J. Web Based Commun.* **2015**, *11*, 170–187. [CrossRef]

7. Rosa, K.D.; Shah, R.; Lin, B.; Gershman, A.; Frederking, R.E. Topical Clustering of Tweets. In Proceedings of the ACM SIGIR: SWSM, Beijing, China, 28 July 2011.

8. Curiskis, S.A.; Drake, B.; Osborn, T.R.; Kennedy, P.J. An evaluation of document clustering and topic modelling in two online social networks: Twitter and Reddit. *Inf. Process. Manag.* **2020**, *57*, 102034. [CrossRef]

9. Lloyd, S.P. Least Squares Quantization in PCM. *IEEE Trans. Inf. Theor.* **1982**, *28*, 129–137. [CrossRef]

10. Pedregosa, F.; Varoquaux, G.; Gramfort, A.; Michel, V.; Thirion, B.; Grisel, O.; Blondel, M.; Prettenhofer, P.; Weiss, R.; Dubourg, V.; et al. Scikit-learn: Machine Learning in Python. *J. Mach. Learn. Res.* **2011**, *12*, 2825–2830.

11. Salton, G.; McGill, M.J. *Introduction to Modern Information Retrieval*; McGraw-Hill, Inc.: New York, NY, USA, 1986.

12. Bellman, R. Dynamic programming. *Science* **1966**, *153*, 34–37. [CrossRef] [PubMed]

13. Mikolov, T.; Chen, K.; Corrado, G.; Dean, J. Efficient Estimation of Word Representations in Vector Space. *arXiv* **2013**, arXiv:1301.3781.

14. Mikolov, T.; Sutskever, I.; Chen, K.; Corrado, G.; Dean, J. Distributed Representations of Words and Phrases and their Compositionality. *Adv. Neural Inf. Process. Syst.* **2013**, *26*, 1421. [CrossRef]

15. Pennington, J.; Socher, R.; Manning, C.D. GloVe: Global Vectors for Word Representation. In Proceedings of the Empirical Methods in Natural Language Processing (EMNLP), Doha, Qatar, 25–29 October 2014; pp. 1532–1543.

16. Bengio, Y.; Ducharme, R.; Vincent, P.; Janvin, C. A Neural Probabilistic Language Model. *J. Mach. Learn. Res.* **2003**, *3*, 1137–1155.

17. Vaswani, A.; Shazeer, N.; Parmar, N.; Uszkoreit, J.; Jones, L.; Gomez, A.N.; Kaiser, L.; Polosukhin, I. Attention Is All You Need. *CoRR* **2017**, *30*, 3058.

18. Devlin, J.; Chang, M.W.; Lee, K.; Toutanova, K. BERT: Pre-training of Deep Bidirectional Transformers for Language Understanding. In Proceedings of the 2019 Conference of the North American Chapter of the Association for Computational Linguistics: Human Language Technologies, Volume 1 (Long and Short Papers), Minneapolis, MN, USA, 2–7 June 2019; pp. 4171–4186. [CrossRef]

19. Wang, A.; Singh, A.; Michael, J.; Hill, F.; Levy, O.; Bowman, S.R. GLUE: A Multi-Task Benchmark and Analysis Platform for Natural Language Understanding. *arXiv* **2018**, arXiv:1804.07461.

20. Wang, A.; Pruksachatkun, Y.; Nangia, N.; Singh, A.; Michael, J.; Hill, F.; Levy, O.; Bowman, S.R. SuperGLUE: A Stickier Benchmark for General-Purpose Language Understanding Systems. *Adv. Neural Inf. Process. Syst.* **2019**, *32*, 1828.

21. Reimers, N.; Gurevych, I. Sentence-BERT: Sentence Embeddings using Siamese BERT-Networks. *arXiv* **2019**, arXiv:1908.10084.

22. Slonim, N.; Friedman, N.; Tishby, N. Unsupervised Document Classification Using Sequential Information Maximization. In Proceedings of the 25th Annual International ACM SIGIR Conference on Research and Development in Information Retrieval, SIGIR'02, Tampere, Finland, 11–15 August 2002; pp. 129–136. [CrossRef]

23. Lang, K. NewsWeeder: Learning to Filter Netnews. In Proceedings of the 12th International Machine Learning Conference (ML95), Tahoe City, CA, USA, 9–12 July 1995.

24. Connor, R.C.H.; Cardillo, F.A.; Moss, R.; Rabitti, F. Evaluation of Jensen-Shannon Distance over Sparse Data. In Proceedings of the SISAP, A Coruna, Spain, 2–4 October 2013.

25. Tishby, N.; Pereira, F.C.; Bialek, W. The information bottleneck method. *arXiv* **2000**, arXiv:physics/0004057.

26. Slonim, N. The Information Bottleneck: Theory and Applications. Ph.D. Thesis, Hebrew University of Jerusalem, Jerusalem, Israel, 2002.

27. Cover, T.M.; Thomas, J.A. *Elements of Information Theory (Wiley Series in Telecommunications and Signal Processing)*; Wiley-Interscience: Hoboken, NJ, USA, 2006.

28. Slonim, N.; Tishby, N. Agglomerative Information Bottleneck. In Proceedings of the Advances in Neural Information Processing Systems, Denver, CO, USA, 29 November–4 December 1999; Solla, S., Leen, T., Müller, K., Eds.; Volume 12.

29. Zhang, J.A.; Kurkoski, B.M. Low-complexity quantization of discrete memoryless channels. In Proceedings of the 2016 International Symposium on Information Theory and Its Applications (ISITA), Monterey, CA, USA, 3 October–2 November 2016; pp. 448–452.

30. Chou, P. Optimal partitioning for classification and regression trees. *IEEE Trans. Pattern Anal. Mach. Intell.* **1991**, *13*, 340–354. [CrossRef]

31. Banerjee, A.; Merugu, S.; Dhillon, I.S.; Ghosh, J. Clustering with Bregman Divergences. *J. Mach. Learn. Res.* **2005**, *6*, 1705–1749.

32. Kurkoski, B.M. On the relationship between the KL means algorithm and the information bottleneck method. In Proceedings of the SCC 2017, 11th International ITG Conference on Systems, Communications and Coding, Hamburg, Germany, 6–9 February 2017; pp. 1–6.

33. Slonim, N.; Aharoni, E.; Crammer, K. Hartigan's K-Means versus Lloyd's K-Means: Is It Time for a Change? In Proceedings of the Twenty-Third International Joint Conference on Artificial Intelligence, IJCAI'13, Beijing, China, 3–9 August 2013; pp. 1677–1684.

34. Greene, D.; Cunningham, P. Practical Solutions to the Problem of Diagonal Dominance in Kernel Document Clustering. In Proceedings of the 23rd International Conference on Machine Learning (ICML'06), New York, NY, USA, 25–29 June 2006; pp. 377–384.

35. Zhang, X.; Zhao, J.; LeCun, Y. Character-level Convolutional Networks for Text Classification. *Adv. Neural Inf. Process. Syst.* **2015**, *28*, 456. [CrossRef]

36. Vinh, N.X.; Epps, J.; Bailey, J. Information Theoretic Measures for Clusterings Comparison: Variants, Properties, Normalization and Correction for Chance. *J. Mach. Learn. Res.* **2010**, *11*, 2837–2854.
37. Meilă, M. Comparing clusterings—An information based distance. *J. Multivar. Anal.* **2007**, *98*, 873–895. [CrossRef]
38. Rosenberg, A.; Hirschberg, J. V-Measure: A Conditional Entropy-Based External Cluster Evaluation Measure. In Proceedings of the 2007 Joint Conference on Empirical Methods in Natural Language Processing and Computational Natural Language Learning (EMNLP-CoNLL), Prague, Czech Republic, 28–30 June 2007; pp. 410–420.
39. Arthur, D.; Vassilvitskii, S. K-Means++: The Advantages of Careful Seeding. In Proceedings of the Eighteenth Annual ACM-SIAM Symposium on Discrete Algorithms, New Orleans, Louisiana, 7–9 January 2007; pp. 1027–1035.
40. Spencer, N. *Essentials of Multivariate Data Analysis*; Taylor & Francis: Abingdon, UK, 2013; p. 95.

Article

Minimum-Integer Computation Finite Alphabet Message Passing Decoder: From Theory to Decoder Implementations towards 1 Tb/s

Tobias Monsees [1,*], Oliver Griebel [2], Matthias Herrmann [2], Dirk Wübben [1], Armin Dekorsy [1] and Norbert Wehn [2]

[1] Department of Communications Engineering, University of Bremen, 28359 Bremen, Germany
[2] Microelectronic Systems Design Research Group, TU Kaiserslautern, 67663 Kaiserslautern, Germany
* Correspondence: tmonsees@ant.uni-bremen.de; Tel.: +49-421-218-62407

Abstract: In Message Passing (MP) decoding of Low-Density Parity Check (LDPC) codes, extrinsic information is exchanged between Check Nodes (CNs) and Variable Nodes (VNs). In a practical implementation, this information exchange is limited by quantization using only a small number of bits. In recent investigations, a novel class of Finite Alphabet Message Passing (FA-MP) decoders are designed to maximize the Mutual Information (MI) using only a small number of bits per message (e.g., 3 or 4 bits) with a communication performance close to high-precision Belief Propagation (BP) decoding. In contrast to the conventional BP decoder, operations are given as discrete-input discrete-output mappings which can be described by multidimensional LUTs (mLUTs). A common approach to avoid exponential increases in the size of mLUTs with the node degree is given by the sequential LUT (sLUT) design approach, i.e., by using a sequence of two-dimensional Lookup-Tables (LUTs) for the design, leading to a slight performance degradation. Recently, approaches such as Reconstruction-Computation-Quantization (RCQ) and Mutual Information-Maximizing Quantized Belief Propagation (MIM-QBP) have been proposed to avoid the complexity drawback of using mLUTs by using pre-designed functions that require calculations over a computational domain. It has been shown that these calculations are able to represent the mLUT mapping exactly by executing computations with infinite precision over real numbers. Based on the framework of MIM-QBP and RCQ, the Minimum-Integer Computation (MIC) decoder design generates low-bit integer computations that are derived from the Log-Likelihood Ratio (LLR) separation property of the information maximizing quantizer to replace the mLUT mappings either exactly or approximately. We derive a novel criterion for the bit resolution that is required to represent the mLUT mappings exactly. Furthermore, we show that our MIC decoder has exactly the communication performance of the corresponding mLUT decoder, but with much lower implementation complexity. We also perform an objective comparison between the state-of-the-art Min-Sum (MS) and the FA-MP decoder implementations for throughput towards 1 Tb/s in a state-of-the-art 28 nm Fully-Depleted Silicon-on-Insulator (FD-SOI) technology. Furthermore, we demonstrate that our new MIC decoder implementation outperforms previous FA-MP decoders and MS decoders in terms of reduced routing complexity, area efficiency and energy efficiency.

Keywords: LDPC code; decoding; finite alphabet message passing; information bottleneck; implementation efficiency

Citation: Monsees, T.; Griebel, O.; Herrmann, M.; Wübben, D.; Dekorsy, A.; Wehn, N. Minimum-Integer Computation Finite Alphabet Message Passing Decoder: From Theory to Decoder Implementations towards 1 Tb/s. *Entropy* **2022**, *24*, 1452. https://doi.org/10.3390/e24101452

Academic Editor: Syed A. Jafar

Received: 26 August 2022
Accepted: 8 October 2022
Published: 12 October 2022

Publisher's Note: MDPI stays neutral with regard to jurisdictional claims in published maps and institutional affiliations.

1. Introduction

Beyond 5G and 6G wireless communication systems, target peak data rates of 100 Gb/s to 1 Tb/s with processing latencies between 10–100 ns [1]. For such high data rate and low latency requirements, the implementation of a Forward Error Correction (FEC) decoder, which is one of the most complex and computationally intense components in the baseband processing chain, is a major challenge [2]. Low-Density Parity Check (LDPC) codes [3] are FEC codes with capacity approaching error correction performance [4] and are part of many communication standards, e.g., DVB-S2x, Wi-Fi, and 3GPP 5G-NR. In contrast to other competitive FEC codes, like Polar and Turbo codes, the decoding of LDPC codes is dominated by data transfers [2] making very high-throughput decoders in advanced silicon technologies challenging, especially from routing and energy efficiency perspectives. For example, in a state-of-the-art 14 nm silicon technology, the transfer of 8 bits on a 1 mm wire costs about 1 pJ, whereas the cost of an 8 bit integer addition is only 10 fJ, which is two orders of magnitude less than the wiring energy cost. During Message Passing (MP) decoding, two sets of nodes, the Check Node (CN) and Variable Node (VN), iteratively exchange messages over the edges of a bipartite graph (Tanner graph of the LDPC code). High-throughput decoding can be achieved by mapping the Tanner graph one-to-one onto hardware, i.e., dedicated processing units are instantiated for each node and the edges of the Tanner graph are hardwired. Unrolling and pipelining the decoding iterations can further boost the throughput towards 1 Tb/s [5], called unrolled full parallel (FP) decoders in the following. However, FP decoders imply large routing challenges, since every edge in the Tanner graph corresponds to $2 \cdot I \cdot n_E$ wires, with I being the number of decoding iterations and n_E being the quantization-width of the exchanged messages. Moreover, to enable good error correction performance, the Tanner graph exhibits limited locality and regularity, which makes efficient routing even more difficult. This problem is even exacerbated in advanced silicon technologies, as routing scales much worse than transistor density [6].

Finite Alphabet Message Passing (FA-MP) decoding has been investigated as a method to mitigate the routing challenges in FP LDPC decoders to reduce the bit-width, i.e., the quantization-width n_E, of the exchanged messages and, thus, the number of necessary wires [7–9]. In contrast to conventional MP decoding algorithms like the Belief Propagation (BP) and its approximations, i.e., Min-Sum (MS), Offset Min-Sum (OMS) and Normalized Min-Sum (NMS) [10], FA-MP use non-uniform quantizers and the node operations are derived by maximizing MI between exchanged messages. Nodes in state-of-the-art FA-MP decoders have to be implemented as Lookup-Tables (LUTs). Since the size of the LUT exponentially increases with the node degree and n_E, investigations were performed to decompose this multidimensional LUT (mLUT) into a chain or tree with only two-input LUTs (denoted as sequential LUT (sLUT) in this paper) yielding only a linear dependency of the node degree but at the cost of a decreased communications performance [11,12]. The Minimum-LUT (Min-LUT) decoder [13] approximates the CN update by a simple minimum search and can be implemented as Minimum-mLUT (Min-mLUT) or Minimum-sLUT (Min-sLUT), i.e., with mLUT or sLUT for VNs, respectively. Other approaches, e.g., Mutual Information-Maximizing Quantized Belief Propagation (MIM-QBP) [14–16] and Reconstruction-Computation-Quantization (RCQ) [17,18], are adding non-uniform quantizers and reconstruction mappings to the outputs and inputs of the nodes, respectively, and performing the standard functional operations inside the nodes, e.g., additions for VNs and minimum search for CNs. The reconstruction mappings generally increase the bit resolution required for node internal representation and processing. It can be shown that this approach is equivalent in terms of error correction performance compared to the mLUT, if the internal quantization after the reconstruction mapping is sufficiently large.

Based on the framework of MIM-QBP and RCQ, the proposed MIC decoder [19] realizes CN updates by a minimum search and VN updates by integer computations that are designed to realize the information maximizing mLUT mappings either exactly or approximately. In this paper, we provide more detailed explanations, extend the discussion to irregular LDPC codes and present a comprehensive implementation analysis. The new contributions of this paper (*Notation:* Random variables are denoted by sans-serif letters x, random vectors by bold sans-serif letters $\mathbf{x}$, realizations by serif letters x and vector-valued realizations by bold serif letters $\boldsymbol{x}$. Sets are denoted by calligraphic letters $\mathcal{X}$. The distribution $p_{\mathsf{x}}(x)$ of a random variable x is abbreviated as $p(x)$. $\mathsf{x} \to \mathsf{y} \to \mathsf{z}$

denotes a Markov chain, and $\mathbb{R}$, $\mathbb{Z}$, $\mathbb{F}_2$ denotes the real numbers, integers and Galois field 2, respectively.) are summarized as follows:

- We provide a novel criterion for the resolution of internal node operations to ensure that the MIC decoder can always replace the information maximizing VN mLUT exactly;
- we show that this MIC decoder has the same communication performance compared to an MI maximizing Min-mLUT decoder;
- we make an objective comparison between different FA-MP decoder implementations (Min-mLUT, Min-sLUT, MIC) in an advanced silicon technology and compare them with a state-of-the-art MS decoder for throughput towards 1 Tb/s;
- we show that our MIC decoder implementation outperforms state-of-the-art FP decoders in terms of routing complexity, area efficiency and energy efficiency and enables the processing of larger block sizes in state-of-the-art FP decoders since the routing complexity is largely reduced.

The remainder of this paper is structured as follows: Section 2 reviews the system model, conventional decoding techniques for LDPC codes such as BP and NMS decoding, and Information Bottleneck (IB) based quantization. Section 3 describes the Min-mLUT and Min-sLUT decoder design for regular and irregular LDPC codes. In Section 4, we introduce the proposed MIC decoder and, in Section 5, we discuss the MIC decoder implementation along with a detailed comparison with state-of-the-art FP MP decoders. Finally, Section 6 concludes the paper.

2. Preliminaries

This section briefly reviews the transmission model, conventional decoding techniques for LDPC codes, and the quantizer design based on IB.

2.1. Transmission Model

The transmission model is shown in Figure 1. An information word $u \in \mathbb{F}_2^K$ is encoded into the codeword $c \in \mathbb{F}_2^N$ via a binary LDPC code [3] of rate $R = \frac{K}{N}$. The Binary Phase Shift Keying (BPSK) modulated vector $x = 1 - 2c$ is transmitted over an Additive White Gaussian Noise (AWGN) channel leading to the received vector $y \in \mathbb{R}^N$ given by $y = x + n$ with AWGN n of variance σ_n^2. A particular LDPC code is defined via a sparse parity check matrix $H \in \mathbb{F}_2^{M \times N}$. The Tanner graph [20] of an LDPC code is a visual representation of its parity check matrix H and consists of a CN for each parity check equation χ_m with $m = 1, ..., M$ and a VN for each codebit c_n with $n = 1, ..., N$. An edge connects VN n and CN m if and only if $H_{m,n} = 1$. The degree of a node is determined by the number of connected edges. Furthermore, the fraction of edges that is connected to a node of a specific degree is characterized by the edge-degree distributions

$$\lambda(\xi) = \sum_{d_V \in \mathcal{D}_V} \lambda_{d_V} \xi^{d_V - 1} \quad \text{and} \quad \rho(\xi) = \sum_{d_C \in \mathcal{D}_C} \rho_{d_C} \xi^{d_C - 1} \tag{1}$$

where λ_{d_V} is the fraction of edges that are connected to VNs of degree $d_V \in \mathcal{D}_V$, and ρ_{d_C} denotes the fraction of edges that is connected to CNs of degree $d_C \in \mathcal{D}_C$.

$u \in \mathbb{F}_2^K$ → | LDPC | → $c \in \mathbb{F}_2^N$ → | BPSK | → $x \in \mathcal{X}^N$ → | AWGN | → $y \in \mathbb{R}^N$ → | Quantizer | → $z \in \mathcal{Z}^N$ → | Decoder | → $\hat{u} \in \mathbb{F}_2^K$

Figure 1. Transmission model for transmission of LDPC encoded messages over an AWGN channel with quantization prior to FEC decoding.

2.2. Iterative Decoding via Belief-Propagation (BP)

LDPC codes are usually decoded by iterative BP, where *extrinsic information* for each codebit c_n is propagated along the edges of the resulting Tanner graph. Figure 2 shows the CN χ_1 that generates *extrinsic information* for the VN c_n by processing the incoming

Variable Node to Check Node (VN-to-CN) messages from the other VNs connected to CN χ_1, i.e., c_1 and c_2. For BP decoding, we define the Variable Node to Check Node (VN-to-CN) messages as $L_{n \to m} \in \mathbb{R}$ and the Check Node to Variable Node (CN-to-VN) messages as $L_{n \leftarrow m} \in \mathbb{R}$. In the first iteration, all messages are initialized by the channel LLRs

$$L_{n \to m}^{(0)} = L(y_n) = \frac{2}{\sigma_n^2} y_n \, . \tag{2}$$

In iteration i, a CN m generates *extrinsic information* for the connected VNs $n \in \mathcal{M}_m$ via the box-plus operation

$$L_{n \leftarrow m}^{(i)} = 2 \operatorname{arctanh} \left(\prod_{j \in \mathcal{M}_m \setminus n} \tanh \left(\tfrac{1}{2} L_{j \to m}^{(i-1)} \right) \right) , \quad \forall n \in \mathcal{M}_m \, . \tag{3}$$

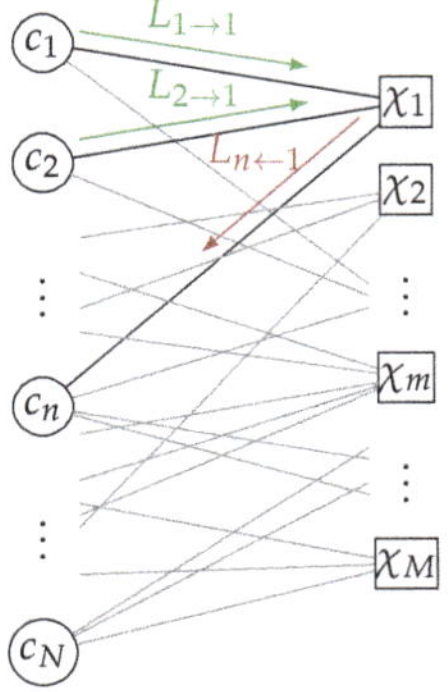

Figure 2. Illustrative example of a CN update on a Tanner graph. The CN χ_1 generates the CN-to-VN message $L_{n \leftarrow 1}$ for the VN c_n based on the VN-to-CN messages $L_{1 \to 1}$ and $L_{2 \to 1}$ from VN c_1 and c_2, respectively.

In case of Normalized Min-Sum (NMS) decoding, the CN update (3) is approximated by

$$L_{n \leftarrow m}^{(i)} \approx \gamma \left(\prod_{j \in \mathcal{M}_m \setminus n} \operatorname{sign} \left(L_{j \to m}^{(i-1)} \right) \right) \min_{j \in \mathcal{M}_m \setminus n} \left| L_{j \to m}^{(i-1)} \right| , \quad \forall n \in \mathcal{M}_m \, . \tag{4}$$

where γ is the normalization factor. In the case of $\gamma = 1$, (4) is the CN update of the MS decoder.

In similar fashion, a VN n generates *extrinsic information* for the connected CNs $m \in \mathcal{N}_n$ by adding the corresponding LLRs

$$L_{n \to m}^{(i)} = L(y_n) + \sum_{v \in \mathcal{N}_n \setminus m} L_{n \leftarrow v}^{(i)} , \quad \forall m \in \mathcal{N}_n \, . \tag{5}$$

The final bit decision $\hat{c}_{n,\text{BP}}^{(i)}$ at iteration i is determined by

$$\hat{c}_{n,\text{BP}}^{(i)} = \frac{1}{2} \left(1 - \operatorname{sign} \left(L(y_n) + \sum_{v \in \mathcal{N}_n} L_{n \leftarrow v}^{(i)} \right) \right) . \tag{6}$$

2.3. Information Bottleneck Based Quantizer Design

For the design of our proposed MIC decoder, we utilize MI maximizing quantization to design an information optimized processing chain that uses only quantizer labels instead of real valued representations [12]. To that end, we first review the principle idea of the MI based quantizer design approach. The considered system model is visualized in Figure 3. The observed signal $y \in \mathcal{Y}$ is mapped to a compressed representation $z \in \mathcal{Z}$ via the scalar quantization function $Q : \mathcal{Y} \rightarrow \mathcal{Z}$. The objective is to find a quantizer function $Q^\star$ that maximizes MI $I(x; z)$ between the relevant source $x \in \mathcal{X}$ and the quantizer output $Q(y) = z \in \mathcal{Z}$ under the condition that the three random variables form a Markov chain $x \rightarrow y \rightarrow z$. Given the joint distribution $p(x, y) = p(y|x)p(x)$, the mapping of the information maximizing quantizer $Q^\star$ is determined by solving the optimization problem

$$Q^\star = \underset{Q}{\arg\max}\, I(x; z) \quad \text{s.t.} \quad |\mathcal{Z}| = 2^{n_Q} < |\mathcal{Y}| \tag{7}$$

where the number of possible quantizer outputs is set to 2^{n_Q}. The optimization problem in (7) is a special case of the Information Bottleneck Method (IBM) [12,21–23]. The optimal solution is a deterministic quantization function where the conditional probability of the quantizer output z given the relevant source x is

$$p(z|x) = \sum_{y \in \mathcal{Y}_z} p(y|x) \tag{8}$$

with $\mathcal{Y}_z = \{y \in \mathcal{Y} \mid Q^\star(y) = z\}$ as the set of observed signals y that are mapped to one specific quantizer output z. Since the maximum of (7) depends only on the cardinality of $\mathcal{Z}$, we utilize a convenient signed integer based representation $\mathcal{Z} = \{-\frac{2^{n_Q}}{2}, ..., -1, 1, ..., \frac{2^{n_Q}}{2}\}$ that simplifies the MIC decoder processing. For the special case where the relevant source x is a binary random variable (i.e., $|\mathcal{X}| = 2$), the algorithm that finds the optimal quantizer via dynamic programming has been derived in [24]. We denote the LLRs of the quantizer output $z \in \mathcal{Z}$ by

$$L(z) = \log\left(\frac{p(z|x = +1)}{p(z|x = -1)}\right). \tag{9}$$

An important property of the MI maximizing quantizer for binary input is that any two different sets of LLRs $\mathcal{L}_{z'} = \{L(y) \,|\, y \in \mathcal{Y}_{z'}\}$ and $\mathcal{L}_{z''} = \{L(y) \,|\, y \in \mathcal{Y}_{z''}\}$ for $z', z'' \in \mathcal{Z}$ and $z' \neq z''$ are separated by a single threshold [19,24,25]. This property will be exploited in the design of the MIC decoder in Section 4.

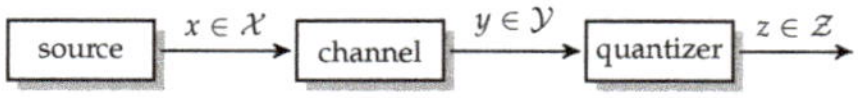

source $\xrightarrow{x \in \mathcal{X}}$ channel $\xrightarrow{y \in \mathcal{Y}}$ quantizer $\xrightarrow{z \in \mathcal{Z}}$

Figure 3. Considered system model for quantizer design.

3. LUT Decoder Design

This section describes the design of the LUT decoder that is optimized via Discrete Density Evolution (DDE) [11] to maximize *extrinsic information* between the codebits and its messages, under the assumption that the Tanner graph is cycle free. In contrast to the BP algorithm, the LUT is optimized to process the quantizer labels z in (7) directly and the bit resolution of the message exchange on the Tanner graph is limited to n_E bits, e.g., 3 or 4 bits. Furthermore, we exploit the signed integer-based representation to simplify the CN update by using the label-based minimum search [13]. In the Min-mLUT decoder design, the VN update functions are optimized to maximize MI. For the Min-sLUT decoder design, the VN update is decomposed into a sequence of two-dimensional updates that generally results in a MI loss compared to the Min-mLUT decoder design.

In the following, we review the calculation of the CN and VN distributions for each iteration that are required for the design of the MI maximizing VN update. As illustrated in Figure 4, we omit the iteration index i and consider messages of an arbitrary CN and

VN for CN degrees $d_C \in \mathcal{D}_C$ and VN degrees $d_V \in \mathcal{D}_V$ to calculate the distributions that are required for the Min-mLUT design.

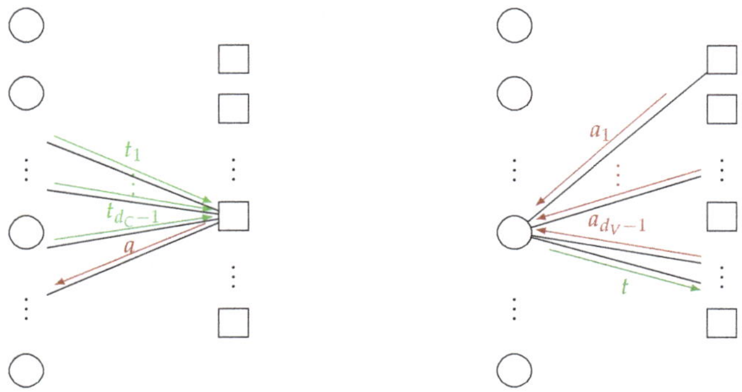

(a) CN update   (b) VN update

Figure 4. Illustrative example for generation of extrinsic information in case of LUT decoding using discrete messages. (**a**) visualizes a CN that generates the CN-to-VN message a based on incoming VN-to-CN messages $t_1, ..., t_{d_C-1}$. In (**b**), a VN generates the VN-to-CN message t based on incoming CN-to-VN messages $a_1, ..., a_{d_V-1}$.

3.1. Check Node LUT Design

The LUT decoder design is based on discrete alphabets $\mathcal{Z}$, $\mathcal{T}$ and $\mathcal{A}$ for the channel information, the VN-to-CN and the CN-to-VN messages, respectively. For the first iteration, the VN-to-CN messages t_j for $j = 1, ..., d_V - 1$ are initialized by the signed integer valued channel information, i.e., $t_j = z_j \in \mathcal{Z}$. The distribution of the $d_C - 1$ VN-to-CN messages $t = [t_1, ..., t_{d_C-1}] \in \mathcal{T}^{(d_C-1)}$ and an arbitrary codebit c of a check equation χ is [11]

$$p_{d_C}(t|c) = \left(\frac{1}{2}\right)^{d_C-2} \sum_{b:\oplus b=c} \prod_{j=1}^{d_C-1} p(t_j|b_j) \tag{10}$$

with $\oplus b = b_1 \oplus ... \oplus b_{d_C-1}$ as the modulo 2 sum of connected codebits. The VN-to-CN messages t_j are processed by a CN update function that generates quantized output messages $a \in \mathcal{A}$ that are represented only by n_E bits.

Given the distribution in (10), the CN update (We keep the node degrees d_C or d_V as index of random variables to indicate that the distribution changes with the corresponding degrees.) $f_{d_C}(\mathbf{t}_{d_C}) = \mathsf{a}_{d_C}$ that maximizes MI is determined by the solution of the quantization problem for binary input ($\mathsf{c} \to \mathbf{t}_{d_C} \to f_{d_C}(\mathbf{t}_{d_C}) = \mathsf{a}_{d_C}$)

$$f_{d_C}^{\mathrm{MI}} = \underset{f_{d_C}}{\arg\max}\; I\left(\mathsf{c}; \mathbf{t}_{d_C}\right) \text{ s.t. } |\mathcal{A}| = 2^{n_E} \quad \text{for } d_C \in \mathcal{D}_C. \tag{11}$$

As discussed in Section 2.3, the optimal solution of (11) is found via dynamic programming.

However, we utilize the minimum update [13] as a CN update for all iterations as an approximation of the MI maximizing CN update in (11). We observed that the output of the minimum update is quite close to the optimal IB update. As visualized for a degree 3 CN in Figure 5, the difference between the optimal IB CN and the minimum update can be interpreted as an additive *correction* LUT where only a small fraction of entries are nonzero. For the label-based minimum search, the CN update rule reads

$$a = f_{d_C}^{\min}(t) = \left(\prod_{j=1}^{d_C-1} \mathrm{sign}(t_j)\right) \min\{|t_1|, ..., |t_{d_C-1}|\}. \tag{12}$$

If the CN update function is given, the conditional distribution of the CN-to-VN messages $a \in \mathcal{A} = \mathcal{T}$ is

$$p_{d_C}(a|c) = \sum_{t \in \mathcal{T}^{(d_C-1)} : f_{d_C}^{\min}(t)=a} p_{d_C}(t|c) \quad \text{for} \ \ d_C \in \mathcal{D}_C. \tag{13}$$

In the design via DDE, the connections between VNs and CNs are considered on average by the degree distribution [26]. Hence, the design considers only the marginal CN-to-VN message distribution $p(a|c)$ that includes averaging over all possible CN degrees by

$$p(a|c) = \sum_{d_C \in \mathcal{D}_C} p_{d_C}(a|c)\rho_{d_C}. \tag{14}$$

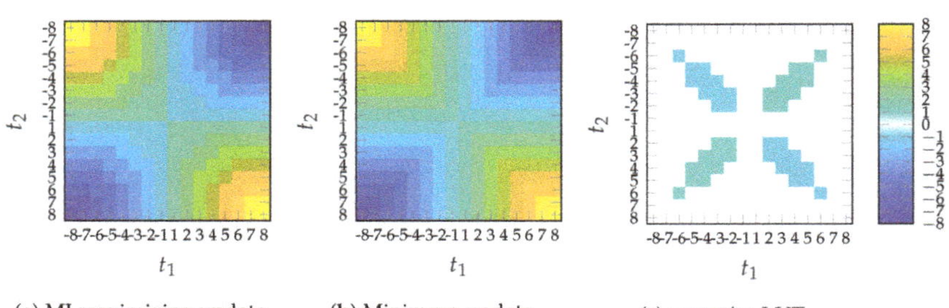

  (a) MI maximizing update   (b) Minimum update   (c) *correction* LUT

Figure 5. Graphical representation of a discrete CN update using $n_E = 4$ bit input messages t_1 and t_2 and a color-coded output message $a \in \mathcal{A} = \{-4, ..., -1, 1, ..., 4\}$. Subfigure (**a**) shows the MI maximizing update $f_3^{MI}(t_1, t_2)$ and subfigure (**b**) the minimum update $f_3^{\min}(t_1, t_2)$. The difference $f_3^{MI}(t_1, t_2) - f_3^{\min}(t_1, t_2)$ in subfigure (**c**) contains only a few non-zero elemets and can be interpreted as a *correction* LUT.

3.2. Variable Node LUT Design

For designing the VN update, we require the joint distribution of the discrete channel information $z \in \mathcal{Z}$ together with the CN-to-VN messages $a_m \in \mathcal{A}$ combined in $\mathbf{a} = [z, a_1, ..., a_{d_V-1}] \in \mathcal{Z} \times \mathcal{A}^{d_V-1} = \mathcal{V}$ and a codebit c [11]

$$p_{d_V}(\mathbf{a}|c) = p(z|c) \prod_{m=1}^{d_V-1} p(a_m|c) \tag{15}$$

where $p(a_m|c) = p(a|c)$ for $m = 1, ..., d_{V,\max} - 1$ and $\mathcal{V}$ is the set of all possible states of the vector $\mathbf{a}$, i.e., $|\mathcal{V}| = 2^{n_Q+(d_V-1)n_E}$. Given the distribution (15), the individual degree-dependent VN update $g_{d_V}(\mathbf{a}_{d_V}) = \mathsf{t}_{d_V}$ that maximizes MI $I(c; \mathsf{t}_{d_V})$ is determined as the solution of the optimization problem $(c \rightarrow \mathbf{a}_{d_V} \rightarrow g_{d_V}(\mathbf{a}_{d_V}) = \mathsf{t}_{d_V})$

$$g_{d_V}^{MI} = \underset{g_{d_V}}{\arg\max} \ I\big(c; \mathsf{t}_{d_V}\big) \ \text{s.t.} \ |\mathcal{T}| = 2^{n_E} \quad \text{for} \ \ d_V \in \mathcal{D}_V. \tag{16}$$

The parameter n_E defines the bit-width of the messages exchanged between VN and CN and controls the complexity of the message exchange. The optimization problem in (16) is the channel quantization problem for binary input (Section 2.3). The optimal solution is a deterministic input–output relation that can be stored as a d_V dimensional LUT with $2^{n_Q+(d_V-1)n_E}$ entries, e.g., for $d_V = 6$ and $n_E = n_Q = 4$, we have approximately 16.8 million entries. Furthermore, the communication performance can be increased by considering the degree distribution in the design of the node updates [13,26]. The gain in communication performance generally depends on the degree distribution and the message resolution n_E [13]. However, a comparison of the different design approaches in [13,26] is beyond the scope of this paper. The distribution of the VN-to-CN messages for the next iteration in (10) is

$$p_{d_V}(t|c) = \sum_{a \in \mathcal{V}\,:\,g_{d_V}^{\text{MI}}(a)=t} p(a|c) \quad \text{for } d_V \in \mathcal{D}_V. \tag{17}$$

Again, the marginal distribution is determined by averaging over all possible VN degrees, i.e.,

$$p(t_j|c_j) = p(t|c) = \sum_{d_V \in \mathcal{D}_V} p_{d_V}(t|c)\lambda_{d_V}, \quad \text{with } j = 1, ..., d_{C,\text{max}}. \tag{18}$$

In case of a regular LDPC code, there is only one possible degree for all VNs and CNs, i.e., the summation term in (14) and (18) vanishes but all other steps remain the same.

For the design of the MI maximizing Min-mLUT decoder, we start with an initial VN-to-CN distribution $p(t_j|c_j)$ and iterate over (10), (13)–(18) and declare convergence if $I(\text{c};\text{t})$ approaches the maximum value of one bit for binary input after I number of iterations.

3.3. Sequential LUT Design

For the sequential design approach sLUT, the node update is split into a sequence of degree two updates that are optimized independently to maximize MI. This approach serves as an approximation of the mLUT design described in Section 3.2 and reduces the number of possible memory locations within each update. In general, multidimensional optimization without decomposition conserves more MI compared to a design that decomposes the optimization problem into a sequence of two-dimensional updates [11,12] or more general nested tree decompositions [13].

4. Minimum-Integer Computation Decoder Design

The MI maximizing Min-mLUT decoder realizes the discrete VN updates by LUTs with $2^{n_Q+(d_V-1)n_E}$ entries leading to prohibitively large implementation complexity. Nevertheless, determining these multidimensional LUTs in the laboratory is feasible with sufficient computing resources. Thus, the idea is to search for the MI maximizing mLUTs but implement the corresponding discrete functions by relatively simple operations in order to avoid performance degradations. As visualized in Figure 6, the computational domain framework [14,16] replaces the VN update by an operation that is decomposed into

(i)   mappings ϕ_V and ϕ of the n_E-bit CN-to-VN messages a_m and n_Q-bit channel information z into *node internal* n_R-bit signed integers with $n_R \geq n_E$ and $n_R \geq n_Q$, respectively;

(ii)   execution of integer additions for n_R-bit signed integers;

(iii)   threshold quantization Q_V to n_E bits determining the VN-to-CN message t.

For the MIC decoder design, we derive a criterion for sufficient internal node resolution n_R such that the mLUT mapping is replaced exactly. Note that the information maximizing mLUT is generated offline and is replaced by an integer function that replaces its functionality exactly or approximately during execution.

To keep the notation simple, we omit the dependency on the iteration index i and node degree in this section.

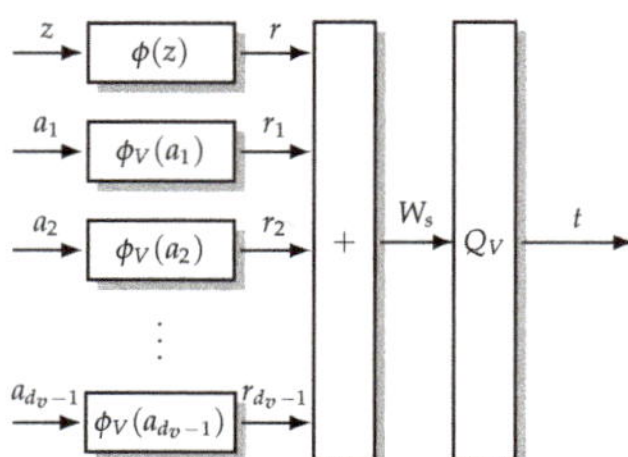

Figure 6. VN update for computational domain framework [14,16]. The n_Q-bit channel information $z \in \mathcal{Z}$ and the n_E-bit CN-to-VN messages $a_1, ..., a_{d_v-1} \in \mathcal{A}$ are transformed to n_R-bit signed integers. This transformation generally increases the required bit resolution for the representation, i.e., $n_\mathrm{R} \geq n_\mathrm{Q}$ and $n_\mathrm{R} \geq n_\mathrm{E}$. The internal signed integers are summed and quantized back into a n_E-bit VN-to-CN message $t \in \mathcal{T}$.

4.1. Equivalent LLR Quantizer

To motivate the integer calculation of the MIC approach, we review the connection between the equivalent LLR quantizer and the VN update of the BP algorithm. Analogous to the VN update of the BP algorithm in (5), the LLR of the combined message vector $a \in \mathcal{V}$ equals the addition of the LLRs of the channel output z and of the individual messages a_m, i.e., for every possible combination $a \in \mathcal{V}$, the LLR of the combined message is

$$L(a) = \log\left(\frac{p(a|c=0)}{p(a|c=1)} \right) = L(z) + \sum_{m=1}^{d_V-1} L(a_m) \ . \tag{19}$$

The LLRs $L(a_m)$ of the individual messages are determined by (14) during DDE. As described in Section 2.3, the information maximizing quantizer for binary input separates the LLR $L(a)$ by using a $|\mathcal{T}| - 1$ threshold quantizer $Q_\mathrm{L} : \mathbb{R} \to \mathcal{T}$, i.e., the relation

$$t = g^{\mathrm{MI}}(a) = Q_\mathrm{L}(L(a)) = Q_\mathrm{L}\left(L(z) + \sum_{m=1}^{d_V-1} L(a_m) \right) \tag{20}$$

can be determined that achieves the same output as the information optimal mLUT in (16). However, to ensure that (20) produces the same output as the information optimal mLUT, calculations over real numbers are required. In the next subsection, we show that we can exploit (20) to find a calculation that requires only a finite resolution. We also provide a condition to limit the resolution that is required for exact calculation of the information optimal mLUT.

4.2. Computations over Integers

The VN update structure using the computational domain framework is visualized in Figure 6. As suggested by [14,16], a possible choice for the integer mappings $\phi_v(m)$ and $\phi_{ch}(z)$ is given by scaling and rounding the corresponding LLRs $L(m)$ and $L(z)$, respectively. In addition to [14,16], we provide further insights on the optimal choice of the scaling factor based on the relation between the VN update of the BP algorithm and the MI maximizing quantizer design. More precisely, based on the established relation in (20), we define an integer mapping for the channel information z and the CN-to-VN messages a_m in order to replace the computations over real numbers by computations over signed integers (With $\lfloor \cdot \rceil$ as round to nearest integer (away from 0 if fraction part is .5))

$$g^{\mathrm{MIC}}(a) = Q_V(W_s(a)) = Q_V\left(\underbrace{\lfloor sL(z) \rceil}_{r=\phi(z)\in\mathcal{R}} + \sum_{m=1}^{d_V-1} \underbrace{\lfloor sL(a_m) \rceil}_{r_m=\phi_V(a_m)\in\mathcal{R}_V} \right) = Q_V\left(r + \sum_{m=1}^{d_V-1} r_m \right) . \tag{21}$$

Compared to (20), the LLRs $L(z)$ and $L(a_m)$ have been multiplied by a non-negative scaling factor s and quantized to the next n_R-bit signed integer r and r_m, respectively. Subsequently, the sum of integers is limited again to n_E bits by threshold quantizer Q_V. We can interpret the scaling and rounding operation also directly as a mapping of signed integer messages z and a_m to n_R-bit signed integer messages

$$r = \phi(z) \in \mathcal{R} \quad \text{and} \quad r_m = \phi_V(a_m) \in \mathcal{R}_V \tag{22}$$

that requires n_R bits for the representation, depending on the scaling factor s.

In the following, we show that we can always find a threshold quantizer $Q_V : \mathcal{W} \to \mathcal{T}$ that maps the summation $W_s(\boldsymbol{a})$ into a VN-to-CN message $t \in \mathcal{T}$ that is identical to the VN-to-CN message of the information optimal VN update in (20), i.e., $t = g^{\mathrm{MIC}}(\boldsymbol{a}) = g^{\mathrm{MI}}(\boldsymbol{a})$. First, we consider the set of messages $\mathcal{A}_t = \{\boldsymbol{a} \in \mathcal{V} : g^{\mathrm{MI}}(\boldsymbol{a}) = t \in \mathcal{T}\}$ that are mapped into a specific output t via the information maximizing VN update $g^{\mathrm{MI}}(\boldsymbol{a})$ in (16). Thus, we can identify a corresponding set of integers $\mathcal{W}_t = \{W_s(\boldsymbol{a}) \in \mathcal{W} : \boldsymbol{a} \in \mathcal{A}_t\}$. By varying the scaling factor s, we can always find a scaling value $s^\star \leq \frac{d_V}{\Delta_{\min}}$ such that the sets of integer values $\mathcal{W}_t$ for all $t \in \mathcal{T}$ are *non-overlapping intervals*, i.e.,

$$[D_{t'}, E_{t'}] \cap [D_{t''}, E_{t''}] = \varnothing \quad \forall t'', t' \in \mathcal{T} , \tag{23}$$

with $D_t = \min \mathcal{W}_t$ and $E_t = \max \mathcal{W}_t$. Condition (23) ensures that any two different clusters t' and t'' can be separated by a simple threshold operation. The value $\Delta_{\min}$ is the minimum separation between the LLRs $L(\boldsymbol{a})$ of the elements of any two neighbouring clusters in (20) and is always larger than zero since Q_L is a threshold quantizer. If we consider a scaled version of the LLRs $sL(\boldsymbol{a})$ with any real valued scaling factor $s > 1$, we can always find a threshold quantizers $Q_{L,s}$ that achieves the same output as the information optimal mLUT. Scaling the LLRs $L(\boldsymbol{a})$ by a factor of $\frac{d_V}{\Delta_{\min}}$ ensures that the minimum separation between any two neighbouring clusters is d_V. Since the influence of the rounding operation can be bounded by $-\frac{d_V}{2} \leq W_s(\boldsymbol{a}) - sL(\boldsymbol{a}) < \frac{d_V}{2}$, scaling with a factor of at least $\frac{d_V}{\Delta_{\min}}$ ensures that any two neighbouring clusters $\mathcal{W}_t$ and $\mathcal{W}_{t+1}$ are separated by at least one integer and, thus, condition (23) is satisfied. Hence, we can always find a corresponding integer function $g^{\mathrm{MIC}}(\boldsymbol{a})$ in (21) that generates exactly the same output as $g^{\mathrm{MI}}(\boldsymbol{a})$ in (20).

Furthermore, an approximate integer calculation is found if the integer valued range of ϕ and ϕ_V are limited to n_R-bits

$$\max_z |\phi(z)| < 2^{n_R} < 2^{n_R^\star} , \quad \max_a |\phi_V(a)| < 2^{n_R} < 2^{n_R^\star} \tag{24}$$

where $n_R^\star = \lceil \log_2(\lfloor s^\star L_{\max} \rfloor) \rceil + 1$ is the bit resolution that is required for exact representation if the largest magnitude of the individual LLRs in (20) is $L_{\max}$. If condition (23) is not fulfilled, we select the output cluster that maximizes MI. If (24) is satisfied, the required bit resolution of the summation $W_s(\boldsymbol{a})$ in (21) is limited by

$$n_W = \lceil \log_2\left(d_V(2^{(n_R-1)} - 1)\right) \rceil + 1 . \tag{25}$$

To consider the influence of this new mapping in the design of subsequent iterations, we also update the VN-to-CN distribution in (17).

We note that the MIC design approach can also be applied for the design of CN operations and can also be used to generate exact or approximate representations of nested tree decompositions similar to the sLUT method. However, the corresponding investigations are beyond the scope of this paper.

Illustrative Example for MIC Calculations

To illustrate the proposed MIC approach, we consider the design of a VN node update for a $(d_V=3, d_C=6)$ regular LDPC codes at iteration $i = 1$ with design $E_b/N_0 = 2.5$ dB. Figure 7a shows the equivalent LLR quantizer (20) with 2^{n_E} *non-overlapping* clusters on the real number line, i.e., $\mathcal{T} = \{-4, ..., -1, 1, ..., 4\}$. E.g. all LLRs $L(a)$ between 0 and 1.1 are mapped into cluster $t = 1$. The threshold values are shown by dashed lines in Figure 7a. Additionally, Figure 7b–d show the output of the integer addition in (21) on the x-axis and the output clusters of the optimal mLUT on the y-axis if the scaling factor is set to $s = 10$, $s = 3$, and $s = 1$, respectively.

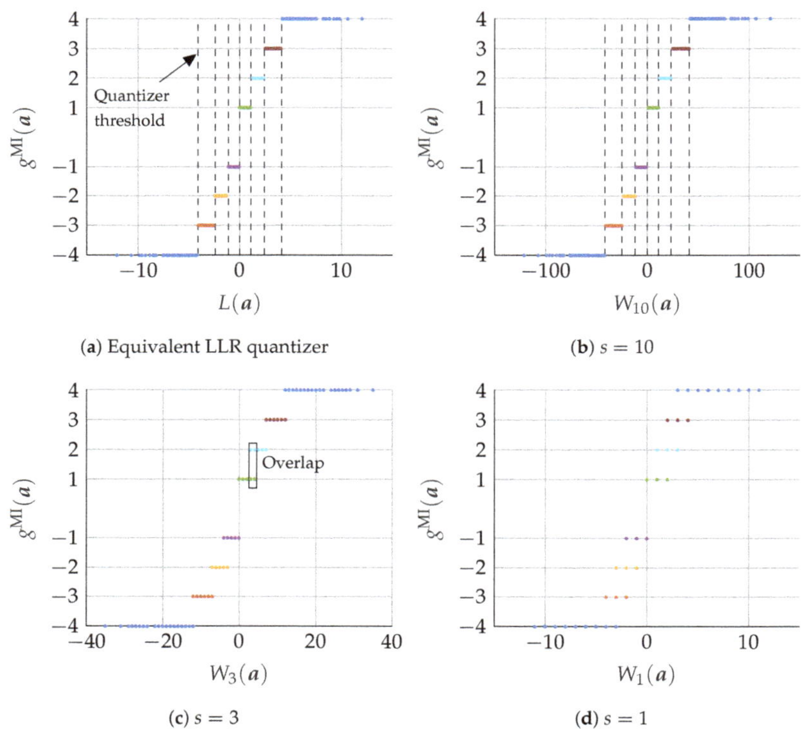

Figure 7. Visualization of the relationship between the result of the calculation in the computational domain and the assignment to mutual information maximizing mLUT mapping. Subfigure (**a**) shows the addition the real valued LLRs of (20) on the x-axis and the mutual information maximizing mLUT assignment of (16) on the y-axis. In Subfigure (**b**)–(**d**), the values on the x-axis are replaced by the corresponding integer additions of (21) for different scaling factors $s \in \{1, 3, 10\}$.

In the case of $s = 10$, all output clusters are separated by using seven integer thresholds, which are indicated by dashed lines in Figure 7b. In this case, the integer computation fully replaces the original mLUT functionality by using only signed integers of low-range. To clarify the example, the numeric values of the corresponding LLRs and integer mappings of (19) and (21) for $s = 10$ are shown in Table 1. For example, the quantized receive message $z = 2$ corresponds to an LLR of $L(z) = 1.56$ leading to the n_R-bit signed integer message $r = \phi(z) = \lfloor 15.6 \rceil = 16$. After summation of r and r_m, all results $12 \leq W_{10}(a) \leq 23$ are again mapped back to the n_E message $t = 2$.

Table 1. Numeric values of integer based VN update g_{10}^{MIC} with scaling parameter $s = 10$.

Cluster index z, a, t	± 4	± 3	± 2	± 1
Channel LLR $L(z)$	± 5.07	± 2.90	± 1.56	± 0.49
Integer mapping $\phi(z)$	± 51	± 29	± 16	± 5
Message LLR $L(a)$	± 3.46	± 2.08	± 1.02	± 0.25
Integer mapping $\phi_V(a)$	± 35	± 21	± 10	± 3
Interval $\mathcal{W}_t$	$\pm[121, 42]$	$\pm[41, 25]$	$\pm[23, 12]$	$\pm[11, 1]$

For $s = 3$ and $s = 1$, the integer range is further reduced, but the original mLUT functionality cannot be represented exactly since some integer additions are mapped to more than one output cluster of the original mLUT (e.g., some values of $W_s(a)$ are mapped into cluster $t = 1$ and $t = 2$ as highlighted in Figure 7c). If some values of $W_s(a)$ are assigned to more than one cluster of the information maximizing mLUT mapping, a merging of these values into a single cluster is required. This merging generally leads to an inevitable loss of information. In order to find a corresponding threshold quantizer for this case, we select the output cluster that minimizes the information loss under the condition that (23) is fulfilled.

4.3. FER Results

In this section, we discuss the communication performance of the proposed MIC decoder for an irregular LDPC code from the 802.11n standard [27] of length $N = 648$ with rate $R = 0.75$ and edge degree distributions $\lambda(\xi) = 0.2083\xi^1 + 0.3333\xi^2 + 0.25\xi^3 + 0.2083\xi^5$ and $\rho(\xi) = \frac{1}{3}\xi^{13} + \frac{2}{3}\xi^{15}$. The realization of the MIC decoder is characterized by three quantization parameters and specified by $\text{MIC}(n_E, n_Q, n_R)$. In contrast, the Min-mLUT decoder with label based minimum operation as CN update has only two parameters and is denominated by $\text{Min-mLUT}(n_E, n_Q)$. Figure 8 shows the Frame Error Rate (FER) performance of Min-mLUT and MIC for $n_E = n_Q = 3$ and $I = 10$ iterations, but varying resolution of internal messages $n_R \in \{4, 5, 6, 7, 11\}$ for MIC.

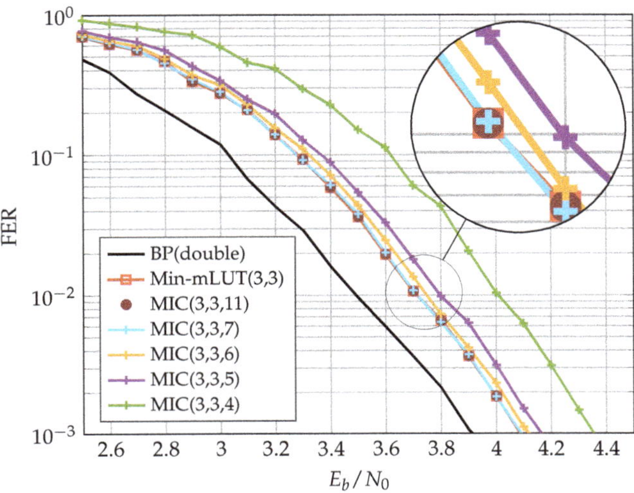

Figure 8. FER performance of $n_E = n_Q = 3$ bit Min-mLUT and MIC decoders using different internal message resolutions n_R for VN update.

The BP decoder with double precision serves as our benchmark simulation. The Min-mLUT decoder with $n_E = n_Q = 3$ bit quantization for the message exchange and channel information results in a minor performance degeneration of only 0.2 dB at a FER of 10^{-3} w.r.t. the benchmark simulation. In comparison, the proposed MIC decoder that replaces the VN update of the Min-mLUT decoder by using the computational domain framework with internal messages of size $n_R = 4$ results in a loss of 0.25 dB compared to the Min-mLUT decoder. The performance gain of the MIC decoder by using $n_R = 5$ compared to $n_R = 4$ is around 0.1 dB. The MIC decoder with $n_R = 7$ has basically identical FER performance compared to the Min-mLUT decoder. If $n_R = 11$, the MIC decoder represents the mLUT functionality exactly by meeting the criterion (23), but the gain in communication performance compared to the MIC decoder with $n_R = 7$ is negligible. Additionally, MIC decoding does not require LUTs with up to 262k entries for each iteration.

5. Finite Alphabet Message Passing (FA-MP) Decoder Implementation

In this section, we investigate the implementation complexity of different LUT-based FA-MP decoders in terms of area, throughput, latency, power, area efficiency, and energy efficiency and compare them with a state-of-the-art Normalized Min-Sum (NMS) decoder. As already stated, we focus on unrolled full parallel (FP) decoder architectures that enable throughput towards 1 Tb/s. The architecture template is shown in Figure 9. Input to the decoder are compressed messages z from the channel quantizer. The decoder uses two-phase decoding. Hence, each iteration consists of two stages: one stage comprises M Check Node Functional Units (CFUs) and the second stage N Variable Node Functional Units (VFUs). The stages are connected by hardwired routing networks, which implement the edges of the Tanner graph. Since the decoding iterations are unrolled, the decoder consists of $2 \cdot I$ stages. Deep pipelining is applied to increase the throughput. For more details on this architecture, the reader is referred to [5].

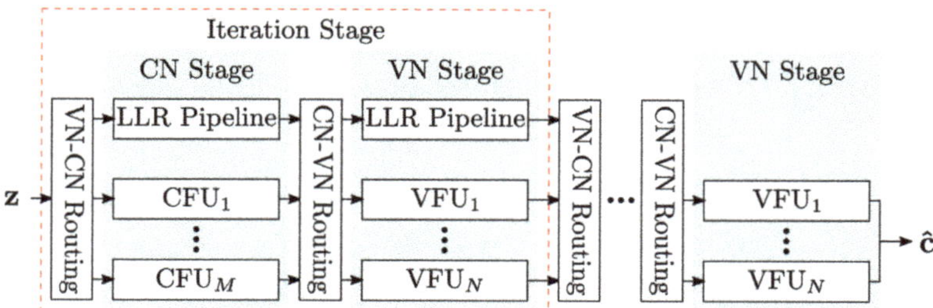

Figure 9. Unrolled full-parallel decoding architecture.

In FP decoders that use the NMS algorithm, node operations are implemented as additions and minimum searches on uniformly quantized messages [5]. In contrast, node functionality in Finite Alphabet (FA) decoders is implemented as LUTs. Implementing a single LUT as memory is impractical in Application-Specific Integrated Circuit (ASIC) technologies since the area and power overhead would be too large. Hence, a single LUT is transformed into n_E Boolean functions $b : B^{inp} \rightarrow B$ with inp being the number of inputs of the LUT, which is the node degree multiplied by n_E. b can consist of up to 2^{inp} product terms if b is represented in sum-of-product form. State-of-the-art logic synthesis tools try to minimize b such that it can be mapped onto a minimum number of gates. Despite this optimization, the resulting logic can be very large for higher node degrees and/or n_E, making this approach unsuitable for efficient FP decoder implementation. It was shown in [7] that the mLUT can be decomposed into a set of two-input sLUTs arranged in a tree structure, which largely reduces the resulting logic at the cost of a small degradation in error correction performance. To compare these approaches with our new decoder, we implemented four different types of FP decoders:

- NMS decoder with extrinsic message scaling factor of 0.75;
- Two LUT-based decoders: in these decoders, we implemented the VN operation by LUTs and the CN operations by a minimum search on the quantized messages. The latter corresponds to the CN Processor implementation of [7]. The LUTs are implemented either as a single LUT (mLUT), or as a tree of two-input LUTs (sLUT);
- Our new MIC decoder in which the VN is replaced by the new update algorithms, presented in the previous section.

For MIC and LUT based decoders, we investigated message quantization $n_E = 3$ and $n_E = 4$. The reference is an NMS decoder with $n_E = 4$ and $n_E = 5$, respectively. For all decoders, the channel and message quantization were set to be identical, i.e., $n_E = n_Q$. We used a different code for our implementation investigation than in the previous sections. This code has a larger block size, which implies increased implementation complexity. The code is a $(816, 406)$ regular LDPC code with $d_V = 3$ and $d_C = 6$ and the number of decoding iterations is $I = 8$.

We applied a Synopsys Design Compiler and IC Compiler II for implementation in a 28 nm Fully-Depleted Silicon-on-Insulator (FD-SOI) technology under worst-case Process, Voltage and Temperature (PVT) conditions (125 °C, 0.9 V for timing, 1.0 V for power). A process with eight metal layers was chosen. Metal layers 1 to 6 are used for routing, with metals 1 and 2 mainly intended for standard cells. The metal layers 7 and 8 are only used for power supply. Power numbers were calculated with back-annotated wiring data and input data for a FER of 10^{-4}. All designs were optimized for high throughput with a target frequency of 1 GHz during synthesis and back-end. To assess the routing congestion, we fixed the utilization to 70 % for all designs as a constraint. The utilization specifies the ratio between logic cell area and total area (=logic cell area plus routing area). Thus, by fixing this parameter, all designs have the same routing area available in relation to their logic cell area.

5.1. FER Performance of Implemented FA-MP Decoders

Figures 10 and 11 show the FER performance for the different decoders. We compare the NMS decoder with the MIC decoder and the two LUT-based decoders. The LUTs of the FA-MP decoders are elaborated to a design Signal-to-Noise-Ratio (SNR) optimized at an FER of 10^{-4}. It should be noted that this may result in an error floor behavior below the target FER. This phenomenon can be mitigated by selecting a larger design SNR at the cost of decreased performance in the waterfall region [13]. For comparison, we also added the BP performance with double precision floating point number representation.

In the previous section, we showed that, for the $(648, 486)$ code, the MIC decoder achieves the same error correction performance as the Min-mLUT decoder for $n_R = 7$. A similar observation was made for the $(816, 406)$ code considered here. In our implementation comparison, we reduced n_R such that the MIC's FER stays below that of the NMS at the target FER of 10^{-4}. In this way, we obtained an $n_R = 5$, which yields a small degradation in the MIC FER compared to the Min-sLUT and Min-mLUT decoders, but outperforms the NMS decoder. We observe that the MIC and Min-mLUT decoders with one bit smaller message quantization n_E have better error correction capability than the NMS decoder at the target FER. In addition, due to the low message quantization and the resulting low dynamic range, the NMS runs into an error floor below FER 10^{-4}.

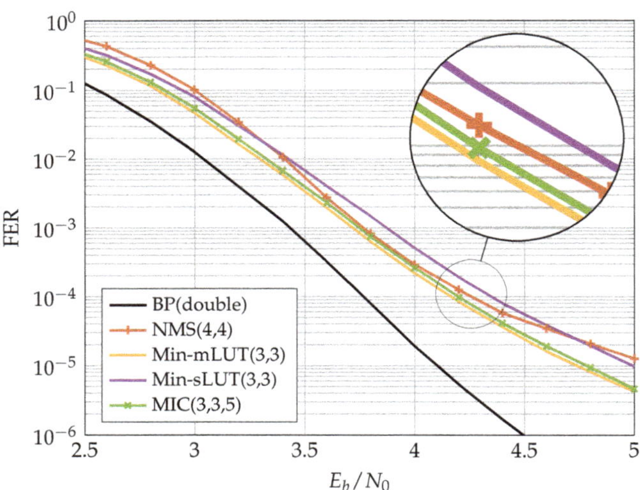

Figure 10. Communication performance of $n_{\mathrm{E}} = 3$ bit FA-MP decoders.

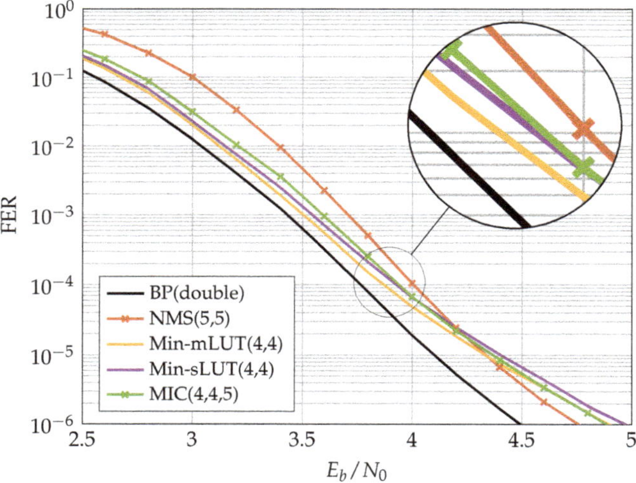

Figure 11. Communication performance of $n_{\mathrm{E}} = 4$ bit FA-MP decoders.

5.2. FD-SOI Implementation Results

Table 2 shows the implementation results for MIC(3,3,5), Min-mLUT(3,3), Min-sLUT(3,3) and NMS(4,4) decoders, whereas Table 3 shows the implementation results for MIC(4,4,5), Min-mLUT(4,4), Min-sLUT(4,4) and NMS(5,5) decoders. As already stated, we fixed the target frequency to 1 GHz and the utilization to 70% for all decoders. Maximum achievable frequency f, final utilization, area A and power consumption P were extracted from the final layout data. From these data, we can derive the important implementation metrics: throughput, latency, area efficiency and energy efficiency. Since the decoders are pipelined, the coded decoder throughput T is $f \cdot N$. The latency is $1/f \cdot 26$ (each iteration consists of three pipeline stages, decoder input and output are also buffered, yielding $8 \cdot 3 + 2 = 26$ pipeline stages in total). The area efficiency is defined as T/A and the energy efficiency as P/T.

The Min-mLUT decoder has the largest area, the worst area efficiency, and the worst energy efficiency. We see an improvement in these metrics for the Min-sLUT at the cost of a slightly decreased error correction performance. The difference in the implementation metrics largely increases when $n_E = 3$ changes to $n_E = 4$. The area increases by a factor of 10 for the Min-mLUT(4,4), but only by a factor of 2.7 for the Min-sLUT(4,4) decoder. Moreover, we had to reduce the utilization to 50 % to achieve a routing convergence for the Min-mLUT(4,4) decoder. The large area increase is explainable with the increase of the LUT sizes from 512 to 4096 entries per LUT when increasing n_E from 3 to 4. Moreover, the frequency largely breaks down, yielding a very low area efficiency and energy efficiency. The Min-sLUT decoders scale better with increasing n_E. Both Min-sLUT decoders outperform the corresponding NMS decoders in throughput and efficiency metrics.

The MIC decoder has the best implementation metric numbers in all cases. It outperforms all other decoders in throughput, area, area efficiency and energy efficiency while having the same or even slightly improved error correction performance compared to the other decoders. It can also be seen that the MIC decoder has a lower routing complexity compared to the Min-sLUT and the NMS decoder. We observe a large drop in the frequency from 595 MHz down to 183 MHz (70 % decrease) when comparing NMS(4,4) with NMS(5,5) under the utilization constraint of 70 %. The large drop in the frequency is explainable with the increased routing complexity for the given routing area constraint that yields longer wires and corresponding delays. This problem is less severe for the Min-sLUT, where the frequency drops from 670 MHz to 492 MHz (27 % decrease). The MIC achieves the highest frequency for all cases and drops from 775 MHz to 633 MHz (18% decrease), only. This shows that the MIC scales much better with increasing n_E.

It should be noted that the CFU implementation is identical for the MIC, Min-mLUT and Min-sLUT decoders. Compared to the corresponding NMS, the CFU implementation is less complex [19] due to: (i) a 1 bit smaller message quantization, (ii) the omission of the scaling unit, and (iii) the omission of the sign-magnitude to two's complement conversion. Hence, the CFU complexity of the FA-MP is always lower than that of the NMS independent of the respective CN degree. Moreover, in contrast to the NMS decoder, the messages from the CFUs to the VFUs are transmitted in sign-magnitude representation via the routing network which reduces the toggling rate and thus the average power consumption.

Table 2. Post-layout results of FA-MP decoders with $n_E = n_Q = 3$, $n_R = 5$.

	MIC	Min-mLUT	Min-sLUT	NMS
n_E, n_Q	3	3	3	4
E_b/N_0 @ FER 10^{-4} [dB]	4.20	4.16	4.35	4.26
Utilization [%]	70	68	71	71
Frequency [MHz]	775	662	670	595
Coded Throughput [Gb/s]	633	540	547	486
Area [mm^2]	2.73	4.23	2.86	3.04
Area Efficiency [Gb/s/mm^2]	231.6	128	190	159.7
Latency [ns]	33.5	39.3	35.8	43.7
Power [W]	4.49	5.07	4.38	4.39
Energy Efficiency [pJ/bit]	7.10	9.4	8.0	9.0

Figure 12 shows the layout of the MIC and the NMS decoder in the same scale. Each color represents one iteration stage, which is composed of CFUs, VFUs, and the routing between the nodes (see also Figure 9). When comparing the same iteration stages (same color) of the two decoders, we can observe that the iteration stages in the MIC decoder are smaller than the corresponding iteration stages in the NMS decoder, although the frequency of the MIC decoder is more than three times higher compared to the NMS decoder. This shows once again that the MIC has a lower implementation complexity, especially from a routing perspective.

Table 3. Post-layout results of FA-MP decoders with $n_E = n_Q = 4$, $n_R = 5$.

	MIC	Min-mLUT	Min-sLUT	NMS
n_E, n_Q	4	4	4	5
E_b/N_0 @ FER 10^{-4} [dB]	3.94	3.87	3.93	4.01
Utilization [%]	69	49	66	69
Frequency [MHz]	633	267	492	183
Coded Throughput [Gb/s]	516	218	401	149
Area [mm^2]	3.66	40.51	7.82	3.99
Area Efficiency [Gb/s/mm^2]	141.1	5.4	51.3	37.4
Latency [ns]	41.1	97.2	48.0	142.0
Power [W]	5.61	11.85	8.68	2.25
Energy Efficiency [pJ/bit]	10.9	54.3	21.6	15.1

(a) NMS(5,5), (3.99 mm^2) (b) MIC(4,4,5), (3.66 mm^2)

Figure 12. Layout of decoders in the same scale; each color indicates one iteration stage from dark red (first iteration) to dark blue (eighth iteration).

Our analysis shows that the new MIC approach largely improves the implementation efficiency and exhibits better scaling compared to the state-of-the-art sLUT and NMS implementations of FP decoder architectures. This enables the processing of larger block sizes, which is mainly due to the reduced routing complexity. Larger block sizes improve the error correction capability and further increase the throughput of FP architectures.

6. Conclusions

This paper provides a detailed investigation of the Minimum-Integer Computation (MIC) decoder for regular and irregular Low-Density Parity Check (LDPC) codes. The MIC decoder utilizes the computational domain framework to realize Variable Node (VN) updates by an equivalent low-range signed integer computation and Check Node (CN) updates by a minimum search. For the VN update, we provide further insights for the design of an Mutual Information (MI) maximizing signed integer computation. To discuss implementation issues on FA-MP decoding architectures, we exemplified this on different LUT-based decoder designs. Furthermore, we compared MIC to state-of-the-art Normalized Min-Sum (NMS) decoder implementations to show the improvement in area efficiency and energy efficiency.

Author Contributions: Conceptualization, T.M., O.G., M.H., D.W., A.D. and N.W.; Funding acquisition, A.D., D.W. and N.W.; Investigation, T.M., D.W., O.G. and M.H.; Software, T.M., O.G. and M.H.; Validation, T.M., O.G., M.H., D.W. and N.W.; Visualization, T.M.; Writing—original draft, T.M., O.G., M.H., D.W. and N.W.; Writing—review and editing, T.M., O.G., M.H., D.W., A.D. and N.W. All authors have read and agreed to the published version of the manuscript.

Funding: This research was funded by the German Ministry of Education and Research (BMBF) projects "FunKI" (grants 16KIS1180K and 16KIS1185) and "Open6GHub" (grants 16KISK016 and 16KISK004).

Institutional Review Board Statement: Not applicable.

Informed Consent Statement: Not applicable.

Data Availability Statement: Not applicable.

Conflicts of Interest: The authors declare no conflict of interest.

References

1. Saad, W.; Bennis, M.; Chen, M. A Vision of 6G Wireless Systems: Applications, Trends, Technologies, and Open Research Problems. *IEEE Netw.* **2020**, *34*, 134–142. [CrossRef]
2. Kestel, C.; Herrmann, M.; Wehn, N. When Channel Coding Hits the Implementation Wall. In Proceedings of the IEEE 10th International Symposium on Turbo Codes Iterative Information (ISTC 2018), Hong Kong, China, 3–7 December 2018. [CrossRef]
3. Gallager, R. Low-Density Parity-Check Codes. *IRE Trans. Inf. Theory* **1962**, *8*, 21–28. [CrossRef]
4. MacKay, D. Good error-correcting codes based on very sparse matrices. *IEEE Trans. Inf. Theory* **1999**, *45*, 399–431. [CrossRef]
5. Schläfer, P.; Wehn, N.; Alles, M.; Lehnigk-Emden, T. A New Dimension of Parallelism in Ultra High Throughput LDPC Decoding. In Proceedings of the IEEE Workshop on Signal Processing Systems (SIPS 2013), Taipei, Taiwan , 16–18 October 2013; pp. 153–158. [CrossRef]
6. IEEE. International Roadmap for Devices and Systems, 2021 Update, More Moore. 2021. Available online: https://irds.ieee.org/images/files/pdf/2021/2021IRDS_MM.pdf (accessed on 8 August 2022)
7. Balatsoukas-Stimming, A.; Meidlinger, M.; Ghanaatian, R.; Matz, G.; Burg, A. A Fully-Unrolled LDPC Decoder based on Quantized Message Passing. In Proceedings of the IEEE Workshop on Signal Processing Systems (SIPS 2015), Hangzhou, China, 14–16 October 2015. [CrossRef]
8. Ghanaatian, R.; Balatsoukas-Stimming, A.; Müller, T.C.; Meidlinger, M.; Matz, G.; Teman, A.; Burg, A. A 588-Gb/s LDPC Decoder Based on Finite-Alphabet Message Passing. *IEEE Trans. Very Large Scale Integr. (VLSI) Syst.* **2018**, *26*, 329–340. [CrossRef]
9. Lee, J.S.; Thorpe, J. Memory-efficient decoding of LDPC codes. In Proceedings of the International Symposium on Information Theory (ISIT 2005), Adelaide, SA, Australia, 4–9 September 2005; pp. 459–463. [CrossRef]
10. Chen, J.; Fossorier, M. Density Evolution for Two Improved BP-Based Decoding Algorithms of LDPC Codes. *IEEE Commun. Lett.* **2002**, *6*, 208–210. [CrossRef]
11. Romero, F.J.C.; Kurkoski, B.M. LDPC Decoding Mappings that Maximize Mutual Information. *IEEE J. Sel. Areas Commun.* **2016**, *34*, 2391–2401. [CrossRef]
12. Lewandowsky, J.; Bauch, G. Information-Optimum LDPC Decoders based on the Information Bottleneck Method. *IEEE Access* **2018**, *6*, 4054–4071. [CrossRef]
13. Meidlinger, M.; Matz, G.; Burg, A. Design and Decoding of Irregular LDPC Codes Based on Discrete Message Passing. *IEEE Trans. Commun.* **2020**, *68*, 1329–1343. [CrossRef]
14. Kang, P.; Cai, K.; He, X.; Li, S.; Yuan, J. Generalized Mutual Information-Maximizing Quantized Decoding of LDPC Codes With Layered Scheduling. *IEEE Trans. Veh. Technol.* **2022**, *71*, 7258–7273. [CrossRef]
15. He, X.; Cai, K.; Mei, Z. On Mutual Information-Maximizing Quantized Belief Propagation Decoding of LDPC Codes. In Proceedings of the IEEE Global Communications Conference (GLOBECOM 2019), Waikoloa, HI, USA, 9–13 December 2019. [CrossRef]
16. He, X.; Cai, K.; Mei, Z. Mutual Information-Maximizing Quantized Belief Propagation Decoding of LDPC Codes. *arXiv* **2019**, arXiv:1904.06666. Available online: https://arxiv.org/abs/1904.06666 (accessed on 7 October 2022).
17. Wang, L.; Wesel, R.D.; Stark, M.; Bauch, G. A Reconstruction-Computation-Quantization (RCQ) Approach to Node Operations in LDPC Decoding. In Proceedings of the IEEE Global Communications Conference (GLOBECOM 2020), Taipei, Taiwan, 7–11 December 2020. [CrossRef]
18. Wang, L.; Terrill, C.; Stark, M.; Li, Z.; Chen, S.; Hulse, C.; Kuo, C.; Wesel, R.D.; Bauch, G.; Pitchumani, R. Reconstruction-Computation-Quantization (RCQ): A Paradigm for Low Bit Width LDPC Decoding. *IEEE Trans. Commun.* **2022**, *70*, 2213–2226. [CrossRef]
19. Monsees, T.; Wübben, D.; Dekorsy, A.; Griebel, O.; Herrmann, M.; Wehn, N. Finite-Alphabet Message Passing using only Integer Operations for Highly Parallel LDPC Decoders. In Proceedings of the IEEE 23rd International Workshop on Signal Processing Advances in Wireless Communication (SPAWC 2022), Oulu, Finland, 4–6 July 2022. [CrossRef]

20. Tanner, R. A recursive approach to low complexity codes. *IEEE Trans. Inf. Theory* **1981**, *27*, 533–547. [CrossRef]
21. Kurkoski, B.M. On the Relationship Between the KL Means Algorithm and the Information Bottleneck Method. In Proceedings of the 11th International ITG Conference on Systems, Communications and Coding (SCC), Hamburg, Germany, 6–9 February 2017.
22. Tishby, N.; Pereira, F.C.; Bialek, W. The Information Bottleneck Method. In Proceedings of the 37th Annual Allerton Conference on Communication, Control, and Computing, Monticello, IL, USA, 22–24 September 1999; pp. 368–377.
23. Hassanpour, S.; Wübben, D.; Dekorsy, A. Overview and Investigation of Algorithms for the Information Bottleneck Method. In Proceedings of the 11th International Conference on Systems, Communications and Coding (SCC), Hamburg, Germany, 6–9 February 2017.
24. Kurkoski, B.M.; Yagi, H. Quantization of Binary-Input Discrete Memoryless Channels. *IEEE Trans. Inf. Theory* **2014**, *60*, 4544–4552. [CrossRef]
25. Burshtein, D.; Pietra, V.D.; Kanevsky, D.; Nadas, A. Minimum Impurity Partitions. *Ann. Stat.* **1992**, *20*, 1637–1646. [CrossRef]
26. Stark, M.; Wang, L.; Bauch, G.; Wesel, R.D. Decoding Rate-Compatible 5G-LDPC Codes With Coarse Quantization Using the Information Bottleneck Method. *IEEE Open J. Commun. Soc.* **2020**, *1*, 646–660. [CrossRef]
27. IEEE. IEEE Standard for Information Technology—Local and Metropolitan Area Networks—Specific Requirements—Part 11: Wireless LAN Medium Access Control (MAC) and Physical Layer (PHY) Specifications Amendment 5: Enhancements for Higher Throughput. In *IEEE Std 802.11n-2009 (Amendment to IEEE Std 802.11-2007 as amended by IEEE Std 802.11k-2008, IEEE Std 802.11r-2008, IEEE Std 802.11y-2008, and IEEE Std 802.11w-2009)*; IEEE: Piscatvey, NJ, USA, 2009; pp. 1–565. [CrossRef]

Article

Information Bottleneck Signal Processing and Learning to Maximize Relevant Information for Communication Receivers

Jan Lewandowsky [1,*], **Gerhard Bauch** [2] **and Maximilian Stark** [2]

[1] Fraunhofer Institute for Communication, Information Processing and Ergonomics, Fraunhoferstraße 20, 53343 Wachtberg, Germany

[2] Institute of Communications, Hamburg University of Technology, Eißendorfer Straße 40, 21073 Hamburg, Germany; bauch@tuhh.de (G.B.); maximilian.stark@tuhh.de (M.S.)

* Correspondence: jan.lewandowsky@fkie.fraunhofer.de; Tel.: +49-228-9435-731

Abstract: Digital communication receivers extract information about the transmitted data from the received signal in subsequent processing steps, such as synchronization, demodulation and channel decoding. Technically, the receiver-side signal processing for conducting these tasks is complex and hence causes bottleneck situations in terms of power, delay and chip area. Typically, many bits per sample are required to represent and process the received signal in the digital receiver hardware accurately. In addition, demanding arithmetical operations are required in the signal processing algorithms. A popular recent trend is designing entire receiver chains or some of their crucial building blocks from an information theoretical perspective. Signal processing blocks with very simple mathematical operations can be designed to directly maximize the relevant information that flows through them. At the same time, a strong quantization reduces the number of bits processed in the receiver to further lower the complexity. The described system design approach follows the principle of the information bottleneck method. Different authors proposed various ideas to design and implement mutual information-maximizing signal processing units. The first important aim of this article is to explain the fundamental similarities between the information bottleneck method and the functionalities of communication receivers. Based on that, we present and investigate new results on an entire receiver chain that is designed following the information bottleneck design principle. Afterwards, we give an overview of different techniques following the information bottleneck design paradigm from the literature, mainly dealing with channel decoding applications. We analyze the similarities of the different approaches for information bottleneck signal processing. This comparison leads to a general view on information bottleneck signal processing which goes back to the learning of parameters of trainable functions that maximize the relevant mutual information under compression.

Keywords: information bottleneck; mutual information; machine learning

Citation: Lewandowsky, J.; Bauch, G.; Stark, M. Information Bottleneck Signal Processing and Learning to Maximize Relevant Information for Communication Receivers. *Entropy* **2022**, *24*, 972. https://doi.org/10.3390/e24070972

Academic Editor: Jerry D. Gibson

Received: 10 May 2022
Accepted: 12 July 2022
Published: 14 July 2022

1. Introduction

As Claude Elwood Shannon postulated in [1], the "fundamental problem of communication is that of reproducing at one point either exactly or approximately a message selected at another point". Hence, in information theoretical terms, the most essential task of any digital communication receiver is to provide a maximum possible amount of mutual information on the transmitted data to its user. One could loosely say that the message selected at the transmitter is relevant to the receiver. Assuming that the possible transmitted messages are chosen from a finite set with certain probabilities at the transmitter, this introduces a quite intuitive understanding of a discrete relevant random variable in the communications context. The transmitted data sequence is relevant to the user of the receiver, but the latter can typically only observe a degraded version of this sequence in the form of a noisy and disturbed received signal.

Interestingly, the concept of a relevant random variable was also introduced by Tishby et al. in [2] in a very generic information theoretical setup termed the information bottleneck method. Conceptually, this method is not directly linked to the communication problem considered above. The elementary idea of the information bottleneck method is to compress an observed random variable Y to a compressed variable T using a compression rule. The compression rule is tailored to preserve relevant mutual information $I(X;T)$, where $I(X;T) \leq I(X;Y)$. In this problem formulation, X is a chosen random variable of interest. This variable defines which features of the observation Y are relevant and should be preserved under the invoked compression. Hence, this concept can be understood as defining relevance through another variable [2]. The information bottleneck method already has numerous very successful applications, such as in image and speech processing, astronomy and neuroscience [3–6]. A comprehensive tutorial on the information bottleneck method, its applications in source coding and its connections to inference and representation learning problems can be found in [7]. In addition, a survey on the applicability and usefulness of the information bottleneck for machine learning is provided in [8].

In the past few years, the information bottleneck method gained massive popularity in the communications community. Various authors brought up different ideas to design and implement receiver subsystems and other parts of communcation systems using the information bottleneck principle of maximizing the preserved relevant information. The applications studied in the communications context lead from the design of channel output quantizers [9–11] over the decoding of low-density parity-check (LDPC) codes [12–31] and polar codes [32–34] to entire receiver chains that include channel estimation and detection [35–38]. Moreover, the information bottleneck method has been applied in joint source–channel coding, forwarding and relaying applications [39–48] and in distributed sensor networks [49–53] successfully. Related works with a focus on inference with the distributiveness of data among multiple nodes and network learning aspects include [54–56].

It needs to be noted that some of the aforementioned works do not explicitly state themselves to be instances of the information bottleneck method. However, their fundamental ideas are in line with the key principle of the preservation of relevant information under compression. Please also note that even though the references provided above are numerous, we cannot claim to have mentioned all applications of the information bottleneck method and the related principles in communications here. Facing the huge variety of research on applications of the information bottleneck method in communications above, we note that a number of techniques have been proposed to apply information bottleneck signal processing lately.

In this article, as a first contribution, we aim to give a quite general introduction to the ideas of receiver-side information bottleneck signal processing. For that purpose, we explain the general idea of the information bottleneck method and link it to the fundamental task of a communication receiver. To illustrate the applicability of the presented ideas to real-world communication receivers, we present and investigate a strongly quantized iterative receiver that is entirely designed with the information bottleneck method and compare its performance to a conventional receiver chain. The presented information bottleneck receiver implements all signal processing operations using lookup tables that are designed with an information bottleneck algorithm. Surprisingly, the presented quantized receiver that is designed with the information bottleneck method performs just as well as a double-precision reference receiver that employs state-of-the-art signal processing algorithms and uses much more costly signal processing operations.

Some very fundamental ideas of information bottleneck signal processing were presented already in 2008 for LDPC decoding by Kurkoski et al. [12]. In fact, a special focus lies on the application of information bottleneck signal processing for the coarsely quantized message passing decoding of LDPC codes in the available literature [12–31]. It needs to be appreciated that various authors have proposed very interesting ideas to solve this problem using the paradigm of maximizing the preserved relevant information under quantization with very few bits per message.

The literature describes quantized decoders that replace the classical node operations of LDPC decoders completely with lookup tables [12,14–16,18,20,21,23]. In addition, the hybrid min-LUT approach from [17,22] uses lookup tables only for the variable node operations and a simple arithmetical operation for the check node operations. Moreover, computational domain approaches that pair relatively simple arithmetical operations with mutual information-maximizing quantizers to implement the node operations of LDPC decoders were studied in [24,25,30,31]. Finally, the idea to learn and implement mutual information-maximizing node operations using neural networks was studied in [28,29]. We note that the design goals of the mentioned methods for LDPC decoding are similar and mainly rely on the information bottleneck idea of maximizing the preserved relevant information. However, what differs is how the mutual information-maximizing operations for the LDPC decoders are designed and implemented.

Therefore, as a second contribution, we discuss what the approaches from the literature have in common, compare them and, based on that, draw some novel conclusions on information bottleneck signal processing in general. Our conclusions break down information bottleneck signal processing to the learning of parameters of trainable functions. The parameters are tuned to maximize the relevant information under compression.

This article is structured as follows. The next section first provides an introduction to the information bottleneck method and its application in coarsely quantized information bottleneck signal processing units. A quite general comparison of the information bottleneck method and a communication system is used to explain the applicability of the information bottleneck method for the design of quantized communication receivers. Afterwards, we substantiate the presented ideas by developing and analyzing a particular information bottleneck receiver structure for an LDPC-encoded transmission over a fading channel. This receiver implements all signal processing operations designed with the information bottleneck method using lookup tables.

Section 3 then starts by recalling and comparing other approaches to information bottleneck signal processing for LDPC decoders from the literature that do not use lookup tables. From this analysis, we draw the conclusion that a unified view on lookup table-based and other approaches is considering information bottleneck signal processing as learning the optimum parameters of trainable functions that maximize the relevant information. Finally, Section 4 summarizes and concludes this article.

2. The Information Bottleneck Method and Coarsely Quantized Information Bottleneck Signal Processing

This section gives an overview of the information bottleneck method and explains its connection to the fundamental design purpose of communication receivers (i.e., extracting relevant information on the transmitted data from the received signal). From that, the idea of information bottleneck signal processing is derived and explained using several examples.

2.1. The Information Bottleneck Method

The information bottleneck method is a quite generic information theoretical setup that was introduced by Tishby et al. in [2]. The basic setup consists of three discrete random variables X, Y and T with realizations $x \in \mathcal{X}, y \in \mathcal{Y}$ and $t \in \mathcal{T}$. These variables follow the Markov relation $X \rightarrow Y \rightarrow T$ and interact as illustrated in Figure 1 and explained in the following.

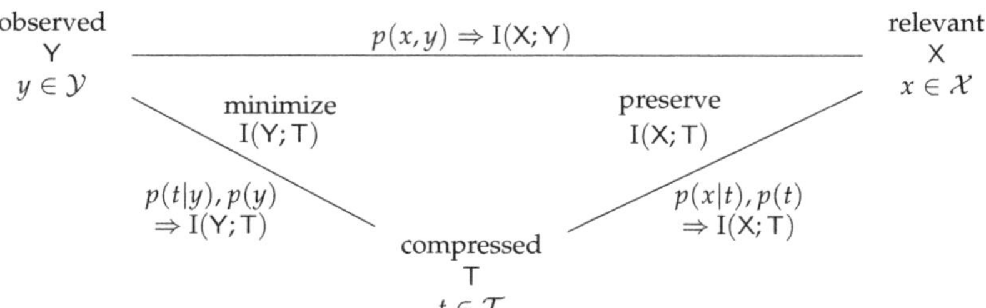

Figure 1. Illustration of the information bottleneck method. The variables X, Y and T form a Markov chain X → Y → T and are termed the relevant, observed and compressed variables, respectively. The fundamental principle is to minimize I(Y; T) while preserving I(X; T).

The random variable Y is considered to be observed and, therefore, termed the observed random variable in the information bottleneck problem setup. The baseline model is that observing the realizations $y \in \mathcal{Y}$ of Y could deliver information that could be classified into relevant and irrelevant information. In order to define which features of Y are considered relevant, the so-called relevant random variable X is introduced. As a result, the relevance is defined through another variable, and the relevant information that X and Y share is

$$I(X;Y) = \sum_{x \in \mathcal{X}} \sum_{y \in \mathcal{Y}} p(x,y) \log \frac{p(x,y)}{p(x)p(y)} = D_{\mathrm{KL}}\{p(x,y) \,|\, p(x)p(y)\}, \tag{1}$$

where $p(x,y)$ is the joint distribution and $p(x)$ and $p(y)$ are the marginal distributions of X and Y, respectively. $D_{\mathrm{KL}}\{.\,|\,.\}$ in Equation (1) is the Kullback–Leibler divergence.

The aims of the information bottleneck method are now twofold:

1. Conduct a lossy compression of the realizations $y \in \mathcal{Y}$ to a compressed realization $t \in \mathcal{T}$ to yield a compact compressed representation T of the observation Y. The information theoretical notion of such a compression is the minimization of the compression information I(Y; T) (i.e., the transmission rate, relating to rate–distortion theory).

2. While conducting the compression mentioned above, preserve the relevant information $I(X; T) \leq I(X; Y)$.

The goals mentioned above are contradictory. Typically, a strong compression will limit the possibility to keep the preserved relevant information I(X; T) above a desired lower bound. Similarly, aiming for a certain minimum amount of $I(X; T) \leq I(X; Y)$ typically provides a lower bound to I(Y; T). As a result, an optimum rule to compress Y onto T in the information bottleneck sense describes a trade-off between achieving a minimum possible compression information I(Y; T) that also allows keeping the preserved relevant information I(X; T) above a desired minimum level.

In technical terms, the compression rule that maps Y onto T is described as a conditional probability distribution $p(t|y)$. This conditional distribution provides the probabilities of a certain $t \in \mathcal{T}$ for a given $y \in \mathcal{Y}$ and hence describes a possibly stochastic mapping of y onto t.

The information bottleneck problem can be understood as the optimization problem of finding a suitable conditional probability distribution $p(t|y)$ with the desired characteristics of minimizing I(Y; T) while preserving I(X; T) for the Markov chain X → Y → T. Tishby et al. proposed finding an optimum mapping $p(t|y)$ using the Lagrange method in [2] and introduced the Lagrangian

$$\mathcal{L}(p(t|y)) = I(Y;T) - \beta I(X;T) \tag{2}$$

which has to be minimized over the set of all valid conditional probability distributions $p(t|y)$. The Lagrangian multiplier $\beta \geq 0$ is a trade-off parameter that allows tuning the aforementioned trade-off between the compression and preservation of relevant information. For $\beta = 0$, the focus is only on compression, and the preservation of $I(X;T)$ is not taken into account, while $\beta \to +\infty$ aims to maximize the preserved relevant information.

In [2], Tishby et al. also derived a set of equations characterizing $p(t|y), p(t)$ and $p(x|t)$ such that

$$p(t|y) = \frac{p(t)}{Z(y,\beta)} \exp(-\beta D_{\mathrm{KL}}\{p(x|y)\,|\,p(x|t)\}) \tag{3}$$

$$p(t) = \sum_{y \in \mathcal{Y}} p(t|y)p(y) \tag{4}$$

$$p(x|t) = \frac{1}{p(t)} \sum_{y \in \mathcal{Y}} p(t|y)p(x,y), \tag{5}$$

where $Z(y,\beta)$ is a normalization function that guarantees that $p(t|y)$ is a valid conditional distribution.

In general, the optimization problem of finding a $p(t|y)$ solution that minimizes the Lagrangian from Equation (2) is neither concave nor convex [2]. Nevertheless, the mentioned equations (Equations (3–5)) naturally suggest an iterative algorithm termed the iterative information bottleneck algorithm to obtain $p(t|y)$ for a given joint distribution $p(x,y)$, a trade-off parameter β and an intended cardinality $|\mathcal{T}|$ of the compression variable T that is also described in [2]. This algorithm provably converges to at least a local minimum of the Lagrangian in Equation (2). Details on its convergence are discussed in [3]. In addition to the iterative information bottleneck algorithm from [2], many other information bottleneck algorithms appeared in the literature (e.g., [57,58]). These algorithms can be understood as the work horses of the information bottleneck method, as they can determine the compression mapping $p(t|y)$ for a given $p(x,y)$, β and a desired cardinality $|\mathcal{T}|$. The information bottleneck algorithms also deliver the distributions $p(x|t)$ and $p(t)$, and thus $p(x,t) = p(x|t)p(t)$ according to Equations (4) and (5) as side products.

Figure 2 provides an overview of the inputs taken and the outputs delivered by an information bottleneck algorithm.

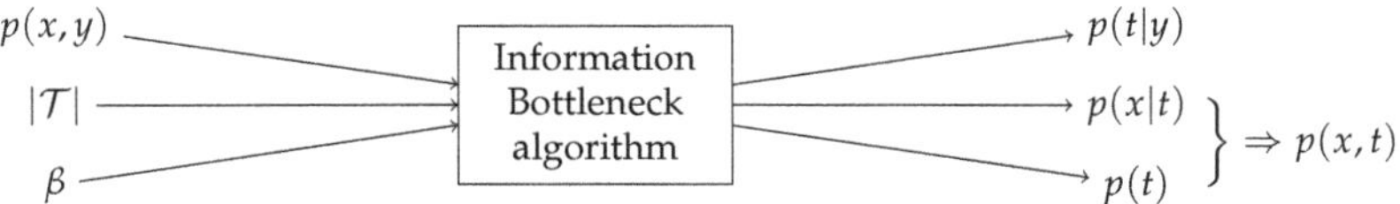

Figure 2. Overview of the inputs taken and the outputs delivered by an information bottleneck algorithm.

An important notion of the resulting conditional probability distribution $p(t|y)$ is that it clusters the event space $\mathcal{Y}$ of the observed random variable Y into clusters $\mathcal{Y}_t$, $t \in \{0,1,\ldots,|\mathcal{T}|-1\}$. This concept covers deterministic mappings of y onto t (i.e., $p(t|y) \in \{0,1\}\forall(y,t)$) that result in a hard clustering of $\mathcal{Y}$ and also probabilistic mappings that describe a soft clustering. For the latter, the realizations y can be contained in various clusters $\mathcal{Y}_t$ with different probabilities.

In the very important special case of aiming to preserve a maximum desired amount of $I(X;T)$ (i.e., $\beta \to +\infty$ for a given cardinality $|\mathcal{T}|$), the clustering of $\mathcal{Y}$ described by $p(t|y)$ becomes a hard clustering [58]. In this case, it is easy to limit the compression information $I(Y;T)$ by a proper choice of the cardinality $|\mathcal{T}|$ of the compressed representation. In the context of this article, this cardinality determines the number of bits required to process the compressed realizations $t \in \mathcal{T}$ in the digital hardware of a communication receiver. The remainder of this article will only focus on deterministic compression mappings, as these are most practical to be applied in digital communication receivers.

It will be important in the following that a deterministic mapping $p(t|y)$ can also be interpreted as a deterministic input/output relation of a system with input $y \in \mathcal{Y}$ and output $t \in \mathcal{T}$. Such a system can be implemented, for example, as a lookup table that holds the respective $t \in \mathcal{T}$ for each possible $y \in \mathcal{Y}$ if the input space $\mathcal{Y}$ has a manageable cardinality. As we will discuss later in Section 3, the lookup table interpretation is just one possibility for implementing deterministic information bottleneck compression mappings in communication receivers. We want to mention here that we only formally consider discrete observed variables Y in this article. The reason for this is that every digital communication receiver needs to quantize the continuous received signal with a limited number of bits per sample before conducting further digital signal processing steps, as will be explained in more detail later.

Before we continue, an important concept from [35] shall be revisited. As the mutual information relation between X, Y and T in the information bottleneck method is quite abstract, it is reasonable to introduce a compact graph notation that allows one to visualize the intended mutual information relations between the variables involved in an information bottleneck problem. In [35], information bottleneck graphs were introduced for that purpose.

Information bottleneck graphs are extended factor graphs that aim to compactly visualize the intended mutual information relations of the information bottleneck method. In an information bottleneck graph, a compression mapping $p(t|y)$ designed with an information bottleneck algorithm is visualized as a trapezoid node that is labeled with the respective relevant random variable. The compressed variable is connected to the shortest side of the trapezoid. The other connected variables form the observation. This concept allows one to cover compression mappings with a single scalar input y or multiple scalar inputs y_n, $n \in \{0, 1, \ldots, N-1\}$. Figure 3 shows an example for a compression mapping $p(t|y)$ with one scalar input on the left. On the right of the figure, an example for a compression mapping $p(t|y_0, y_1, y_2, y_3, y_4)$ with five scalar inputs is shown. A compression mapping $p(t|y_0, y_1, \ldots, y_{N-1})$ with N scalar inputs can equivalently be seen as a compression mapping $p(t|\mathbf{y})$ for a row vector $\mathbf{y} = [y_0, y_1, \ldots, y_{N-1}]$. In this case, the observed random variable Y is a random vector, but the principle of the information bottleneck method remains unchanged, and the compression mapping $p(t|\mathbf{y})$ can still be designed with the available information bottleneck algorithms. Problems arise, however, for a large number of inputs N, as the number of possible conditions $\mathbf{y}$ might grow massively with N. Nonetheless, Figure 3 illustrates that the intended information relations can be visualized very compactly in information bottleneck graphs. Information bottleneck graphs will be used extensively in Section 2.3.

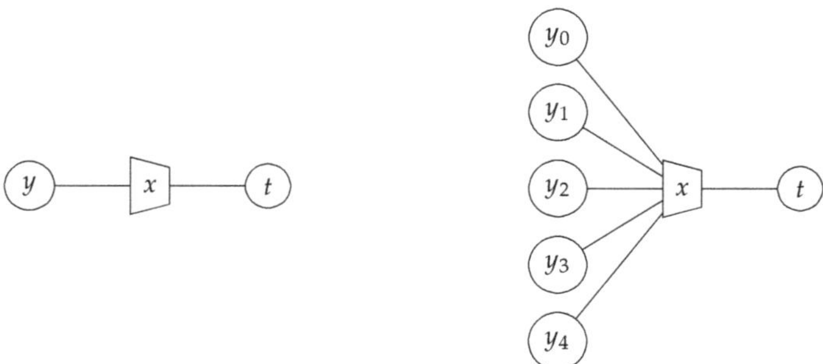

Figure 3. Examples of information bottleneck graphs. The trapezoid nodes correspond to the compression mappings $p(t|y)$ and $p(t|y_0, y_1, y_2, y_3, y_4)$, respectively. Both are designed to preseve I(X; T).

2.2. General View on Information Bottleneck Signal Processing for Receiver Design

Conceptually, the purpose of the information bottleneck method is intuitively related to the most famous fundamental problem of communications cited at the beginning. In order to explain this connection, Figure 4 compares the information bottleneck method at the top with a model view of a communication system at the bottom.

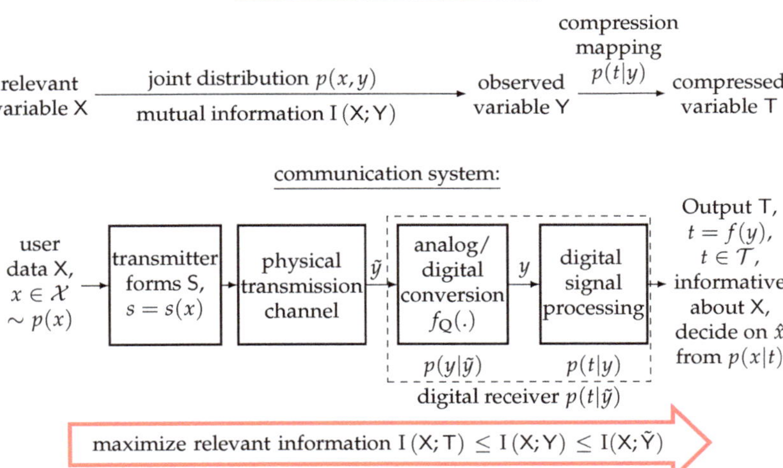

Figure 4. Illustration of the information bottleneck design idea of a communication system. The digital communication receiver shall be designed such that the relevant information I(X; T) is maximized from end to end.

As shown in the figure and discussed above, the transmitted user data X can be considered to be relevant, and the task is to estimate the realization x after the transmitter has transformed it into a transmit signal $s(x)$ to prepare it for the transmission over a physical transmission channel. The transmission of $s(x)$ then results in the received realization $\tilde{y}$, which is typically continuous.

As shown in Figure 4, we assume that $\tilde{y}$ is quantized to a discrete representation y by an analog-to-digital converter with the quantization function $f_Q(.)$ in the receiver (i.e., $y = f_Q(\tilde{y})$). The quantized representation y is then used for further digital signal processing steps to estimate x. Digital communication receivers often have to process huge blocks of the received samples at once due to the transmitter-side channel coding and modulation techniques that spread information on a transmitted bit x over a huge block of transmitted symbols. To obtain a simple notation that is coherent with the information bottleneck notation used above, we do not explicitly highlight the fact the variables X, Y and T could be random vectors in the communication system from Figure 4. However, one should keep in mind that a communication receiver typically estimates a vector **x** of transmitted bits from a long vector $\tilde{\mathbf{y}}$ of channel observations. In this case, the application of $f_Q(.)$ has to be understood as an element-wise application (i.e., $y_k = f_Q(\tilde{y}_k)$, with the time index k enumerating subsequent samples).

The amount of information present about X in the quantized received signal Y is bounded by the capacity of the quantized output transmission channel $p(y|x)$ at the upper bound, which is given by

$$C = \max_{p(x)} I(X;Y) \leq \max_{p(x)} I(X;\tilde{Y}). \qquad (6)$$

Hence, the task of the communication receiver is to extract the information on the relevant X and to provide the best possible estimate $\hat{x}$ at its output. Clearly, such an estimate can be obtained from a receiver output variable T that fulfils I(X; T) $\rightarrow$ max. As a result, the further receiver-side digital signal processing modeled by $p(t|y)$ should maximize

$I(X;T)$. Then, $p(x|t)$ can provide an estimate of the transmitted data x, for example, using the maximum a-posteriori criterion:

$$\hat{x} = \arg\max_{x} p(x|t). \tag{7}$$

Please note that $p(x|t)$ is inherently obtained when the information bottleneck method is applied according to Equation (4). This already indicates that it is possible to estimate x based on a compressed representation t of the quantized received signal y.

An interesting fact in this context is that in information theoretical terms, all signal processing of Y can only preserve or lower the information on X as a direct result of the data processing inequality [59] (i.e., $I(X;T) \leq I(X;Y)$). Due to the typically very advanced channel coding and modulation steps conducted at the transmitter, obtaining an estimate $\hat{x}$ for x, for example, based on the straightforward application of the maximum likelihood criterion via the equation

$$\hat{x} = \arg\max_{x} p(y|x) \tag{8}$$

is often prohibitively complex. The reason for this is that the transmitter spreads information on the user data x over huge blocks of transmitted symbols to protect it against transmission errors, as already mentioned above. In addition, it might include control data in the transmitted data stream to simplify the receiver-side detection process, for example, for channel estimation. As a result, the digital communication receiver has to aggregate the information on the user data X by undoing the transmitter-side encoding and modulation process using suitable algorithms composed in a signal processing chain to process the received y.

In order to overcome the complexity of a straightforward maximum likelihood sequence detection, it is common practice to separate and distribute receiver-side signal processing tasks such as synchronization, channel estimation, detection and channel decoding over concatenated signal processing blocks that exchange messages or signals. Unfortunately, the performance of such signal processing algorithms still relies on sufficient precision. The number of bits per sample has to be large enough to adequately represent the received signal in the digital hardware without a degrading quantization distortion. In addition, the required signal processing operations are often computationally demanding.

It is well known that a coarse quantization (i.e., a low resolution of the analog-to-digital converter) can help to reduce the implementation efforts of communication receivers, as it directly reduces the number of bits needed to process the signal in the digital hardware. In addition, simplified arithmetical operations can reduce the implementation efforts of optimum or close-to-optimum signal processing algorithms. However, both typically result in performance degradations.

Therefore, the fundamental idea of information bottleneck signal processing is to build subptimum receiver components that, despite being strongly quantized, are designed to preserve the maximum possible amount of relevant information and only use simple arithmetical operations. This design idea inherently minimizes the inevitable performance degradation resulting from the quantization and from using potentially suboptimal signal processing operations.

Information bottleneck signal processing has already been applied to various receiver-side signal processing tasks in the past, such as quantized detection and channel estimation [35–38]. In the following, we will analyze a receiver chain that applies information bottleneck signal processing extensively. In this receiver, we consider the mutual information preserving input/output relations of $p(t|y)$ to be implemented as lookup tables that replace the conventional signal processing operations.

2.3. An Example of Information Bottleneck Receiver Design with Iterative Detection and Decoding

In this section, we illustrate the idea of information bottleneck signal processing by considering data transmission with binary phase shift keying (BPSK) modulation over a frequency flat block fading channel. The presented study continues and extends the investigation from our conference publication [60].

We consider data transmission over a frequency-flat block fading channel. The complex-valued continuous received signal at time instance k in the symbol clock is given by

$$\tilde{y}_k = \tilde{h}s_k + \tilde{n}_k = \tilde{y}_k^{\mathrm{re}} + j\tilde{y}_k^{\mathrm{im}}, \tag{9}$$

where $\tilde{h}$ is a complex channel coefficient from a zero-mean circularly complex Gaussian process that is modelled as a constant for B symbol durations and $s_k \in \{-1, +1\}$ is the transmitted BPSK symbol. The noise sample $\tilde{n}_k$ is also a realization from a complex zero-mean white Gaussian process with variance $\sigma_{\tilde{n}}^2$. Please note that in our notation, we use a tilde to indicate that a signal has not been quantized and hence is continuous, and the superscripts re and im distinguish real and imaginary parts, respectively.

For clarity, Figure 5 provides an overview of the considered transmitter that is explained in the following.

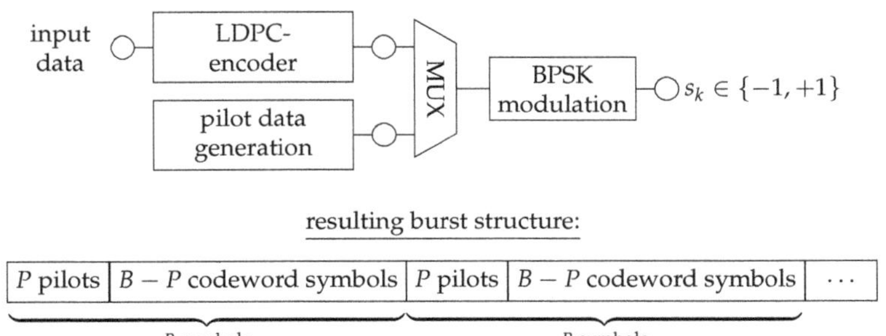

Figure 5. Overview of the considered transmitter. P pilot bits are multiplexed into a length $N^{\mathrm{LDPC}} >> B$ LDPC codeword periodically with a distance of $B - P$ bits, and the resulting data stream is modulated using BPSK modulation.

The transmitter transmits codewords from a regular LDPC code with a codeword length $N^{\mathrm{LDPC}} >> B$ over the channel using BPSK modulation. In order to simplify the receiver-side detection and decoding, the transmitter multiplexes P known pilot symbols into a transmitted codeword with a distance of $B - P$ symbols such that each block of B symbols weighted with the same channel coefficient $\tilde{h}$ starts with a few known pilot symbols. The resulting burst structure is also sketched at the bottom of Figure 5. These pilots shall serve as a reference signal to enable a simple receiver-side channel estimation.

Figure 6 shows a conventional receiver structure for the considered transmitter that uses classical signal processing operations and does not employ any quantization beyond the numerical limits of double-precision at the top. This receiver shall serve as a reference receiver that illustrates the performance of the considered data transmission scheme if no quantization is applied at all. It initially employs minimum mean squared error (MMSE) channel estimation based on the inserted pilot symbols. Based on the obtained channel estimate, it then performs a matched soft demodulation. The term matched reflects that the demodulator takes into account the uncertainty of the estimated channel coefficient. The demodulator then delivers log-likelihood ratios (LLRs) and provides them to a belief propagation decoder for the decoding of the LDPC code. This decoder performs i_{max} decoding iterations. After one such decoding round, the receiver has estimates of the transmitted codeword bits. The reliability of the different codeword bits can be judged by analyzing the absolute magnitudes of the decision LLRs from the decoder. In order to further improve the detection and decoding quality, the receiver then iteratively employs decision feedback (FB) of the N_{FB} most reliable decision bits within a block to virtually enlarge the number of available pilot symbols used for channel estimation. The considered receiver implements a hard-decision turbo channel estimation scheme. Its decision feedback loop is repeated i_{FB} times, and then the final decision bits are provided.

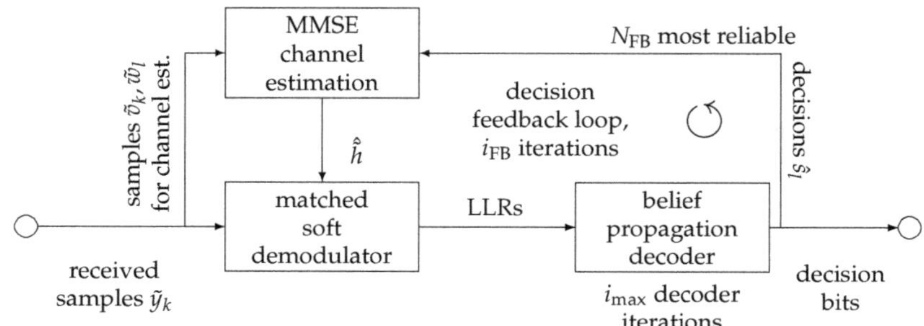

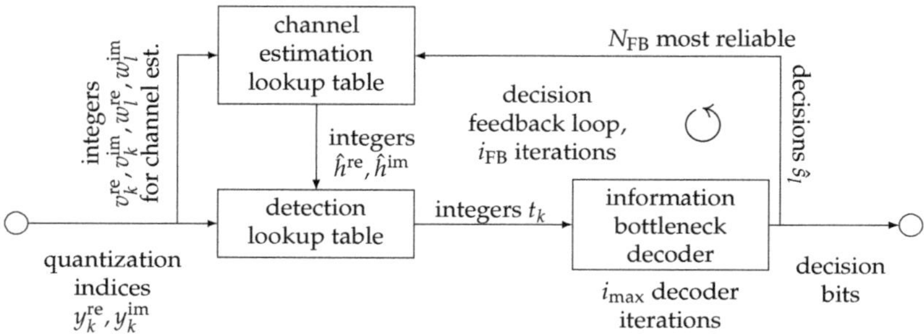

Figure 6. Conventional and information bottleneck receiver chains for LDPC-encoded data transmission over a frequency-flat fading channel. The conventional receiver uses quasi-continuous received samples $\tilde{y}_k$ with double-precision and processes them in state-of-the-art detection and decoding algorithms. The information bottleneck receiver works on quantization indices and implements all signal processing using information bottleneck lookup tables.

Underneath the conventional receiver chain in Figure 6, the proposed information bottleneck receiver chain is depicted. At first glance, both receiver chains look similar. The most important difference is, however, that the information bottleneck receiver chain does not have access to the continuous received samples $\tilde{y}_k$. Instead, it only processes quantization indices from a coarse channel output quantizer, which quantizes the real and imaginary parts of the received signal to quantization indices (i.e., $y_k^{re}, y_k^{im} \in \{0, 1, \ldots, 2^q - 1\}$). Based on these observed quantization indices, the rest of the signal processing chain for channel estimation, detection and LDPC decoding can also be developed using the information bottleneck method. This will be explained in detail in the following.

2.3.1. Information Bottleneck Channel Estimation

In the considered conventional receiver chain, the task of the channel estimation is to obtain a reliable estimate $\hat{\tilde{h}}$ of $\tilde{h}$ from the knowledge of P pilot symbols and their respective received samples. We denote these received samples by $\tilde{v}_0, \tilde{v}_1, \ldots, \tilde{v}_{P-1}$. The MMSE channel estimate is then given by

$$\hat{\tilde{h}} = \frac{1}{P + \sigma_{\tilde{n}}^2/\sigma_{\tilde{h}}^2} \sum_{k=0}^{P-1} \tilde{v}_k s_k^{P}, \tag{10}$$

where $s_k^{\mathrm{p}} \in \{-1, +1\}$ are the known pilot symbols.

It is obvious, that the evaluation of Equation (10) requires high-precision arithmetic. In a digital receiver, a high quantization resolution is required such that the quantized received samples adequately approximate the continuous $\tilde{v}_k$ to obtain a good estimate with high accuracy. As mentioned above, we assume that the conventional reference receiver is implemented using double floating-point precision and therefore does not suffer from any mentionable quantization loss.

In the information bottleneck receiver, however, the aim is to directly process the q-bit integer indices $v_k^{\mathrm{re}} = f_{\mathrm{Q}}(\tilde{v}_k^{\mathrm{re}})$ and $v_k^{\mathrm{im}} = f_{\mathrm{Q}}(\tilde{v}_k^{\mathrm{im}})$. The function $f_{\mathrm{Q}}(.)$ describes a channel output quantizer that is also designed to preserve the maximum relevant information with the information bottleneck method as explained in [35]. Please note that the design of the channel output quantizer $f_{\mathrm{Q}}(.)$ is not discussed in detail in this article. However, we want to mention that it requires handling continuous observed random variables with the information bottleneck method, which has not been discussed in this article so far. A simple method to deal with continuous observations for the quantizer design problem is approximating the continuous variables as very finely quantized discrete variables. More details on the quantizer design with a desired maximum preservation of relevant information can be found in [35,61].

The quantization function $f_{\mathrm{Q}}(.)$ conducts threshold decisions on the real and imaginary parts of the received signal $\tilde{y}_k$. It delivers quantization indices from the set $\{0, 1, \ldots, 2^q - 1\}$. This reflects an analog-to-digital conversion with q bits per sample in the real and imaginary parts, respectively. The obtained unsigned integers neither approximate the continuous received samples nor can they be used to obtain a channel estimate with an arithmetic rule such as that in Equation (10).

Essentially, the task of channel estimation is to extract information on the unknown channel coefficient $\tilde{h}$ from P pairs of integers $(v_k^{\mathrm{re}}, v_k^{\mathrm{im}})$ that correspond to the coarsely quantized received samples for the pilot symbols. Please note that due to the considered channel model and the fact that the pilot symbols are BPSK symbols, it is possible to handle v_k^{re} and v_k^{im} independently. In the following, the signal processing of v_k^{im} will not be discussed, as it is equivalent to that of v_k^{re}.

Note that the quantizer $f_{\mathrm{Q}}(.)$ leads to the fact that up to the distortion caused by the receiver noise and the influence of $s_k \in \{-1, +1\}$, the real part of the continuous channel coefficient $\tilde{h}$ is also observed in a quantized manner. Therefore, the integer h^{re} is the index of the quantization region the continuous $\tilde{h}^{\mathrm{re}}$ falls into; that is, $h^{\mathrm{re}} = f_{\mathrm{Q}}(\tilde{h}^{\mathrm{re}})$. From an information theoretical perspective, the task of a pilot-based channel estimator is to map the unsigned integers $v_0^{\mathrm{re}}, v_1^{\mathrm{re}}, \ldots, v_{P-1}^{\mathrm{re}}$ onto $\hat{h}_{\mathrm{FW}}^{\mathrm{re}}$ such that the mutual information $\mathrm{I}(\hat{\mathsf{H}}_{\mathrm{FW}}^{\mathrm{re}}; \mathsf{H}^{\mathrm{re}})$ is maximized. Please note that we add the index forward (FW) to distinguish the channel estimation from the one conducted in the feedback (FB) iterations, which will be discussed later.

An intuitive approach to designing an information bottleneck equivalent to the pilot-based channel estimation in the forward path of the conventional receiver is to design an information bottleneck compression mapping $p\left(\hat{h}_{\mathrm{FW}}^{\mathrm{re}} | v_0^{\mathrm{re}}, v_1^{\mathrm{re}}, \ldots, v_{P-1}^{\mathrm{re}}\right)$ which aims to maximize $\mathrm{I}(\hat{\mathsf{H}}_{\mathrm{FW}}^{\mathrm{re}}; \mathsf{H}^{\mathrm{re}})$. However, designing and implementing this compression mapping as a lookup table would typically result in prohibitive complexity, since the cardinality of the observation random variable $(\mathsf{V}_0^{\mathrm{re}}, \mathsf{V}_1^{\mathrm{re}}, \ldots, \mathsf{V}_{P-1}^{\mathrm{re}})$ is $2^{q \cdot P}$. For example, with $q = 5$ for bit quantization and $P = 6$ pilot symbols, this would result in a lookup table with more than 10^9 entries. This cardinality exceeds the afordable runtime and space complexities of all available information bottleneck algorithms, as these algorithms typically have to handle a dense matrix representation of $p(x, y)$ that consists of $|\mathcal{X}||\mathcal{Y}|$ real numbers. Moreover, the resulting lookup table would be prohibitively large.

In order to cope with the resulting complexity, a particularly useful feature of information bottleneck compression mappings is that they can be concatenated to reduce the complexity. Such a concatenation is shown in an information bottleneck graph in the upper part of Figure 7 for the considered channel estimation scheme for the pilot-based forward channel estimation. As shown, the task of processing P quantized samples is split into a series of $P-1$ concatenated compression mappings, each of which processes two inputs. This way, the space complexity of the component compression mappings that we implement as lookup tables is drastically reduced to a concatenation of lookup tables with $2^{2q} + (P-2)2^{q^{ce}+q}$ entries, where q^{ce} denotes the number of bits needed to represent the intermediate results $\hat{h}_m^{re}$ from Figure 7. Please note that this number of bits is a design choice, and it can be adjusted by the choice of the cardinality of the compression variable of the information bottleneck algorithm that is used. For simplicity, all compression mappings shall use the same output bit width q^{ce} such that all intermediate results $\hat{h}_m^{re}$ and the final output $\hat{h}_{FW}^{re}$ are from the same set $\left\{0, 1, \ldots, 2^{q^{ce}} - 1\right\}$. With $q^{ce} = 5$ and $P = 6$ pilot symbols, as considered before, the overall size of the lookup tables to implement the concatenated scheme from Figure 7 is reduced from roughly 10^9 to 5120.

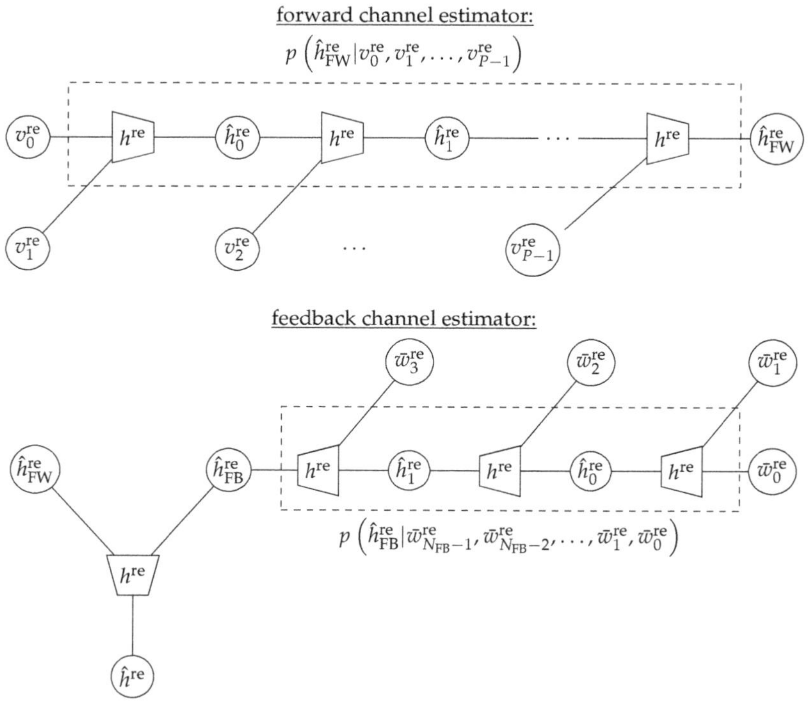

Figure 7. Information bottleneck graphs of the forward and feedback channel estimation schemes for the information bottleneck receiver. The upper part of the figure shows the forward channel estimator consisting of $P-1$ two-input information bottleneck compression mappings. The lower part shows the feedback channel estimator with a similar structure to process $N_{FB} = 4$ inputs.

All compression mappings appearing inside the forward channel estimator from Figure 7 preserve information on the same variable H^{re}. The design of the compression mappings from Figure 7 requires feeding the joint distributions

$$p\left(h^{re}, v_0^{re}, v_1^{re} | s_0^P, s_1^P\right) = p\left(h^{re}, v_0^{re} | s_0^P\right) p\left(h^{re}, v_1^{re} | s_1^P\right) \frac{1}{p(h^{re})} \tag{11}$$

and

$$p\left(h^{\mathrm{re}}, \hat{h}^{\mathrm{re}}_{m-1}, v^{\mathrm{re}}_{m+1} | s^{\mathrm{P}}_0, s^{\mathrm{P}}_1, \dots, s^{\mathrm{P}}_{m+1}\right) =$$

$$p\left(h^{\mathrm{re}}, \hat{h}^{\mathrm{re}}_{m-1} | s^{\mathrm{P}}_0, s^{\mathrm{P}}_1, \dots, s^{\mathrm{P}}_m\right) p\left(h^{\mathrm{re}}, v^{\mathrm{re}}_{m+1} | s^{\mathrm{P}}_{m+1}\right) \frac{1}{p(h^{\mathrm{re}})} \quad \text{for } m \geq 1. \quad (12)$$

to an information bottleneck algorithm.

One question that is still open is how to obtain the joint distributions $p\left(h^{\mathrm{re}}, v^{\mathrm{re}}_m | s^{\mathrm{P}}_m\right)$ needed to evaluate Equations (11) and (12). Recall that h^{re} and v^{re}_m are quantization indices of $\tilde{h}^{\mathrm{re}}$ and $\tilde{v}^{\mathrm{re}}_m$, respectively. Hence, $h^{\mathrm{re}} = f_{\mathrm{Q}}(\tilde{h}^{\mathrm{re}})$ and $v^{\mathrm{re}}_m = f_{\mathrm{Q}}(\tilde{v}^{\mathrm{re}}_m)$, where $f_{\mathrm{Q}}(.)$ characterizes the threshold decisions of the channel output quantizer. The quantizer $f_{\mathrm{Q}}(.)$ for $\tilde{h}^{\mathrm{re}}$ and $\tilde{v}^{\mathrm{re}}$ can also be described by $p(h^{\mathrm{re}}|\tilde{h}^{\mathrm{re}})$ and $p(v^{\mathrm{re}}_m|\tilde{v}^{\mathrm{re}}_m)$. This allows expressing the joint distribution $p\left(h^{\mathrm{re}}, v^{\mathrm{re}}_m | s^{\mathrm{P}}_m\right)$ as

$$p(h^{\mathrm{re}}, v^{\mathrm{re}}_m | s^{\mathrm{P}}_m) = \int\limits_{-\infty}^{+\infty} \int\limits_{-\infty}^{+\infty} p\left(\tilde{h}^{\mathrm{re}}, \tilde{v}^{\mathrm{re}}_m | s^{\mathrm{P}}_m\right) p(h^{\mathrm{re}}|\tilde{h}^{\mathrm{re}}) p(v^{\mathrm{re}}_m|\tilde{v}^{\mathrm{re}}_m)\, \mathrm{d}\tilde{h}^{\mathrm{re}}\, \mathrm{d}\tilde{v}^{\mathrm{re}}_m. \quad (13)$$

In this integral, the factors $p\left(h^{\mathrm{re}}|\tilde{h}^{\mathrm{re}}\right)$ and $p(v^{\mathrm{re}}_m|\tilde{v}^{\mathrm{re}}_m)$ gather all the probability masses of the continuous $\tilde{h}^{\mathrm{re}}$ and $\tilde{v}^{\mathrm{re}}_m$, which are mapped onto the same pair $(h^{\mathrm{re}}, v^{\mathrm{re}}_m)$ by application of the channel output quantizer on these variables. Therefore, they determine the integration area for a particular pair $(h^{\mathrm{re}}, v^{\mathrm{re}}_m)$. The distribution $p\left(\tilde{h}^{\mathrm{re}}, \tilde{v}^{\mathrm{re}}_m | s^{\mathrm{P}}_m\right)$ for the considered channel model is a multivariate Gaussian distribution. Both components have a zero mean. As a result, the multivariate Gaussian distribution $p\left(\tilde{h}^{\mathrm{re}}, \tilde{v}^{\mathrm{re}}_m, | s^{\mathrm{P}}_m\right)$ is fully characterized by the covariance matrix

$$\mathbf{C}_{\tilde{H}^{\mathrm{re}}, \tilde{V}^{\mathrm{re}}_m | S^{\mathrm{P}}_m} = \frac{\sigma^2_{\tilde{h}}}{2} \begin{bmatrix} 1 & s^{\mathrm{P}}_m \\ s^{\mathrm{P}}_m & 1 \end{bmatrix} + \frac{\sigma^2_{\tilde{n}}}{2} \begin{bmatrix} 0 & 0 \\ 0 & 1 \end{bmatrix}. \quad (14)$$

The integral from Equation (13) with the considered Gaussian distributions and covariance matrices characterized by Equation (14) is easily solved numerically using the method described in [62]. This allows one to obtain the joint distributions $p(h^{\mathrm{re}}, v^{\mathrm{re}}_m | s^{\mathrm{P}}_m)$ and hence construct the concatenated structure of information bottleneck compression mappings from the top of Figure 7 using Equations (11) and (12). We apply the KL-means algorithm from [57] to construct the proposed information bottleneck channel estimator, as this algorithm can be highly parallelized and is very fast.

The design process of the information bottleneck channel estimator delivers the joint distribution $p\left(\hat{h}^{\mathrm{re}}, h^{\mathrm{re}}\right) = p\left(h^{\mathrm{re}}|\hat{h}^{\mathrm{re}}\right) p\left(\hat{h}^{\mathrm{re}}\right)$, which has to be used further to construct an information bottleneck detection lookup table. Its design is explained in the next subsection.

A very similar concept of information bottleneck channel estimation can also be used in the feedback iterations of the iterative information bottleneck receiver structure from Figure 6 to implement a turbo-like channel estimation scheme. After decoding the LDPC code, the receiver can identify the most reliable bit decisions $\hat{s}_l$ and their corresponding received samples w^{re}_l.

Let w^{re}_l, $l \in \{0, 1, \dots, N_{\mathrm{FB}} - 1\}$ denote the quantized integer indices y^{re}_k, which correspond to the most reliable bit decisions in the decoded codeword. Moreover, let $\hat{s}_l$ denote the respective hard decision BPSK symbols corresponding to the respective decision bits. The task of an information bottleneck feedback channel estimator is to extract the information on h^{re} from $\hat{s}_l$ and w^{re}_l. Please note that the transformation

$$\bar{w}^{\mathrm{re}}_l = \begin{cases} w^{\mathrm{re}}_l & \hat{s}_l = +1 \\ 2^q - w^{\mathrm{re}}_l - 1 & \hat{s}_l = -1. \end{cases} \quad (15)$$

undoes the influence of $\hat{s}_l$ and makes all $\tilde{w}_l^{\text{re}}$ look like quantization indices received for the transmitted symbol $s_l = +1$ if the decision $\hat{s}_l$ is correct.

The integers $\tilde{w}_l^{\text{re}}$ can be used in the feedback channel estimation scheme shown in the bottom part of Figure 7. The design of the concatenated compression mappings is completely equivalent to the one for the forward channel estimator above, with the only difference being that one assumes the transmitted symbol to be $s_l = +1$ due to the transformation from Equation (15). In fact, this transformation is a simple way to make the feedback channel estimation lookup tables independent of the symbol decision and thus keep them minimal in size. However, we want to mention that this method only works for BPSK modulation. Higher-order modulation schemes are not studied in this article. The presented information bottleneck channel estimation could be implemented for higher-order modulation schemes equivalenty, but this might also require feeding the symbol decisions $\hat{s}_l$ to the feedback channel estimation lookup tables.

The final compression mapping $p(\hat{h}^{\text{re}} | \hat{h}_{\text{FW}}^{\text{re}}, \hat{h}_{\text{FB}}^{\text{re}})$ combines the relevant information on h^{re} from the forward and feedback channel estimator lookup tables to finally deliver the integer $\hat{h}^{\text{re}}$ that is highly informative about h^{re}. The obtained $\hat{h}^{\text{re}}$ can be used for detection in the next step.

2.3.2. Information Bottleneck Detection

In a conventional receiver with high precision, the obtained MMSE channel estimate $\hat{h}$ from Equation (10) can be used for soft demodulation. The LLRs of the soft demodulator are given by

$$
L^{\text{ch}}(s_k) = \frac{1}{1 + \frac{1}{P + \sigma_{\hat{h}}^2 / \sigma_{\hat{h}}^2}} \cdot \frac{4\,\text{Re}\left\{\hat{h}^* \tilde{r}_k\right\}}{\sigma_{\hat{h}}^2},
\tag{16}
$$

where $\text{Re}\{.\}$ denotes the real part of a complex number and $.^*$ denotes the conjugate complex. Using these channel LLRs, it is easy to apply the LLR-based belief propagation decoding algorithm in the conventional receiver at the top of Figure 6.

When the information bottleneck channel estimator from the preceding section is applied, it delivers the pair $\left(\hat{h}^{\text{re}}, \hat{h}^{\text{im}}\right)$ of unsigned integers. Moreover, the quantizer delivers a pair $(y_k^{\text{re}}, y_k^{\text{im}})$ for each symbol s_k. The task of an information bottleneck demodulator equivalent for the considered channel is extracting relevant information on s_k from $\left(\hat{h}^{\text{re}}, \hat{h}^{\text{im}}\right)$ and $(y_k^{\text{re}}, y_k^{\text{im}})$.

Figure 8 shows the information bottleneck graph performing this task. In this graph, one first successively combines $\hat{h}^{\text{re}}, y_k^{\text{re}}$ and $\hat{h}^{\text{im}}, y_k^{\text{im}}$ to obtain t_k^{re} and t_k^{im}. The compression mappings $p\left(t_k^{\text{re}} | \hat{h}^{\text{re}}, y_k^{\text{re}}\right)$ and $p\left(t_k^{\text{im}} | \hat{h}^{\text{im}}, y_k^{\text{im}}\right)$ are identical, as their inputs are identically distributed. To design $p\left(t_k^{\text{re}} | \hat{h}^{\text{re}}, y_k^{\text{re}}\right)$ with an information bottleneck algorithm, one needs the joint probability distribution

$$
p\left(s_k, \hat{h}^{\text{re}}, y_k^{\text{re}}\right) = \sum_{h^{\text{re}}=0}^{2^q - 1} p(h^{\text{re}}, y_k^{\text{re}} | s_k) \underbrace{\frac{p\left(h^{\text{re}}, \hat{h}^{\text{re}}\right)}{p(h^{\text{re}})}}_{p\left(\hat{h}^{\text{re}} | h^{\text{re}}\right)} p(s_k).
\tag{17}
$$

In Equation (17), the distributions $p\left(h^{\text{re}}, \hat{h}^{\text{re}}\right)$ and $p(h^{\text{re}})$ are known from the design of the channel estimation lookup table. The distribution $p(h^{\text{re}}, y_k^{\text{re}} | s_k)$ is again an integral of a multivariate Gaussian distribution, where

$$
p(h^{\text{re}}, y_k^{\text{re}} | s_k) = \int_{-\infty}^{+\infty} \int_{-\infty}^{+\infty} p(\tilde{h}^{\text{re}}, \tilde{y}_k^{\text{re}} | s_k)\, p(h^{\text{re}} | \tilde{h}^{\text{re}})\, p(y_k^{\text{re}} | \tilde{y}_k^{\text{re}})\, \mathrm{d}\tilde{h}^{\text{re}}\, \mathrm{d}\tilde{y}_k^{\text{re}}.
\tag{18}
$$

Again, the integral can be solved using the method from [62]. The required covariance matrix is obtained analogously with Equation (14).

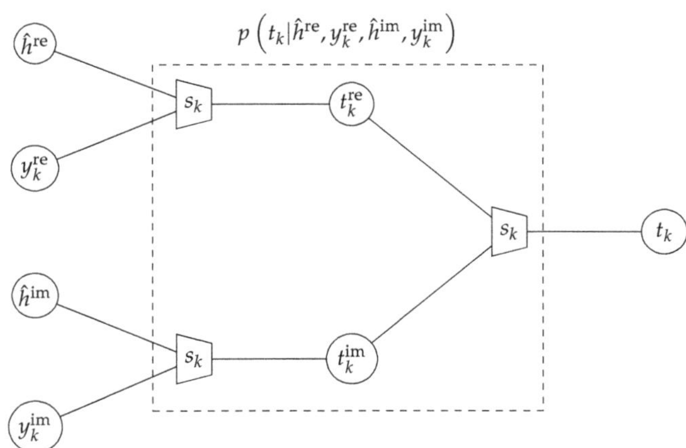

Figure 8. Information bottleneck graph of the detection scheme for the information bottleneck receiver. The detection scheme first extracts information on s_k from $(y_k^{\mathrm{re}}, \hat{h}^{\mathrm{re}})$ and $(y_k^{\mathrm{im}}, \hat{h}^{\mathrm{im}})$ independently. Afterwards, it yields a an integer output t_k from t_k^{re} and t_k^{im} that is informative about s_k using the mapping $p(t_k|t_k^{\mathrm{re}}, t_k^{\mathrm{im}})$.

An important note is that the design of the final concatenated information bottleneck compression mapping $p(t_k|t_k^{\mathrm{re}}, t_k^{\mathrm{im}})$ inherently delivers the distributions $p(s_k|t_k)$ and $p(t_k)$ and, therefore, $p(s_k, t_k)$. Reviewing Figure 4, the concatenation of the transmitter, channel, quantizer and the proposed concatenation of information bottleneck compression mappings therefore corresponds to a discrete input, discrete output transmission scheme. This discrete input, discrete output transmission scheme delivers $t_k \in \{0, 1, \ldots, 2^{q^{\mathrm{det}}} - 1\}$ for each transmitted modulation symbol s_k at the receiving end, where $2^{q^{\mathrm{det}}}$ is the output cardinality of the information bottleneck algorithm used to design $p(t_k|t_k^{\mathrm{re}}, t_k^{\mathrm{im}})$. The statistical properties of (s_k, t_k) and the preserved relevant mutual information $\mathrm{I}(S_k; T_k)$ are completely determined by $p(s_k, t_k)$, which describes a discrete source and channel model for the considered data transmission scheme. This model is needed to construct the iterative information bottleneck LDPC decoder from Figure 6. The design of this LDPC decoder is sketched in the following section. More details can be found in [18,61].

2.3.3. Information Bottleneck LDPC Decoder Design

As briefly mentioned above, the design of an information bottleneck LDPC decoder requires knowledge of the joint probability distribution $p(s_k, t_k)$. Luckily, this distribution is delivered inherently by the information bottleneck algorithm used to design the detection scheme from the previous section.

This probability distribution is iteratively processed in a discretized density evolution algorithm [12,18,61] to construct a coarsely quantized information bottleneck LDPC decoder. The density evolution algorithm is needed to determine the input distributions of the information bottleneck algorithms used to design the lookup tables that implement the variable and the check node operations of a message passing LDPC decoder in the subsequent decoder iterations. The lookup tables only process and deliver integer indices and therefore are fundamentally different in their operation than conventional LDPC decoders, which process real-valued LLRs.

Figure 9 gives an overview of the notation used to describe the design of the information bottleneck LDPC decoder. We consider a regular LDPC code with a variable node degree d_{v} and check node degree d_{c}. In decoder iteration i, the variable nodes pass quantized messages (i.e., unsigned integers $y_k^{(i)}$) to the check nodes. The check nodes generate outgoing integer messages $t_{c \rightarrow v}^{(i)}$ for all their connected edges from these incoming messages. This process is illustrated for one particular target edge at the top of Figure 9. The check node operation is designed such that the mutual information shared between the outgoing message and the codeword bit that this message represents in the LDPC codeword is maximized with the information bottleneck method. This codeword bit is denoted by x in the upper part of Figure 9. The other $d_c - 1$ codeword bits connected to the check node are denoted by $b_0, b_1, \ldots, b_{d_c - 2}$.

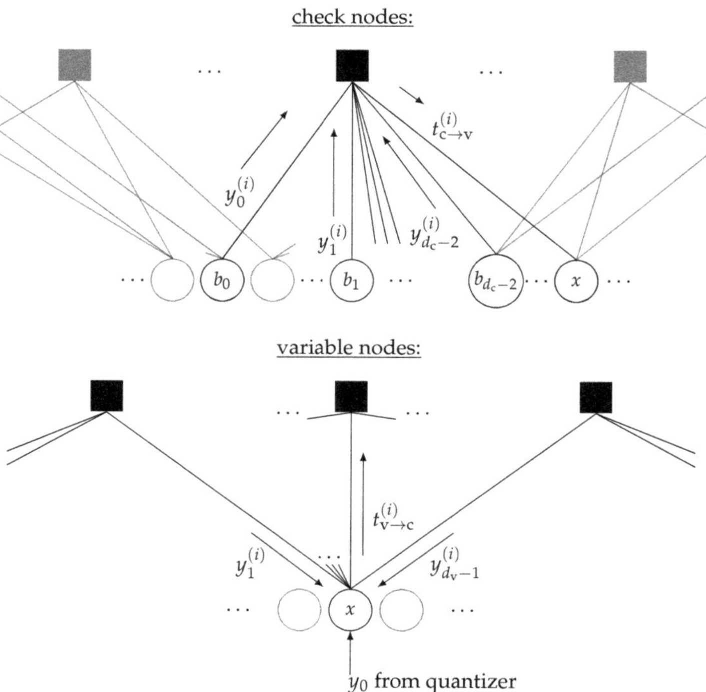

Figure 9. Message generation of a check node and a variable node in an iterative information bottleneck LDPC decoder. The nodes generate integer messages for their connected edges.

As the variable nodes pass the received messages from the channel to the check nodes in the first decoding iteration, in principle, the check node operation could be designed in a straightforward manner by feeding the joint probability distribution

$$p\left(x, y_0^{(i)}, y_1^{(i)}, \ldots, y_{d_c-2}^{(i)}\right) = \sum_{\substack{(b_0, b_1, \ldots, b_{d_c-2}): \\ x = b_0 \oplus b_1 \oplus \ldots \oplus b_{d_c-2}}} \prod_{k=0}^{d_c-2} p(b_k, y_k) \tag{19}$$

to an information bottleneck algorithm. This approach, however, suffers from intractable complexity, as the number of input configurations $(y_0^{(i)}, y_1^{(i)}, \ldots, y_{d_c-2}^{(i)})$ grows exponentially with the node degree d_c. As a result, the authors of [12] proposed splitting the check node operation into a series of concatenated two-input operations, exactly as was performed in Section 2.3.1 for the channel estimation scheme to reduce the complexity. In this way, a series of concatenated two-input lookup tables can be designed with an information bottleneck algorithm that processes unsigned integers and aims to maximize the preserved relevant information $I(X; T_{c \to v}^{(i)})$.

The information bottleneck algorithm used for the design of the check node operation finally delivers $p(x, t_{c \to v}^{(i)})$. Based on this distribution, a very similar process can be used to design the variable node operation depicted in the bottom part of Figure 9. The variable nodes create outgoing messages $t_{v \to c}^{(i)}$ that shall be highly informative of their corresponding codeword bit x (i.e., $I(X; T_{v \to c}^{(i)}) \to \max$). To obtain extrinsic information on the codeword bit x, the variable nodes process $d_v - 1$ messages from the connected check nodes and a message y_0 from the channel. In the considered data transmission scheme, the channel messages correspond to $t_k \in \{0, 1, \ldots, 2^{q^{ce}} - 1\}$ from the detection scheme described in Section 2.3.2. As a result, the design of the variable node operation with the information bottleneck method processes the joint probability distribution $p(x, y_0)$, which is equivalent to $p(s_k, t_k)$ and $p(x, t_{c \to v}^{(i)})$ from the design process of the check node operation. For the equations needed to obtain all involved joint probability distributions, the reader is asked to refer to [18,61]. Most importantly, the information bottleneck algorithm applied to the variable node design delivers $p(x, t_{v \to c}^{(i)})$. This distribution has to be processed to design the check node operation for the next decoder iteration $i + 1$, which will in turn deliver $p(x, t_{c \to v}^{(i+1)})$ for the check nodes in the next decoding iteration. This naturally suggests an iterative algorithm to determine lookup tables used as node operations in an information bottleneck LDPC decoder. This algorithm is iterated for a desired number of $i_{\max}$ iterations of the constructed LDPC decoder, and the lookup tables constructed for the variable and the check node operations in each iteration of the decoder are stored.

Finally, an important note on the proposed receiver design shall be made. Obviously, the joint probability distributions that are processed for the decoder design and also for the channel estimation and detection scheme studied before depend on the channel conditions. As a result, the reader might expect that the lookup tables designed with the information bottleneck method need to be adjusted to the signal-to-noise ratio or, equivalently, to the current E_b / N_0 on the channel. As the design process of the detection, channel estimation and especially the LDPC decoder is computationally quite complex and involves numerous information bottleneck algorithms, this would clearly question the practical use of the proposed receiver design. However, interestingly, the proposed receiver design provides excellent performance, even with information bottleneck lookup tables that are used being mismatched to the actual channel conditions. This allows conducting the entire receiver design process offline to store the resulting lookup tables and to use them in a receiver with coarse quantization and very simple operations only. We will analyze the performance of the resulting receiver and compare it to several reference receivers in the following to prove our statement on excellent performance.

2.3.4. Comparison of Iterative Receiver Performances

In this section, we investigate the bit error rate performance of the proposed information bottleneck receiver and compare it to the one of several reference receivers. As mentioned above, the proposed information bottleneck receiver from Figure 6 was designed for a manually chosen design E_b / N_0 and was not matched to the actual E_b / N_0 on the channel in the results presented next. The lookup tables were constructed offline with the information bottleneck method.

The bit error rate performances of the considered receivers are compared in Figure 10 for a number of $i_{FB} = 0$ and $i_{FB} = 5$ feedback iterations of the decision feedback channel estimation. The applied LDPC code was a length of 8000 $(d_v, d_c) = (3, 6)$ regular LDPC code from [63]. In the studied scenario, we considered a block fading channel that was constant for $B = 35$ symbol durations before a new independent channel coefficient was drawn, which weighted the transmitted symbols according to Equation (9). The first $P = 3$ symbols transmitted for each of these blocks were the known pilot symbols for the initial channel estimation.

The quantization bit widths of the information bottleneck receiver were chosen as follows. We used $q = 5$ bits per sample for channel output quantization in the real and imaginary parts of the received signal, respectively, $q^{\text{ce}} = 8$ bits for channel estimation and $q^{\text{det}} = 5$ bits for the detection lookup table.

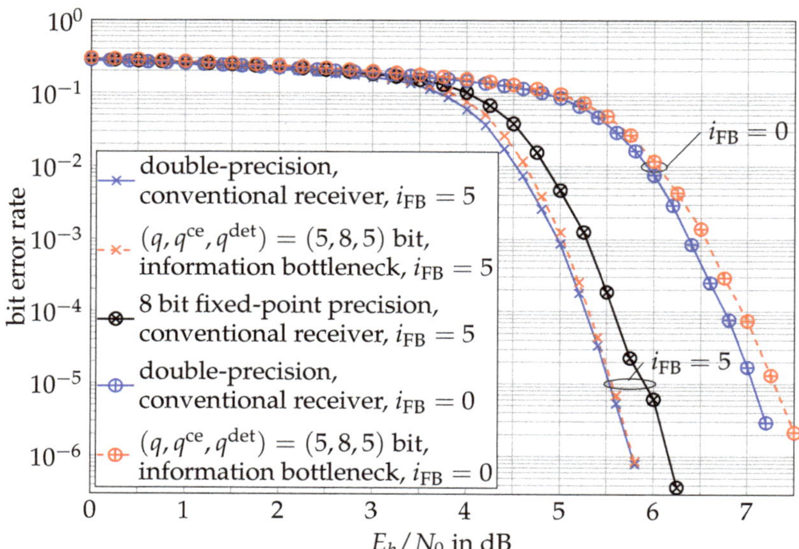

Figure 10. Bit error rates of different quantized receivers and the conventional receiver from Figure 6, which did not suffer from a quantization loss at all. The quantized information bottleneck receiver with $q = 5$ bit channel output quantization, $q^{\text{ce}} = 8$ bit channel estimation and $q^{\text{det}} = 5$ bit detection and LDPC decoding met the performance of the double-precision reference receiver for $i_{\text{FB}} = 5$ feedback iterations.

We also used $q^{\text{det}} = 5$ bit messages in the information bottleneck LDPC decoder. The decoder applied $i_{\text{max}} = 25$ iterations in each of the outer feedback iterations for the channel estimation. In each outer iteration, the $N_{\text{FB}} = 20$ most reliable bit decisions from this decoder in each block were fed to the feedback channel estimator illustrated in Figure 7 to supplement the channel estimation. The red curves in Figure 10 show the bit error rate performance of the proposed information bottleneck receiver with these parameters for $i_{\text{FB}} = 0$ and $i_{\text{FB}} = 5$ outer iterations of the decision feedback loop for channel estimation. A comparison of the receiver's performance for $i_{\text{FB}} = 0$ and $i_{\text{FB}} = 5$ outer iterations of the channel estimation feedback loop clearly illustrates the gain resulting from improving the initial channel estimate based on the decision feedback from the channel decoder.

The blue curves in Figure 10 correspond to the conventional receiver chain from Figure 6 with the respective numbers of feedback iterations $i_{\text{FB}} = 0$ and $i_{\text{FB}} = 5$ and also used $i_{\text{FB}} = 25$ iterations of the conventional belief propagation LDPC decoder. We want to stress again that this receiver does not suffer from any mentionable quantization loss.

Despite that, it can clearly be observed that the performance of the quantized information bottleneck receiver chain that only processed unsigned integers and replaced all conventional arithmetical operations in the signal processing algorithms with simple lookup operations for $i_{\text{FB}} = 5$ iterations was tremendously close to the one of the conventional reference receiver. It is also observable that without feedback iterations (i.e., $i_{\text{FB}} = 0$), the loss of the information bottleneck receiver was slightly higher. The reason for this is that all of the lookup tables in the information bottleneck receiver were designed only for a single design E_b / N_0, which we optimized to yield the best performance for the case with $i_{FB} = 5$ feedback iterations. It is, however, possible to also tune the design E_b/N_0 to minimize the performance gap with respect to the non-quantized receiver without feedback iterations to achieve a negligible loss over E_b/N_0 for $i_{FB} = 0$ [61].

The numbers of bits $(q, q^{ce}, q^{det}) = (5, 8, 5)$ used in the information bottleneck signal processing units were determined by carefully analyzing which settings resulted in the optimum performance. While for the detection and decoding stage we found $q^{det} = 5$ bit processing sufficient, during our investigations, we noted that the channel estimation appeared to require a larger bit width of $q^{ce} = 8$ bits to perform that close to the non-quantized reference system. A possible intuitive explanation is that other than for all the detection and decoding parts of the receiver, the relevant variable of the channel estimator $H^{re} = f_Q(\tilde{H}^{re})$ is not binary. Instead, with a q-bit channel output quantizer $f_Q(.)$ in place, it can take 2^q different values. This intuitively suggests that in order to achieve close-to-optimum performance, the channel estimation stage requires a larger bit width to preserve the relevant information.

As another reference, the bit error rate of an 8-bit fixed-point implementation of the conventional receiver is included in Figure 10 for $i_{FB} = 5$ feedback iterations in black. This receiver used the Q5.3 fixed-point format in the detection and channel estimation stages and the Q4.4 fixed-point format for the messages in the belief propagation decoder. In this notation, $Qm.n$ denotes using m bits for the integer part and n bits for the fractional part of the processed fixed-point numbers. In order to not disadvantage the conventional receiver chain, all other possible combinations of 8-bit fixed-point formats were also tested, but the chosen selection offered the best performance. The 8-bit fixed-point conventional receiver was clearly outperformed by the information bottleneck receiver. This is particularly interesting because the information bottleneck receiver only used 8-bit integers in its channel estimation stage and shorter 5-bit integers in all other detection and decoding stages.

So far, Figure 10 shows that the performance of the proposed receiver was very close to that of the conventional receiver with double-precision if $i_{FB} = 5$ feedback iterations were performed. It is, however, still unclear whether or not the same information bottleneck receiver suffers from performance degradation if fewer feedback iterations are performed.

As a result, Figure 11 shows the bit error rates of both receivers for all $i_{FB} \in \{0, 1, 2, 3, 4, 5\}$. The figure illustrates that the performance of the information bottleneck receiver can compete with the double-precision conventional receiver for all investigated numbers of feedback iterations i_{FB}. The figure also indicates that the gains of the feedback loop were very significant in the first three feedback iterations for both receivers and then became less significant. It was found that using more than five feedback iterations hardly improved the bit error rate performance of any investigated receiver.

As the most important result so far, we can summarize that the information bottleneck design of the proposed receiver yielded a quantized information bottleneck receiver with very simple and homogenous operations that could deliver performance practically identical to that of a conventional receiver with double-precision arithmetic. This receiver employs advanced signal processing concepts, such as decision feedback for quantized channel estimation and quantized LDPC decoding. All of the signal processing operations used in the receiver were designed using the same principle of maximizing the flow of relevant information through the quantized signal processing operations of the receiver with the information bottleneck method.

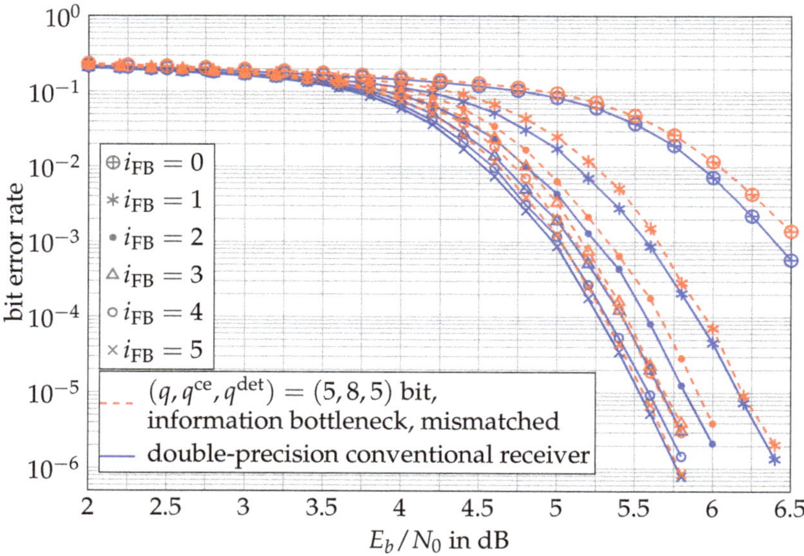

Figure 11. Bit error rates of different receiver implementations as a function of the number i_{FB} of feedback iterations. The quantized information bottleneck receiver with $q = 5$-bit channel output quantization, $q^{\text{ce}} = 8$-bit channel estimation and $q^{\text{det}} = 5$-bit detection and LDPC decoding offered performance similar to the double-precision conventional receiver for all investigated i_{FB}.

3. Parameter Learning of Trainable Functions to Maximize the Relevant Information

In the information bottleneck receiver presented above, as we have often emphasized, all signal processing operations were implemented as lookup tables. As mentioned in the introduction, however, there exist several other approaches to implementing information bottleneck signal processing units in the literature, especially in the context of LDPC decoding. In this section, our goal is to provide an overview of the lookup table approach, the computational domain methods used in [24,25,30,31] and the neural network approach from [28,29]. All can be seen as mappings $t = f_\theta(\mathbf{y})$ implied by a trainable function $f_\theta(\mathbf{y})$, with parameters θ that are trained to maximize the preserved relevant information $I(X; T)$.

This view on information bottleneck signal processing is depicted in Figure 12 and was introduced previously in [64,65].

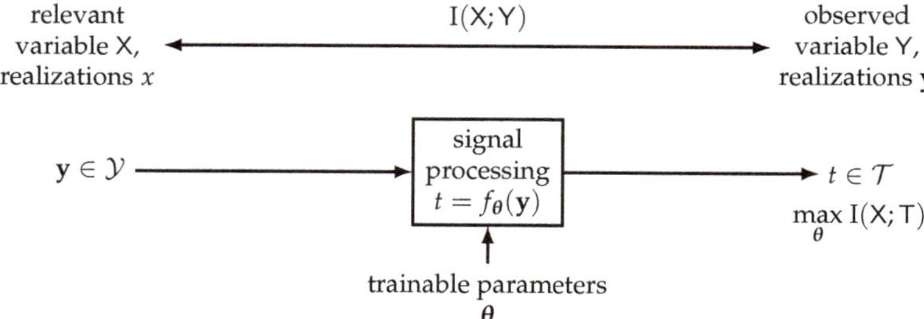

Figure 12. General illustration of an information bottleneck signal processing unit. The signal processing unit consists of a trainable function with parameters θ that can be learned to maximize $I(X; T)$.

3.1. Lookup Tables

As mentioned in Section 2.1, it is intuitive that any deterministic compression mapping $p(t|y)$ for discrete t and y can be implemented in a lookup table. For that purpose, one just has to store the respective $t \in \mathcal{T}$ for all possible $y \in \mathcal{Y}$. Of course, this also holds if the realizations of Y are random vectors $\mathbf{y} = [y_0, y_1, \ldots, y_{N-1}]$.

For simplicity, we assume that $y_n \in \{0, 1, \ldots, |\mathcal{Y}|_n - 1\}$ in this section (i.e., the y_n in vector $\mathbf{y}$ are unsigned integers). Please note that the elements of the event space are totally irrelevant for the mutual information $I(X; Y)$, as this mutual information only depends on the probabilities implied by $p(x, \mathbf{y}), p(x)$ and $p(\mathbf{y})$. We can easily understand a lookup table holding the t for each possible $\mathbf{y}$ as a trainable function $f_\theta(\mathbf{y})$ with parameters $\boldsymbol{\theta}$.

Consider a software implementation of a lookup table consisting of a row vector $\boldsymbol{\theta} = [\theta_0, \theta_1, \ldots, \theta_{|\mathcal{Y}|-1}]$. This vector holds the outcome t for every possible input vector $\mathbf{y}$ at a certain position that is termed the address. To obtain the output t, the input $\mathbf{y}$ has to be mapped onto the address of the corresponding output t in the vector. Let the function implementing this address transformation be denoted by $\xi(\mathbf{y})$. It is not important which address transformation $\xi(\mathbf{y})$ is used as long as it maps the input vector $\mathbf{y}$ onto the address of the corresponding $t \in \mathcal{T}$ uniquely. Then, the lookup table delivers $t = f_\theta(\mathbf{y}) = \theta_{\xi(\mathbf{y})}$ by just accessing the vector $\boldsymbol{\theta}$ at the calculated address.

With this quite formal description of a simple lookup table, it is clear that the problem of maximizing the preserved relevant information $I(X; T)$ with $t \in \mathcal{T}$ and $\mathbf{y} \in \mathcal{Y}$ is formally equivalent to determining a vector of the optimum discrete parameters:

$$\boldsymbol{\theta}^{\mathrm{IB}}_{\beta \to \infty} = \arg \max_{\boldsymbol{\theta}} I(X; f_\theta(Y)). \tag{20}$$

Therefore, we can summarize that any deterministic mapping of $\mathbf{y}$ onto t that is implemented in a lookup table and determined to maximize the preserved relevant information with an information bottleneck algorithm can equivalently be interpreted as a trainable function $f_\theta(\mathbf{y})$ with a maximum of $|\mathcal{Y}|$ discrete parameters $\theta_{\xi(\mathbf{y})} \in \mathcal{T}$. The number of parameters $|\mathcal{Y}|$ is identical to all possible input configurations $\mathbf{y} \in \mathcal{Y}$. This allows representing arbitrary mappings in this setup and has the consequence that a lookup table can implement every possible input/output relation for discrete $\mathbf{y}$ and t, including one that maximizes $I(X; T)$ globally. This holds independent of the probability distribution $p(x, \mathbf{y})$. Yet, we note that finding an optimum mapping with the available information bottleneck algorithms is not guaranteed. However, for discrete $\mathbf{y} \in \mathcal{Y}$ and discrete $t \in \mathcal{T}$, we can conclude that no other mapping can offer more flexibility than a lookup table. Therefore, in theory, a lookup table can be seen as an optimum choice in terms of providing the possibility to maximize $I(X; f_\theta(Y))$ with properly tuned parameters $\boldsymbol{\theta}$. The way to learn these parameters in the receiver presented above was feeding the joint probability distributions $p(x, \mathbf{y})$ to the information bottleneck algorithms.

Unfortunately, however, problems arise if the number N of scalar inputs to be processed by a lookup table is large, as was the case, for example, for the node operations in the information bottleneck LDPC decoders from Section 2.3.3. These operations have to process a huge number of scalar input variables, and the number of possible input configurations of a lookup table scales exponentially with the number of scalar inputs. However, we have seen that in the studied cases, it was possible to split the signal processing operations designed as information bottleneck lookup tables into concatenated two-input tables to reduce the complexity. This, however, results in a concatenation of several lossy two-input lookup tables. Hence, the authors of [24,25] argued reasonably that such a splitting procedure likely results in losses of the preserved relevant information in comparison with an input/output mapping that can process all scalar inputs consolidated in the input vector $\mathbf{y}$ at once.

3.2. Computational Domain Technique

The authors of [24,25] proposed an interesting idea to implement check and variable node operations of LDPC decoders that, just as with the information bottleneck approaches based on lookup tables described above, aims at maximizing the preserved relevant information under compression. Their approach is called mutual information-maximizing quantized belief propagation decoding. We briefly recall this approach that is illustrated in Figure 13 in the following.

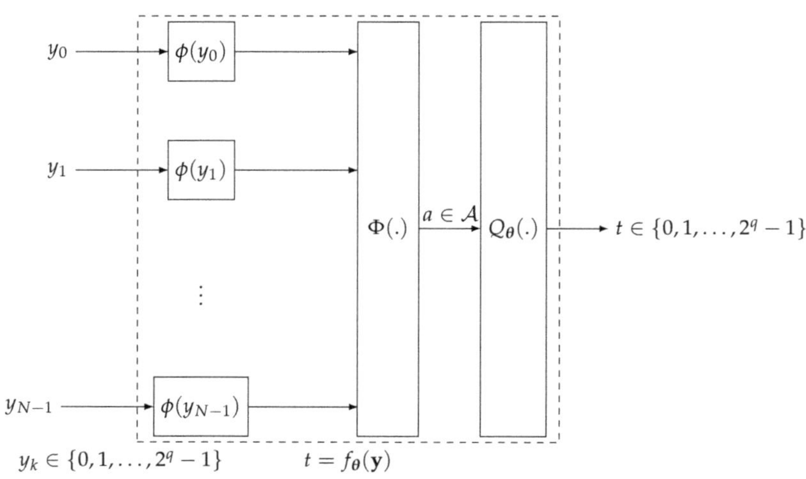

Figure 13. Illustration of the computational domain approach for information bottleneck-like LDPC decoding from [24,25]. Figure adapted from [24,25] with minor adaptions of the notation.

The authors of [24,25] distinguished between variable and check node operations in an LDPC decoder. Here, however, we study a general input/output relation that maps an incoming vector $\mathbf{y} = [y_0, y_1, \ldots, y_{N-1}]$ onto an outgoing t. For simplicity, we assume the most practical approach of all y_n and t to be from the same set $\{0, 1, \ldots, 2^q - 1\}$ of q-bit unsigned integers. The computational domain approach consists of three subsequent steps [24,25]:

1. Use a predefined reconstruction function $\phi(.)$ to transfer the incoming messages y_n to numbers $\phi(y_n)$ in a computational domain $\mathbb{D}$;
2. Use a function $\Phi : \mathbb{D}^N \longrightarrow \mathcal{A}$ to process the numbers in the computational domain and to map them onto a single number $a \in \mathcal{A}$;
3. Apply a scalar quantizer $Q_\theta(.)$ with $2^q - 1$ ordered thresholds $\theta = [\theta_0, \theta_1, \ldots, \theta_{2^q-2}]$ on a that quantizes $a \in \mathcal{A}$ back to the set $\{0, 1, \ldots, 2^q - 1\}$.

In [24,25], certain reconstruction functions for the check and variable nodes of LDPC decoders and also reasonable functions $\Phi(.)$ for these applications, which we do not recall here in detail for brevity, are proposed.

The idea of maximizing the preserved relevant mutual information in the information bottleneck sense is applied in the design of the scalar quantizer $Q_\theta(.)$ in the computational domain approach from [24,25]. The thresholds $\theta = [\theta_0, \theta_1, \ldots, \theta_{2^q-2}]$ of this quantizer are chosen such that $I(X; T)$ is maximized.

Of course, one can understand the quantization thresholds stored in vector $\theta = [\theta_0, \theta_1, \ldots, \theta_{2^q-2}]$ as parameters of a function implementing the node operation

$$t = f_\theta(\mathbf{y}) = Q_\theta(\Phi(\phi(y_0), \phi(y_1), \ldots, \phi(y_{N-1}))), \tag{21}$$

such that the design of the node operation ends up in the exact same problem formulation as that given in Equation (20) (that is, the learning parameters θ of a trainable function $f_\theta(\mathbf{y})$ such that $I(X; f_\theta(Y)) \longrightarrow \max$).

The major difference is that the number of parameters that are tuned to maximize $I(X; f_\theta(Y))$ can be drastically reduced in the computational domain approach with respect to the lookup table. Essentially, only $2^q - 1$ quantization thresholds instead of 2^{qN} entries of a lookup table need to be learned and optimized to maximize $I(X; T)$. Interestingly, the computational domain approach employs arithmetical operations in the function $\Phi(.)$. In return, it needs much fewer parameters to preserve significant amounts of $I(X; T)$ than the lookup table approach. We note, however, that the choice of the reconstruction function $\phi(.)$ and the function $\Phi(.)$ as well as their abilities to preserve significant amounts of $I(X; T)$ are problem-specific and depend on $p(x, \mathbf{y})$.

It has to be appreciated that the proposed method from [24,25] offers excellent bit error rate performance for LDPC decoding.

To illustrate this fact, Figure 14 compares the bit error rate performances of the computational domain approach and two lookup table-based information bottleneck decoders for the $(d_v, d_c) = (3, 6)$ regular LDPC code that was already used in Section 2.3.4. For these simulations, all quantized decoders exchanged $q = 4$-bit messages in the iterative message passing process for their respective decoding. The bit width used internally to represent $a \in \mathcal{A}$ in the computational domain decoders was 10 bits for the check nodes and variable nodes. The shown results for the computational domain decoder were adopted from [24]. The bit error rate curves refer to data transmission with BPSK modulation over an additive white Gaussian noise channel. The number of decoder iterations was $i_{\max} = 50$ for all investigated decoders.

The relatively small node degrees of the applied code allow constructing lookup table-based information bottleneck decoders with and without splitting the node operation into two-input operations. We want to mention that the decoder without splitting has an impractically large memory demand but shall be used as a benchmark here. For the relatively small node degrees of the $(3, 6)$ regular LDPC code used here, the lookup tables in the decoder without splitting already had more than 300 million entries [28].

As can be seen, the computational domain decoder performed just as well as a lookup table-based decoder without internal splitting of the node operations. Both quantized decoders effectively reached the performance of the non-quantized belief propagation decoder with double-precision up to a negligible gap over E_b / N_0.

As can also be seen, splitting the node operation of the information bottleneck decoder into a series of two-input operations offered slightly worse performance but yielded much smaller lookup tables in return. The number of lookup table entries used for this decoder was just 384,000 and therefore drastically reduced to a tractable number in comparison with the lookup table-based decoder without two-input splitting of the node operations.

Again, we want to stress here that the information bottleneck decoders and the computational domain decoder [24] were entirely constructed offline and used mismatched to the E_b / N_0 on the channel. For reference, the bit error rate of the well-known min-sum decoder with double-precision is also shown in Figure 14. This decoder was outperformed by all decoders designed with the information bottleneck principle of maximizing the preserved relevant information.

Finally, we note that the computational domain approach is well suited for LDPC decoding. Thus far, to the best of our knowledge, studies of computational domain approaches that maximize the relevant information for other applications of information bottleneck signal processing such as the channel estimation studied in Section 2.3.1, cannot be found in the literature.

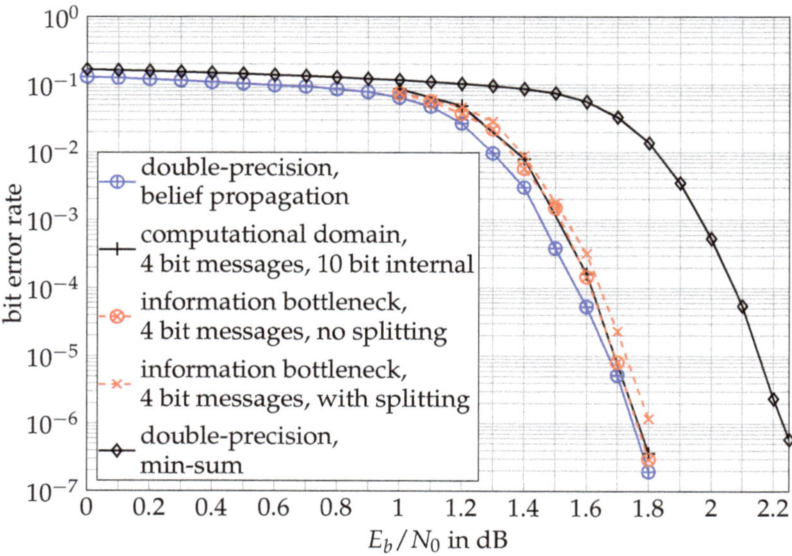

Figure 14. Bit error rates of several LDPC decoders for data transmission with BPSK over an additive white Gaussian noise channel. The applied code was a $(3,6)$ regular LDPC code. All decoders conducted $i_{\mathrm{max}} = 50$ decoding iterations. The computational domain approach refers to [24].

3.3. Neural Networks

Another idea to implement information bottleneck signal processing for LDPC decoding was introduced in [28]. In order to get rid of the need for implementing very large lookup tables to process the incoming messages of a check or variable node in one shot (i.e., without splitting the node operation), neural networks were trained to maximize the preserved relevant information. Learning the parameters of such neural networks was conducted in a supervised manner in [28] first. To accomplish this, as a first step, lookup tables that did not use two-input table splitting were constructed with the information bottleneck method for the variable and check node operations. Afterwards, different neural network structures were trained to mimic the resulting node operations. The key is, of course, that the trained networks have much fewer parameters θ than there are entries in the original lookup tables.

We want to recall a certain neural network structure proposed to implement the node operations in [28] here. This neural network directly inputs the binary representations of the integer messages exchanged to implement the node operations of the check or the variable nodes. Moreover, this network is iteration-aware [28], meaning that the decoder iteration i is used as an additional input to the network together with the incoming messages of a node. In this way, only one network for the check nodes and one network for the variable nodes are required to implement all node operations for all iterations of the decoder.

For the $(3,6)$ regular LDPC code considered in the prior section, this architecture results in a neural network-based decoder with 25,564 parameters of the involved neural networks. Recall again that the lookup table-based information bottleneck decoder without split node operations required more than 300 million lookup table entries, and the one with splitting required 384,000 lookup table entries (for more quantitative details, see Table I from [28]).

Figure 15 shows the bit error rate results of the neural network method for implementation of the node operations from [28]. The simulation setup, the used LDPC code and the number of decoding iterations were the same as those for the computational domain approach from the prior section.

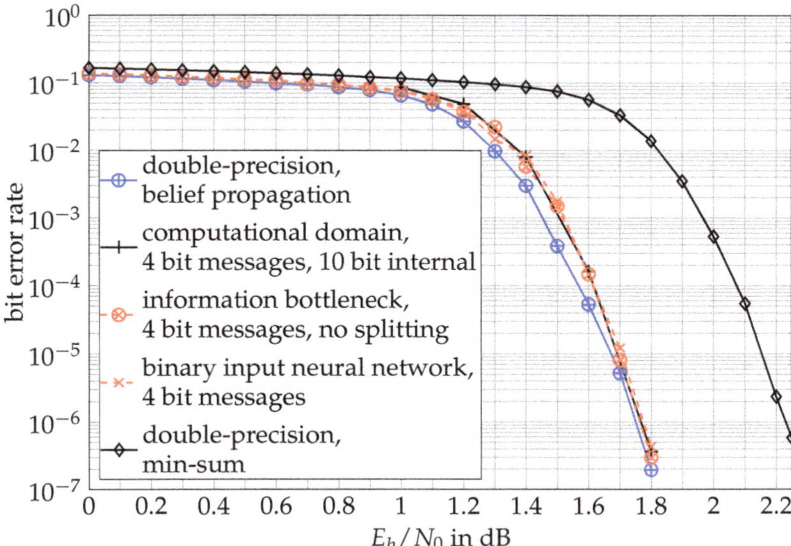

Figure 15. Bit error rates of several LDPC decoders for data transmission with BPSK over an additive white Gaussian noise channel. The applied code was a $(3, 6)$ regular LDPC code. All decoders conducted $i_{\max} = 50$ decoding iterations. The computational domain approach refers to [24].

As can clearly be seen, the neural network implementation of the node operations achieved a practically identical performance to the lookup table-based information bottleneck decoder without split node operations as well as the computational domain decoder studied before. All these decoders approach the performance of the double-precision belief propagation decoder with quite different approaches to implement the node operations which, however, all are motivated by the information bottleneck idea of maximizing the preserved relevant information under quantization.

The supervised learning approach to train a neural network that shall maximize $I(X; T)$ presented in [28] still requires designing huge lookup tables used to generate the training data as an intermediate step. Hence, the idea was also extended to unsupervised learning in [29] such that the parameters of the neural networks involved were directly trained to maximize the preserved relevant information with standard gradient methods that are commonly used to train the parameters of neural networks. This eliminates the intermediate step of designing lookup tables with impractical sizes for decoder implementation. Moreover, it directly leads back to the problem formulation from Equation (20), which again illustrates that this neural network-based approach of information bottleneck signal processing can also be understood as the learning parameters of a trainable function that shall maximize the preserved relevant information.

However, if the number N of scalar inputs consolidated in the input vector $\mathbf{y}$ is large, representing $p(x, \mathbf{y})$ in closed form to calculate $p(x, t)$ and from that $I(X; T)$ is also infeasible. As a way out, a number of samples $(x, \mathbf{y})$ can be used to determine the corresponding $t = f_{\boldsymbol{\theta}}(\mathbf{y})$ for all samples and, from that, estimate $p(x, t)$ and $I(X; T)$.

Finally, we want to mention that obtaining an accurate estimate of $I(X; T)$ in this unsupervised learning procedure typically requires a vast amount of samples $(x, \mathbf{y})$ if the number of scalar inputs consolidated in $\mathbf{y}$ is large. This problem is well known in the machine learning community and often referred to as the curse of dimensionality. However, this number of samples only influences the construction complexity of the node operations in the LDPC decoder but not their implementation complexity.

3.4. Further Discussion, Other Approaches and Future Work

The discussion of the lookup table approach, the computational domain approach and the neural network approach to information bottleneck signal processing reveals the key principles of information bottleneck signal processing. These are choosing appropriate functions $f_\theta(\mathbf{y})$ with trainable parameters θ and determining the values of these parameters that maximize the relevant information. The functions used should only use a small number of simple arithmetical operations, and the number of parameters should also be small for complexity reasons.

As was discussed above, lookup tables have strengths regarding flexibility, as for discrete inputs $\mathbf{y}$ and discrete outputs t, they can implement arbitrary input/output relations $t = f_\theta(\mathbf{y})$. However, they often need too many parameters if the number of inputs is large. The studied computational domain approach requires much fewer parameters but needs a sophisticated choice for the reconstruction function $\phi(.)$ and the function $\Phi(.)$ for the signal processing problem of interest, (e.g., LDPC decoding). Neural network approaches offer an almost endless choice of network architectures. These are determined by the number of layers, the types of the layers and their respective activation functions. All these aspects influence their implementation complexity and the number of parameters. However, neural networks enable realizing almost arbitrary mappings with proper architectures and a reasonable number of parameters. Moreover, they can be trained efficiently using gradient-based algorithms.

In future work, it will be interesting to further study flexible, trainable functions for information bottleneck signal processing units. To illustrate that many more options exist, we briefly mention a quite different approach from [64,65] as an example. This approach is based on efficient nearest neighbor search algorithms in graphs. These search algorithms are used to implement trainable functions $f_\theta(\mathbf{y})$ that aim to maximize $I(X;T)$ in [64,65].

The mappings $t = f_\theta(\mathbf{y})$ in [64,65] use a small number of distance calculations between the incoming vector $\mathbf{y}$ and some trained parameter vectors θ_t to determine the system output t such that $I(X;T) \to$ max. An exemplary nearest neighbor search is illustrated in Figure 16.

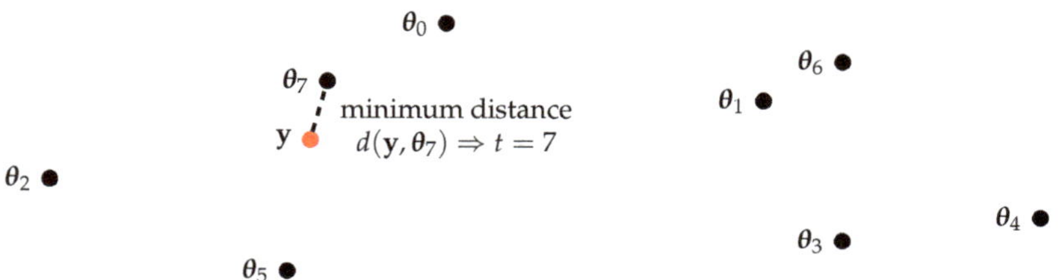

Figure 16. Illustration of a mapping $t = f_\theta(\mathbf{y})$ based on the nearest neighbor search. The output t is the index of the nearest neighbor θ_t of $\mathbf{y}$. It can be found by using graph-based algorithms efficiently.

In the figure, the parameters $\theta = [\theta_0, \theta_1, \ldots, \theta_7]$ reflect the positions of the points labeled θ_t in the two-dimensional plain. In addition, an exemplary two-dimensional input vector $\mathbf{y}$ is shown as a red dot. The outcome of the operation $t = f_\theta(\mathbf{y})$ is the integer index t of an approximate or exact nearest neighbor θ_t of the input vector $\mathbf{y}$ under some arbitrary distance measure $d(\mathbf{y}, \theta_t)$. In Figure 16, the distance is the Euclidean distance, and the exact nearest neighbor is considered for illustration purposes. Please note that it is not required to calculate all possible distances $d(\mathbf{y}, \theta_t)$ to determine the system output t. Instead, either an exact or approximate nearest neighbor search using very efficient search algorithms in graph structures of the parameter vectors θ_t can be applied [64,65]. The parameters involved in nearest neighbor search-based mappings were trained using genetic algorithms in [64,65]. Genetic algorithms are parameter optimization algorithms which are inspired by the natural evolution of the species [66].

Clearly, this realization of a mapping $t = f_\theta(\mathbf{y})$, which aims to maximize $I(X; T)$, differs a lot from the lookup table-based method, the computational domain approach and also the neural network method studied above. Yet, the results from [64,65] prove it to be quite powerful in terms of the preservation of $I(X; T)$, with few parameters for distance-based quantization and demodulation in communication receivers. This illustrates that there is a lot of potential in finding different flexible and simple trainable functions $f_\theta(\mathbf{y})$ to design information bottleneck signal processing units for different signal processing problems in the future. In addition, efficient training of the parameters involved offers interesting research directions.

4. Conclusions

In this article, we first gave an overview of the information bottleneck method and explained how it can be linked to the fundamental task of a digital communication receiver. The application of the information bottleneck method for receiver-side signal processing in this context effectively allows building quantized signal processing units that aim to maximize the relevant information that flows through them. This concept is fundamentally different from conventional quantized signal processing approaches, which typically aim to minimize an expected error measure, such as the mean squared error. Based on the principle of maximizing the preserved relevant information under quantization, we presented and investigated an iterative receiver structure for a frequency-flat fading channel that employs advanced signal processing concepts for iterative LDPC decoding, decision feedback-aided channel estimation and detection. All signal processing units in this receiver were designed using the information bottleneck principle of maximizing the preserved relevant information under quantization. The corresponding signal processing operations applied in the constructed receiver were implemented as static lookup tables that were designed offline with the information bottleneck method. Despite this fact, the designed receiver did not suffer from any mentionable performance degradation in comparison with a conventional receiver with double-precision signal processing and belief propagation decoding.

After having studied the fundamental idea and the lookup table-based implementation of information bottleneck signal processing units in the receiver implementation, other methods for learning and implementing mutual information-maximizing signal processing units for LDPC decoders were also recalled from the literature and investigated. Our main conclusion is that the considered approaches to information bottleneck signal processing, including the one based on lookup tables, effectively aim to learn the parameters θ of trainable functions $f_\theta(\mathbf{y})$ that are tuned to maximize the preserved relevant information $I(X; f_\theta(Y))$ under a constraint for the cardinality of the value set of $f_\theta(\mathbf{y})$. We note that the ways to learn the respective parameters differ depending on the kind of function that is designed to preserve the relevant information. While the lookup table based approaches typically employ information bottleneck algorithms, the computational domain approach from [24,25] uses a quantizer design algorithm to determine the optimum quantization thresholds. The neural network-based approaches from [28,29] mimic lookup tables or directly use gradient-based algorithms to maximize the preserved relevant information. Even more ideas that use nearest neighbor search algorithms to implement information bottleneck signal processing units and genetic algorithms to tune their parameters were proposed in the literature [64,65]. Finally, what all these approaches have in common is the learning of trainable parameters to maximize the preserved relevant information under quantization.

A lot of very interesting open research questions on information bottleneck signal processing arise from the problem of finding powerful and, at the same time, simple trainable functions for different signal processing applications that allow preserving huge amounts of relevant information. Moreover, exploring the methods for efficient training of the involved parameters, such as using genetic algorithms or gradient-based schemes, is interesting.

Author Contributions: Conceptualization, J.L., G.B. and M.S.; methodology, J.L. and M.S.; software, J.L. and M.S.; validation, J.L., G.B. and M.S.; formal analysis, J.L., G.B. and M.S.; investigation, J.L., G.B. and M.S.; resources, J.L.; writing—original draft preparation, J.L.; writing—review and editing, J.L., G.B. and M.S.; visualization, J.L., G.B. and M.S.; supervision, G.B.; project administration, G.B. All authors have read and agreed to the published version of the manuscript.

Funding: This research received no external funding.

Institutional Review Board Statement: Not applicable.

Informed Consent Statement: Not applicable.

Data Availability Statement: Not applicable.

Conflicts of Interest: The authors declare no conflict of interest.

References

1. Shannon, C.E. A Mathematical Theory of Communication. *Bell Syst. Tech. J.* **1948**, *27*, 379–423. [CrossRef]
2. Tishby, N.; Pereira, F.C.; Bialek, W. The information bottleneck method. In Proceedings of the 37th Allerton Conference on Communication and Computation, Monticello, NY, USA, 22–24 September 1999; 368–377.
3. Slonim, N. The Information Bottleneck: Theory and Applications. Ph.D. Dissertation, Hebrew University, Jerusalem, Israel, 2002.
4. Slonim, N.; Somerville, R.; Tishby, N.; Lahav, O. Objective classification of galaxy spectra using the information bottleneck method. *Mon. Not. R. Astron. Soc.* **2001**, *323*, 270–284. [CrossRef]
5. Bardera, A.; Rigau, J.; Boada, I.; Feixas, M.; Sbert, M. Image segmentation using information bottleneck method. *IEEE Trans. Image Process.* **2009**, *18*, 1601–1612. [CrossRef]
6. Buddha, S.; So, K.; Carmena, J.; Gastpar, M. Function identification in neuron populations via information bottleneck. *Entropy* **2013**, *15*, 1587–1608. [CrossRef]
7. Zaidi, A.; Agueri, I.-E.; Shamai, S. On the information bottleneck problems, connections, applications and information theoretic views. *Entropy* **2020**, *22*, 151. [CrossRef]
8. Goldfeld, Z.; Polyanskiy, Y. The information bottleneck problem and its applications in machine learning. *IEEE J. Sel. Areas Inf. Theory* **2020**, *1*, 19–38. [CrossRef]
9. Zeitler, G.; Low-precision analog-to-digital conversion and mutual information in channels with memory. In Proceedings of the 2010 48th Annual Allerton Conference on Communication, Control, and Computing (Allerton'2010), Monticello, NY, USA, 29 September–1 October 2010; pp. 745–752.
10. Zeitler, G.; Singer, A.C.; Kramer, G. Low-precision A/D conversion for maximum information rate in channels with memory. *IEEE Trans. Commun.* **2012**, *60*, 2511–2521. [CrossRef]
11. Kurkoski, B.M.; Yagi, H. Quantization of binary-input discrete memoryless channels. *IEEE Trans. Inf. Theory* **2014**, *60*, 4544–4552. [CrossRef]
12. Kurkoski, B.M.; Yamaguchi, K.; Kobayashi, K. Noise thresholds for discrete LDPC decoding mappings. In Proceedings of the 2008 IEEE Global Telecommunications Conference (GLOBECOM'2008), New Orleans, LA, USA, 30 November–4 December 2008; pp. 1–5.
13. Meidlinger, M.; Balatsoukas-Stimming, A.; Burg, A.; Matz, G. Quantized message passing for LDPC codes. In Proceedings of the 2015 49th Asilomar Conference on Signals, Systems and Computers (ACSSC'2015), Pacific Grove, CA, USA, 8–11 November 2015; pp. 1606–1610.
14. Lewandowsky, J.; Bauch, G. Trellis based node operations for LDPC decoders from the information bottleneck method. In Proceedings of the 2015 9th International Conference on Signal Processing and Communication Systems (ICSPCS'2015), Cairns, Australia, 14–16 December 2015; pp. 1–10.
15. Romero, F.J.C.; Kurkoski, B.M. LDPC decoding mappings that maximize mutual information. *IEEE J. Sel. Areas Commun.* **2016**, *34*, 2391–2401. [CrossRef]
16. Lewandowsky, J.; Stark, M.; Bauch, G. Optimum message mapping LDPC decoders derived from the sum-product algorithm. In Proceedings of the 2016 IEEE International Conference on Communications (ICC'2016), Kuala Lumpur, Malaysia, 22–27 May 2016; pp. 1–5.
17. Meidlinger, M.; Matz, G. On irregular LDPC codes with quantized message passing decoding. In Proceedings of the 2017 IEEE 18th International Workshop on Signal Processing Advances in Wireless Communications (SPAWC'2017), Sapporo, Japan, 3–6 July 2017; pp. 1–5.
18. Lewandowsky, J.; Bauch, G. Information-optimum LDPC decoders based on the information bottleneck method. *IEEE Access* **2018**, *6*, 4054–4071. [CrossRef]
19. Stark, M.; Lewandowsky, J.; Bauch, G. Information-optimum LDPC decoders with message alignment for irregular codes. In Proceedings of the 2018 IEEE Global Communications Conference (GLOBECOM'2018), Abu Dhabi, United Arab Emirates, 9–13 December 2018; pp. 1–6.

20. Stark, M.; Lewandowsky, J.; Bauch, G. Information-bottleneck decoding of high-rate irregular LDPC codes for optical communication using message alignment. *Appl. Sci.* **2018**, *8*, 1884. [CrossRef]
21. Stark, M.; Bauch, G.; Lewandowsky, J. Decoding of non-binary LDPC codes using the information bottleneck method. In Proceedings of the 2019 IEEE International Conference on Communications (ICC'2019), Shanghai, China, 20–24 May 2019; pp. 1–6.
22. Meidlinger, M.; Matz, G.; Burg, A. Design and decoding of irregular LDPC codes based on discrete message passing. *IEEE Trans. Commun.* **2019**, *68*, 1329–1343. [CrossRef]
23. Lewandowsky, J.; Bauch, G.; Tschauner, M.; Oppermann, P. Design and evaluation of information bottleneck LDPC decoders for digital signal processors. *IEICE Trans. Commun.* **2019**, *E102-B*, 1363–1370. [CrossRef]
24. He, X.; Cai, K.; Mei, Z. Mutual information-maximizing quantized belief propagation decoding of regular LDPC codes. *arXiv* **2019**, arXiv:1904.0666.
25. He, X.; Cai, K.; Mei, Z. On mutual information-maximizing quantized belief propagation decoding of LDPC codes. In Proceedings of the 2019 IEEE Global Communications Conference (GLOBECOM'2019), Waikoloa, HI, USA, 9–13 December 2019; pp. 1–6.
26. Stark, M.; Bauch, G.; Wang, L.; Wesel, R. Information bottleneck decoding of rate-compatible 5G-LDPC codes. In Proceedings of the 2020 IEEE International Conference on Communications (ICC'2020), Taipei, Taiwan, 7–11 December 2020; pp. 1–6.
27. Stark, M.; Bauch, G.; Wang, L.; Wesel, R. Decoding rate-compatible 5G-LDPC codes with coarse quantization using the information bottleneck method. *IEEE Open J. Commun. Soc.* **2020**, *1*, 646–660. [CrossRef]
28. Stark, M.; Lewandowsky, J.; Bauch, G. Neural information bottleneck decoding. In Proceedings of the 2020 14th International Conference on Signal Processing and Communication Systems (ICSPCS'2020), Adelaide, Australia, 14–16 December 2020; pp. 1–7.
29. Stark, M.; Machine Learning for Reliable Communication under Coarse Quantization. Ph.D. Dissertation, Hamburg University of Technology, Hamburg, Germany, 2021.
30. Mohr, P.; Bauch, G. Coarsely Quantized Layered Decoding Using the Information Bottleneck Method. In Proceedings of the 2021 IEEE International Conference on Communications (ICC'2021), Montreal, QC, Canada, 14–23 June 2021; pp. 1–6.
31. Wang, L.; Terrill, C.; Stark, M.; Li, Z.; Chen, S.; Hulse, C.; Kuo, C.; Wesel, R.; Bauch, G. Reconstruction-Computation-Quantization (RCQ): A Paradigm for Low Bit Width LDPC Decoding. *IEEE Trans. Commun.* **2022**, *70*, 2213–2226. [CrossRef]
32. Stark, M.; Shah, A.; Bauch, G. Polar code construction using the information bottleneck method. In Proceedings of the 2018 IEEE Wireless Communications and Networking Conference Workshops (WCNCW'2018), Barcelona, Spain, 15–18 April 2018; pp. 7–12.
33. Shah, S.A.A.; Stark, M.; Bauch, G. Design of quantized decoders for polar codes using the information bottleneck method. In Proceedings of the 12th International ITG Conference on Systems, Communications and Coding 2019 (SCC'2019), Rostock, Germany, 11–14 February 2019; pp. 1–6.
34. Shah, S.A.A.; Stark, M.; Bauch, G. Coarsely Quantized Decoding and Construction of Polar Codes Using the Information Bottleneck Method. *Algorithms* **2019**, *12*, 192. [CrossRef]
35. Lewandowsky, J.; Stark, M.; Bauch, G. Information bottleneck graphs for receiver design. In Proceedings of the 2016 IEEE International Symposium on Information Theory (ISIT'2016), Barcelona, Spain, 10–15 July 2016; pp. 1–5.
36. Lewandowsky, J.; Stark, M.; Bauch, G. Message alignment for discrete LDPC decoders with quadrature amplitude modulation. In Proceedings of the 2017 IEEE International Symposium on Information Theory (ISIT'2017), Aachen, Germany, 25–30 June 2017; pp. 2925–2929.
37. Lewandowsky, J.; Stark, M.; Mendrzik, R.; Bauch, G. Discrete channel estimation by integer passing in information bottleneck graphs. In Proceedings of the 2017 11th International ITG Conference on Systems, Communications and Coding (SCC'2017), Hamburg, Germany, 6–9 February 2017; pp. 1–6.
38. Bauch, G.; Lewandowsky, J.; Stark, M; Oppermann, P. Information-optimum discrete signal processing for detection and decoding. In Proceedings of the 2018 IEEE 87th Vehicular Technology Conference (VTC Spring), Porto, Portugal, 3–6 June 2018; pp. 1–6.
39. Hassanpour, S.; Monsees, T.; Wübben, D.; Dekorsy, A. Forward-aware information bottleneck-based vector quantization for noisy channels. *IEEE Trans. Commun.* **2020**, *68*, 7911–7926. [CrossRef]
40. Hassanpour, S.; Wübben, D.; Dekorsy, A. Forward-aware information bottleneck-based vector quantization: Multiterminal extensions for parallel and successive retrieval. *IEEE Trans. Commun.* **2021**, *69*, 6633–6646. [CrossRef]
41. Zeitler, G.; Koetter, R.; Bauch, G.; Widmer, J. Design of network coding functions in multihop relay networks. In Proceedings of the 5th International Symposium on Turbo Codes and Related Topics (ISTC'2008), Lausanne, Switzerland, 1–5 September 2008; pp. 249–254.
42. Zeitler, G.; Koetter, R.; Bauch, G.; Widmer, J. On quantizer design for soft values in the multiple-access relay channel. In Proceedings of IEEE International Conference on Communications (ICC'2009), Dresden, Germany, 14–18 June 2009; pp. 1–5.
43. Winkelbauer, A.; Matz, G. Joint network-channel coding for the asymmetric multiple-access relay channel. In Proceedings of IEEE International Conference on Communications (ICC'2012), Ottawa, ON, Canada, 10–15 June 2012; pp. 2485–2489.
44. Winkelbauer, A.; Matz, G. Joint network-channel coding in the multiple-access relay channel: Beyond two sources. In Proceedings of 5th International Symposium on Communications, Control and Signal Processing, Rome, Italy, 2–4 May 2012; pp. 1–5.
45. Kern, D.; Kühn, V. Practical aspects of compress and forward with BICM in the 3-node relay channel. In Proceedings of 20th International ITG Workshop on Smart Antennas (WSA), Munich, Germany, 9–11 March 2016; pp. 1–7.

46. Kern, D.; Kühn, V. On compress and forward with multiple carriers in the 3-node relay channel exploiting information bottleneck graphs. In Proceedings of 11th International ITG Conference on Systems, Communications and Coding (SCC'2017), Hamburg, Germany, 6–9 February 2017; pp. 1–6.
47. Kern, D.; Kühn, V. On implicit and explicit channel estimation for compress and forward relaying OFDM schemes designed by information bottleneck graphs. In Proceedings of 21th International ITG Workshop on Smart Antennas (WSA'2017), Berlin, Germany, 15–17 March 2017; pp. 1–6.
48. Chen, D.; Kühn V. Alternating information bottleneck optimization for the compression in the uplink of C-RAN. In Proceedings of the 2016 IEEE International Conference on Communications (ICC'2016), Kuala Lumpur, Malaysia, 22–27 May 2016; pp. 1–7.
49. Stark, M.; Lewandowsky, J.; Bauch, G. Iterative message alignment for quantized message passing between distributed sensor nodes. In Proceedings of the 2018 IEEE 87th Vehicular Technology Conference (VTC Spring'2018), Porto, Portugal, 3–6 June 2018; pp. 1–6.
50. Steiner, S.; Kühn, V. Distributed compression using the information bottleneck principle. In Proceedings of the 2021 IEEE International Conference on Communications (ICC'2021), Montreal, QC, Canada, 14–23 June 2021; pp. 1–6.
51. Steiner, S.; Kühn, V.; Stark, M.; Bauch, G. Reduced-complexity greedy distributed information bottleneck algorithm. In Proceedings of the 2021 IEEE Statistical Signal Processing Workshop (SSP'2021), Rio de Janeiro, Brazil, 11–14 July 2021; pp. 361–365.
52. Steiner, S.; Kühn, V.; Stark, M.; Bauch, G. Reduced-complexity optimization of distributed quantization using the information bottleneck principle. *IEEE Open J. Commun. Soc.* **2021**, *2*, 1267–1278. [CrossRef]
53. Steiner, S.; Aminu, A.D.; Kühn, V. Distributed quantization for partially cooperating sensors using the information bottleneck method. *Entropy* **2022**, *24*, 438. [CrossRef] [PubMed]
54. Aguerri, I.-E.; Zaidi, A. Distributed variational representation learning. *IEEE Trans. Pattern Anal. Mach. Intell.* **2021**, *43*, 120–138. [CrossRef] [PubMed]
55. Moldoveanu, M.; Zaidi, A. On In-network learning. A comparative study with federated and split learning. In Proceedings of the 2021 IEEE 22nd International Workshop on Signal Processing Advances in Wireless Communications (SPAWC'2021), Lucca, Italy, 27–30 September 2021; pp. 221–225.
56. Moldoveanu, M.; Zaidi, A. In-network learning for distributed training and inference in networks. In Proceedings of the 2021 IEEE Globecom Workshops (GC Wkshps'2021), Madrid, Spain, 7–11 December 2021; pp. 1–6.
57. Zhang, J.A.; Kurkoski, B.M. Low-complexity quantization of discrete memoryless channels. In Proceedings of the 2016 International Symposium on Information Theory and Its Applications (ISITA), Monterey, CA, USA, 30 October–2 November 2016; pp. 448–452.
58. Hassanpour, S.; Wübben. D.; Dekorsy, A. On the equivalence of double maxima and KL-means for information bottleneck-based source coding. In Proceedings of the 2018 IEEE Wireless Communications and Networking Conference (WCNC), Barcelona, Spain, 15–18 April 2018; pp. 1–6.
59. Cover, T.M.; Thomas, J.A. *Elements of Information Theory*, 2nd ed.; Wiley: New York, NY, USA, 2006.
60. Lewandowsky, J.; Stark, M.; Bauch, G. A discrete information bottleneck receiver with iterative decision feedback channel estimation. In Proceedings of the 2018 IEEE 10th International Symposium on Turbo Codes and Iterative Information Processing (ISTC'2018), Hong Kong, China, 25–29 November 2018; pp. 1–5.
61. Lewandowsky, J. The Information Bottleneck Method in Communications. Ph.D. Dissertation, Hamburg University of Technology, Hamburg, Germany, 2020.
62. Genz, A. Numerical computation of multivariate normal probabilities. *J. Comput. Graph. Stat.* **1992**, *1*, 141–149.
63. MacKay, D. J. C. Encyclopedia of Sparse Graph Codes. Available online: http://www.inference.phy.cam.ac.uk/mackay/codes/data.html (accessed on 14 November 2017).
64. Lewandowsky, J.; Dongare, S.J.; Adrat, M; Schrammen, M.; Jax, P. Optimizing parametrized information bottleneck compression mappings with genetic algorithms. In Proceedings of the 2020 14th International Conference on Signal Processing and Communication Systems (ICSPCS'2020), Adelaide, Australia, 14–16 December 2020; pp. 1–8.
65. Lewandowsky, J.; Dongare, S.J.; Martín Lima, R.; Adrat, M.; Schrammen, M.; Jax, P. Genetic Algorithms to Maximize the Relevant Mutual Information in Communication Receivers. *Electronics* **2021**, *10*, 1434. [CrossRef]
66. Coley, D. A. *An Introduction to Genetic Algorithms for Scientists and Engineers*; World Scientific Publishing Co., Inc.: Singapore, 1998.

MDPI
St. Alban-Anlage 66
4052 Basel
Switzerland
www.mdpi.com

Entropy Editorial Office
E-mail: entropy@mdpi.com
www.mdpi.com/journal/entropy